Springer Advanced Texts in Chemistry

Springer
New York
Berlin
Heidelberg
Barcelona
Budapest
Hong Kong
London
Milan
Paris
Santa Clara
Singapore
Tokyo

Springer Advanced Texts in Chemistry

Series Editor: Charles R. Cantor

John F. Robyt

Essentials
of Carbohydrate
Chemistry

With 370 Illustrations

 Springer

John F. Robyt
Laboratory of Carbohydrate Chemistry and Enzymology
Department of Biochemistry and Biophysics
Iowa State University
Ames, IA 50011
USA

Series Editor:
Charles R. Cantor
Center for Advanced Biotechnology
Boston University
Boston, MA 02215, USA

The cover illustrations depict a computer model of a proteoglycan structure, constructed by Jun-yong Choe of the Department of Biochemistry and Biophysics, Iowa State University. Teresa Larsen of The Scripps Research Institute rendered the model using AVS and custom modules created by Thomas J. Macke. © 1997 by T. Larsen, TSRI.

Library of Congress Cataloging-in-Publication Data
Robyt, John F., 1935–
     Essentials of carbohydrate chemistry / John F. Robyt.
        p.   cm.—(Springer advanced texts in chemistry)
     Includes bibliographical references (p.   –   ) and index
     ISBN 0-387-94951-8 (alk. paper)
     1. Carbohydrates.   I. Title.   II. Series.
QD321.R667   1998
572'.56—dc21   97-19019

Printed on acid-free paper.

Production coordinated by Impressions Book and Journal Services, Inc., and managed by Bill Imbornoni; manufacturing supervised by Johanna Tschebull.
Typeset by Impressions Book and Journal Services, Inc., Madison, WI.
Printed and bound by Maple-Vail Book Manufacturing Group, York, PA.
Printed in the United States of America.

9 8 7 6 5 4 3 2 1

ISBN 0-387-94951-8 Springer-Verlag New York Berlin Heidelberg   SPIN 10556320

Series Preface

New textbooks at all levels of chemistry appear with great regularity. Some fields such as basic biochemistry, organic reaction mechanisms, and chemical thermodynamics are well represented by many excellent texts, and new or revised editions are published sufficiently often to keep up with progress in research. However, some areas of chemistry, especially many of those taught at the graduate level, suffer from a real lack of up-to-date textbooks. The most serious needs occur in fields that are rapidly changing. Textbooks in these subjects usually have to be written by scientists actually involved in the research that is advancing the field. It is not often easy to persuade such individuals to set time aside to help spread the knowledge they have accumulated. Our goal, in this series, is to pinpoint areas of chemistry where recent progress has outpaced what is covered in any available textbooks, and then seek out and persuade experts in these fields to produce relatively concise but instructive introductions to their fields. These should serve the needs of one semester or one quarter graduate courses in chemistry and biochemistry. In some cases the availability of texts in active research areas should help stimulate the creation of new courses.

Charles R. Cantor

Preface

There has been interest in research on carbohydrates at Iowa State University for nearly 100 years. A course on carbohydrate chemistry has been offered for over 60 years. When my predecessor, Professor Dexter French, joined the faculty in 1947, he started teaching a graduate course on carbohydrate chemistry every summer semester. When I joined the faculty in 1964, we team taught the course until his death in 1981. I have continued to teach it every other year. *Essentials of Carbohydrate Chemistry* has developed, in part, from this course. Besides covering the traditional subjects of carbohydrate structure, reactions, and organic chemical modifications, subjects not usually covered in texts on carbohydrates such as history of carbohydrates, sweetness, cyclodextrins, glycoproteins, biosynthesis, and biodegradation are presented. The book ends with general considerations involved in the separation, purification, and structure determination of carbohydrates. Many carbohydrate transformations and reactions are treated from the perspective of organic reaction mechanisms and, where appropriate, have been integrated with analytical procedures for the qualitative and quantitative determination of carbohydrates. Numerous examples of the roles that carbohydrates play in living organisms are included in nearly every chapter, and enzyme catalyzed reactions are given an emphasis not usually found in carbohydrate texts.

The book has been written for students beginning the study of carbohydrate chemistry who have some background in organic chemistry. It should also have appeal to those wanting a review of carbohydrate chemistry from either an elementary or an advanced level, and the many structures, literature citations, and suggestions for further study should serve as a helpful reference.

John F. Robyt

Acknowledgments

I wish to thank my wife, Lois, for reading the entire manuscript and making helpful suggestions, my research associate, Rupendra Mukerjea, for reading the manuscript and carefully checking the many structures, and Henry Zobel who also read the manuscript and made helpful suggestions and comments. I also wish to acknowledge my colleagues Herb Fromm, Don Graves, Rich Honzatko, Dave Metzler, and Bernie White in the Department of Biochemistry and Biophysics who have supported me and encouraged me during the writing of the manuscript.

Contents

Chapter 3
Transformations   48

Chapter 4
Modifications   76

Chapter 8
Cyclodextrins   245

Chapter 9
Glycoconjugates   262

Chapter 10
Biosynthesis   290

Contents

Contents   XV

Chapter 11
Biodegradation

Chapter 12
Determinations

Appendix A
Primer on Carbohydrate Nomenclature

Appendix B
Primer on Enzyme Names and Their Catalyzed Reactions

Index

Chapter 1

Beginnings

1.1 Carbohydrates and Their Involvement in Life Processes

Carbohydrates are the most widely distributed, naturally occurring organic compounds on Earth. They include some of the first organic compounds to have their structures determined. As such, carbohydrates helped establish and bridge the disciplines of organic chemistry and biochemistry. They played a particularly important role in the development of the field of optical isomerism.

Much earlier, however, carbohydrates played a key role in the establishment and evolution of life on the Earth by making a direct link between the energy of the sun and chemical energy. Carbohydrates are produced during the process of photosynthesis, in which the energy from the sun is converted into chemical energy by combining carbon dioxide with water to form carbohydrates and molecular oxygen (reaction 1.1). Thus light energy is stored as chemical energy in the form of carbohydrates in plants. Photosynthesis and the formation of carbohydrate is discussed in more detail in Chapter 10. The energy stored in carbohydrates is harvested by nonphotosynthesizing organisms in the processes known as *glycolysis* and *respiration*.

$$6\,CO_2 \;+\; 6\,H_2O \;\xrightarrow[\text{photosynthesis}]{h\nu \;\text{(light energy from the sun)}}\; C_6H_{12}O_6 \;+\; 6\,O_2 \qquad (1.1)$$

Current geochemical theory suggests that molecular oxygen was in very low concentration in the atmosphere of the ancient earth, and carbohydrates were broken down by the process of anaerobic glycolysis (reaction 1.2) to give energy in the form of adenosine triphosphate (ATP), a phosphorylated and purine-

1

substituted carbohydrate ribose derivative (see Fig. 1.7). This is discussed in more detail in Chapter 11.

$$C_6H_{12}O_6 \xrightarrow{\text{anaerobic glycolysis}} 2\ CH_3-\overset{\overset{\displaystyle OH}{|}}{CH}-COOH + \text{energy} \qquad (1.2)$$
$$\text{L-lactic acid} \qquad \text{(2 ATP)}$$

Only a limited amount of energy was available for living organisms via anaerobic glycolysis (2 ATP per $C_6H_{12}O_6$), and life remained simple and primitive. As the concentration of atmospheric molecular oxygen increased due to microbial photosynthesis, mechanisms evolved for the complete oxidation of carbohydrates by molecular oxygen to give $CO_2 + H_2O$ (respiration, reaction 1.3), which gave much greater amounts of energy (38 ATP per $C_6H_{12}O_6$). This increase in available energy led to an explosion in the number and complexity of organisms.

$$C_6H_{12}O_6 + 6\,O_2 \xrightarrow{\text{respiration}} 6\,CO_2 + 6\,H_2O + \begin{array}{c}\text{energy}\\ \text{(38 ATP)}\end{array} \qquad (1.3)$$

Today it is estimated that 3.4×10^{14} kg (4×10^{11} tons) of carbohydrate are biosynthesized each year on the Earth by plants and photosynthesizing bacteria. Carbohydrates make up an important constituent in the human diet and provide a high proportion (50–60%) of the calories consumed. Carbohydrates are also utilized as structural materials in plants, animals, and microorganisms.

1.2 The Nature of Carbohydrates

Reaction 1.1 indicates that the carbohydrates synthesized in photosynthesis have the empirical formula $C_6H_{12}O_6$. It was found in the nineteenth century that carbohydrates in general have the formula $C_n(H_2O)_n$. They were therefore thought to be hydrates of carbon and hence were called *carbohydrates.* Later it was found that carbohydrates in fact contain hydroxyl groups and carbonyl groups and are polyhydroxy aldehydes or ketones. Still later it was recognized that compounds need not be aldehydes or ketones or have the empiric formula of a hydrate of carbon to be a carbohydrate; rather, compounds could be derived from polyhydroxy aldehydes or ketones and have properties of a carbohydrate. Thus the modern definition of a carbohydrate is that it is a polyhydroxy aldehyde or ketone, or a compound that can be derived from them by any of several means including (1) reduction to give sugar alcohols; (2) oxidation to give sugar acids; (3) substitution of one or more of the hydroxyl groups by various chemical groups, for example, hydrogen [H] may be substituted to give deoxysugars, and amino group [NH_2 or acetyl-NH] may be substituted to give amino sugars; (4) derivatization of the hydroxyl groups by various moieties, for example, phosphoric acid to give phospho sugars, or sulfuric acid to give sulfo sugars, or reaction of the hydroxyl groups with alcohols to give saccharides, oligosaccharides, and polysaccharides.

All of these types of carbohydrates play important roles in the scheme of life. For example, the phospho esters of carbohydrates are intermediates in the metabolism of carbohydrates that produces $CO_2 + H_2O +$ energy in the process of respiration. The major compound in the conversion and interchange of energy is the phospho sugar derivative ATP (see Fig. 1.7 for the structure of ATP), sometimes called the "molecule of energy currency." The sugar alcohols and sugar acids are involved in diverse functions of living systems. For example, the sugar alcohol sorbitol is found in some fruits, where it imparts a distinctive sweet taste, and the sugar alcohol ribitol is found in the vitamin riboflavin. Sugar acids are found in plant, animal, and bacterial polysaccharides, such as pectin, hyaluronate, alginate, and bacterial capsules (see Chapter 6), where they impart important functional properties of acids and anions. Carbohydrates are found in combined forms as oligosaccharides and polysaccharides that have the distinctive properties of high molecular weight polymers (Chapter 6). Carbohydrates are also found combined with proteins and lipids, and there is a growing appreciation for their role in protein solubility, protein folding, protein turnover, cell-surface receptors, cell-cell recognition, cellular differentiation, and immunological recognition (see Chapter 9).Carbohydrates make up the backbones of RNA and DNA.

1.3 Occurrence of Carbohydrates

Carbohydrates are widely distributed on the Earth in many different forms and in substances and materials. These materials have long been recognized and exploited and used to improve human life. Some examples of notable and familiar carbohydrates follow:

Cellulose is an abundant carbohydrate of commercial and biological importance, found in all plants as the major structural component of the plant cell wall. When we think of cellulose, we usually think of trees and wood, although this is a relatively impure form of cellulose mixed with many other components such as hemicelluloses and lignin. The cellulose in wood is used to make paper, which is a type of refined cellulose. The fluffy fiber found in the cotton boll is the purest naturally occurring form of cellulose (see Fig. 1.1A).

Starch is another abundant carbohydrate of commercial and biological importance. It is found in the leaves, stems, roots, seeds, and tubers of plants, where it serves to store energy captured in photosynthesis. Nonphotosynthesizing organisms eat and digest the parts of plants that are high in starch as a source of the energy stored in the chemical bonds of the starch molecule. Maize kernels (see Fig. 1.1B) are notably high in starch. Cereal grains such as maize, wheat, rye, and rice, and potato tubers (Fig. 1.1) are well known as important agricultural commodities that are high in starch. It is less widely appreciated that other vegetables (beans, peas, and sweet potatoes) and fruits (bananas) are also high in starch.

Figure 1.1. Important agricultural commodities that are high in material or food carbohydrates: **A,** cotton bol; **B,** maize; **C,** field of wheat; **D,** rice plants; **E,** potatoes.

Chitin is related to cellulose in structure. It makes up the major organic component of the exoskeletons of arthropods such as insects, crabs, and so forth. This is the largest class of organisms, outside of bacteria, and comprises some 900,000 known species, more than are found in all other classes of organisms put together and therefore represents a substantial amount of carbohydrate on the Earth.

Murein is related to chitin and is the structural component of the cell wall of all species of bacteria. Bacteria represent an extremely large group of organisms that is very widely distributed and comprises a substantial amount of carbohydrate. Although there are other kinds of carbohydrates in the bacterial cell wall, it is murein that is the major unifying structural component. The other carbohydrates are more diverse and distinctive in the various species, and they impart distinctive characteristics to the individual bacterial species, often serving to set them apart from each other.

Sucrose is a naturally occurring carbohydrate, chiefly found in sugar cane and sugar beet but also found in many other plants, especially fruits. It also occurs in honey produced by bees and in the sap of the maple tree, in sorghum, and in certain date and palm trees. Its particular attribute is its sweetness. Although many carbohydrates are sweet, it is sucrose that has become the sugar of commerce, primarily because of the ease of obtaining it in high purity from sugar cane or sugar beets. It has long been used by man as a food, sweetening agent, and has had special significance in ancient religious ceremonies (see Chapter 2).

Lactose is a naturally occurring carbohydrate found in the milk of mammals. It serves as the principal source of carbohydrate and energy for their young. It is much less sweet than sucrose and has a relatively bland taste.

α,α-*Trehalose* is a carbohydrate found in yeasts and fungi and is the major sugar in the hemolymph fluid of insects, where it serves as a source of energy.

The following shows what happens when these various carbohydrates are heated in aqueous solution with acid:

$$\text{Cellulose} + H_2O \xrightarrow{H^+} \text{glucose}$$

$$\text{Starch} + H_2O \xrightarrow{H^+} \text{glucose}$$

$$\text{Chitin} + H_2O \xrightarrow{H^+} N\text{-acetyl glucosamine}$$

$$\text{Murein} + H_2O \xrightarrow{H^+} N\text{-acetyl glucosamine} + N\text{-acetyl muramic acid}$$

$$\text{Sucrose} + H_2O \xrightarrow{H^+} \text{glucose} + \text{fructose}$$

$$\text{Lactose} + H_2O \xrightarrow{H^+} \text{glucose} + \text{galactose}$$

$$\alpha,\alpha\text{-Trehalose} + H_2O \xrightarrow{H^+} \text{glucose}$$

Glucose is a carbohydrate found in many materials in combined forms such as cellulose, starch, sucrose, lactose, and α,α-trehalose. It is also found in the free state in a number of materials such as honey, grapes, and raisins. Glucose plays an important role in the blood of all animals, where it serves as an immediate source of energy and as a stabilizer of the osmotic pressure of the blood. It further serves as the precursor for the formation of glycogen and fat.

1.4 Asymmetry and the Structures of Carbohydrates

We have seen that different carbohydrates occur naturally, and many are composed wholly or in part of glucose. So, what are the structures of carbohydrates and how do they differ from each other?

We begin by asking what is the smallest compound that fulfills the definition of a carbohydrate. The smallest carbohydrate would have to have three carbons, an aldehyde or keto group, and two hydroxyl groups. There are three compounds that fulfill that definition. They have the following structures and names:

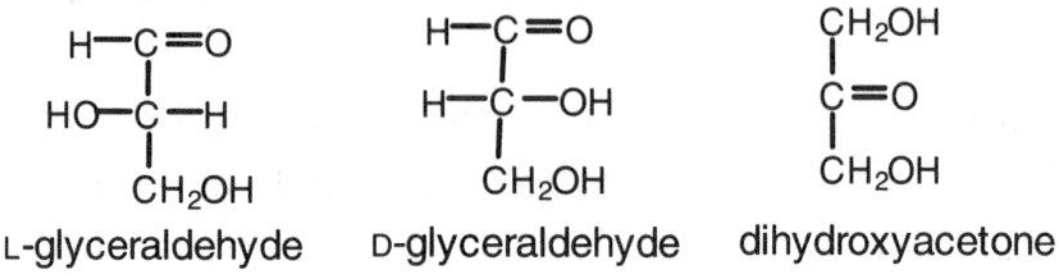

L-glyceraldehyde   D-glyceraldehyde   dihydroxyacetone

The two aldehyde compounds have asymmetric, or chiral, centers, with the hydroxyl group on the chiral center either to the *right* or to the *left,* which have been designated as "D" or "L," respectively. The three compounds have the same empirical formula, $C_3H_6O_3$, but are distinct, with different chemical and physical properties. The two D- and L-structures cannot be superimposed. D-Glyceraldehyde rotates plane polarized light (the D-line of sodium) to the right and has a specific optical rotation $[\alpha]_D^{25}$ at 25°C of +8.7°, and L-glyceraldehyde rotates plane polarized light to the left with a specific optical rotation at 25°C of −8.7°. D- and L-Glyceraldehyde have been selected as the configurational reference standards for carbohydrates. D-Glyceraldehyde, however, is the only one of the two that occurs naturally to any significant extent, and it occurs primarily as the 3-phosphate.

Almost all of the naturally occurring carbohydrates have the D-configuration. It is not understood why and how carbohydrates with only the D-configuration were selected when the formation of carbohydrates first occurred on the Earth. The configuration of the 20 naturally occurring α-amino acids is opposite (L-configuration) to that of naturally occurring carbohydrates. The L-α-amino acids can have their structures related to D- and L-glyceraldehyde and hence to the configuration of carbohydrates. If one replaces the hydroxyl group at the asymmetric carbon of L-glyceraldehyde with an amino group and oxidizes the aldehyde group to a carboxyl group, the α-amino acid L-serine would be produced.

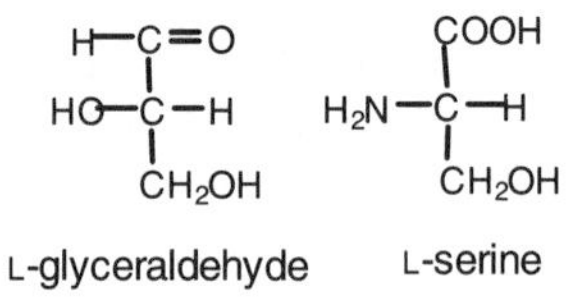

L-glyceraldehyde   L-serine

Again, the actual mechanism for the formation and selection of the L-amino acids is not understood. It might have been that ammonia reacted with the keto group of dihydroxy acetone to form the imine, the primary hydroxyl group was oxidized, and the imine was reduced stereoselectively to give L-serine (reaction 1.4); or ammonia might have reacted with the keto group of a breakdown product of carbohydrates, pyruvic acid, to give an imine that was stereoselectively reduced to give L-alanine (reaction 1.5).

$$
\begin{array}{c}
\underset{\substack{\text{dihydroxyacetone}}}{\overset{\displaystyle CH_2OH}{\underset{\displaystyle CH_2OH}{C=O}}} + NH_3 \xrightarrow{\quad \overset{H_2O}{\diagup}\quad} \underset{\displaystyle CH_2OH}{\overset{\displaystyle CH_2OH}{HN=C}} \xrightarrow{\text{oxidation}} \underset{\displaystyle CH_2OH}{\overset{\displaystyle COOH}{HN=C}}
\end{array}
\qquad (1.4)
$$

$$
\underset{\substack{\text{L-serine}}}{\overset{\displaystyle COOH}{\underset{\displaystyle CH_2OH}{H_2N-C-H}}} \xleftarrow[\substack{\text{stereoselective} \\ \text{reduction}}]{2\,H^{(+)} + 2e^{(-)}}
$$

$$
\underset{\substack{\text{pyruvic acid}}}{\overset{\displaystyle COOH}{\underset{\displaystyle CH_3}{C=O}}} + NH_3 \xrightarrow{\quad\overset{H_2O}{\diagup}\quad} \underset{\displaystyle CH_3}{\overset{\displaystyle COOH}{HN=C}} \xrightarrow[\substack{\text{stereoselective} \\ \text{reduction}}]{2\,H^{(+)} + 2e^{(-)}} \underset{\substack{\text{L-alanine}}}{\overset{\displaystyle COOH}{\underset{\displaystyle CH_3}{H_2N-C-H}}}
\qquad (1.5)
$$

Using paper and pencil, all of the potential structures of the D- and L-configuration of the carbohydrates can be drawn by systematically inserting an asymmetric carbon in D- and L-glyceraldehyde between the aldehyde group and the adjacent chiral carbon, with the hydroxyl group of the new carbon to the right for one compound and to the left for the other compound (see Fig. 1.2). We can see that D-glyceraldehyde gives two structures and L-glyceraldehyde gives two structures, each with four carbons. The D- and L-structures are the mirror images of each other and are called *enantiomers*. Thus a D-carbohydrate is defined by the fact that the hydroxyl group is to the right on the asymmetric carbon atom furthest from the most oxidized carbon atom, the first or reference carbon (aldehydo, keto, or carboxyl group), as is found in D-glyceraldehyde. Similarly, all of the structures obtained from L-glyceraldehyde (in which the chiral carbon that is the most remote from the most oxidized carbon is to the left) would be L-sugars. It is obvious that if we know the structure of any D-carbohydrate, we can draw the structure of its corresponding L-carbohydrate by drawing its mirror image or enantiomer. This designation and nomenclature was proposed by Rosanoff in 1906 [1,2], although it was first recognized and used in the determination of the structures of carbohydrates by Emil Fischer in 1885. Fischer assigned compounds as "*d*" (for *dextrorotatory*) if the compound rotated plane polarized light to the right and "*l*" (for *levorotatory*) if the compound rotated plane polarized light to the left. Later, as we will see in Chapter 2, Fischer assigned the D- and L- designations to absolute structures.

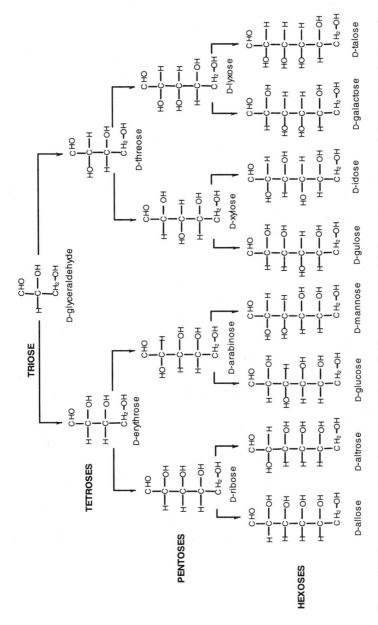

Figure 1.2. Structures and names of the D-family of aldoses starting with the D-triose and progressing systematically to the eight possible D-hexoses by adding a single asymmetric center with the two possible configurations to each structure.

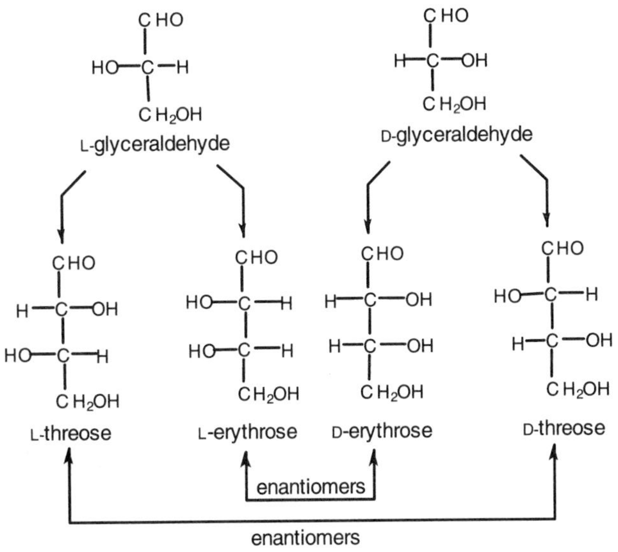

Carbohydrates with three or more carbons are called, successively, *trioses, tetroses, pentoses, hexoses, heptoses, octoses,* and so on.

Figure 1.2 shows the structures and gives the names of the D-family of carbohydrate aldehydes (called *aldoses*) that can be derived from D-glyceraldehyde, through the hexoses, by sequentially adding a new hydroxyl carbon between the aldehyde group and the adjacent hydroxy carbon. As noted above, the addition of a single asymmetric center to one of the compounds gives two new compounds, with the hydroxyl to the right and to the left, respectively. Thus D-glyceraldehyde gives two D-tetroses; the two D-tetroses give four D-pentoses; and the four D-pentoses give eight D-hexoses. If we consider that each asymmetric center can have two configurations, one to the right and the other to the left, the number of possible isomers is equal to 2^n, where n is the number of asymmetric centers. This would give us the total number of isomers containing both the D- and the L-configurations. Hexoses, with four asymmetric centers, have a total number of isomers equal to 2^4, or 16 D- and L-isomers. The number of D- (or L-) hexoses is half that number, or eight isomers. Thus the number of D- or L-aldoses is given by 2^{n-1}.

In a similar manner, the family of ketoses with the D-configuration can be drawn from dihydroxy acetone. The first ketose with an asymmetric center is a tetrose in which the configuration of the asymmetric center could be either to the left or to the right. If it is to the right, we have by definition a D-sugar. A second asymmetric center could be added, between the keto carbon and the hydroxyl carbon, with its configuration either to the left or to the right, giving two new D-ketopentoses. The structures and names for the D-keto-sugars through the hexoses are given in Fig. 1.3.

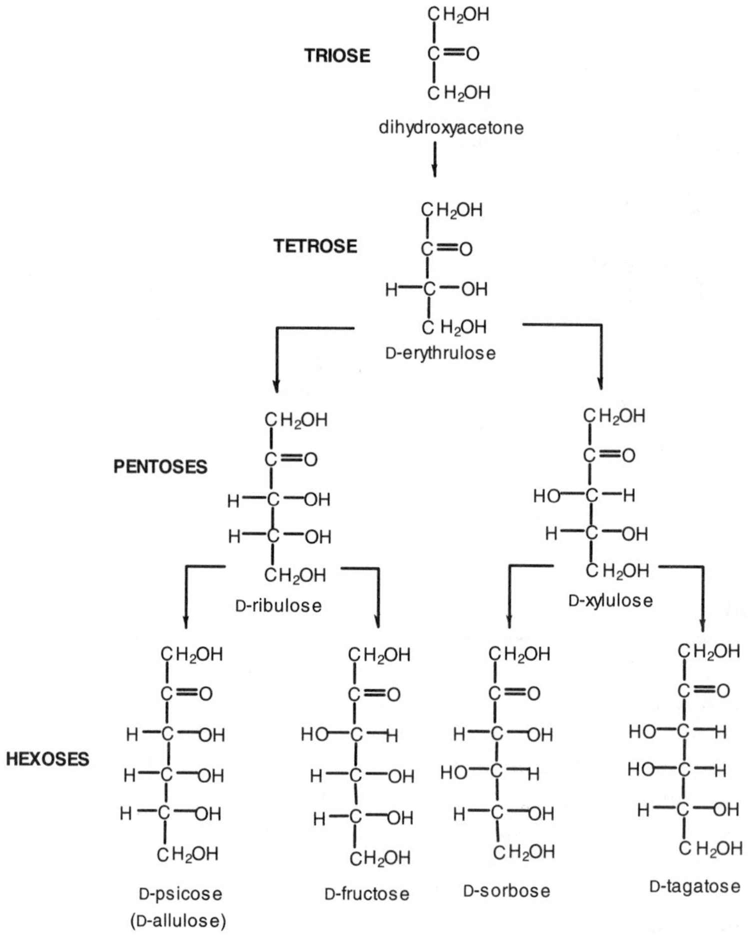

Figure 1.3. Structures and names of the D-family of ketoses starting with the triose and progressing systematically to the four possible D-hexoses by adding a single asymmetric center with the two possible configurations to each structure.

1.5 Remembering the Structures of Carbohydrates

A common problem in learning carbohydrate chemistry is the difficulty in remembering the structures and names of carbohydrates, as they differ from each other only by the configuration of the asymmetric carbon atoms. The trick is not to *memorize* the structures but to *learn* them by developing a "system." The structure of D-glyceraldehyde is easy, since it has only one asymmetric carbon atom and it is to the right, by definition, because it has the D-configuration. The two tetroses are also easy, since the new asymmetric carbon has the hydroxyl group to the right in one structure and to the left in the other structure, and the last asymmetric carbon is to the right for both structures because they both are "D." The pentoses present more of a problem. But only two of the four possible D-isomers are of any importance as naturally occurring carbohydrates. They are D-ribose and D-xylose. Neither D-arabinose nor D-lyxose occur naturally to any extent. This is very fortunate because both D-ribose and D-xylose have structures that are easy to remember. D-Ribose has all of its asymmetric hydroxyl groups to the right, and D-xylose has the D-hydroxyl group to the right, the next hydroxyl up is to the left, and the next one is to the right. Thus, starting with the most remote hydroxyl group, the hydroxyl groups of D-xylose alternate: right, left, right.

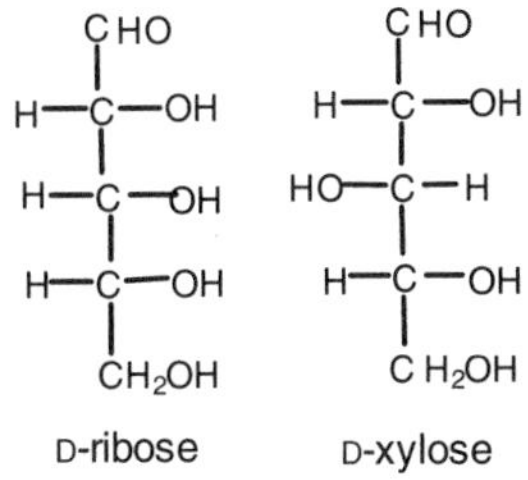

D-ribose   D-xylose

The structures of the aldohexoses would appear even more difficult to remember than the aldopentoses as there are eight isomeric forms. But only four of the eight D-isomers are of any biological or economic importance: D-allose, D-glucose, D-mannose, and D-galactose. D-Allose is very easy to remember, as D-ribose, it has all of its hydroxyl groups to the right, and furthermore its name is *all*ose. D-Glucose has a structure that some would say is difficult to remember because there doesn't appear to be any simple pattern relating the structures of the four asymmetric carbons. But, if we were to remember the structure of any carbohydrate, it should be D-glucose because it constitutes 99.95% of the carbohydrates on the Earth, either in combined or derivatized form. Actually it is easy to remember, but first we must consider the structures of the two other important aldohexoses, D-mannose and D-galactose. D-Mannose has a point of symmetry between carbons 3 and 4 in which the bottom two hydroxyl groups are to the right and the next two are to the left. D-Galactose and D-allose on the other hand have a plane of symmetry between carbons 3 and 4 so that the bottom two asymmetric carbons are mirror images of the top two asymmetric carbons.

D-mannose   D-galactose   D-allose

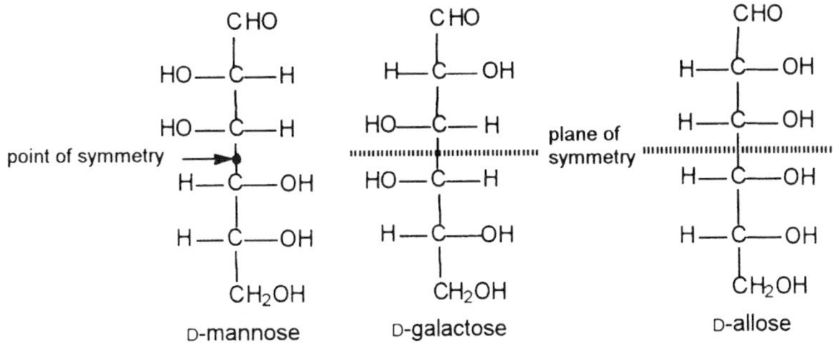

D-Glucose is a specific isomer at a single carbon of D-mannose, D-galactose, and D-allose. It is a special isomer of D-mannose, in which the configuration of the position-2 carbon is opposite that of D-glucose. It is so special that Emil Fischer gave a special name to the isomers; he called them *epimers,* and this concept played an important role in the arguments that Fischer made in determining their structures (see Section 2.3 in Chapter 2).

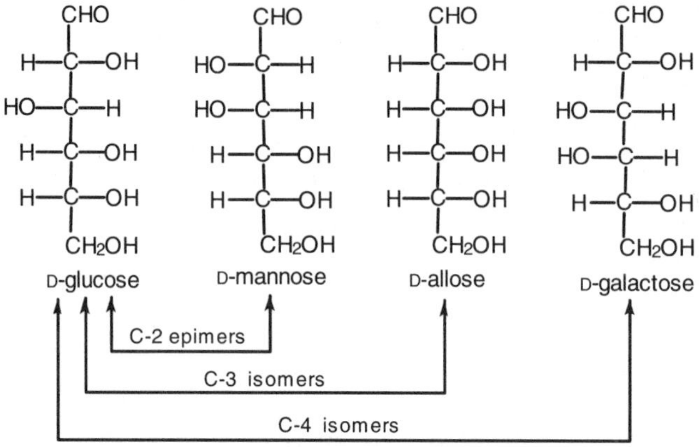

D-glucose   D-mannose   D-allose   D-galactose

C-2 epimers

C-3 isomers

C-4 isomers

D-Glucose is also an isomer of D-allose at C-3 and an isomer of D-galactose at C-4. In recent years, some organic chemists and biochemists have obscured and blurred the concept of epimers by stating that D-allose and D-galactose are also epimers of D-glucose. Because the isomerization at C-2 occurs readily, as shown in reactions 1.6 and 1.7, the isomers at C-2 are given the special name *epimers.* Thus the character of D-glucose and D-mannose at C-2 is very special and sets them apart from D-allose and D-galactose.

Carbohydrates that are epimers of each other at the 2-position can be relatively easily interconverted by treating them at pH 11 and 25°C (alkaline isomerization). The transformation of one epimer into the other goes though the 2-keto compound, and the three isomeric carbohydrates are formed. Thus D-glyceraldehyde is converted into dihydroxyacetone and L-glyceraldehyde, resulting in an equilibrium mixture. Likewise D-glucose can be converted into D-fructose and D-mannose. By using higher concentrations of base and higher temperatures, D-glucose can be converted into D-allose, D-galactose, and many other products (see section 3.2 in Chapter 3). It has been reported that over 100 products were produced by treating D-glucose under alkaline conditions at higher temperatures. These products included the various 2-, 3-, 4-, and 5-keto-sugars, the isomers occurring from the conversion of the respective keto groups into the two hydroxyl isomers, as well as oxidation products.

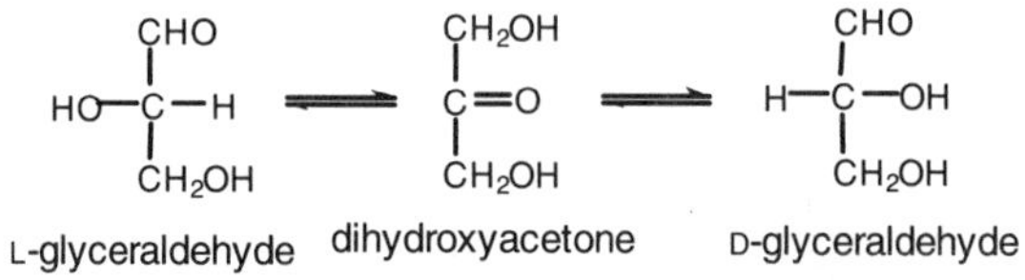

L-glyceraldehyde   dihydroxyacetone   D-glyceraldehyde   (1.6)

Alkaline isomerization at pH 11 and 25° C

$$
\begin{array}{ccc}
\text{CHO} & \text{CH}_2\text{OH} & \text{CHO} \\
\text{H}-\text{C}-\text{OH} & \text{C}=\text{O} & \text{HO}-\text{C}-\text{H} \\
\text{HO}-\text{C}-\text{H} & \text{HO}-\text{C}-\text{H} & \text{HO}-\text{C}-\text{H} \\
\text{H}-\text{C}-\text{OH} & \text{H}-\text{C}-\text{OH} & \text{H}-\text{C}-\text{OH} \\
\text{H}-\text{C}-\text{OH} & \text{H}-\text{C}-\text{OH} & \text{H}-\text{C}-\text{OH} \\
\text{CH}_2\text{OH} & \text{CH}_2\text{OH} & \text{CH}_2\text{OH} \\
\text{D-glucose} & \text{D-fructose} & \text{D-mannose}
\end{array}
$$

(1.7)

Alkaline isomerization at pH 11 and 25° C

1.6 Derived Carbohydrate Structures

Many of the carbohydrates that we have considered do not occur naturally. Instead, they are found as derived or combined compounds and often play important roles in living systems. We will consider the structures of some of these compounds in this section.

a. Sugar Alcohols

Sugar alcohols are produced when the carbonyl group of either the aldehyde or ketone group is reduced. The resulting products are polyhydroxy compounds. They retain some of the properties of carbohydrates, such as high water solubility, sweetness, optical activity, and reactions of secondary and primary alcohols. When either dihydroxy acetone or D-glyceraldehyde are reduced, a single sugar alcohol, *glycerol,* is produced. When D-glucose is reduced, we get *D-glucitol* (commonly called *D-sorbitol*). When D-fructose is reduced, we get two sugar alcohols, D-glucitol and *D-mannitol.* D-Glucitol is widely distributed in plants and is especially found in fruits and berries such as apples, apricots, cherries, and pears. It is also found in seaweed and algae. D-Mannitol has widespread occurrence in plants and is found especially in plant exudates and seaweeds. *Xylitol* occurs naturally in three fruits: strawberries, raspberries, and plums. It has an unusual sweet taste and gives these fruits some of their characteristic flavor. It further imparts a cooling effect in the mouth. *Ribitol* occurs naturally as part of the vitamin riboflavin (vitamin B$_2$). It also occurs in the capsular polysaccharides of *Pneumococcus* bacteria and in teichoic acids that are found in the cell wall of several gram-positive bacteria (see Chapter 6). *Erythritol* is found in certain algae, lichens, and grasses. A 1,4-dithio derivative of erythritol and *threitol* have been synthetically produced for use as efficient reducing agents for disulfide linkages, which are found primarily in proteins. See Figure 1.4 for the structures of some common sugar alcohols.

b. Sugar Acids

Three kinds of sugar acids can be obtained from the common aldo carbohydrates. The first occurs from the oxidation of the aldehyde group to give an "*onic acid.*

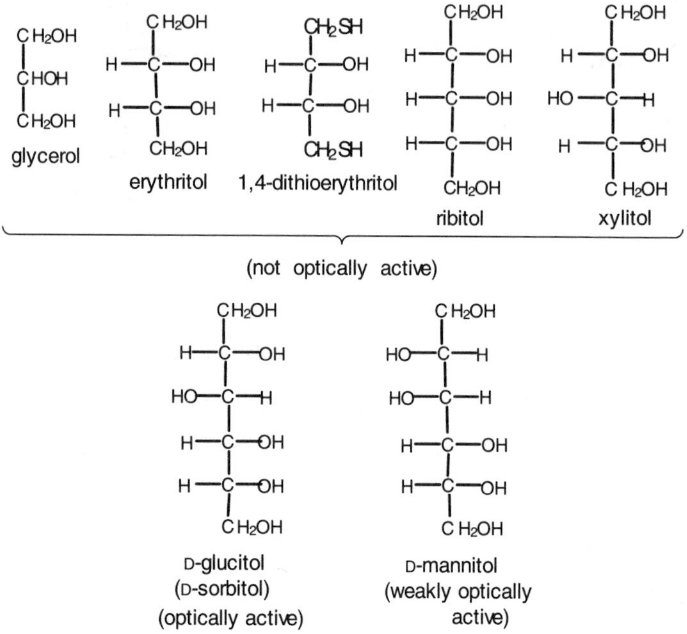

Figure 1.4. Structures and names of some common sugar alcohols.

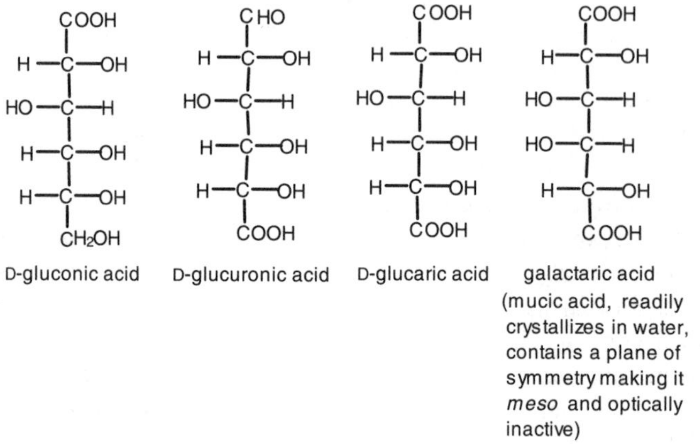

Figure 1.5. Structures and names of some common sugar acids.

The second occurs from the oxidation of the primary alcohol group to give a "*uronic acid.* The third occurs from the oxidation of both the aldehyde and the primary alcohol groups to give an "*aric acid.* Structures of these sugar acids are illustrated in Fig. 1.5 using D-glucose as the parent compound. A common aric acid is obtained from the oxidation of D-galactose when this sugar is heated with nitric acid to produce *galactaric acid.* It is commonly called *mucic acid* and is a *meso* compound that has internal symmetry and does not rotate plane polarized light. It also readily crystallizes and is not water soluble. The uronic acids occur in combined form in many polysaccharides and will be considered in Chapter 6.

c. Deoxy Sugars

When a hydroxyl group of a carbohydrate is substituted by a hydrogen atom, a deoxy sugar results. This can theoretically occur for any of the hydroxyl groups in a carbohydrate. There are, however, only a limited number of naturally occurring deoxy sugars. When D-ribose is substituted in the 2-position, 2-deoxy-D-ribose results. This is the carbohydrate constituent of DNA (deoxyribonucleic acid), the genetic polymer found in chromosomes. The other common hydroxyl replacement is at C-6 of the hexoses. When the C-6 hydroxyl group of D-glucose is replaced, we obtain 6-deoxy-D-glucose, which is commonly called D-quinivose. D-Quinivose occurs as a glycoside in many plant species (see section 3.6 of Chapter 3). The replacement of the hydroxyl group at C-6 of D-mannose results in 6-deoxy-D-mannose, commonly called D-rhamnose. Rhamnose is also known to occur naturally as the L-isomer. L-Rhamnose is found as a glycoside in many plants and occurs free in poison sumac and poison ivy. It is also found as a constituent of extracellular, gram-negative bacterial polysaccharides, such as the *Salmonella* O-antigen discussed in Chapter 6. The replacement of the hydroxyl group at C-6 of D-galactose gives 6-deoxy-D-galactose, commonly called D-fucose. D-Fucose is found in plants and makes up various cardiac glycosides. L-Fucose occurs in polysaccharides found in seaweeds, eggs of sea urchins, frog spawn, and gum tragacanth. D- and L-Rhamnose and D- and L-fucose are frequently found as carbohydrate constituents of glycoproteins. See Fig. 1.6 for the structures of these deoxy sugars.

A number of 3,6-dideoxyhexoses have also been found in extracellular gram-negative bacterial polysaccharides. Of the eight possible isomers, four have been isolated and identified. We will consider their structures in Chapter 6 when we discuss the bacterial polysaccharides. Synthetic deoxy sugars have been made and serve as enzyme inhibitors and/or as probes in studying the function of biological systems. One particular group of synthetic deoxy carbohydrates, the 2,3-dideoxy-D-ribonucleotide triphosphates, are used in the sequencing of DNA by the Sanger, or chain termination, method.

d. Amino Sugars

When the hydroxyl group is replaced by an amino group, we get an amino sugar. As with the deoxy sugars, theoretically any hydroxyl group can be replaced. In nature, however, as with the deoxy sugars, we find only a limited number of replacements by amino groups. Sometimes it is a replacement by an *N*-acetyl-amino group. The most commonly occurring amino sugars are 2-amino-2-deoxy-D-glucose (commonly called D-glucosamine), *N*-acetyl-2-amino-2-deoxy-D-glucose,

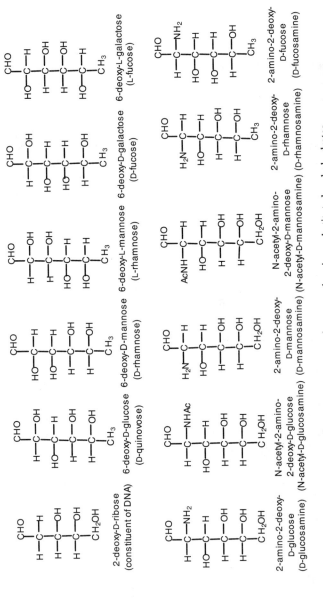

Figure 1.6. Structures and names of some common deoxy- and amino-substituted carbohydrates.

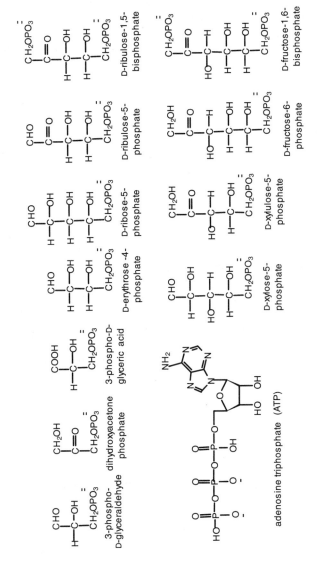

Figure 1.7.　Structures and names of some common and metabolically important carbohydrate phosphates.

2-amino-2-deoxy-D-galactose (commonly called D-galactosamine), the *N*-acetyl derivative, and substitution of an amino or *N*-acetyl-amino group on C-2 of the 6-deoxy analogues of D- or L-mannose and D- or L-galactose (giving D- or L-rhamnosamine and D- or L-fucosamine). These amino sugars are also frequently found as carbohydrate constituents of glycoproteins (see Chapter 9). See Fig. 1.6 for the structures of these amino sugars. Sugars with amino groups in other positions that are not naturally occurring can also be synthesized (see Chapter 4) and sometimes serve as enzyme inhibitors and/or probes in the study of the function of biological systems.

e. Sugar Phosphates

Many of the simple trioses, tetroses, and pentoses do not occur naturally in the free state but are commonly found as phosphate-ester derivatives. The phospho-esters are important intermediates in the breakdown and synthesis of carbohydrates by living organisms. D-Glucose is converted into D-fructose-1,6-bisphosphate that is then cleaved in half to give D-glyceraldehyde-3-phosphate and dihydroxyacetone phosphate (see Chapter 11). D-Erythrose is found as the 4-phosphate in the pentose-phosphate pathway of carbohydrate degradation and in the photosynthetic process. D-Ribose-5-phosphate, D-ribulose-5-phosphate, D-xylose-5-phosphate, and D-xylulose-5-phosphate are found in the pentose phosphate pathway as well as in the photosynthetic pathway (see Chapter 10). D-Ribulose-1,5-bisphosphate is the direct intermediate to which CO_2 is added in the photosynthetic pathway. D-Ribose-5-phosphate also is the precursor of RNA (ribonucleic acid) and DNA (deoxyribonucleic acid). See Fig. 1.7 for the structures of these common sugar phosphates.

In addition to the naturally occurring deoxy, amino, and phospho sugars, other kinds of esters, ethers, and substituted carbohydrates (carbohydrates with hydroxyl and/or hydrogen replaced by other groups), can be chemically or enzymatically synthesized (see Chapter 4).

f. Condensed Sugar Products

We will defer the discussion of the structures of these carbohydrates to Chapter 3, the section on the formation of saccharides, after we have further discussed the intramolecular reactions of pentoses and hexoses in Chapters 2 and 3.

1.7 Literature Cited

1. M. A. Rosanoff, *J. Am. Chem. Soc.,* **28** (1906) 114–116.
2. C. S. Hudson, *Adv. Carbohydr. Chem.,* **3** (1948) 12–22.

Chapter 2

Developments

2.1 Carbohydrates in the History and Development of Human Culture

The basic carbohydrates of cellulose, starch, and sucrose were known and used by humans from very ancient times. They have had a major impact on the development of culture by providing clothing, food, shelter, and writing materials. The following are some examples.

Cellulose. Papyrus was a material used as a medium for writing from the very earliest times (4000 B.C.E.), especially in Egypt. It was prepared as thin films from the outer bark of a reedy plant known as *cyperus papyrus.* Papyrus was a cellulosic material that was glued together with a paste of wheat starch to give it body and the ability to hold ink. The manufacture of papyrus for writing became a major industry in Egypt, and the Egyptians exported it throughout the Mediterranean and Europe. It was used in Europe until about the tenth century C.E., when it was replaced by paper. The ancient scrolls, such as the *Dead Sea Scrolls,* were written on papyrus.

Archaeological evidence indicates that cotton was being used in India to make fabric and string as early as 3000 B.C.E. Around 1500 B.C.E., India apparently was the center of a cotton industry that spread to Persia (Iran), China, and Japan.

The writings of Herodotus (500 B.C.E.) and Pliny the Elder (60 C.E.) indicate that cotton fiber was known to the Greeks and Romans. Columbus (1492 C.E.) found it used by the natives of the New World. In the conquest of Mexico and Peru, cotton cloth was discovered to be in use, and ancient Peruvian tombs had mummies covered with cotton strips, dating 6000–5000 B.C.E.

Flax is a fiber plant that was cultivated in the stone age. Flaxseed was probably first gathered for food. The fiber is a relatively high-quality cellulose that is used to make linen. It was grown in ancient Egypt for the manufacture of linen cloth. Linen was also produced by the Chaldeans, Babylonians, Phoenicians,

19

Greeks, and Romans. The use of flax spread through the European continent and reached America with the early colonists. It was grown in Massachusetts in 1630, where it was spun into yarn and woven into cloth.

Starch. Foods containing starch were items in the diet of early humans. Large quantities of grains and vegetables were consumed. As mentioned above, papyrus was sized with starch pastes.

Paper for writing was produced in China around 100 C.E. The process used fibrous material from the bark of trees combined with old rags and fish nets. A Chinese document, dated 312 C.E., contained starch as a sizing agent. By about 600 C.E. papermaking had spread to Japan where it was made from the bark of the mulberry tree by Buddhist monks. From the Far East, the use of paper traveled westward, following the caravan routes. Between 700 and 1300 C.E., paper was coated with starch paste. The papers were sized with a thin starch paste to provide a surface that could absorb ink. They were then covered with unmodified starch to give thickness, rigidity, and strength. Rice, wheat, and barley starch granules have been identified in these early papers [1].

The Romans used starch to whiten cloth and powder hair. Roman writings of over 2,000 years ago recorded a procedure for the preparation of starch [1].

> The plant material containing the starch (bulbs or roots) is macerated in water, or grain is steeped in water for several days and then macerated. The solid material is allowed to settle and the water is poured off. The solid is resuspended in fresh water and then pressed through cloth. The white liquid that comes through is allowed to stand. A white solid settles to the bottom, the clear liquid is poured off, and the white solid (the starch) is air dried in the sun.

This procedure is very similar to that used today.

About 300 C.E., starch was widely used to stiffen cloth. The starch was often mixed with other materials to produce different colors in the cloth. Colored starches (especially yellow and red) were also used as cosmetics. Starch was modified by heating with vinegar to create syrups. Starch was also deliberately modified by the action of enzymes. Around 975 C.E., an Arabian pharmacologist described the treatment of starch with saliva to produce a poultice (that he called "artifical honey") for the treatment of wounds [2]. From around 1350 C.E., starch was a common material of commerce. A story has been told that in 1821 the accidental discovery of the production of soluble starch occurred in a Dublin textile mill where starch was used as a sizing agent. A fire occurred in the mill, and after the fire was extinguished, it was noticed that some of the unused starch had been turned brown by the heat and that this material easily dissolved in water to make syrups. The heating of starch was repeated under more controlled conditions, and the product was found to be much more soluble than it was before heating. Heat-treated starch became known as *British gums,* and the process became a common method for modifying starch and obtaining a water-soluble product.

In the early United States (1800 C.E.), the primary source of starch was potatoes. By 1880, there were more than 150 potato starch factories through out the United States. Potatoes were also a source of starch in Holland and Germany. Today, maize has become the primary source of industrial starch used though out the world, although sweet potato starch is important in Japan, and rice starch is important in Europe and Asia. Tapioca (cassava) starch is important in the East Indies, Dominican Republic, and Brazil, and arrowroot starch is produced in the West Indies.

Sucrose. Sugarcane is also an ancient source of carbohydrate. The origins of sugarcane and the extraction of sugar (sucrose) from this grass are obscure. They apparently originated either in northeast India or the South Pacific. There are indications of primitive sugar processing in New Guinea as early as 10,000 B.C.E. and about 6000 B.C.E. in India. Many woody, wild sugarcanes are still found growing in India. These plants have all of the fundamental characteristics of modern cultivated strains, and it is speculated that sugarcane originated in the Indus Valley. Sugar or sucrose is mentioned in very early Hindu sacred books. The word *sucrose* itself is derived from the Sanscrit word *sarkara,* meaning sweet. The sugar culture spread from India to China between 1800 and 1700 B.C.E. and then went westward and eastward. The Arabs spread sugarcane culture into the Mediterranean and developed some of the earliest refining processes from which the preparation of candies and confections developed. These early products reached a high peak in Egypt (ca. 1000 B.C.E.) and were also used in medicines. The high food value of sugar (sucrose) was appreciated, and its use was adopted by many cultures and civilizations.

Crystalline sugar from cane was known in India around 300 C.E., and the use of sugar spread via the caravan routes over North Africa into Spain around this time. Sugarcane cannot be grown in Europe because it requires a tropical or semi-tropical climate. Cane sugar was exported to Europe in the fourteenth and fifteenth centuries and used as a costly sweetening agent. By the late fifteenth century, many sugar refineries were in operation in Europe, and the use of sugar had become widespread. The restriction of growing sugarcane in a tropical climate led to a search for other plants that might produce sugar in a temperate climate. This search produced the sugar beet which was cultivated in the late eighteenth century on the European continent. The desire to find alternate sweetening agents also led to the use of honey, grape juice, and raisins.

2.2 Development of Carbohydrate Chemistry

At the end of the eighteenth century (1792), a carbohydrate was isolated from honey that was different from cane sugar (sucrose). Ten years later (1802), a sweet carbohydrate was found in grapes that was also different from sucrose. Almost ten years after that (1811), it was found that acid hydrolysis of starch produced a carbohydrate that was identical to the carbohydrate found in grapes. And ten years after that (1820), it was found that this same crystalline sugar could be obtained from the urine of diabetics and from the acid hydrolysis of cellulose. These sugars were shown to be identical to the sugar in honey that was different from sucrose. In 1838, this sugar was called *glucose.* In 1866, the well-known German chemist August Kekulé changed the name to *dextrose* because it rotated plane polarized light to the right. In 1881, a third sugar was isolated and crystallized from honey. It was called *levulose* because it strongly rotated plane polarized light to the left. In 1890, the German chemist Emil Fischer, a former student of Kekulé's, renamed it *fructose* and changed the name of dextrose back to glucose.

By the middle of the nineteenth century, a number of relatively pure carbohydrates such as sucrose, cellulose from cotton, starch, glucose, fructose, mannose, and lactose were known to the chemists of Europe, especially in Germany, who were studying natural products [3]. These chemists were investigating the physical and chemical properties of carbohydrates. The following is a summary of what was known about carbohydrates in the nineteenth century.

a. Chemical Properties of Carbohydrates, 1860–1880

In the period from 1860 to 1880, the following chemical facts were known about (+)-glucose and (+)-mannose:

1. They reduced Tollen's reagent, $Ag(NH_3)_2^+$, to $Ag°$, indicating that they were aldehydes.
2. They formed pentaacetates, indicating that they had five hydroxyl groups.
3. They could be oxidized with Br_2/H_2O to a monocarboxylic acid.
4. They could be oxidized with nitric acid to give dicarboxylic acids that were optically active.
5. They could be reduced to give a sugar alcohol that gave a hexaacetate.
6. Reduction of the sugar alcohol with HI gave 2-iodohexane, and reduction of the monocarboxylic sugar acid gave hexanoic acid (reaction 2.1), indicating they had a straight chain of carbon atoms.

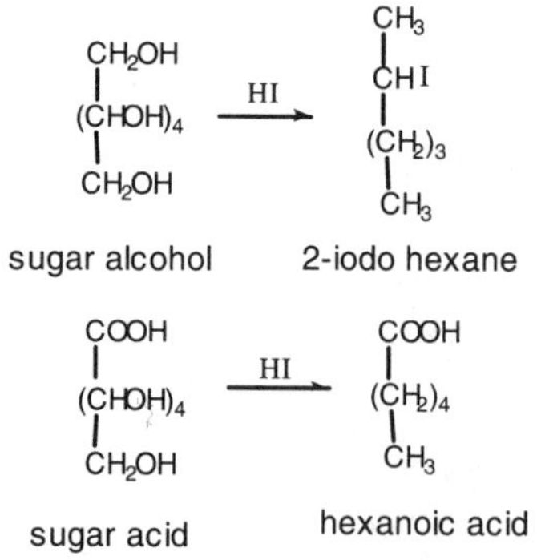

$$
\begin{array}{ccc}
\text{CH}_2\text{OH} & & \text{CH}_3 \\
| & \text{HI} & | \\
(\text{CHOH})_4 & \longrightarrow & \text{CHI} \\
| & & | \\
\text{CH}_2\text{OH} & & (\text{CH}_2)_3 \\
& & | \\
& & \text{CH}_3
\end{array}
$$

sugar alcohol   2-iodo hexane

$$
\begin{array}{ccc}
\text{COOH} & & \text{COOH} \\
| & \text{HI} & | \\
(\text{CHOH})_4 & \longrightarrow & (\text{CH}_2)_4 \\
| & & | \\
\text{CH}_2\text{OH} & & \text{CH}_3
\end{array}
$$

sugar acid   hexanoic acid

(2.1)

7. The empirical formula was $C_6H_{12}O_6$.
8. In 1870, Bayer and Fittig gave the following formulation for a hexose:
   $HO-CH_2-CH(OH)-CH(OH)-CH(OH)-CH(OH)-CHO$
9. The Kiliani cyanohydrin reaction added cyanide to aldehydes giving two isomeric hydroxy acids with the chain length increased by one carbon.
10. (+)-Arabinose was isolated from beet pulp in 1868 and found to be an aldopentose.

11. In 1878, Emil Fischer synthesized phenylhydrazine for his thesis at the University of Munich. In 1884, he further discovered that carbohydrates gave crystalline phenylosazones in which two phenylhydrazines reacted with the aldehyde group and the carbon adjacent to the aldehyde group (reaction 2.2)

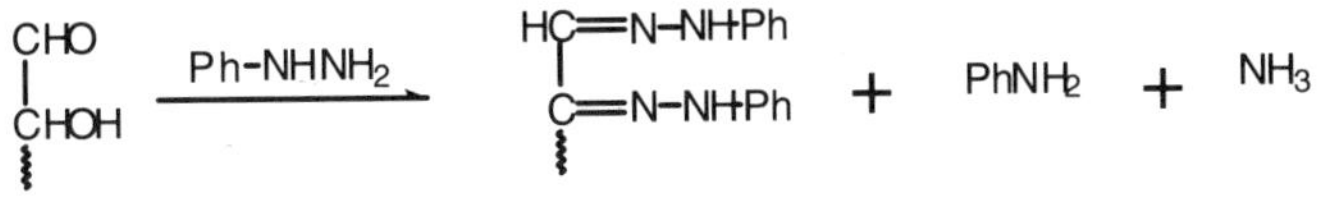

$$
\begin{array}{c}
\text{CHO} \\
| \\
\text{CHOH} \\
\vdots
\end{array}
\xrightarrow{\text{Ph-NHNH}_2}
\begin{array}{c}
\text{HC}=\text{N-NHPh} \\
| \\
\text{C}=\text{N-NHPh} \\
\vdots
\end{array}
+ \quad \text{PhNH}_2 \quad + \quad \text{NH}_3
\qquad (2.2)
$$

b. Fischer's Demonstration of the Structures of Glucose, Mannose, Arabinose, and Fructose

Fischer had taken a position at the University of Würzburg in 1885 and set out in 1888 to determine the structure of glucose. He made the following observations, arguments, and deductions after performing the indicated experiments.

1. (+)-Glucose is an aldohexose. The stereochemical theories of van't Hoff and Le Bel predicted that (+)-glucose would have four chiral centers and that it would be one of 2^4, or 16, possible stereoisomers. Of the 16 possible isomers, there were eight pairs of enantiomers. Fischer realized that he could simplify the problem by limiting the structure of glucose to a pair of enantiomers. He, therefore, considered only eight enantiomers by arbitrarily placing the C-5 hydroxyl group to the right, giving what is now called the D-family of carbohydrates.

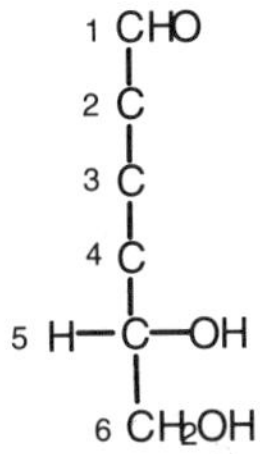

$$
\begin{array}{c}
1\ \text{CHO} \\
| \\
2\ \text{C} \\
| \\
3\ \text{C} \\
| \\
4\ \text{C} \\
| \\
5\ \text{H}-\text{C}-\text{OH} \\
| \\
6\ \text{CH}_2\text{OH}
\end{array}
$$

Fischer was well aware that this arbitrary decision had only a 50% chance of being correct. He also realized that if he was wrong, the correct structure would simply be the mirror image of the structure that he determined. The actual configuration of C-5 had to await 60 years before it was determined by X-ray crystallography.

2. (+)-Glucose, (+)-mannose, and (–) fructose all reacted with excess phenylhydrazine to give the same crystalline phenylosazone. This indicated to Fischer that the configurations of the three sugars at C-3, C-4, and C-5 were identical, and that glucose differed from mannose at *only* C-2. Fischer called the two sugars *epimers*. (–) Fructose was shown to be a ketone at C-2, instead of an aldehyde, and its configuration at C-3, C-4, and C-5 was the same as that of (+)-glucose and (+)-mannose, since all three of these sugars gave the same phenylosazone.

3. (+)-Arabinose reacted with HCN to give two monocarboxylic acids (A and B in reaction 2.3) that were mirror images. When (+)-mannose was oxidized with bromine-water, the product was found to be identical to one of the monocarboxylic acids (B) that was formed by reaction of (+)-arabinose with HCN.

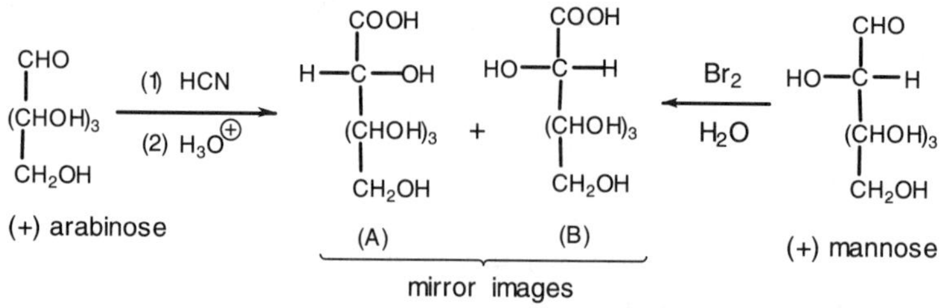

$$(2.3)$$

mirror images

Further, when the two acids were separated and the Fischer-Kiliani synthesis was performed, (+)-glucose and (+)-mannose were obtained. Fischer's proof then depended, in part, on the relationship between the structure of D-glucose and D-arabinose, and Fischer's problem was to determine the configurations of C-2 and C-3 of D-arabinose.

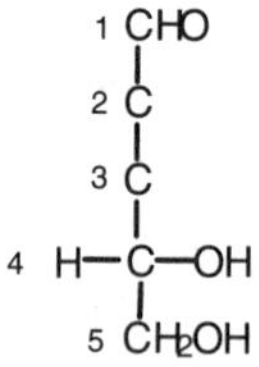

4. Oxidation of (+)-arabinose with warm nitric acid gave an optically active aric acid. Fischer deduced that the oxidation of one of the C-2 isomers of D-arabinose would also give an optically active aric acid, but oxidation of the other C-2 isomer of D-arabinose would give an optically inactive aric acid. Fischer argued that regardless of the configuration of C-3, the configuration of C-2 *had* to be the *left* to give an optically active aric acid because, if it was to the *right*, an optically inactive aric acid would result as shown by the reactions in 2.4:

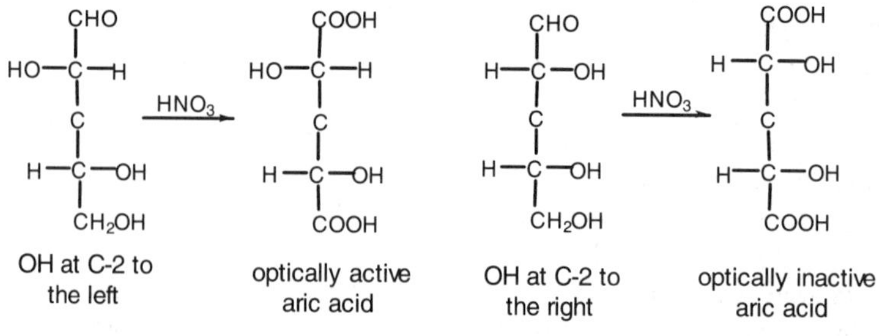

$$(2.4)$$

5. Because D-glucose and D-mannose had the same configurations at C-3, C-4, and C-5 and were each formed from D-arabinose by the Fischer-Kiliani syn-

thesis, they were epimers; the configurations at C-2 had to be opposite. So, the reaction of D-arabinose (structure G in reaction 2.5) in the Fischer-Kiliani synthesis gave two epimeric structures, H and J in reaction 2.5.

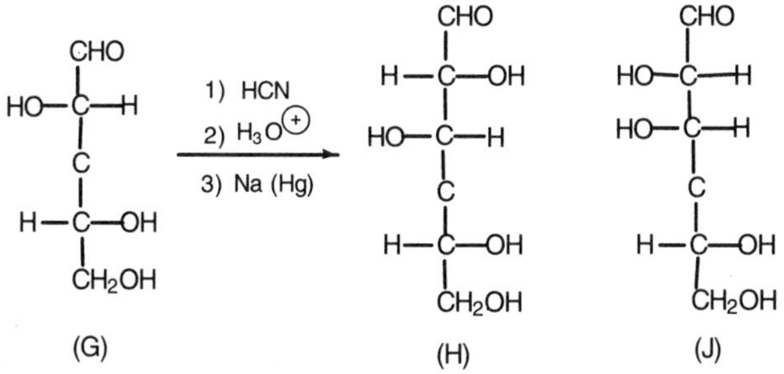

(2.5)

6. The problem now resolved into determining the configuration of C-4 for structures H and J (reaction 2.5). Fischer reasoned that because both D-glucaric acid and D-mannaric acid were optically active, the configuration of the OH group at C-4 had to be to the *right* (structure K below) because, if it were to the *left,* the aric acid of structure H would be optically inactive due to intramolecular symmetry (structure L, below).

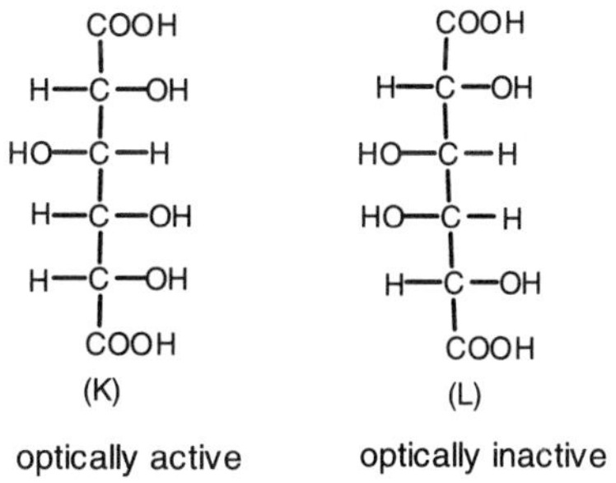

So, the structures for D-glucose and D-mannose were the following:

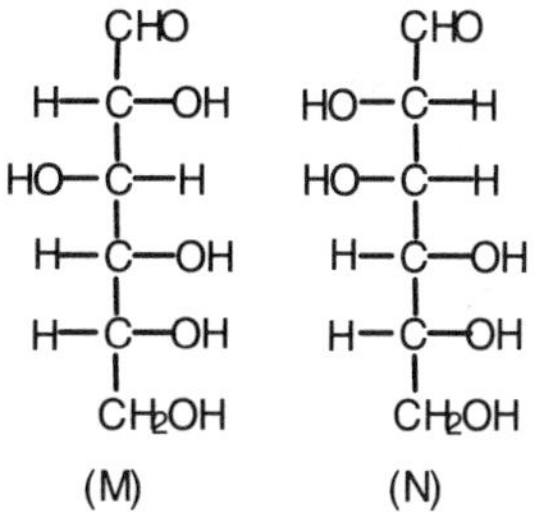

The problem, however, remained as to which one was D-glucose and which one was D-mannose.

7. Fischer contemplated the problem and reasoned that D-glucaric acid could be obtained by nitric acid oxidation of two different hexoses, structure M above

and L-gulose (structure P, below). D-Mannaric acid (structure N′, below), on the other hand, could be obtained only by the nitric acid oxidation of one hexose (structure N in reaction 2.6).

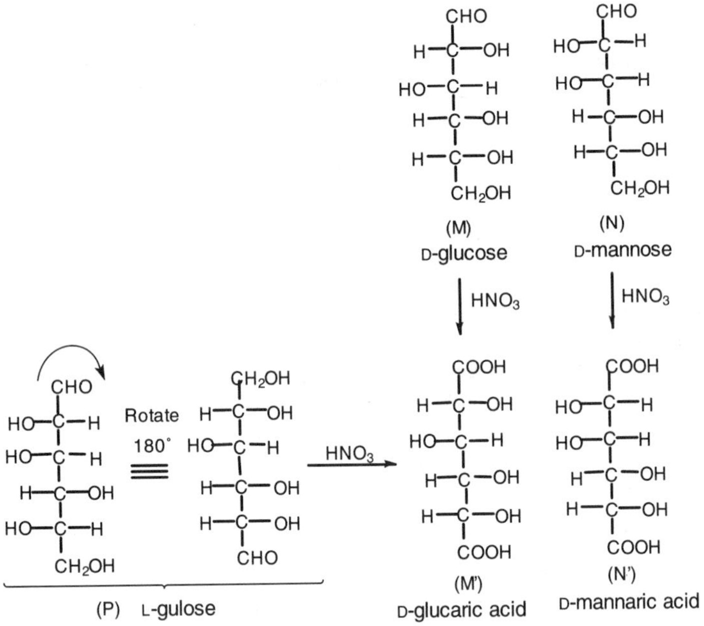

(2.6)

Conversion of D-glucose and L-gulose to D-glucaric acid

So, the structure for D-glucose was M and D-mannose was N, shown below:

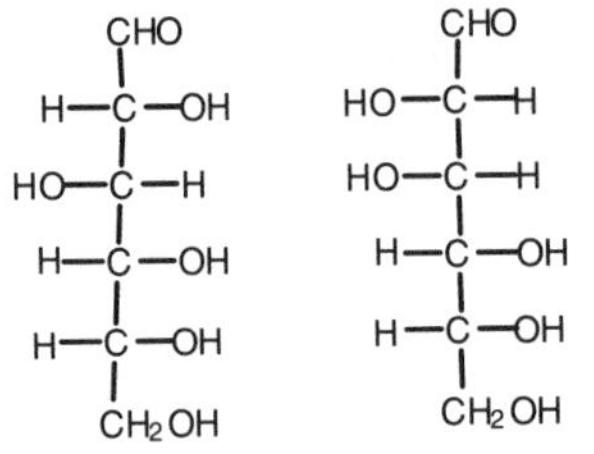

(M) D-glucose (N) D-mannose

Fischer confirmed the structures by synthesizing L-gulose and showing that its oxidation with nitric acid did indeed give D-glucaric acid, the same compound that was obtained by oxidizing D-glucose [4]. This also meant that the structure for D-arabinose was Q and D-fructose was R, shown below:

By 1891, Fischer had established the structures of these carbohydrates and published the results in three papers [4–6]. He continued his studies for several more years, determining the structures of all of the eight D-aldohexoses and four

D-aldopentoses. He correlated the structures of the hexoses and pentoses with the structures of the tartaric acids by reducing the size of the sugar chains one carbon at a time using the Ruff degradation (see Chapter 4, section 4.5.e.i). He performed two Ruff degradations followed by nitric acid oxidation on the hexoses, and one Ruff degradation followed by nitric acid oxidation on the pentoses [7]. The absolute configuration of (+)-tartaric acid was determined 60 years later by Bijvoet et al. [8,9]. From this the configurations of (–)-tartaric acid and *meso*-tartaric acid could be deduced. Thus, the absolute structures at C-5 that Fischer obtained by Ruff degradations and nitric acid oxidations on D-glucose, D-mannose, D-arabinose, and D-xylose were confirmed, and it turned out that Fischer had chosen the correct configuration for C-5 when he made his apparently arbitrary decision to place the hydroxyl group to the right.

Following his ingenious experiments, using only chemical reactions, crystallizations, optical rotation, and his brilliant deductions, Fischer was awarded the Nobel prize in chemistry in 1902 for his monumental studies on the structures of carbohydrates.

The configurations of the members of the family of D-aldoses, from D-glyceraldehyde to the eight D-hexoses, are shown in Fig. 1.2 of Chapter 1. All of the D-sugars in this figure can be shown to be derived from D-glyceraldehyde by using the Fischer-Kiliani cyanohydrin synthesis. In 1917, Wohl and Momber [10] confirmed this by synthesizing D-glucose from (+)-D-glyceraldehyde.

2.3 Cyclic Structures for the Carbohydrates

For many, it seemed that the structures of the hexoses and pentoses had been definitively demonstrated by Emil Fischer, and that the chapter on the structures of carbohydrates was closed. However, there were certain troubling facts that were not entirely consistent with the structures proposed by Fischer.

1. The carbohydrates failed to undergo certain reactions that were typical of aldehydes. Although the hexoses and pentoses were readily oxidized at C-1 under mild alkaline conditions, they did not give a positive result with the Schiff test (reaction with basic fuchsin), and they did not form bisulfite addition products, reactions that are typical of aldehydes. The cyanohydrin and phenylhydrazine reactions also went much more slowly than they did for other α-hydroxy aldehydes.

2. Acetylation produced two isomeric pentaacetates. Similarly, the formation of methyl glucoside by reaction of glucose in methanol with hydrogen chloride as a catalyst also produced two isomeric glycosides, each containing only one methyl group. The methyl glycosides had the properties of an acetal in that they did not spontaneously revert to an aldehyde and an alcohol when dissolved in water. The formation of the two isomeric glycosides and the two isomeric pentaacetates was not predicted from Fischer's structures.

Emil Fischer was born on October 9, 1852, and grew up in Euskirchen, a small village near Bonn in western Germany. In 1870, Fischer entered the University of Bonn and studied chemistry under August von Kekulé and Adolf von Baeyer. In 1872, he transferred to the University at Strasbourg with von Baeyer and received a doctorate, working on the structure of fluorescein. In 1878, he qualified for the rank of Privatdozent at the University of Munich, with a thesis on the synthesis of substituted hydrazines. In 1882, at the age of 30, he became a full professor and Director of the Chemical Institute at the University of Erlangen. Then in 1885, he moved to the University of Würzburg as a professor and commenced his studies on the structures of carbohydrates. In 1891, he published three papers describing the structures of glucose, fructose, mannose, and arabinose. In 1892, at age 40, he accepted the Chair of Chemistry at the University of Berlin. At that time, this was the highest professorial chair in chemistry in Germany. In spite of being saddled with heavy administrative responsibilities at the University of Berlin, Professor Fischer made time to continue to work in the laboratory on carbohydrates, developing among other things the glycal synthesis for 2-deoxyglucose and the synthesis of alkyl glycosides. He also began studies on amino acids, proteins, and nucleic acids. The combination of his studies on sugars, polysaccharides, amino acids, proteins, and nucleic acids led him to understand and to establish experimental evidence as to how bifunctional monomer units were combined to give biopolymers of high molecular weight. Emil Fischer was also very interested in the action of enzymes, especially the high degree of specificity that they display. He developed the "lock and key" hypothesis to explain this specificity. His hypothesis has turned out to be essentially correct. Under his leadership and influence, synthetic organic chemistry and theoretical chemistry were united with biochemistry, which became established as a legitimate field of study. It thus can legitimately be claimed that he is the father of both carbohydrate chemistry and biochemistry.

3. Two crystalline forms of D-glucose were obtained. One crystallized at room temperature (20–21°C) and had a melting point of 146°C and specific optical rotation, $[\alpha]_D^{20}$, of +112°; the other crystallized at temperatures above 98°C and had a melting point of 150°C and specific optical rotation of +19°. When these two crystalline D-glucoses were dissolved in water and their optical rotation was measured over time, it was found that the specific optical rotation of the first form slowly dropped to a value of +52.7° and did not change; the specific optical rotation of the second form of D-glucose slowly increased from +19° to give the same constant rotation of 52.7°. Some believed that this suggested D-glucose existed in three forms: α, β, and γ, with the α-form having $[\alpha]_D^{20}$ equal to +112°, the β-form having $[\alpha]_D^{20}$ equal to +19°, and the γ form having $[\alpha]_D^{20}$ equal to +52.7°. The change in the specific optical rotation with time was called *mutarotation* and is illustrated in Fig. 2.1. However, some sugars such as sucrose and the methyl glycosides did not have two crystal forms, and their specific optical rotations did not change on standing in solution.
4. The fact that the formation of the methyl glycosides and the pentaacetates resulted in only two isomeric forms each, suggested that D-glucose and related hexoses and pentoses existed in only two forms and not three.
5. The phenomenon of mutarotation for D-glucose had been recognized since 1846. Also as early as 1856, two different crystalline forms of lactose had been obtained.
6. It was shown that aldehydes would react with alcohols to give *hemiacetals* that could have two isomeric forms as shown in reaction 2.7.

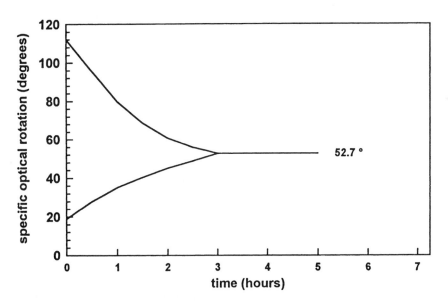

Figure 2.1. Change in the specific optical rotations of water solutions of the two forms of D-glucose (α-D-glucose, [α] = +112° and β-D-glucose, [α] = +19°) with time.

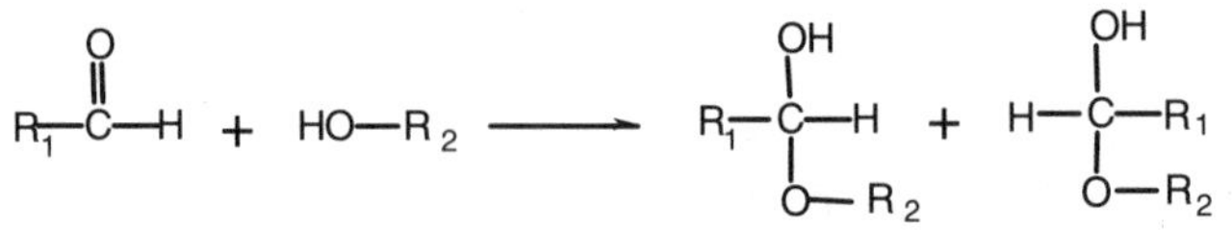

$$R_1\text{--}\overset{\overset{\displaystyle O}{\|}}{C}\text{--}H \ + \ HO\text{--}R_2 \ \longrightarrow \ R_1\text{--}\overset{\overset{\displaystyle OH}{|}}{\underset{\underset{\displaystyle O\text{--}R_2}{|}}{C}}\text{--}H \ + \ H\text{--}\overset{\overset{\displaystyle OH}{|}}{\underset{\underset{\displaystyle O\text{--}R_2}{|}}{C}}\text{--}R_1 \qquad (2.7)$$

hemiacetal enantiomers

These two forms would be opposite in configuration around the new asymmetric center formed from the aldehyde carbon and would have different specific optical rotations. They are called *anomers*.

7. These hemiacetal compounds can react with a second molecule of alcohol and split out water to form stable *acetals* according to reaction 2.8:

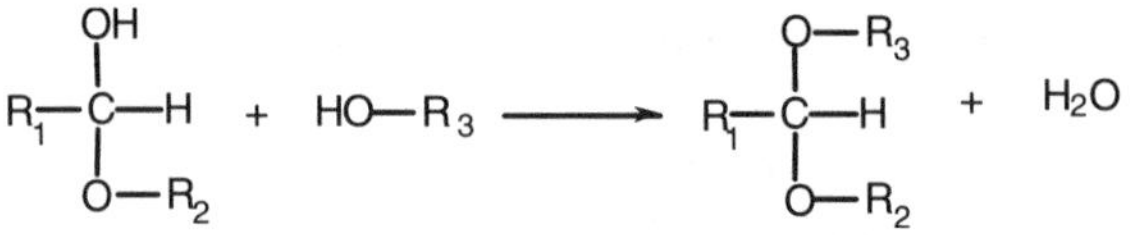

$$R_1\text{--}\overset{\overset{\displaystyle OH}{|}}{\underset{\underset{\displaystyle O\text{--}R_2}{|}}{C}}\text{--}H \ + \ HO\text{--}R_3 \ \longrightarrow \ R_1\text{--}\overset{\overset{\displaystyle O\text{--}R_3}{|}}{\underset{\underset{\displaystyle O\text{--}R_2}{|}}{C}}\text{--}H \ + \ H_2O \qquad (2.8)$$

acetal

Hence, the two anomeric hemiacetal forms can form two isomeric acetals, accounting for the formation of two methyl-D-glycosides and two forms of pentaacetates.

8. Thus it was argued that because the carbohydrates contain both an aldehyde and several alcohol groups, they could form two intramolecular hemiacetals and, hence, two isomeric acetals when they react with another exogenous alcohol such as methanol.

9. The problem was which one of the particular alcohol hydroxyl groups in the carbohydrate structure was reacting with the aldehyde group to form the two hemiacetals.

10. Two structures were proposed, the Cholley structure and the Tollens structure:

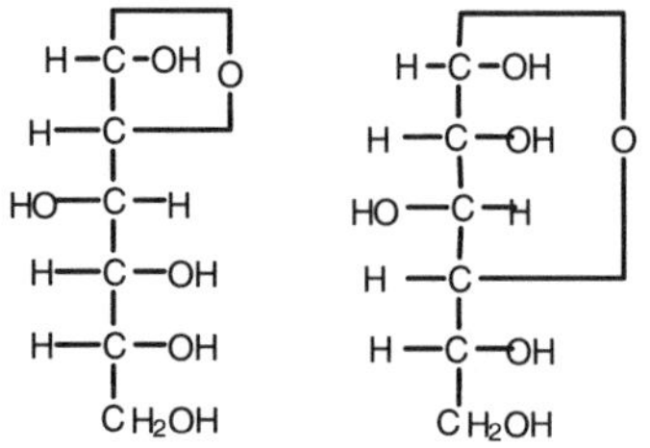

Cholley Structure   Tollen Structure

Most chemists of the time favored the structure proposed by Tollens.

11. It was not until the period from 1920 to 1930 that the size of the rings was definitively determined by the English carbohydrate research group at Birmingham. This group eventually was led by W. Norman Haworth. These researchers found that the hexoses formed six-membered rings preferentially

with the hemiacetal oxygen in the ring to give two anomeric hemiacetals. These were called α- and β-D-glucopyranose, and they had the following Fischer structures:

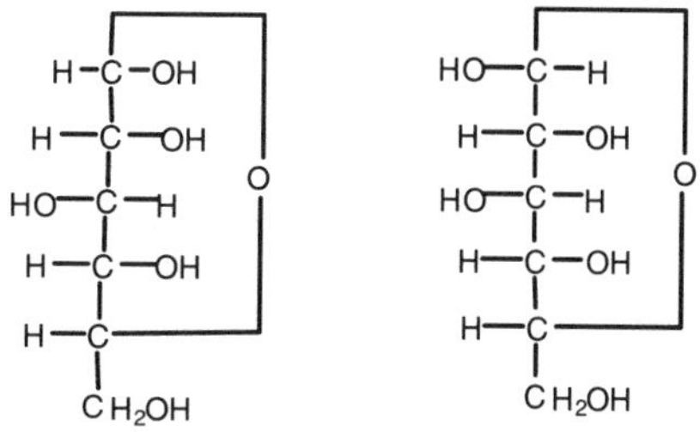

α -D-glucopyranose
Fischer projection

β-D-glucopyranose
Fischer projection

These structures were called *pyranoses* after the six-membered heterocycle, pyran, with oxygen in the ring.

By building models using the known carbon-carbon and carbon-oxygen bond lengths and angles, Haworth realized that the Fischer structures, although correct in terms of the configurations of the chiral centers, were not an accurate description of the cyclic structures. Haworth proposed a more accurate picture of cyclic carbohydrate structures with equal lengths of carbon-carbon and carbon-oxygen bonds, and tetrahedral angles of 109° [11]. This is called the *Haworth projection*. It can be drawn in 12 different orientations. It has become customary to use only one of these orientations whenever possible. This is the so-called *standard orientation* that has C-1, the hemiacetal carbon, to the extreme right, and the ring oxygen in the upper right corner. The numbering of the carbons then proceeds in a clockwise manner.

Sometimes five-membered rings also occur, although they usually are minor structures unless it is impossible to form a six-membered ring. In this case, C-1 is still placed to the extreme right, with the ring oxygen in the upper center. These are called *furanose* rings, after the five-membered oxygen heterocycle, furan.

The standard Haworth formulas for pyranose and furanose rings are the following:

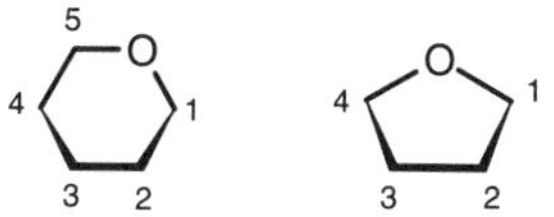

Pyranose   Furanose

Whenever possible, these are the orientations that are used to prevent confusion. There are times, however, when it is desirable to turn the ring over from top to bottom, or from left to right, or to rotate the ring. This especially is required when glycosidic bonds between sugars need to be drawn (see Chapter 3, Figs. 3.10 and 3.11).

32 2: Developments

To convert a Fischer projection into a Haworth projection, the standard ring is drawn. For a D-sugar, the bottom primary hydroxyl group in the Fischer projection is placed *up* at C-5 in the Haworth projection. Then all of the substituents to the right in the Fischer projection are placed *down* on the Haworth ring, and all of the substituents to the left in the Fischer projection are placed *up*. The following examples illustrate the conversion:

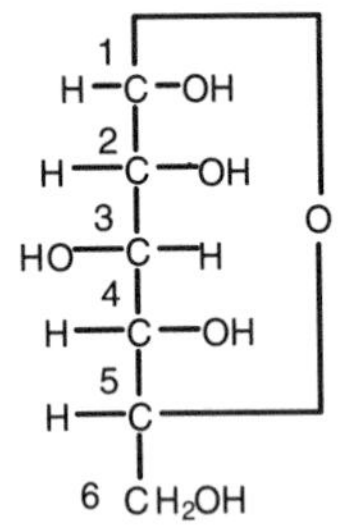

Fischer projection α-D-glucopyranose

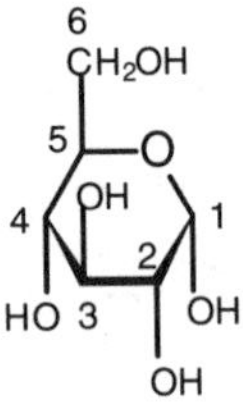

Haworth projection α-D-glucopyranose

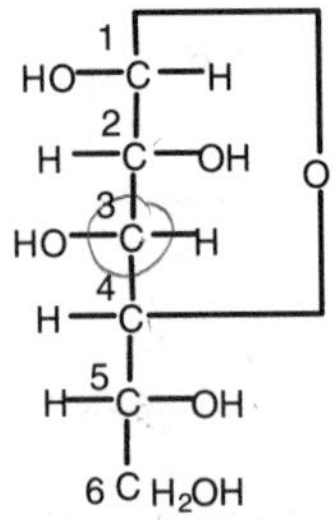

Fischer projection β-D-glucofuranose

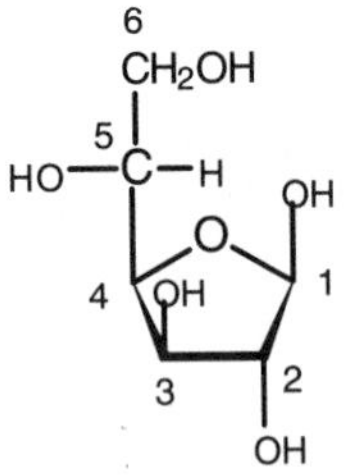

Haworth projection β-D-glucofuranose

The following illustrate some other orientations for α-D-glucopyranose:

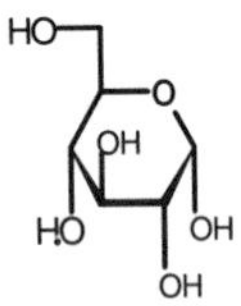

α-D-glucopyranose Standard orientation

I

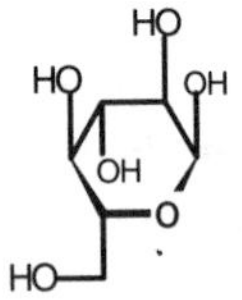

turned over top-to-bottom

II

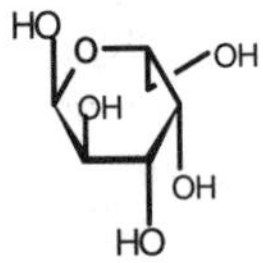

turned over right-to-left

III

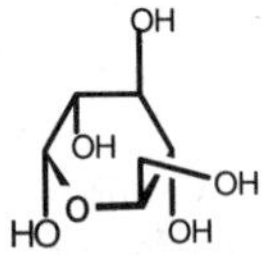

formula III turned over top-to-bottom

IV

One has to be careful not to confuse structures II and III for the mirror images of structure I because the mirror image would be L-glucose, as shown below:

MIRROR

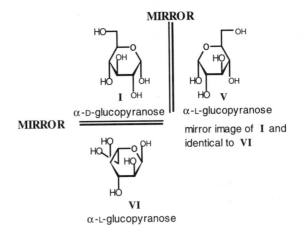

I α-D-glucopyranose

V α-L-glucopyranose
mirror image of I and identical to VI

MIRROR

VI
α-L-glucopyranose

2.4 Naming the Anomeric Forms of Carbohydrates

The formation of two isomeric methyl-D-glucosides required a systematic nomenclature to identify them. In 1909, C. S. Hudson, an American carbohydrate chemist, defined that system. He stated that for D-sugars, the form that is most dextrorotatory of an α-/β- pair of anomers should be named α-D-, and the other form β-D-. For the L-sugars, the more levorotatory member of such a pair should be named α-L-, and the other β-L-. If the pair of D-sugars are not glycosides, and mutarotate, the form that mutarotates downward should be called α-D-, and the form that mutarotates upward should be β-D- [12]. This was called Hudson's rule. The form of D-glucose that has a specific optical rotation of +112° would be α-D-glucose, and the form that has a specific optical rotation of +19° would be β-D-glucose.

These rules were empirical and based on observations of limited samples of compounds. Subsequent to Hudson's rule, it was found that there were exceptions to the rule, and it was proposed that a better system of nomenclature would be based on the absolute configurations in which the hemiacetal hydroxyl that is on the same side (cis) as the hydroxyl group reacting with the aldehyde or ketone is α-, and the hemiacetal hydroxyl that is on the opposite side (trans) is β-.

2.5 Determination of the Size of Carbohydrate Rings

It was 20 years later when the English carbohydrate research group at Birmingham experimentally demonstrated the size of the carbohydrate ring. Hirst and Purves methylated methyl-α-D-xyloside with dimethyl sulfate, removed the methyl acetal group by dilute acid hydrolysis (the methyl ethers that formed by methylating the free hydroxyl alcohol groups were stable to acid hydrolysis), and oxidized the product with nitric acid, after which the structures of the cleavage products were determined [13]. They later extended the study to methyl-α-D-glucoside [14,15]. From these analyses, they deduced the size of the ring. The following summarizes the reactions carried out by periodate for methyl-D-glucoside and gives the structures of the cleavage products.

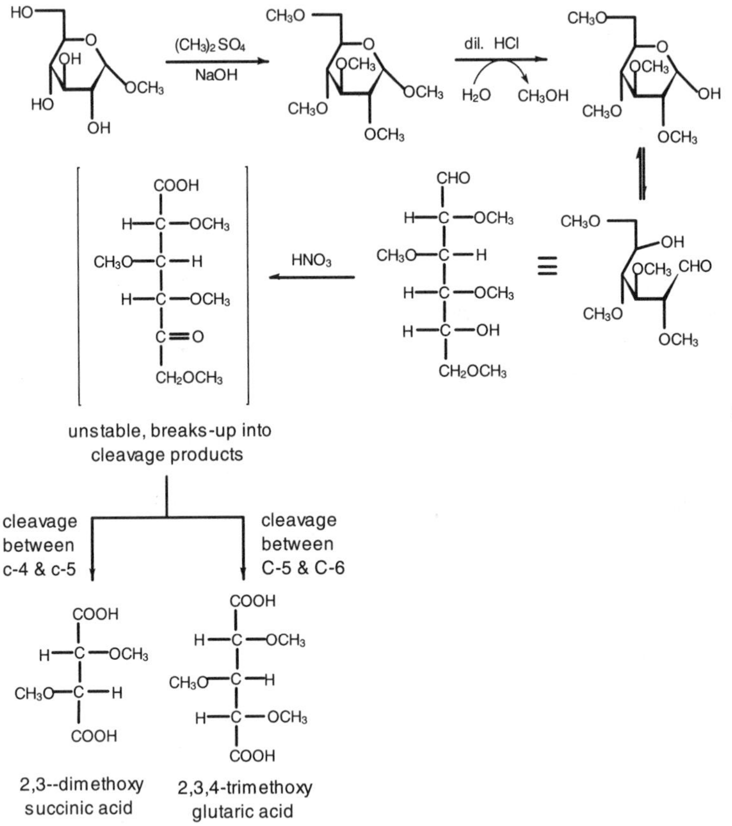

(2.9)

The formation of the four-carbon dimethoxy and five-carbon trimethoxy dicarboxylic acids indicated that the acetal ring of methyl-α-D-glucoside was between C-1 and C-5, and the ring was, therefore, a six-membered pyranoside.

Actually, four methyl-D-glycosides were obtained for D-glucose. Two of them made up the major ring structure for methyl-D-glucoside and both had six-membered rings and were methyl-α-D-glucopyranoside and methyl-β-D-glucopyranoside. The other two were produced in much lesser amounts. Haworth performed a similar study for these minor forms and found that the cleavage products were dimethoxy-D-glyceric acid and dimethoxy-succinic acid, indicating cleavage between C-4 and C-5, and C-3 and C-4. This showed that the minor forms had five membered rings and were methyl-α-D-glucofuranoside and methyl-β-D-glucofuranoside.

The size of the rings were subsequently confirmed by C. S. Hudson and colleagues by using periodate oxidation and the determination of the amount of periodate consumed and the nature of the oxidized products [16,17]. It was shown that periodate oxidation is stoichiometric for the cleavage of carbon-carbon bonds whose carbons each have a hydroxyl group. The oxidation cleaves the bond with one periodate and produces two aldehydes. Periodate oxidation of a carbon bond to a hydroxy methyl group ($-CH_2-OH$) gives formaldehyde, and oxidation of a series of two adjacent carbon bonds between these carbons, each with a hydroxyl group, uses two periodates and gives formic acid from the center carbon and two aldehydes. The periodate oxidations are summarized in reactions 2.10–2.12:

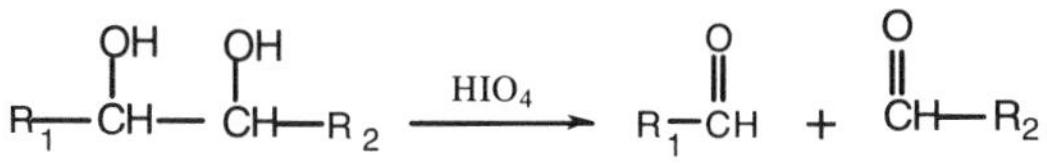

(2.10)

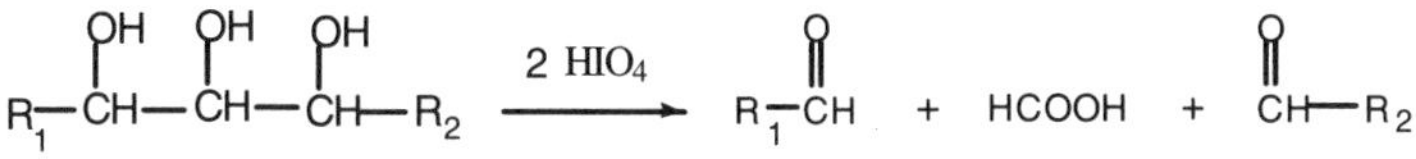

(2.11)

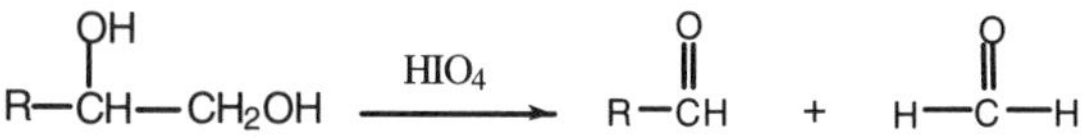

(2.12)

The amounts of periodate consumed and the amount of formic acid and formaldehyde can be quantitatively determined. Thus, for methyl-α- (or β)-D-glucopyranoside, 2 moles of periodate were consumed and 1 mole of formic acid was formed per mole of glucopyranoside. The wavy line indicates the site of periodate carbon bond cleavage.

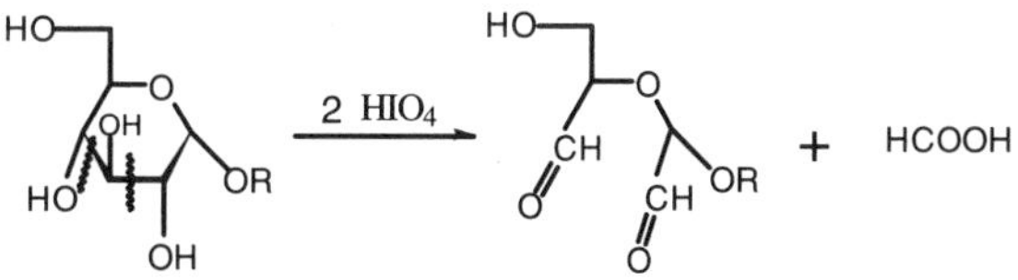

(2.13)

It was thus confirmed that the structures of the major forms of methyl-D-glucoside were six-membered pyranosides. When the minor forms were oxidized, 2 moles of periodate was consumed, but 1 mole of formaldehyde was formed instead of 1 mole of formic acid per mole of glucoside, as shown in the following reaction.

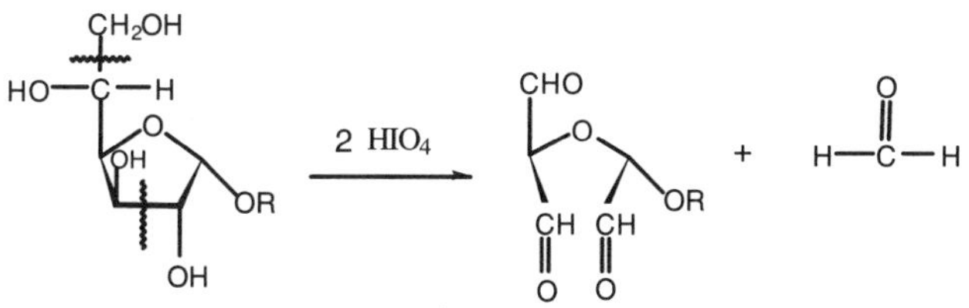

(2.14)

With the development of nuclear magnetic resonance spectroscopy (NMR), the size of the ring in solution and the amounts of each form at equilibrium have been determined [18,19].

2.6 Conformational Structures of Cyclic Sugars

The Fischer and Haworth projections for the carbohydrates do provide the relative spatial configuration of the hydroxyl groups. Of these projections, the Haworth projections represent the cyclic sugars better than the Fischer projections. Three-dimensional model building using the correct bond lengths and bond angles of the tetrahedral carbons showed that the rings are not planar as they appear in the Haworth projection. Further, rotations about the sigma bonds between the carbon atoms and about the carbon-oxygen atoms in the ring can provide an infinite number of shapes for the ring in three dimensions. The shape of the ring and the positions of the hydroxyl groups and hydrogen atoms in relation to the ring is called *conformation.* Of the many possible conformations, there is usually a "favored conformation" that is the shape with the minimum net energy. The various factors that contribute to the favored conformation involve bond-angle strain, bond-torsional strain, steric interactions of the atoms and groups, dipolar interactions, hydrogen-bonding effects, solvation effects, and hydrophobic interactions.

For crystalline sugars, a single, most-favored conformation is usually assumed. For sugars in solution, an equilibrium of different conformations is obtained in which the most favored conformation is found in the greatest amount.

Of the infinite conformations that pyranose sugars can have, there are two extremes that resemble chairs. The two chair conformations are usually nonequivalent, such that one is the most favored conformation and the other is the least favored conformation. There are two types of bonds around each of the carbon atoms in the ring. These bonds are either within the plane of the ring and are called *equatorial bonds,* or they are perpendicular to the plane of the ring and are called *axial bonds* [19].

For the D-hexoses, the C1 conformation (reaction 2.15) is defined as the form in which the primary hydroxyl group ($-CH_2-OH$) is equatorial. It is also the form in which C-1 is down and C-4 is up when the orientation of the ring is in the standard position. The C1 conformation is, thus, sometimes called a $^4C_1$ conformation. The 1C conformation is then the form in which the primary hydroxyl group is perpendicular to the ring, or axial. In this form, the C-1 is up and C-4 is down; this is called a $_4C^1$ conformation. For the pyranose ring forms of the D-pentoses, the designation of the conformation has to use the positions of C-1 and C-4, since there is no primary hydroxyl group.

The most stable or most favorable conformation usually is the one that places the majority of the bulky substituents (for most of the carbohydrates a bulky substituent is a hydroxyl group or a hydroxy methyl group) in an equatorial position, and, likewise, the least favored conformation is the one that places the majority of the bulky groups in an axial position. The placing of the bulky groups in the plane of the ring, or equatorial position, puts the bulky groups as far apart from each other as possible, creating a low-energy form with a minimum of bulky group interactions. Placing the bulky groups perpendicular to the ring, in an axial position, puts the bulky groups as close together as possible, creating a higher energy form with a maximum of interaction. This can be illustrated for β-D-glucopyranose in which all of the bulky groups are equatorial when the molecule in the $^4C_1$ (C1) conformation. An intermediate in the transformation from C1 to 1C is a so-called *boat conformation* in which C-1 has been moved up by rotation around the ring atoms [20].

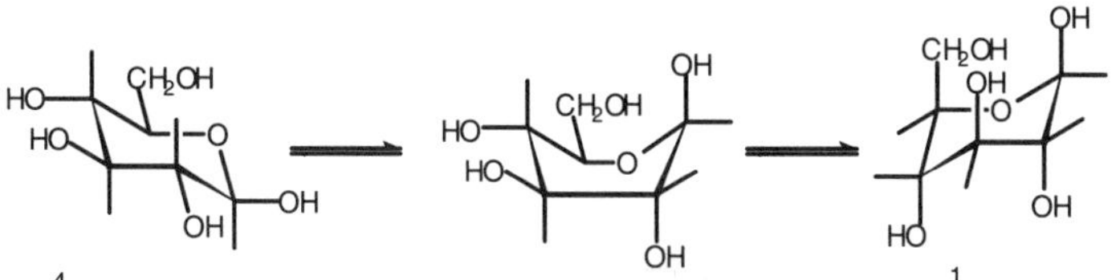

$$^4C_1 \ (C1) \qquad\qquad\qquad\qquad\qquad\qquad\qquad _4C^1 \ (1C) \tag{2.15}$$

conformation of β-D-glucopyranose

a boat confirmation conformation of β-D-glucopyranose

conformation of β-D-glucopyranose

It should be noted that the different conformations, called *conformers,* all have the *same* configurations for the hydroxyl groups. Thus, for β-D-glucopyranose, the β-hemiacetal hydroxyl group is up, C-2 hydroxyl group is down, C-3 hydroxyl group is up, and so forth in the C1, 1C, and boat conformations.

β-D-Glucopyranose is the *only* D-hexose that can have all of the bulky groups in the equatorial position. β-D-mannopyranose has the C-2 hydroxyl group axial, β-D-allopyranose has the C-3 hydroxyl group axial, and β-D-galactopyranose has the C-4 hydroxyl group axial when the conformation is $^4C_1$ (C1). It has been speculated that this is the reason that the evolutionary process selected glucose as the predominant carbohydrate structure on the Earth.

β-D-Xylopyranose also can assume a $^4C_1$ conformation with all of the bulky substituents in an equatorial, but it apparently was not favored in the evolutionary process because it has an odd number of carbon atoms and hence would require an asymmetric metabolism, that is metabolism of a two-carbon fragment and a three-carbon fragment, requiring two distinct metabolic pathways. Whereas a hexose such as D-glucose would require only a single pathway for the metabolism of two three-carbon fragments (see Chapter 11 for the pathway).

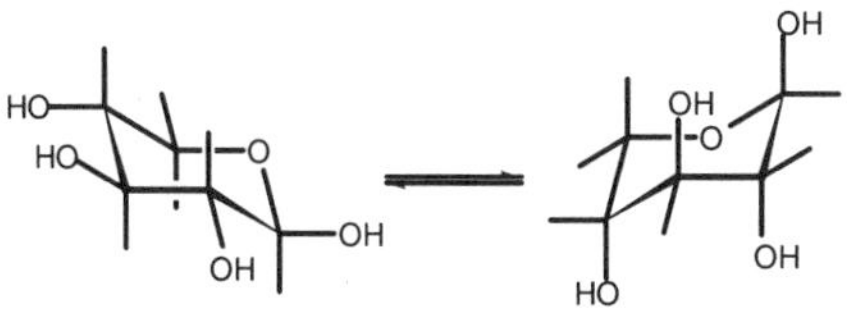

$$^4C_1 \qquad\qquad\qquad\qquad\qquad _4C^1 \tag{2.16}$$

conformation of β-D-xylopyranose

conformation of β-D-xylopyranose

An interesting D-hexose conformation is found in β-D-idopyranose; the $^4C_1$ (C1) conformation has three axial hydroxyl groups, and the $_4C^1$ (1C) conformation has one hydroxyl group axial and a hydroxy methyl group that is axial. These two conformers have about equal energy and hence at equilibrium occur in about equal amounts. D-Idose has never been crystallized, possibly because of the presence of the two conformers in about equal amounts.

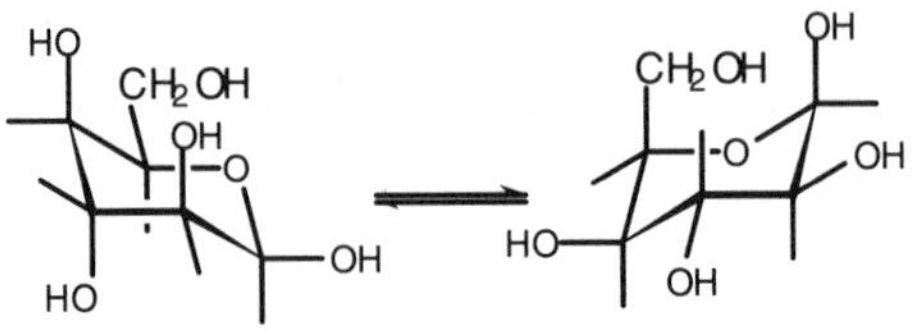

(2.17)

β-D-idopyranose in two conformations with three
and two bulky groups that are axial, giving two
conformations that are of about equal energy

The five-membered furanose rings are also not planar. The furanose ring can have an *envelope conformation,* with four atoms coplanar and the fifth out of the plane (the latter can be any of the four carbons or the oxygen), or it can have a *twist conformation* in which three adjacent atoms are coplanar and the other two adjacent atoms are above and below the plane of the other three atoms. For any furanose ring, there are 10 possible envelope forms and 10 possible twist forms. The energies of the 20 possible forms are all about the same, and rapid interconversions can take place between them.

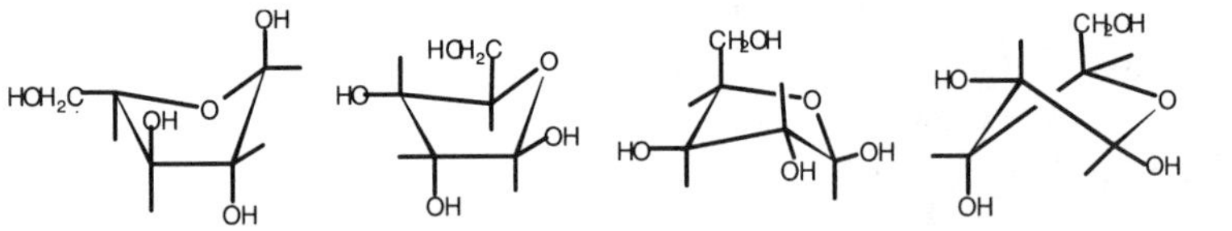

Three envelope conformations and one twist conformation of β-D-xylofuranose

In the case of the hexose lactones, four of the six atoms in the ring (O, C-1, C-2, and C-3) are constrained to coplanarity by the sp2 carbon-oxygen double bond, resulting in a *half-chair conformation* or what is sometimes called a *sofa conformation.*

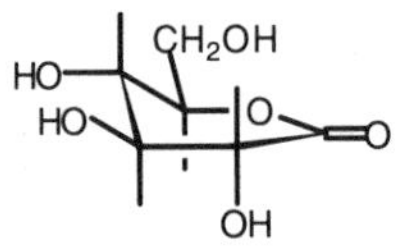

half-chair or sofa conformation
for δ-D-gluconolactone

The conformation of the sugar molecules must be kept in mind because the three-dimensional shape very often plays an important, albeit subtle, role in the biological function of carbohydrates. The Fischer or Haworth formulas are usually easier to draw and can often be used since they convey the necessary stereochemical information to distinguish the various sugars from each other. The conformational formulas, however, are often easier to use when two or more carbohydrate units are joined together by glycosidic linkages (see Chapter 3, section 3.7, and Fig. 3.10). Care has to be used in drawing the conformational for-

mulas so that the various equatorial bonds are in the plane of the ring parallel to
the bonds in the ring to which they are opposed.

2.7 The Literature of Carbohydrate Chemistry

The early studies on carbohydrates in the nineteenth century by Emil Fischer and
others were published mainly in the German general chemical journals such as
Justus Liebigs Annalen Chemie (most often just called *Annalen Chemie*), *Chemis-
che Berichte* (most often called just *Berchite*), and *Zeitschrift für physiologie
Chemie*. With the development of carbohydrate studies in England and the United
States in the early part of the twentieth century, the English researchers published
in the *Journal of the Chemical Society*, and the Americans published in the *Journal of the American Chemical Society* and the *Journal of Biological Chemistry*.

A general book titled *Carbohydrate Chemistry* was published in the United
States in 1948 by Ward W. Pigman and Rudolph M. Goepp, Jr. It consisted of 500
pages and was considered an extensive work. Pigman then published a second
edition in 1957, that was 900 pages, nearly double in size. A three-volume work
titled The *Carbohydrates* was edited by Pigman and Derek Horton in 1972. A
fourth volume was added in 1980. The four volumes consisted of individual chapters by many authors who were expert in the specific topics in the chapters. The
work consisted of over 2,000 pages.

An *Advances in Carbohydrate Chemistry* series was started in 1945 (volume
1), with a new volume appearing every year. In 1969, the name was changed to
Advances in Carbohydrate Chemistry and Biochemistry. The series continues
today and is a compendium of detailed reviews, four to five per volume, of carbohydrate subjects and authored by experts knowledgeable in the subjects. The
work reflects the knowledge of the time and usually includes background and history, often including a biography of individual(s) that contributed a lifetime to the
study of carbohydrates.

With the expansion of scientific studies, which included carbohydrates, in the
middle of the twentieth century, two important works on carbohydrates appeared.
The first was a methods series, *Methods in Carbohydrate Chemistry*. Volume 1
appeared in 1962, and other volumes followed after that. It provided detailed procedures for the isolation, analysis, and synthesis of carbohydrates. Again, each
chapter or section was authored by an expert that was actively working in the field
and knew the important details that were necessary for successfully carrying out
the described procedures. To date, there are nine volumes published in various
years. The principal editor has been Roy Whistler of Purdue University. The volumes, titles, year of publication, and editors are the following:

Vol.	Title	Year	Editors
I	*Analysis and Preparation of Sugars*	1962	R. L. Whistler and M. L. Wolfrom
II	*Reactions of Carbohydrates*	1963	R. L. Whistler and M. L. Wolfrom
III	*Cellulose*	1963	R. L. Whistler

IV	Starch	1964	R. L. Whistler
V	General Polysaccharides	1965	R. L. Whistler
VI	General Carbohydrate Methods	1972	R. L. Whistler and J. N. BeMiller
VII	General Methods, Glycosaminoglycans, and Glycoproteins	1976	R. L. Whistler and J. N. BeMiller
VIII	General Methods	1980	R. L. Whistler and J. N. BeMiller
IX	General Methods	1995	R. L. Whistler and J. N. BeMiller

About the same time, a new international research journal, *Carbohydrate Research*, appeared (volume 1) in 1964. It has become the principal journal for the publication of all aspects of carbohydrate research, such as chemical synthesis and modification; enzymatic synthesis; organic and enzymatic mechanisms involving carbohydrates; carbohydrate metabolism; conformational analysis; isolation of carbohydrates from natural sources; analytical methods; chemistry of monosaccharides, oligosaccharides, and polysaccharides; biological function of carbohydrates; and physical properties. Approximately four to five volumes are published each year.

A more recent international journal, *Journal of Carbohydrate Chemistry,* was introduced in 1982. It requires camera-ready copy and is devoted primarily to the organic and physical chemistry of carbohydrates, such as novel synthetic methods; mechanisms involved in carbohydrate reactions; uses of carbohydrates in the synthesis of natural products, drugs, and antibiotics; use of carbohydrates as synthetic reagents; separation methods as applied to carbohydrate reactions and synthesis; spectroscopic and crystallographic structure studies of carbohydrates; molecular modeling studies; and the chemistry of carbohydrate polymers, oligosaccharides, polysaccharides, and glycoconjugates.

About the same time (1981), another international journal, *Carbohydrate Polymers,* was introduced. Its emphasis is on technical aspects of industrially important polysaccharides. It covers the study and uses of industrial applications of carbohydrate polymers in foods, textiles, paper, wood, adhesives, and pharmaceuticals and includes topics concerning structure and properties; biological and industrial development; analytical methods; chemical, microbiological, and enzymatic modifications; and interactions with other materials.

The newest carbohydrate journal is *Carbohydrate Letters.* Its purpose is to provide fast publication of short notes on carbohydrate chemistry.

Articles on carbohydrates appear in other journals as well. Articles concerning chemical reactions and modifications are often found in the *Journal of Organic Chemistry, Tetrahedron,* and *Tetrahedron Letters.* Articles on biochemical, physiological, and enzymatic reactions of carbohydrates can be found in *Archives of Biochemistry and Biophysics, Biochemistry, Biochimica et Biophysica Acta, Biochemical Journal, Bioscience Biotechnology and Biochemistry* (formerly known as *Agricultural and Biological Chemistry*), *European Journal of Biochemistry, Journal of Biological Chemistry,* and *Starch/Stärke* (formerly known as *Die Stärke*).

Several monographs on specific topics have appeared in the twentieth century.

A Comprehensive Survey of Starch Chemistry, R. P. Walton (1928).

Polarimetry, Saccharimetry, and the Sugars, U.S. Dept. of Commerce C440, F. J. Bates, ed. (1942).

Chemistry and Industry of Starch, R. W. Kerr (1944).

Chemistry and Industry of Starch, R. W. Kerr, 2nd ed. (1950).

Polysaccharide Chemistry, R. L. Whistler (1953).

Recent Advances in the Chemistry of Cellulose and Starch, J. Honeyman (1959).

Industrial Gums, Polysaccharides, and Their Derivatives, 1st edn., R. L. Whistler and J. N. Bemiller, eds. (1959).

Starch Chemistry and Technology, 1st edn., R. L. Whistler and E. F. Paschall, eds. (1965).

Starch and Its Derivatives, 4th edn., J. A. Radley, ed. (1968).

The Carbohydrates, Chemistry and Biochemistry, four volumes, W. Pigman and D. Horton, eds. (1972–1980).

Industrial Gums, Polysaccharides, and Their Derivatives, 2nd edn., R. L. Whistler and J. N. BeMiller, eds. (1973).

Sugar Chemistry, R. S. Shallenberger and G. G. Birch (1975).

Biochemistry of Carbohydrates (MTP International Review of Science, Vol. 5) W. J. Whelan, ed. (1975).

Extracellular Microbial Polysaccharides (ACS Symposium Series 45) P. A. Sandford and A. Laskin (1977).

Microbial Polysaccharides and Polysaccharases, R. C. W. Berkeley, G. W. Gooday, and D. C. Ellwood, eds. (1979).

Cell Surface Carbohydrate Chemistry, R. E. Harmon, ed. (1978).

Advanced Sugar Chemistry, R. S. Shellenberger (1982).

Biology of Carbohydrates, Vol. 1, V. Ginsburg and P. Robbins, eds. (1981); Vol. 2 (1984).

Starch Chemistry and Technology, 2nd edn., R. L. Whistler, J. N. BeMiller, and E. F. Paschall, eds. (1984).

The Polysaccharides, vol. 1, G. O. Aspinall, ed. (1982); Vol. 2 (1983); Vol. 3 (1985).

Sugar: A User's Guide to Sucrose, N. L. Pennington and C. W. Baker (1990).

Carbohydrates as Organic Raw Materials, F. W. Lichtenthaler, ed. (1990).

Biotechnology of Amylodextrin Oligosaccharides (ACS Symposium Series 458) R. B. Friedman, ed. (1991).

Developments in Carbohydrate Chemistry, R. J. Alexander and H. F. Zobel, eds. (1992).

Carbohydrates in Industrial Synthesis, M. A. Clarke, ed. (1992).

Industrial Gums, Polysaccharides, and Their Derivatives, 3rd edn., R. L. Whistler and J. N. BeMiller, eds. (1993).

Starch Chemistry and Technology, 3rd edn., R. L. Whistler and J. N. BeMiller, eds. (1997).

2.8 Chronological Summary of the Uses of Carbohydrates by Humans and the Development of Carbohydrate Chemistry

10,000 B.C.E.	Primitive sugarcane processing in New Guinea.
6000 B.C.E.	Sugarcane culture develops in India.
6000 B.C.E.	Cellulose, as cotton, is used by many cultures.
4000 B.C.E.	Starch is used as an adhesive for papyrus in Egypt.
3000 B.C.E.	Cotton cloth and string are produced in India.
1800 B.C.E.	Sucrose culture spreads from India to China and Mediterranean.
1500 B.C.E.	Cotton cloth from India spreads to Persia, China, and Japan.
1000 B.C.E.	Use of sucrose in candies, confections, and medicines in Egypt.
100 C.E.	Paper is made for writing in China.
300 C.E.	Starch is used by many cultures to stiffen and color cloth.
312 C.E.	Starch is used as a sizing agent for paper in China.
600 C.E.	Papermaking spreads from China to Japan.
700 C.E.	Paper is coated with starch paste to retain ink and provide strength.
975 C.E.	Arabian pharmacologist describes treatment of starch with saliva to produce a poultice for wounds.
1600 C.E.	Many sugar refineries develop in Europe.
1700 C.E.	Sugar beet is developed in Europe for obtaining sucrose.
1792 C.E.	A sugar is isolated from honey that is different from cane sugar (sucrose).
1800 C.E.	Potatoes are used as the primary source of starch in the United States.
1802 C.E.	A sweet sugar is found in grapes that is also different from sucrose.
1808 C.E.	Malus develops plane polarized light and observes optical rotation by carbohydrates.
1811 C.E.	Acid-hydrolyzed starch produces a sweet crystalline sugar.
1820 C.E.	Acid-hydrolyzed cellulose also produces a sweet crystalline sugar.
1821 C.E.	Production of dextrins by heating starch is discovered and leads to the formation of British gums.
1838 C.E.	The sugar from honey, grapes, starch, and cellulose is found to be identical and is called glucose by Dumas.
1858 C.E.	Empirical formula of glucose is determined to be $C_6H_{12}O_6$.
1860 C.E.	Glucose is shown to have several hydroxyl groups, by forming a compound with five acetate groups.

1866 C.E.	Kekulé changes the name of glucose to dextrose because it rotates plane polarized light to the right.
1870 C.E.	Bayer and Fittig propose that the formula for glucose is HO–CH$_2$–CH(OH)–CH(OH)–CH(OH)–CH(OH)–CHO.
1874 C.E.	Le Bel and van't Hoff independently develop the theory of optical rotation by asymmetric tetrahedral carbon atoms with four different groups around each carbon.
1878 C.E.	E. Fischer synthesizes phenylhydrazine.
1881 C.E.	A sugar is crystallized from honey that is different from sucrose and glucose. It is named levulose because it rotates plane polarized light to the left.
1882 C.E.	Sucrose, glucose, fructose, lactose, starch, cellulose, and mannose were organic compounds studied by nineteenth-century chemists.
1884 C.E.	E. Fischer shows that phenylhydrazine reacts with sugars to give crystalline phenylosazones by incorporating two phenylhydrazines.
1885 C.E.	E. Fischer shows that glucose, mannose, and fructose provide the same crystalline phenylosazones.
1886 C.E.	Fructose, galactose, sorbose, and mannose are shown to have the empirical formula $C_6H_{12}O_6$.
1886 C.E.	Kiliani shows that an aldehyde chain can be lengthened by adding HCN, to give two isomeric acids.
1887 C.E.	E. Fischer shows that the Kiliani synthesis can be used to create sugars with one more carbon by reducing the lactone.
1888 C.E.	Glucose is shown to be a six-carbon polyhydroxy aldehyde.
1888 C.E.	E. Fischer embarks on the determination of the structures of carbohydrates.
1890 C.E.	E. Fischer changes the name of dextrose back to glucose, and levulose to fructose.
1891 C.E.	E. Fischer reports the structures of glucose, mannose, fructose, and arabinose.
1892–1900 C.E.	Controversies develop about the structures of the carbohydrates, based on the formation of two methyl-D-glucosides, mutarotation, and other inconsistencies of a polyhydroxy aldehyde or ketone. In this same period, it is proposed that the carbohydrates are intramolecular hemiacetals that can form two isomeric acetals (glycosides).
1892–1900 C.E.	E. Fischer continues his studies of carbohydrates and eventually proposes the nature of biopolymers to be due to the combination of bifunctional monomers.

1902 C.E.	E. Fischer is awarded the Nobel prize in chemistry for his studies on carbohydrates.
1909 C.E.	C. S. Hudson proposes a nomenclature for the two isomeric methyl-D-glucosides and related hemiacetal isomers.
1920–1930 C.E.	The English carbohydrate research group, led by W. Norman Haworth, definitively demonstrate the size of carbohydrate rings to be primarily six-membered pyranoses and propose a six-membered hexagon to represent the carbohydrates.

2.9 Specific Terms and Concepts Used in Carbohydrate Chemistry

Acetal	A compound that results when an alcohol reacts with a hemiacetal, and water is released (see Hemiacetal and Ketal).
Anomers	A configurational isomer that occurs at C-1 when hemiacetal (or hemiketal) ring structures are formed. For example, α- and β-isomers are anomers.
Asymmetric carbon	A tetrahedral carbon that has four different groups substituted at the four valence bonds.
Axial bonds	Covalent bonds that are perpendicular to the plane of a ring (see equatorial bonds).
Chiral	A molecular center that cannot be superimposed onto its mirror image. A chiral center, thus, has handedness. A chiral carbon would have four different groups attached.
Cis-trans isomerism	Atoms or groups are *cis* or *trans* when they, respectively, are on the same or opposite sides of a plane of reference.
Configuration	Refers to the relative orientation or position in space of one group in relation to another. This can be the position of a group around a tetrahedral carbon, or it can refer to groups attached to different carbon atoms (see *Cis-trans* isomerism).
Conformation	Arrangement of atoms in a structure that can be changed in relationship to each other by rotation around single covalent bonds.
Conformational isomers or conformers	Interconvertible geometric shapes that are obtained by rotation about single covalent bonds. Examples are chair and boat forms of a

	single pyranose ring structure, or envelope or twist forms of a single furanose ring. A single structure can have, in theory, an infinite number of conformers.
Diasteromers or diasteroisomers	Stereoisomers that are identical except for one or more chiral centers, for example, D-glucose and D-galactose, D-glucose and D-mannose, D-mannose and D-allose, and so forth.
D-Sugar	Carbohydrate belonging to the D-chiral group in which the OH group on the last, highest-numbered, chiral carbon is to the right in a Fischer projection (see *Fischer projection* and L-Sugar).
Enantiomers	A pair of stereoisomers that are mirror images of each other, for example, D- and L-sugars.
Epimers	Isomers that originally differed from each other at C-2, for example, D-glucose and D-mannose. Now the definition has been expanded to include any pair of diasteromers that differ from each other at only one chiral center. For example, D-galactose and D-allose are epimers of D-glucose and D-mannose.
Equatorial bonds	Bonds that are within the plane of a ring (see Axial bonds).
Fischer projection	A three-dimensional structural arrangement on a flat page, in which a tetrahedral carbon atom is represented by two crossed lines. The horizontal lines represent bonds coming out of the page, and the vertical lines represent bonds going into the page.

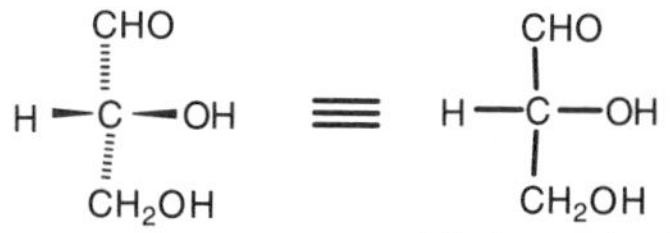

```
   CHO                    CHO
    |                      |
H — C — OH    ≡    H — C — OH
    |                      |
  CH₂OH                  CH₂OH
                   Fischer projection
```

Furanose	A five-membered sugar hemiacetal (or hemiketal) ring.
Glycoside	A special acetal or ketal that results when a hemiacetal or hemiketal reacts with a noncarbohydrate alcohol, and water is released.
Glycosidic linkage	The linkage that results when an alcohol group of a carbohydrate (or a noncarbohydrate alcohol) reacts with a hemiacetal or hemiketal

hydroxyl, and water is released. Glycosidic linkages can be either α or β.

Haworth projection	Representation of pyranose and furanose ring structures as planar hexagons and pentagons, respectively.
Hemiacetal	A compound that results when an alcohol reacts with an aldehyde.
Hemiketal	A compound that results when an alcohol reacts with a ketone.
Ketal	A compound that results when an alcohol reacts with a hemiketal, and water is released (see Acetal).
L-Sugar	A carbohydrate belonging to the L-chiral group in which the OH group on the last, highest-numbered, chiral carbon is to the left in a Fischer projection (see Fischer projection and D-sugar).
Meso compounds	Compounds that contain chiral centers but are superimposable onto their mirror images by virtue of a plane of symmetry. For example, galactaric acid (mucic acid) is a *meso* compound, and it does not rotate plane polarized light due to intramolecular optical compensation (intramolecular mirror images).

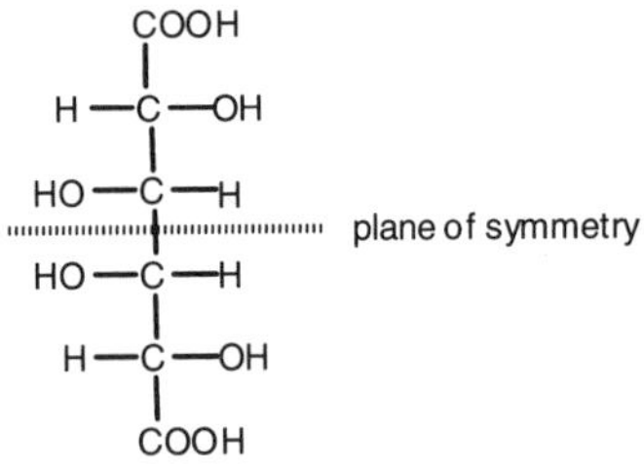

```
            COOH
             |
     H —C—OH
             |
   HO —C—H
·············|·············  plane of symmetry
   HO —C—H
             |
     H —C—OH
             |
            COOH
```

galactaric acid

Optical activity	A compound shows optical activity when it rotates plane polarized light. Compounds that have chiral centers will usually rotate plane polarized light (see *Meso* compounds for exceptions).
Pyranose	A six-membered sugar hemiacetal (or hemiketal) ring.

2.10 Literature Cited

1. R. L. Whistler, in *Starch, Chemistry and Technology,* Vol. 1, pp. 1–10 (R. L. Whistler, E. F. Paschall, J. N. BeMiller, H. J. Roberts, eds.) Academic, New York (1965).
2. As quoted by R. P. Walton, in *A Comprehensive Survey of Starch Chemistry,* Part 1, p. 236, Chemical Catalog, New York (1928).
3. W. Pigman and D. Horton, in *The Carbohydrates,* Vol. 1A, pp. 6–9 (W. Pigman and D. Horton, eds.) Academic, New York (1972).
4. E. Fischer and O. Piloty, *Chem. Ber.,* **24** (1891) 521–533.
5. E. Fischer, *Chem. Ber.,* **24** (1891) 1836–1841.
6. E. Fischer, *Chem. Ber.,* **24** (1891) 2683–2695.
7. E. Fischer, *Chem. Ber.,* **29** (1896) 1377–1384.
8. A. F. Peerdemann, A. J. van Bommel, and J. M. Bijvoet, *Nature,* **168** (1951) 271–273.
9. J. M. Bijvoet, *Endeavour,* **14** (1955) 71–77.
10. A. Wohl and F. Momber, *Chem. Ber.,* **50** (1917) 455–460.
11. H. D. K. Drew and W. N. Haworth, *J. Chem. Soc.* (1926) 2303–2312.
12. C. S. Hudson, *J. Am. Chem. Soc.,* **31** (1909) 66–70
13. E. L. Hirst and C. B. Purves, *123 J. Chem. Soc.* (1923) 1352–1356.
14. E. L. Hirst, *J. Chem. Soc.* (1926) 350–355.
15. W. N. Haworth, E. L. Hirst, and A. Learner, *J. Chem. Soc.* (1927) 1040–1047; 2432–2438.
16. E. L. Jackson and C. S. Hudson, *J. Am. Chem. Soc.,* **59** (1937) 996–999; **61** (1939) 959–961.
17. W. D. Maclay, R. M. Hann, and C. S. Hudson, *J. Am. Chem. Soc.,* **61** (1939) 1660–1664.
18. C. Williams and A. Allerhand, *Carbohydr. Res.,* **56** (1977) 173–179.
19. D. Horton and Z. Walaszek, *Carbohydr. Res.,* **105** (1982) 145–153.
20. W. Pigman and D. Horton, in *The Carbohydrates,* Vol. 1A, pp. 55–64 (W. Pigman and D. Horton, eds.) Academic, New York (1972).

2.11 References for Further Study

"The Fischer cyanohydrin synthesis and the configuration of higher–carbon sugars and alcohols," C. S. Hudson, *Adv. Carbohydr. Chem.,* **1** (1945) 1–36.

"Historical aspects of Emil Fischer's fundamental conventions for writing stereo–formulas in a plane," C. S. Hudson, *Adv. Carbohydr. Chem.,* **3** (1948) 1–22.

"Emil Fischer and his contributions to carbohydrate chemistry," K. Freudenberg, *Adv. Carbohydr. Chem.,* **21** (1966).

E. L. Eliel, *Stereochemistry of Carbon Compounds,* McGraw-Hill, New York (1962).

"Emil Fischer, 1852–1919," E. Farber, in *Great Chemists,* pp. 982–995, Wiley Interscience, New York (1961).

Chapter 3

Transformations

3.1 Mutarotation

In the last chapter, we briefly discussed mutarotation. If one dissolves either pure α-D-glucopyranose ($[\alpha]_D^{20} = +112°$) or pure β-D-glucopyranose ($[\alpha]_D^{20} = +19°$) in water, a complex series of reactions take place to give a mixture of products that are in equilibrium. The $[\alpha]_D^{20}$ of this mixture is $+52.7°$ and represents the resultant optical rotation of five different compounds: 37% α-D-glucopyranose, 67% β-D-glucopyranose, 0.5% α-D-glucofuranose, 0.5% β-D-glucofuranose, and 0.002% of the open-chain free aldehyde [1]. The structures of the five forms of D-glucose in solution at equilibrium are shown in Fig. 3.1. The process of mutarotation gives this equilibrium mixture if the starting compound is any one of the five forms. The four ring structures are transformed into each of the other ring structures through the open-chain form, until the equilibrium amounts are obtained. The process is slow, taking many hours to reach equilibrium in distilled water at 20°C. Both acid and base can catalyze the transformations. Alkali is the better of the two catalysts. Dilute alkali (pH 10) catalyzes the transformation approximately 5,000 times faster than an equivalent amount of acid (pH 4). The transformation is also catalyzed by 2-hydroxypyridine and by the enzyme mutarotase, which is produced by several fungi such as *Penicillium notatum* and *Aspergillus niger* and found in some animal tissues. Catalysis by 2-hydroxypyridine was the first reported example of concerted acid-base catalysis. 2-Hydroxypyridine has acidic (pyridinium ion) and base (phenoxylate ion) groups rigidly held in a favorable position for effecting catalysis. 2-Hydroxypyridine is approximately 7,000 times more effective as a catalyst than the hydroxide ion at pH 10.

Mutarotase catalyzes the conversion 4–5 orders of magnitude faster than 2-hydroxypyridine, Its mechanism, however, is believed to be similar to 2-hydroxypyridine in that it involves both a basic group and an acid group at the active site of the enzyme. The enzyme has a specific D-glucose binding site that stereo-

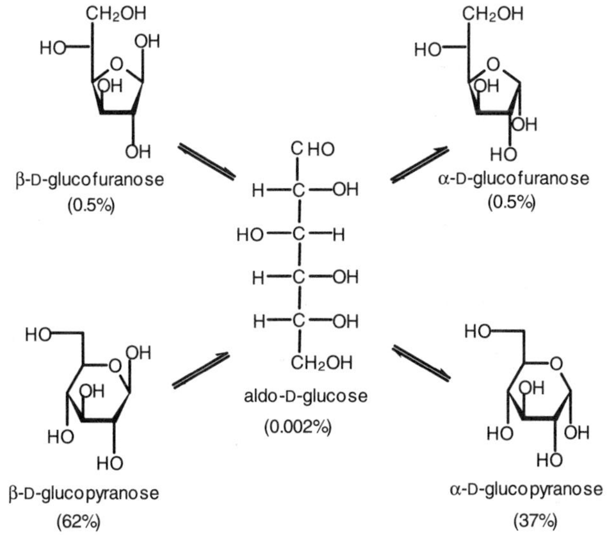

Figure 3.1. Process of mutarotation starting with any one of the five possible structural forms of D-glucose, resulting in an equilibrium mixture at 30°C.

specifically orients the glucose ring in the correct position for a basic group (believed to be a carboxylate ion) and an acidic group (which possibly is a carboxyl group) to effect concerted acid-base catalysis. The mechanisms for the catalysis of mutarotation by alkali, acid, acid/base, and mutarotase are given in Fig. 3.2.

Many of the reactions of carbohydrates that involve C-1 take place with the open-chain, aldehyde form, even though it is present in very small amounts for most carbohydrates. When the open-chain form reacts, one of the ring forms is converted into the open-chain form, until all of the carbohydrate has reacted. Different carbohydrates have different distributions of the various forms when at equilibrium in water solutions. Table 3.1 gives the various distributions for some carbohydrates.

Table 3.1. Distribution of α/β-Pyranose, α/β-Furanose, and Open-Chain Forms of Carbohydrates at Equilibrium in Water Solutions[a]

Carbohydrate	Temp. °C	Pyranose		Furanose		Open-chain
		α	β	α	β	
D-Glucose	31	38.0%	62.0%	0.5%	0.5%	0.002%
D-Mannose	44	65.5%	34.5%	0.6%	0.3%	0.005%
D-Rhamnose	44	65.5%	34.5%	0.6%	0.3%	0.005%
D-Fructose	31	2.5%	65.0%	6.5%	25.0%	0.8%
D-Galactose	31	30.0%	64.0%	2.5%	3.5%	0.02%
D-Idose	31	38.5%	36.0%	11.5%	14.0%	0.20%
D-Xylose	31	36.5%	63.0%	0.3%	0.3%	0.002%
D-Ribose	31	21.5%	58.5%	6.4%	13.5%	0.05%

[a]Data from S. J. Angyal, *Adv. Carbohydr. Chem. Biochem.*, **42** (1984) 63–65.

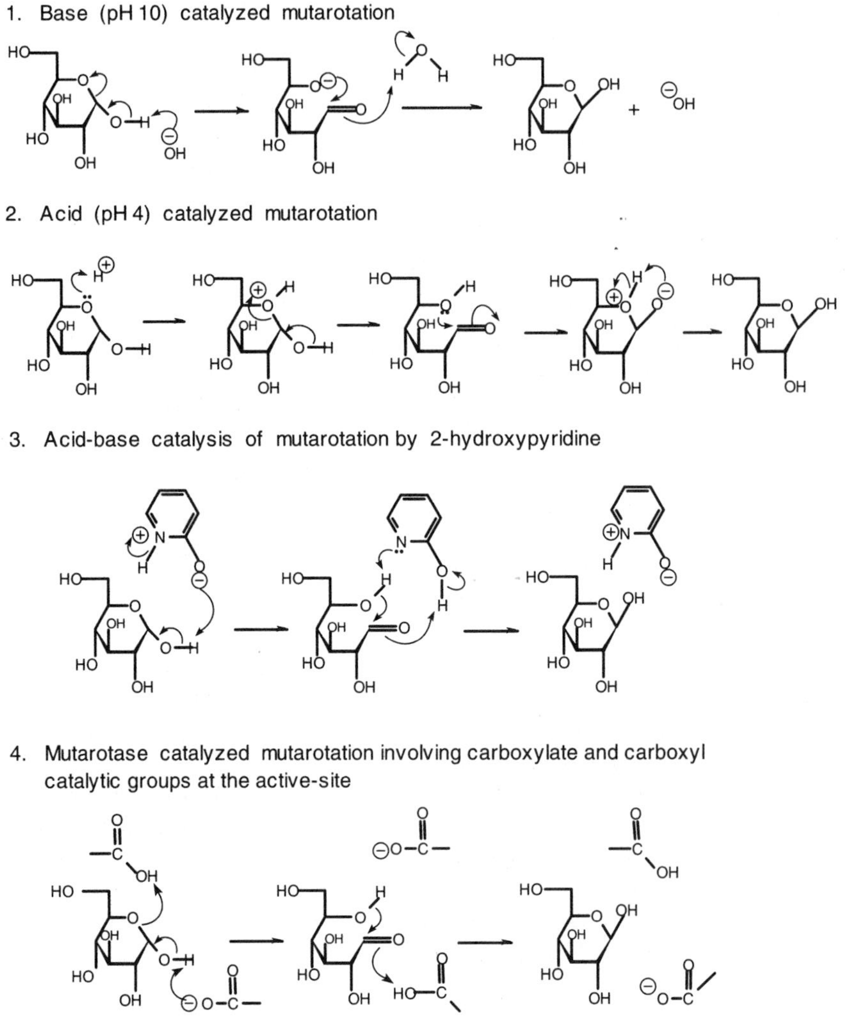

1. Base (pH 10) catalyzed mutarotation

2. Acid (pH 4) catalyzed mutarotation

3. Acid-base catalysis of mutarotation by 2-hydroxypyridine

4. Mutarotase catalyzed mutarotation involving carboxylate and carboxyl catalytic groups at the active-site

Figure 3.2. Mechanisms for the catalysis of the transformation of α-D-glucopyranose into β-D-glucopyranose by alkali, acid, acid/base, and mutarotase.

3.2 Reactions of Carbohydrates with Strong Alkali

Two Dutch chemists, Lobry de Bruyn and Alberda van Eckenstein, collaborated in the study of the effects of alkali on carbohydrates. The reaction with alkali produces epimerization of aldoses and ketoses and aldose-ketose isomerization [2]. At pH values of 11–13 and 20°C, alkali catalyzes the transformation of D-glucose into D-fructose and D-mannose. The transformation most probably takes place by the formation of two enediols, although the enolic forms of the sugars have never been isolated. The transformation produces the epimerization of two aldoses and the formation of a ketose and has been called the *Lobry de Bruyn–Alberda van Eckenstein transformation*. The mechanisms for the transformations, starting with D-glucose, are shown in Fig. 3.3. The hydrogen on C-2 of the aldose is acidic because of the electron-withdrawing power of the carbonyl oxygen. The hydrogen is removed by the hydroxide catalyst. Likewise, the hydrogen on the C-2 hydroxyl group of the *cis*-enediol is acidic because of the electron-withdrawing influence of the carbon-carbon double bond. The hydroxide ion removes this hydrogen to give the ketose; then the hydrogen on C-1 of the ketose becomes acidic because of the electron-withdrawing power of the ketocarbonyl group, and the *trans*-enediol is formed by the hydroxide catalyst. This produces an acidic hydrogen on the C-1 hydroxyl group of the *trans*-enediol to give D-mannose. The transformation could start with either D-mannose or D-fructose to produce the same set of equilibrium products.

There are several enzyme-catalyzed reactions that involve aldo-keto isomerizations of both phosphorylated and nonphosphorylated carbohydrates. The following illustrate some of the important enzyme-catalyzed aldose-ketose isomerizations:

$$\text{D-glucose} \xrightleftharpoons[\text{isomerase}]{\text{glucose}} \text{D-fructose} \tag{3.1}$$

$$\text{D-xylose} \xrightleftharpoons[\text{isomerase}]{\text{xylose}} \text{D-xylulose} \tag{3.2}$$

$$\text{D-glucose-6-phosphate} \xrightleftharpoons[\text{isomerase}]{\text{phosphogluco}} \text{D-fructose-6-phosphate} \tag{3.3}$$

$$\text{3-phospho-D-glyceraldehyde} \xrightleftharpoons[\text{isomerase}]{\text{triose phosphate}} \begin{array}{c}\text{dihydroxyacetone}\\\text{phosphate}\end{array} \tag{3.4}$$

$$\text{D-ribose-5-phosphate} \xrightleftharpoons[\text{isomerase}]{\text{ribose-5-phosphate}} \text{D-ribulose-5-phosphate} \tag{3.5}$$

The commercial enzyme (glucose isomerase) is used to convert high-glucose corn syrups into high-fructose corn syrups that are commonly used as a sweetening agent in soft drinks. This enzyme is actually a xylose isomerase that can catalyze the isomerizations of both D-glucose and D-xylose.

As with mutarotation, acid will catalyze aldo-keto isomerization and epimerization reactions, but it is not as effective as alkali. In fact, carbohydrates usually have maximum stability in slightly acid conditions of pH 3–4.

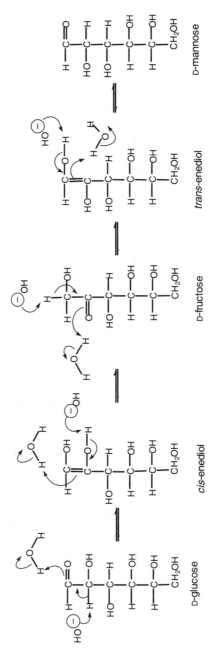

Figure 3.3. Lobry de Bruyn–Alberda van Eckenstein base-catalyzed isomerization.

When the alkaline conditions are stronger (pH > 14) and when the temperature is increased, the keto-enol isomerizations will take place all along the carbohydrate chain to give enediols between carbons 2 and 3, 3 and 4, 4 and 5, and 5 and 6 for hexoses, the consequent keto groups at C-3, C-4, and C-5, aldehyde at C-6, and the isomerization of the hydroxyl groups at C-3, C-4, and C-5, resulting in all of the possible aldohexoses and ketohexoses [3]. These reactions are illustrated in Fig. 3.4.

3.3 Alkaline Dehydration, Fragmentation, and Oxidation Reactions of Carbohydrates

Under even stronger alkaline conditions and higher temperatures, other reactions take place of which dehydrations are the most frequent and important. Wolfrom and colleagues [4,5] studied the alkaline dehydration reactions for the breakdown of D-glucose and D-xylose. The dehydration reactions lead to the formation of 5-(hydroxymethyl) furfural. Hydroxymethyl furfural itself can undergo further dehydration and fragmentation to give levulinic acid as shown in Fig. 3.5.

At higher temperature and strong alkaline conditions, fragmentation reactions occur to give a large number of compounds. A common type of cleavage is a reverse also condensation reaction that can be illustrated for D-fructose and forms D-glyceraldehyde and dihydroxyacetone. In the presence of oxygen a number of oxidation and decarboxylation products occur to give a large number of compounds such as D-glyceric acid, pyruvic acid, DL-lactic acid, formaldehyde, glyoxal, oxalic acid, acetic acid, formic acid, D-erythrose, and saccharinic acids, as shown in Fig. 3.6. Nef [3,6] reported the formation of over 100 different products from strong alkaline decomposition of D-fructose and D-glucose at higher temperatures.

The alkaline process of cleavage of D-fructose to give dihydroxy acetone and D-glyceraldehyde is similar to the important glycolysis reaction catalyzed by the enzyme *aldolase* (see Chapter 11) which reacts with D-fructose-1,6-bisphosphate at 37°C to give 3-phospho-D-glyceraldehyde and dihydroxy acetone phosphate. The comparison of the alkaline and enzyme cleavage and mechanisms of reaction are shown in Fig. 3.7.

3.4 Reactions of Carbohydrates with Strong Acid, and the Qualitative and Quantitative Determination of Carbohydrates

The degradation of carbohydrates by strong acids has been recognized for many years. It was a common elementary school demonstration to add sulfuric acid to sugar (sucrose) to produce carbon. Under more controlled conditions, the reaction of all carbohydrates in aqueous solution with strong acid has been shown to produce the formation of a derivative of furfural as the major product [7].

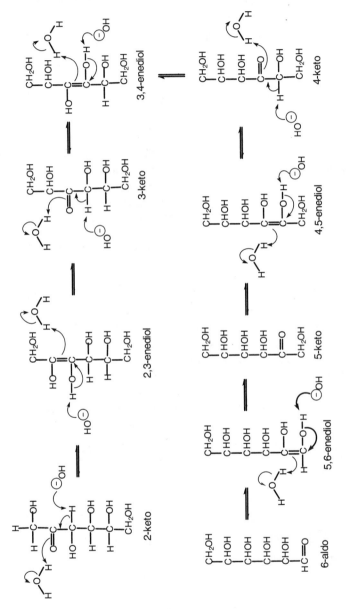

Figure 3.4. Lobry de Bruyn–Alberda van Eckenstein transformations at higher base concentrations and temperatures.

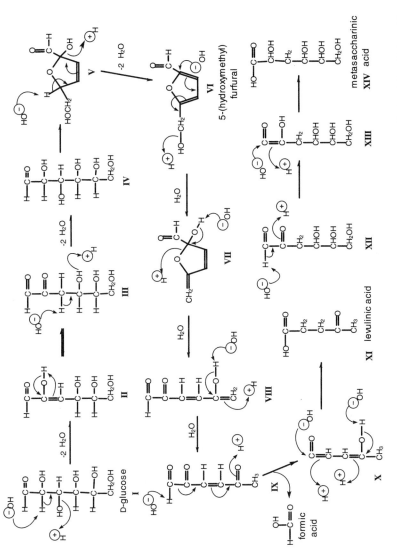

Figure 3.5. Alkaline dehydrogenation reactions and the formation of 5-(hydroxymethyl) furfural, levulinic acid, and saccharinic acid.

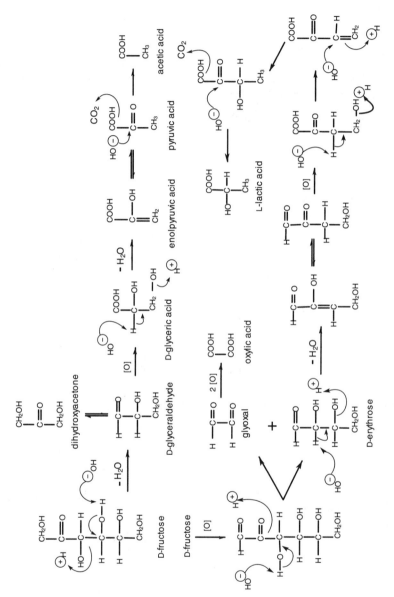

Figure 3.6. Alkaline oxidation and fragmentation reactions of carbohydrates.

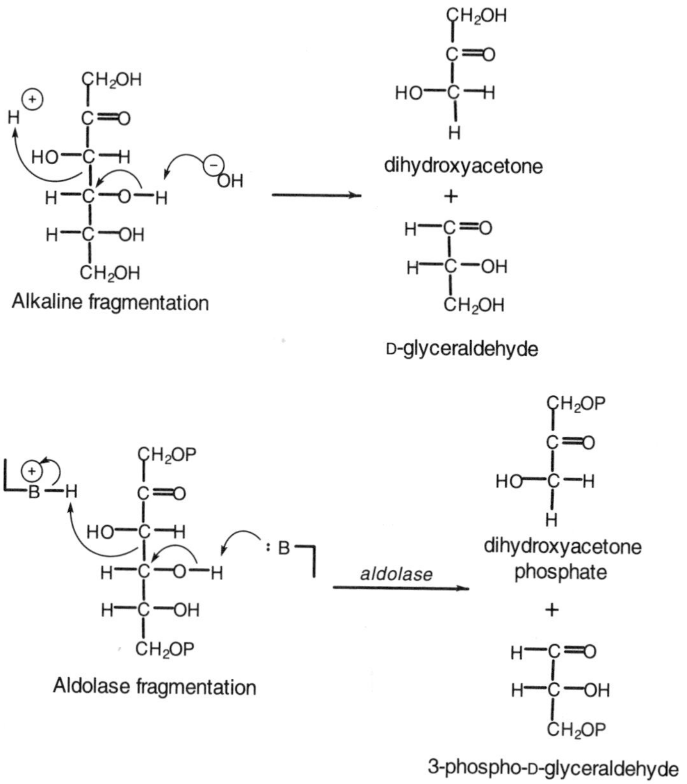

Figure 3.7. Mechanism for the alkaline fragmentation of D-fructose and fragmentation of D-fructose-1,6-bisphosphate by aldolase.

The formation of furfural is an acid-catalyzed dehydration reaction in which 3 moles of water are removed from 1 mole of hexose or pentose. The reactions are shown in Fig. 3.8. A common method for producing 5-(hydroxymethyl) furfural is the reaction of sucrose described by Haworth and Jones [8] in which a 30% solution of sucrose in water is heated to 120–140°C under pressure for 2–3 hr with 0.3% oxalic acid. In this process, the 5-(hydroxymethyl) furfural is formed primarily from the fructose part of sucrose and results in a 57% yield.

Under stronger conditions using concentrated sulfuric acid, both aldohexoses and aldopentoses give furfural derivatives. The mechanisms for the dehydration of D-glucose by strong acid are given in Fig. 3.8. Aldohexoses give 5-(hydroxymethyl) furfural, aldopentoses give furfural, 6-deoxyaldohexoses give 5-methyl furfural, and aldohexuronic acids give 5-carboxy furfural.

A number of well-known color reactions that are specific for carbohydrates depend on the formation of furfural derivatives by the action of strong acid. The furfural that is produced is condensed with a phenolic compound or an aryl amine,

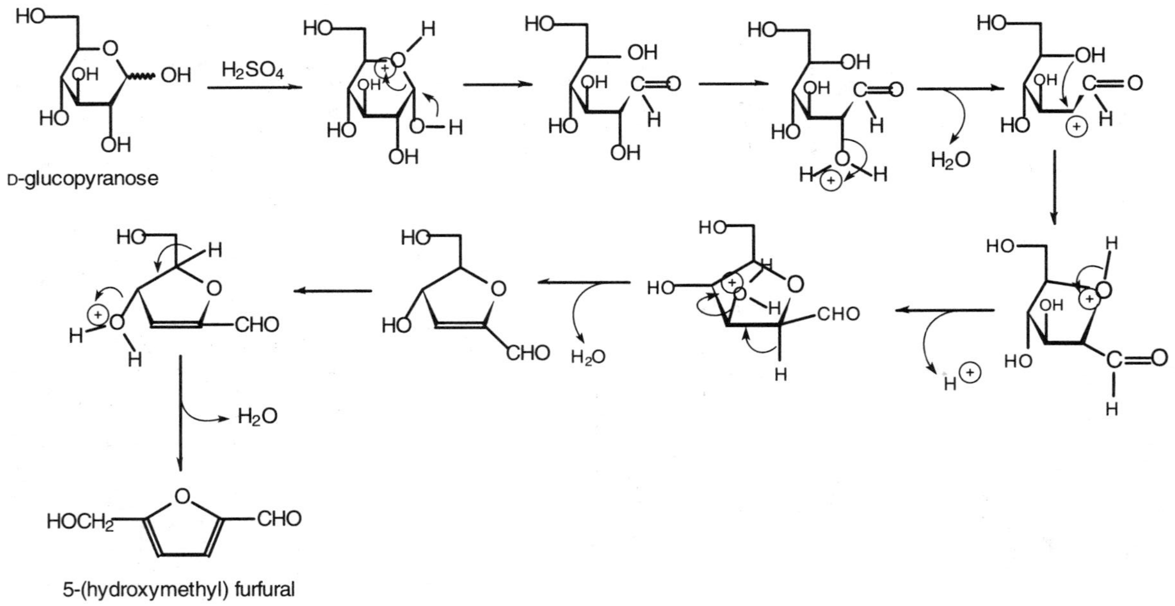

Figure 3.8. Mechanisms in the dehydration of carbohydrates by strong acid.

resulting in a colored product. A general qualitative test for carbohydrates is the *Molisch test* in which the substance to be tested in aqueous solution is treated with sulfuric acid in the presence of α-naphthol to give a characteristic purple color [9]. The color is produced by almost all carbohydrates, with the exception of sugar alcohols (alditols) and 2-amino-2-deoxy sugars. The concentrated sulfuric acid hydrolyzes the glycosidic bonds that are present in oligosaccharides and polysaccharides. The resulting monosaccharides are then dehydrated to produce the furfural derivatives. The furfural compounds condense with α-naphthol, creating the purple color.

Ketoses are dehydrated more rapidly than aldoses to give furfural derivatives that can be condensed with phenols. In the *Seliwanoff test* for ketoses, the aqueous carbohydrate sample is treated with 3 M hydrochloric acid containing resorcinol. A deep-red precipitate is formed, indicating the presence of ketose(s). The faster rate of dehydration of ketoses by HCl versus the dehydration of aldoses permits the specific detection of ketoses [10,11].

When pentoses are heated with concentrated (12 M) HCl, furfural is produced and gives a blue-green color with orcinol and ferric ions. This is called *Bial's test* and is specific for pentoses. Although hexoses do react under the condition of the Bial test, they give only a faint, yellow color that is masked by the blue-green color given by pentoses [12]. 2-Deoxypentoses are dehydrated very slowly in comparison with pentoses and hence do not give a color under the conditions of the test. D-Ribose can, thus, be detected in the presence of 2-deoxy-D-ribose.

The reaction of carbohydrates with concentrated sulfuric acid and a phenol is used to quantitatively determine the amount of carbohydrate present in a solution. An early method used anthrone dissolved in concentrated sulfuric acid [13]. Problems with the stability of the reagent and the relative insolubility of anthrone in water have lead to the replacement of the anthrone–sulfuric acid procedure by the phenol–sulfuric acid procedure [14]. Phenol is miscible with water in all proportions and does not need to be dissolved in the sulfuric acid before use. A micro procedure has been developed in which a 98-sample plate is used with a microplate spectrometer [15]. This permits the use of small volumes of sulfuric acid and sample. Further, a large number of samples can be determined in triplicate relatively easily. The absorbance of 98 samples can be measured in the spectrometer in about 60 seconds. The plate can be either discarded or easily rinsed out and used again. In either the macro or micro procedure, a standard curve has to be prepared using as the standard the carbohydrate that is being determined, for example, D-glucose, sucrose, D-galactose, D-mannose, and so forth. For compounds that contain only one type of monosaccharide (for example, D-glucose), such as starch, maltodextrins, dextran, isomaltodextrins, and cellulose, D-glucose can be used as the standard. When galactans and xylans are being determined, D-galactose and D-xylose, respectively, would be used as standards. Uronic acids are qualitatively and quantitatively determined by substituting an aromatic heterocyclic amine, carbazole, for the phenol. A purple color is obtained that is measured at 535 nm [16].

A summary of the strong acid/phenolic tests for carbohydrates is given in Table 3.2.

3.5 Reducing Reactions of Carbohydrates

Aldehydes in general will reduce an ammoniacal solution of silver ion to metallic silver (*Tollen's test*). Most carbohydrates are aldehydes or ketones. The free aldehyde or ketone usually exists in very small amounts. It is the aldo or keto group, however, that usually reacts, and as it reacts, the other forms (hemiacetals or hemiketals) are converted into the free aldehyde or ketone forms. The conversion is catalyzed by alkali. The oxidation of the aldehyde is conducted under alkaline conditions of pH 10–11 and is catalyzed by metal ions such as Ag^+, Cu^{+2}, or Fe^{+3}.

A number of qualitative tests for "reducing sugars," that is, sugars that have free hemiacetal or hemiketal groups, have been developed. These usually involve an alkaline copper reagent such as Fehling's reagent [17] or Benedict's reagent [18]. Under the alkaline conditions, the cupric ion is kept in solution because it forms a complex with tartarate, citrate, or carbonate ions. The aldehyde form is oxidized to an *onic acid,* and the cupric ion is reduced to cuprous oxide (Cu_2O), which is a brick-red precipitate.

$$
\begin{array}{c}
\text{CHO} \\
| \\
\text{CHOH}
\end{array}
+ \; 2\,Cu^{+2} \; + \; 2\,OH^{\ominus} \longrightarrow
\begin{array}{c}
\text{COOH} \\
| \\
\text{CHOH}
\end{array}
+ \; Cu_2O \; + \; 2\,H^{\oplus}
\qquad (3.6)
$$

The reducing methods have been developed into quantitative procedures for determining the amount of reducing sugars present in a sample. Over 80 methods, using copper as an oxidizing agent, have been devised. A problem with the procedures is that the amount of copper that is reduced is not stoichiometric with the amount of carbohydrate aldehyde oxidized. The amount of copper that is reduced by the reducing carbohydrate is dependent on the degree of alkalinity, temperature, time of heating, type of complexing agent, and structure of the carbohydrate. Nevertheless, the amount of metal ion reduced can be used to quantitatively determine the amount of reducing carbohydrate that is present by carefully controlling the temperature and the time of heating and by preparing a standard curve using various concentrations of the reducing carbohydrate.

Several macro or titration methods involve the heating (usually in a boiling water bath) of an aqueous solution of the carbohydrate for a specified period of time with an alkaline solution of complexed cupric ion. After cooling, an excess amount of sodium iodide solution is added, which is oxidized to iodine by the unreacted cupric ion. The iodine is titrated with a standard solution of sodium thiosulfate (Schoorl method [19]).

$$
2\,Cu^{+2} \; + \; 2\,I^{-1} \; + \; H_2O \longrightarrow Cu_2O \; + \; I_2 \; + \; 2\,H^+
\qquad (3.7)
$$

Table 3.2. Qualitative and Quantitative Determination of Carbohydrates with Strong Acid and Phenols

Name of test	Acid	Phenol	Carbohydrate determined	Type of test	Sensitivity	Color and absorbance	Ref.
Molisch	H_2SO_4	α-Naphthol	All carbohydrates[a]	Qualitative	10 μg/mL	Purple	[9]
Anthrone	H_2SO_4	Anthrone	All carbohydrates[a]	Quantitative	50 μg/mL	Blue-green (600 nm)	[13]
Phenol sulfuric acid	H_2SO_4	Phenol	All carbohydrates[a]	Quantitative	10 μg/mL	Amber (470 nm)	[14]
Seliwanoff	HCl	Resorcinol	Ketoses	Qualitative	20 μg/mL	Red precipitate	[10,11]
Bial's	HCl	Orcinol	Pentoses[b]	Qual./quant.	20 μg/mL	Blue-green (600 nm)	[12]
Dische	H_2SO_4	Carbazole[c]	Uronic acids	Qual./quant.	10 μg/mL	Purple (535 nm)	[16]

[a]All carbohydrates except sugar alcohols, 2-amino and 2-acetamido, and 2-deoxy sugars.
[b]All pentoses except 2-deoxypentoses, which are dehydrated very slowly.
[c]Carbazole is an aromatic heterocyclic amine.

Instead of measuring the amount of cupric ion present after reaction with a reducing carbohydrate, the cupric solution may be titrated directly by the addition of the reducing carbohydrate to a boiling cupric solution until all of the cupric ion is consumed. The end point is determined by using methylene blue as a redox indicator (Lane and Eynon method [19]). Titration methods using ferricyanide as the oxidizing agent have also been devised. The oxidation of the carbohydrate is achieved by using an excess of ferricyanide (reaction 3.8). The amount of ferricyanide remaining after oxidation of the carbohydrate is determined iodometrically, as shown in reaction 3.9 (Hagedorn-Jensen method [20]).

$$\begin{array}{l} CHO \\ | \\ CHOH \\ | \end{array} + 2\ Fe(CN)_6^{-3} + 2\ OH^- \xrightarrow{KCN} \begin{array}{l} COOH \\ | \\ CHOH \\ | \end{array} + 2\ Fe(CN)_6^{-4} + H_2O \qquad (3.8)$$

$$2\ Fe(CN)_6^{-3} + 2\ I^- \longrightarrow 2\ Fe(CN)_6^{-4} + I_2 \qquad (3.9)$$

The amount of reducing carbohydrate is determined for these methods from a standard curve that is generated using an appropriate reducing carbohydrate. These macro procedures are usually used by the starch industries that produce dextrins from acid and/or enzyme hydrolysis of starch.

The need of the medical profession to determine small amounts of glucose present in blood and urine, and the small amounts of reducing sugars in research samples have led to the development of colorimetric procedures. One of these is the Nelson-modified Somogyi procedure [21,22]. It measures the oxidation of reducing carbohydrates by measuring the amount of Cu_2O formed. The cuprous oxide is reacted with an arsenomolybdate reagent that gives a blue-green color whose absorbance is measured at 600 nm. The method can be performed on a semimicroscale, and a quantitative determination obtained for an unknown sample by using a properly prepared standard curve. The procedure is sensitive to about 1 mg/mL of reducing sugar.

The amount of reducing sugar can also be determined colorimetrically by oxidation of the carbohydrate with potassium ferricyanide, as described in reaction 3.8 [23,24]. The reaction is performed in the presence of excess cyanide to keep the ferric ion completely complexed with six cyanide groups. Ferricyanide ion is a bright-yellow color, whereas the reduced ferrocyanide is colorless. As the oxidation occurs there is a loss of ferricyanide ion and a loss of yellow color that can be measured at 420 nm. The reduction of the ferricyanide and loss of color produce a standard curve that has a negative slope. The method is sensitive to 100 μg/mL of reducing sugar, which is 1 order of magnitude below the Somogyi-Nelson procedure.

A modification of the ferricyanide procedure is the Park-Johnson method [25] that adds a ferric chloride solution to the solution containing reduced ferrocyanide. The ferric ion reacts with the ferrocyanide ion to give a deep-blue-colored ferric ferrocyanide, $Fe_4[Fe(CN)_6]_3$, that is measured at 620 nm. The Park-

Johnson modification is about 2 orders of magnitude more sensitive than the Somogyi-Nelson procedure, and 1 order of magnitude more sensitive than the ferricyanide method, measuring 10 μg/mL of reducing sugar.

A micro copper procedure has been developed for use in the microplate spectrophotometer [15]. It is relatively simple, using the copper bicinchoninate reagent [26], and can measure 1 μg/mL of reducing sugar. This is 1 order of magnitude more sensitive than the Park-Johnson method.

Another method involving the reduction of alkaline 3,5-dinitrosalicylate (DNSA) was introduced in 1921 for the determination of glucose in blood and urine [27,28]. It was later adapted for the measurement of reducing groups produced by the action of carbohydrases such as amylases and cellulases [29–31]. Reducing sugars reduce the nitro groups to amines. The method requires very strong alkali (0.4–0.7 M sodium hydroxide) and gives overoxidation, especially for reducing dextrins produced by these enzymes, and also gives nonlinear standard curves [32]. It is popular because of the simplicity of the reagent and the procedure, but the amount of oxidation is dependent on the chain length of the dextrins that are formed and gives an inaccurate, elevated determination of the activity of the enzyme and the amount of reducing sugar present. In contrast, the alkaline copper and ferricyanide methods give linear standard curves for maltodextrins of varying chain length. The amount of absorbance is proportional to the millimolar concentration of reducing sugars; it is not chain-length dependent [24,32]. It should be noted, however, that monosaccharides such as glucose give a standard curve that is not the same as the standard curve given by the maltodextrins. Further, the DNSA method is not very sensitive. It requires an amount of reducing carbohydrate that is an order of magnitude greater than the amount required for the Somogyi-Nelson procedure.

The alkaline–metal ion oxidations of reducing carbohydrates are useful as analytical methods, when carefully standardized, to detect and quantitatively determine reducing sugars. They, however, produce a number of oxidized carbohydrate products, including the onic acids, α-keto onic acids, and other degradation products. They, therefore, are not useful for the preparation of the onic acids. For this purpose, halogens and hypohalites are used (see Chapter 4).

Enzymes have also been used in the oxidation of carbohydrates as a means of determining specific carbohydrates. β-D-Glucopyranose is very specifically oxidized to D-gluco-1,5-lactone by the enzyme *glucose oxidase*. This enzyme is elaborated by several fungi such as *Penicillium notatum* and *Aspergillus niger* and is available from biochemical supply companies. The α-anomer of D-glucose is oxidized only 0.007 times as fast as the β-anomer. However, total glucose can be determined if the enzyme mutarotase is present to catalyze the conversion of α-D-glucopyranose into β-D-glucopyranose. Crude preparations of glucose oxidase contain mutarotase. The quantitative determination of D-glucose is obtained by the use of three enzymes: *glucose oxidase, mutarotase, and peroxidase.* The latter converts one of the products, hydrogen peroxide, into molecular oxygen that oxidizes a dye, *o*-dianisidine, to a purple color that is measured at 525 nm [33].

$$\text{D-glucose} \xrightarrow[\text{mutarotase}]{\textit{glucose oxidase}} \text{D-glucono-1,5-lactone} \quad + \quad H_2O_2$$

$$\begin{array}{c}\text{purple color}\\\text{measured at 525 nm}\end{array} \xleftarrow[\textit{o-dianisidine}]{} [O_2] + H_2O \xleftarrow{} \Big] \textit{peroxidase} \quad (3.10)$$

Another enzyme that has been used to measure glucose is D-*glucose-6-phosphate dehydrogenase.* As its name implies, it oxidizes D-glucose-6-phosphate, utilizing an oxidative coenzyme, $NADP^+$ (see Chapter 10, Fig. 10.2 for the structure and the oxidation-reduction reaction of $NADP^+$). To measure D-glucose, another enzyme, *hexokinase,* and ATP are required. The procedure is relatively expensive, requiring two coenzymes and two enzymes. It is, however, favored by the medical community because of the ease of measuring the reduced coenzyme product, NADPH, at 320 nm [34] (see reaction 3.11).

$$\text{D-glucose} \xrightarrow[\text{ATP}]{\textit{hexokinase}} \text{D-glucose-6-phosphate}$$

$$\begin{array}{cc}\text{NADPH} & + \\\text{measured at 320 nm}\end{array} \quad \begin{array}{c}\text{6-phospho-1,5-D-}\\\text{gluconolactone}\end{array} \xleftarrow{} \Big] \textit{Glc-6-P dehydrogenase} \quad (3.11)$$

3.6 Reactions of Hemiacetals or Hemiketals with Alcohols

The most reactive hydroxyl group of a carbohydrate is the hemiacetal or hemiketal hydroxyl group that is formed when cyclic furanose or pyranose rings are made (reaction 2.7 in Chapter 2). These hydroxyl groups can react with alkyl or aralkyl alcohols and with aryl phenols to split out water and give an acetal or *glycoside* (reaction 2.8 in Chapter 2). The alkyl, aralkyl, or aryl group is called the *aglycone* group, and the sugar residue is called the *glycosyl* group. Glycosides can be formed from five-membered rings to give glycofuranosides, and they can be formed from six-membered rings to give glycopyranosides; and they can have α- or β-configurations.

The simplest method for synthesizing glycosides is the Fischer method [35] which involves reaction of the carbohydrate (usually a monosaccharide) with an anhydrous alkyl or aralkyl alcohol, and hydrogen chloride as a catalyst. Phenols do not react using this method. Higher sugars such as di- or trisaccharides also are not good reactants because the acid usually cleaves the glycosidic linkage between the monosaccharides by alcoholysis. The furanose forms of the sugar react most rapidly and are the initial products (kinetic products), but as the reaction progresses, the pyranosides are formed. At equilibrium the pyranosides are the major products because they are the most thermodynamically stable (reaction 3.12). It is the α-pyranosides that usually predominate, making the method particularly useful for the preparation of this glycoside anomer, which is usually difficult to prepare by other methods. High yields of the α-pyranoside can be obtained by crystallizing and removing it from the mother liquor. Second and third crops of crystals are obtained by further conversion of the other forms to the thermodynamically stable α-pyranoside anomer.

Reaction of D-glucose in methanol proceeds through the α- and β-furanoside forms, through the β-pyranoside form, ultimately to the α-pyranoside form, which becomes the predominant product.

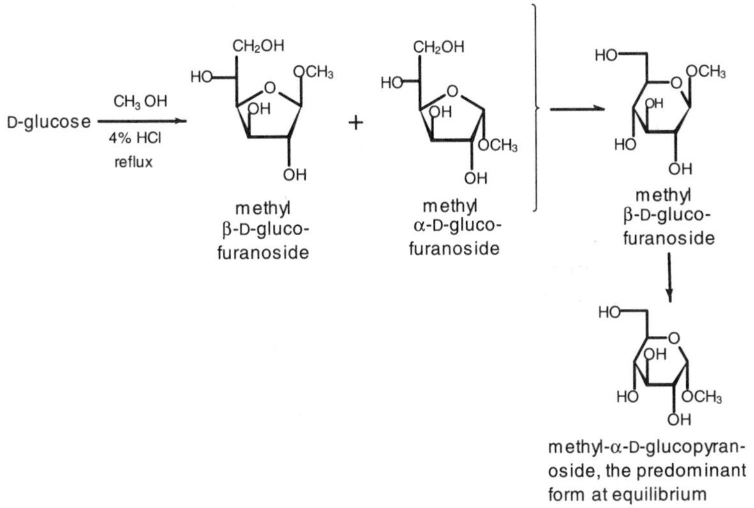

methyl
β-D-gluco-
furanoside

methyl
α-D-gluco-
furanoside

methyl
β-D-gluco-
furanoside

(3.12)

methyl-α-D-glucopyran-
oside, the predominant
form at equilibrium

The formation of α-D-pyranoside as the predominant thermodynamic product is interpreted as due to the phenomenon called the *anomeric effect* that arises from a favorable dipole-dipole interaction that is obtained when the glycosidic oxygen is axial or α at C-1 [36,37]. This configuration places the partial negative charge of the glycosidic oxygen farther from the partial negative charge of the ring oxygen and hence is the most thermodynamically stable form.

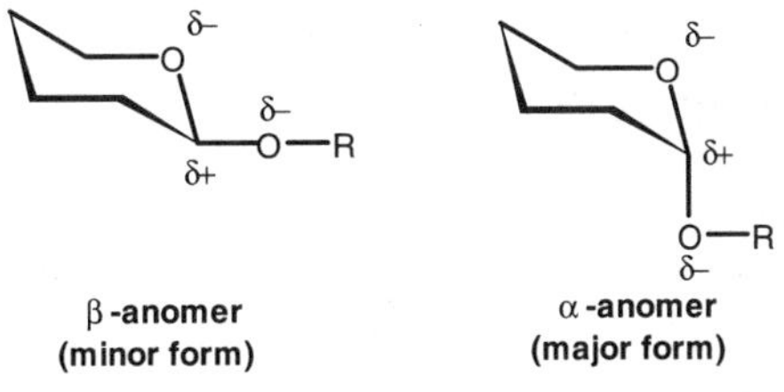

β-anomer
(minor form)

α-anomer
(major form)

Anomeric Effect

There is a *reverse anomeric effect* that occurs when an atom from the aglycone attached to C-1 has a positive or partial positive charge [38]. This is evidenced by pyridinium glycosides or glycosides containing a carbonyl group attached to the anomeric carbon.

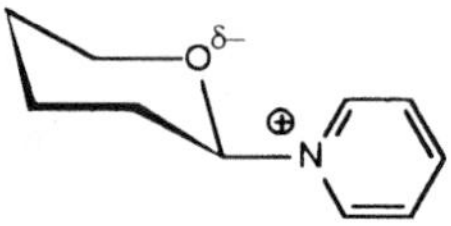

β-anomer
(major form)

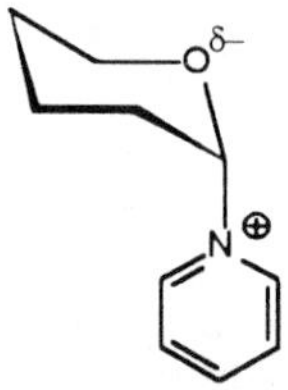

α-anomer
(minor form)

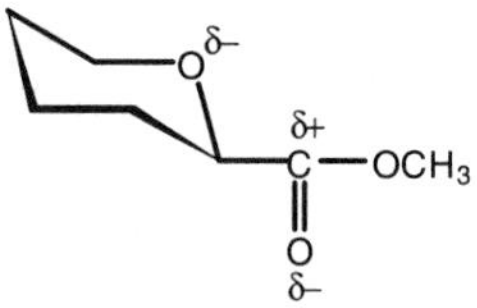

β-anomer
(major form)

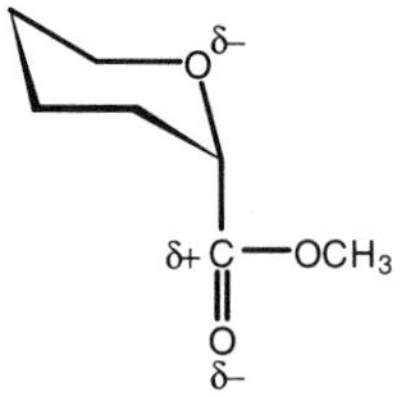

α-anomer
(minor form)

Reverse Anomeric Effect

The formation of glycosides produces a nonreducing carbohydrate because it is no longer possible to form the free aldehyde or ketone from the hemiacetals or hemiketals. Naturally occurring glycosides are found primarily in plants, where they obviously have some biological function. Some of these glycosides have been shown to have medicinal effects on animals, and some have been found to be poisonous. Figure 3.9 gives the structures, names, and properties of some of these naturally occurring glycosides.

3.7 Formation of Glycosidic Linkages to Give Di-, Tri-, and Oligosaccharides

Because carbohydrates themselves are polyalcohols with primary and secondary alcohol groups, the alcohol groups can react with a hemiacetal hydroxyl group to split out water and form a glycoside between two carbohydrate residues, producing a disaccharide. Taking D-glucopyranose as an example, 11 different disaccharides can be formed, among which the α-D-glucopyranose could react with the alcohol group at C-2, C-3, C-4, and C-6 to give four possible reducing disaccharides (see Fig. 3.10). β-D-glucopyranose could react with the same alcohol groups to give another set of four possible reducing disaccharides (Fig. 3.11). Each one of these disaccharides has been observed, although none of them per se are pre-

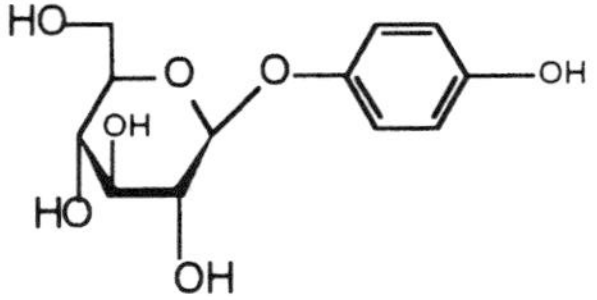

Arbutin is found in the leaves of blueberries, cranberries, and pear tree. It has been used as a diuretic and anti-infective urinary agent.

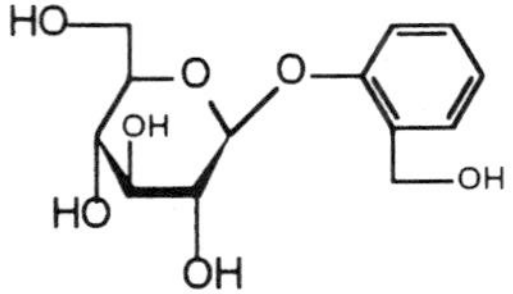

Salicin is found in the bark of willow and poplar. It has been used as an analgesic agent.

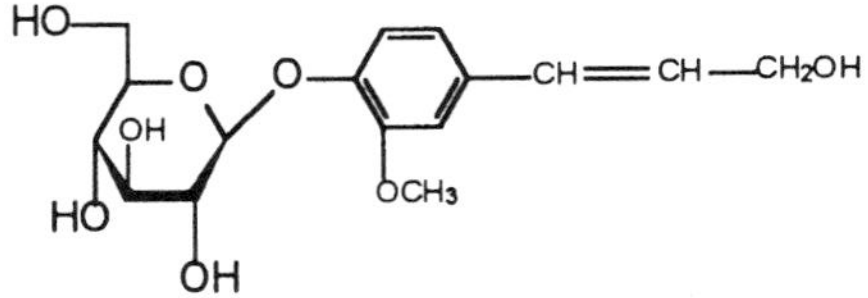

Coniferin is found throughout the *Gymnospermae*. The aglycone is the monomer of lignin and the glycoside may be the precursor to lignin biosynthesis. It is also found in comfrey root, sugar beets, and asparagus.

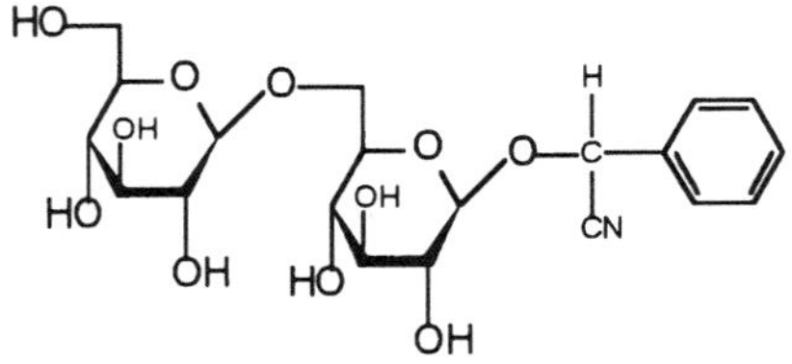

Amygdalin occurs in the seeds of Rosaceae -- bitter almonds, peach, and apricot seeds. It has been suggested as an anti-cancer drug, and has been used in some countries to treat cancers. There, however, is no hard data to substantiate its effectiveness in such treatments.

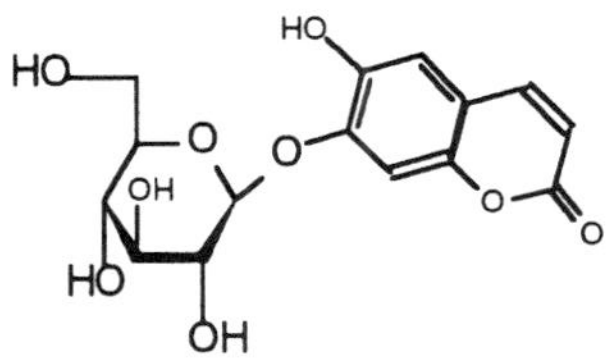

Aesculin is found in the leaves and bark of the horse chestnut. It is used as a skin protectant in sun-tanning preparations.

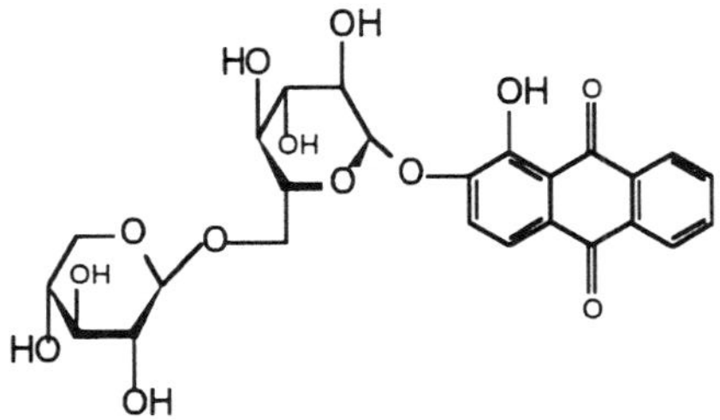

Ruberythric acid is an anthraquinone glycoside found in the madder root. The aglycone is alizarin, which is used in dyeing and staining. The carbohydrate is a disaccharide of β-D-xylopyranose linked 1,6 to β-D-glucopyranose, which is called *primeverose*.

Figure 3.9. Glycosides found in plants, and their sources and uses.

(*continued*)

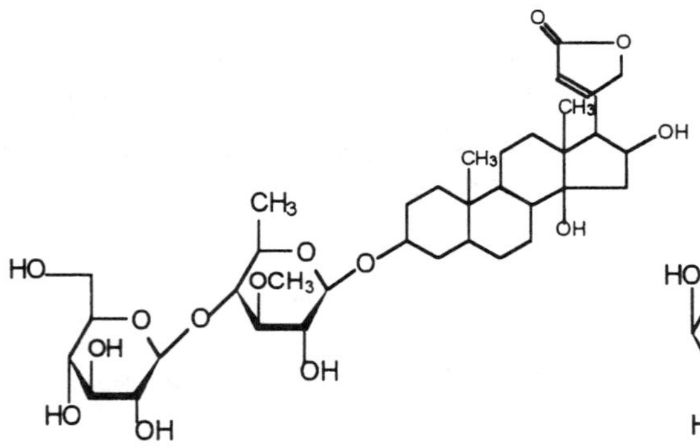

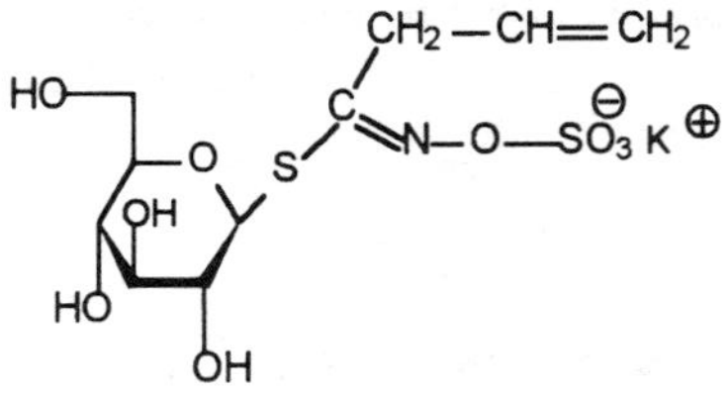

Sinigrin and analogous glycosides are found throughout the *Crucifera*. It is commonly isolated from black mustard seeds and from horseradish root. It is believed to contribute to the hot flavor of these materials.

Digitalin is obtained from the seeds of *Digitalis purpure*. The aglycone is a steroid attached to a disaccharide of β-D-glucopyranose attached to 3-O-methyl-D-quinovose (3-O-methyl-6-deoxy-D-glucose). It is a powerful cardiac stimulant.

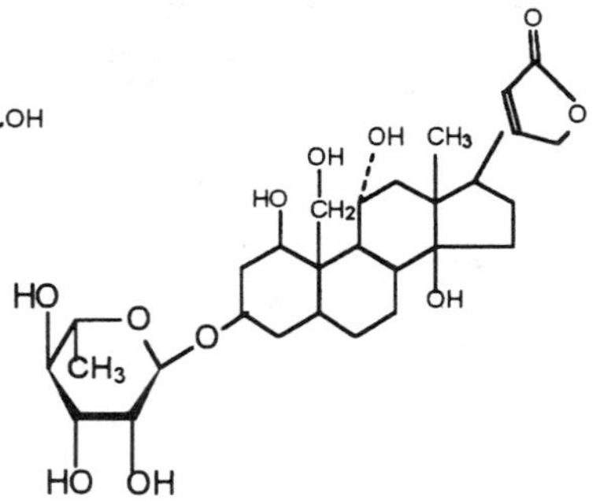

Ouabain is a cardiac glycoside (used as an arrow poison). Even though it is a toxic glycoside, it is used in low concentrations as a heart stimulant. The carbohydrate is α-L-rhamnose.

Figure 3.9. *(continued)*

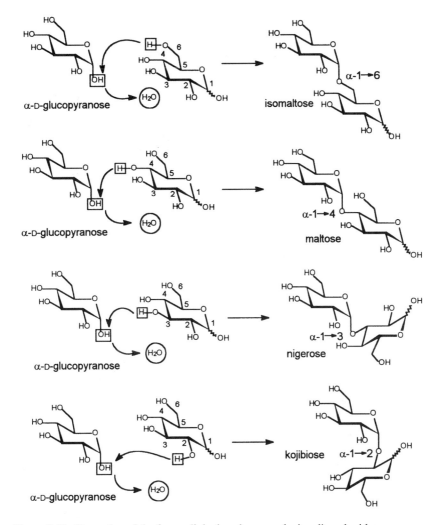

Figure 3.10. Formation of the four α-linked D-glucose reducing disaccharides.

dominant in biological systems. The α- and β-hemiacetal hydroxyl groups of two glucopyranose residues could also react with each other to split out water and form 3 possible nonreducing disaccharides (Fig. 3.12), thus providing a total of 11 possible disaccharides. Of the latter three disaccharides, one of them, α,α-trehalose, occurs in relatively large amounts in specific biological systems. Figures 3.10 and 3.11 show the structures of the eight reducing disaccharides of D-glucopyranose, and Figure 3.12 shows the structures of the three nonreducing trehaloses. The bond between the monosaccharide residues is called a *glycosidic bond*. Each of the 11 disaccharides has distinctive chemical and physical properties that are imparted by the type of glycosidic bond that joins the monosaccharide residues together. Isomaltose, for example, has an α-1 → 6 linkage, maltose has

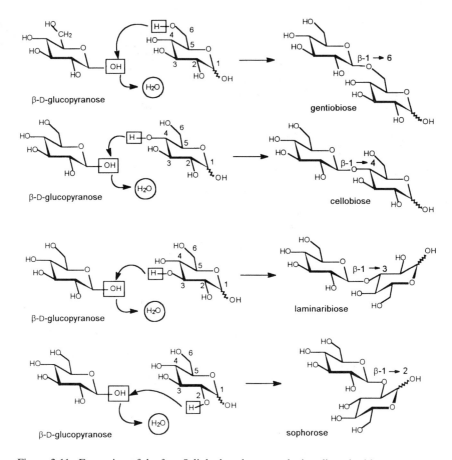

Figure 3.11. Formation of the four β-linked D-glucose-reducing disaccharides.

an α-1 → 4 linkage, gentiobiose has a β-1 → 6 linkage, and cellobiose has a β-1 → 4 linkage, and so on. The trehaloses are nonreducing disaccharides because the two hemiacetal hydroxyl groups react with each other to give an acetal-acetal linkage rather than the acetal linkage between the two monosaccharide residues that occurs for the eight reducing disaccharides.

α,α-Trehalose is a naturally occurring, nonreducing disaccharide that is found in the spores and fruiting bodies of fungi. It is also produced by yeasts. Its major biological occurrence, however, is in insect lymph fluid, where it functions as the major energy source for insects. This function is similar to that of D-glucose in mammalian blood. There are two other disaccharides that occur naturally in significant amounts. One is sucrose, which occurs in all plants, but in especially appreciable amounts in sugarcane and sugar beets. It is a nonreducing, heterodisaccharide that is composed of two different sugar units: α-D-glucopyranose and β-D-fructofuranose are combined together by splitting out water between the

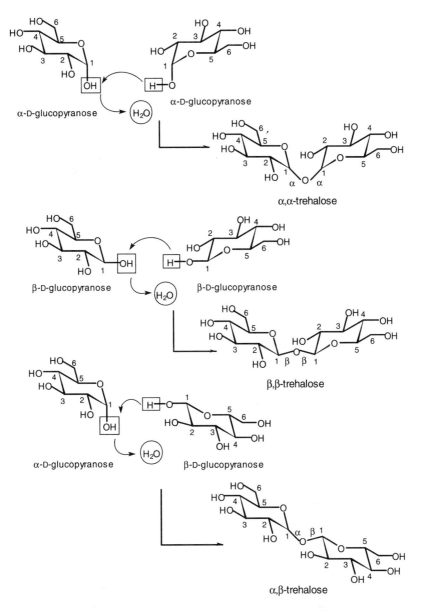

Figure 3.12. Formation of the three possible D-glucose nonreducing disaccharides, the trehaloses.

hemiacetal hydroxyl group of α-D-glucopyranose and the hemiketal hydroxyl group of β-D-fructofuranose. It is, thus, a nonreducing disaccharide with a six-membered ring and a five-membered ring. Its chemical name is α-D-*glucopyranosyl*-β-D-*fructofuranoside,* so it is not surprising that it is usually called by its shorter common name, *sucrose* (see Fig. 3.13). The third naturally occurring dis-

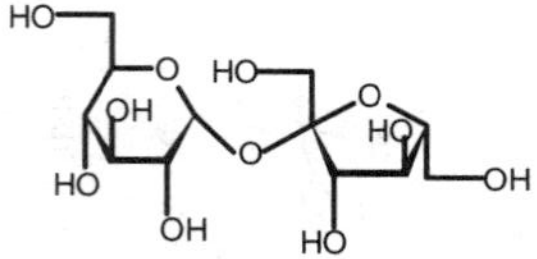

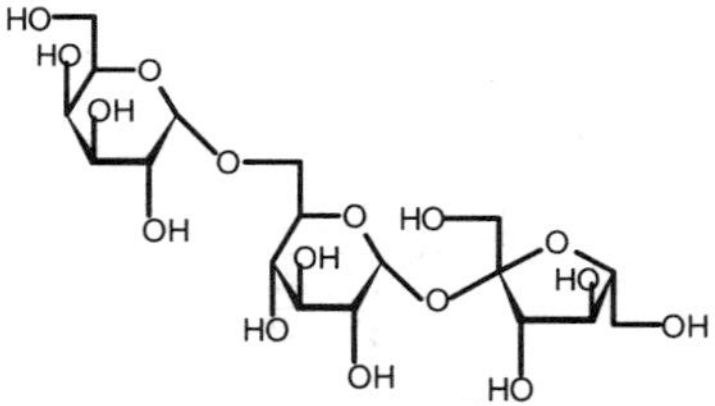

Raffinose is a nonreducing trisaccharide of α-D-galactopyranose linked 1,6 to the glucose part of sucrose. Widely distributed in plants and prepared from beet molasses and cotton seed.

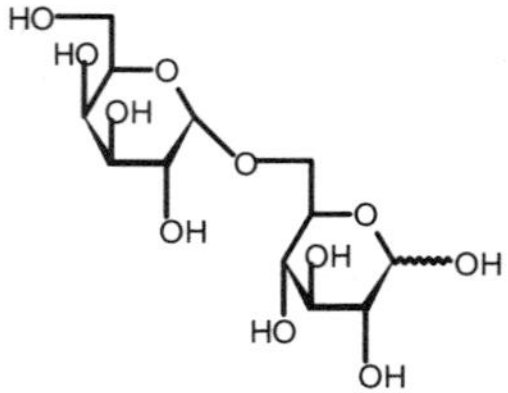

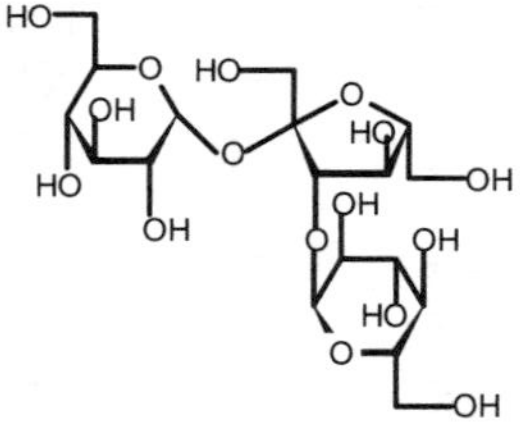

Melezitose is a nonreducing trisaccharide with α-D-glucopyranose linked 1,3 to the fructose part of sucrose. It is present in the sweet exudate of many trees, such as larch, Douglas fir, Virginia pine, poplars, etc.

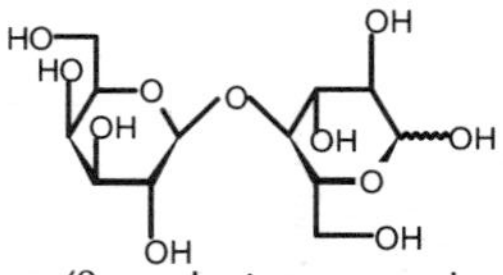

Lactose (β-D-galactopyranosyl-1,4-D-glucopyranose) is a reducing disaccharide found in mammalian milk where it is the source of energy for the young.

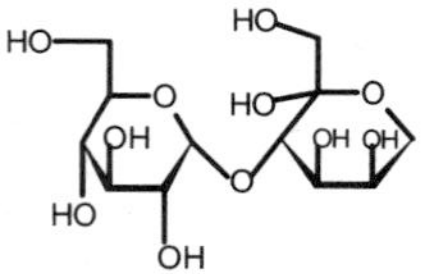

Turanose is a reducing disaccharide with α-D-glucopyranose linked α-1,3 to D-fructopyranose, found as part of the trisaccharide melezitose from which it is formed.

Sucrose (α-D-glucopyranosyl-β-D-fructofuranoside) is found in most plants, but especially in sugar cane and sugar beets.

Melibiose is a reducing disaccharide [α-D-galactopyranosyl-(1,6)-D-glucopyranose] obtained from raffinose by yeast fermentation.

Figure 3.13. Structures of some naturally occurring di- and trisaccharides, their sources, and some properties.

accharide is *lactose,* which is found in the milk of mammals where it serves as the major energy source for the young. It too is a heterodisaccharide in which the β-hemiacetal hydroxyl group of D-galactopyranose splits out water with the C-4 alcohol group of D-glucopyranose to form a reducing disaccharide with a free hemiacetal hydroxyl group on the glucose residue. Its chemical name is β-*D*-*galactopyranosyl-1* → 4-*D*-*glucopyranose.*

There are a number of other naturally occurring di- and trisaccharides that are found in lesser amounts in specific types of plants or are produced by the action of specific microbial enzymes. *Raffinose* is a nonreducing trisaccharide that is formed from sucrose by the addition of α-D-galactopyranose to the C-6 alcohol group of the glucose moiety of sucrose. It is found primarily in sugar beets. *Melez-itose* is a nonreducing trisaccharide with α-D-glucopyranose linked 1 → 3 to the fructose moiety of sucrose; it is found in the sweet exudates of many trees. *Planteose* is a nonreducing trisaccharide with α-D-galactopyranose linked 1 → 6 to the fructose moiety of sucrose. *Melibiose* is a reducing disaccharide with α-D-galactopyranose linked 1 → 6 to D-glucopyranose; it is obtained by the hydrolysis of the fructose moiety of raffinose. *Turanose* is a reducing disaccharide with α-D-glucopyranose linked 1 → 3 to D-fructopyranose and is obtained by hydrolysis of the sucrose linkage of melezitose. *Panose* is a reducing trisaccharide with α-D-glucopyranose linked 1 → 6 to the nonreducing end of maltose; it is synthesized by several microbial enzymes and represents the branch structure found in amy-lopectin. See Fig. 3.13 for the structures of these saccharides.

The actual mechanisms for the formation of the glycosidic bond(s) between the two or more monosaccharide residues cannot be accomplished by the simple Fis-cher type of synthesis that is used for the formation of the alcohol glycosides. Na-ture uses relatively elaborate, enzymatically catalyzed reactions with high-energy, activated monosaccharide units (sometimes di-, tri-, and tetrasaccharide units) to achieve the synthesis of these combined monosaccharides (see Chapter 10 for some of these biological syntheses).

3.8 Literature Cited

1. S. J. Angyal, *Adv. Carbohydr. Chem. Biochem.,* **42** (1984) 63–65.
2. J. C. Speck, Jr., *Adv. Carbohydr. Chem.,* **13** (1958) 63–103.
3. J. U. Nef, *Ann. Chem.,* **376** (1910) 1–12.
4. M. L. Wolfrom, R. D. Schuetz, and L. F. Cavalieri, *J. Am. Chem. Soc.,* **70** (1948) 514–518.
5. M. L. Wolfrom, R. D. Schuetz, and L. F. Cavalieri, *J. Am. Chem. Soc.,* **71** (1949) 3518–3522.
6. J. U. Nef, *Ann. Chem.,* **357** (1907) 294–306; **403** (1913) 204–211.
7. F. H. Newth, *Adv. Carbohydr. Chem.,* **6** (1951) 83–106.
8. W. N. Haworth and W. G. M. Jones, *J. Chem. Soc.,* (1944) 667–672.
9. H. Molisch, *Monatsch. Chem.,* **7** (1886) 108–111.
10. J. H. Roe, *J. Biol. Chem.,* **107** (1934) 15–20.
11. D. J. Gray, *Analyst,* **75** (1950) 314–318.

12. Z. Dische and K. Schwartz, *Mikrochim. Acta,* **2** (1937) 13–18.
13. R. Dreywood, *Ind. Eng. Chem. Anal. Ed.,* **18** (1946) 499–501; E. E. Morse, *Anal. Chem.,* **19** (1947) 1012–1014.
14. M. Dubois, K. A. Gilles, J. K. Hamilton, P. A. Rebers, and F. Smith, *Anal. Chem.,* **28** (1956) 350–353.
15. J. D. Fox and J. F. Robyt, *Anal. Biochem.,* **195** (1991) 93–96.
16. Z. Dische, *J. Biol. Chem.,* **167** (1947) 189–198.
17. H. Fehling, *Ann. Chem.,* **72** (1849) 106–200; **106** (1858) 75–78.
18. S. R. Benedict, *J. Am. Med. Assoc.,* **57** (1911) 1193–1197.
19. F. J. Bates, *Polarimetry, Saccharimetry and the Sugars,* (Circular of the National Bureau of Standards, C440) pp. 192–381, U.S. Government Printing Office, Washington, D.C. (1942).
20. H. C. Hagedorn and B. N. Jensen, *Biochem. Z.,* **135** (1923) 46–51.
21. N. Nelson, *J. Biol. Chem.,* **153** (1944) 357–381.
22. J. F. Robyt and W. J. Whelan, *Starch and Its Derivatives,* p. 432, Chapman & Hall, London (1968).
23. W. S. Hoffman, *J. Biol. Chem.,* **120** (1927) 51–58.
24. J. F. Robyt, R. J. Ackerman, and J. G. Keng, *Anal. Biochem.,* **45** (1972) 517–524.
25. J. T. Park and M. J. Johnson, *J. Biol. Chem.,* **181** (1949) 149–153.
26. S. Waffenschmidt and L. Jaenicke, *Anal. Biochem.,* **165** (1987) 337–340.
27. J. B. Sumner, *J. Biol. Chem.,* **47** (1921) 5–9.
28. J. B. Sumner, *J. Biol. Chem.,* **62** (1925) 287–290.
29. G. Noelting and P. Bernfeld, *Helv. Chim. Acta,* **31** (1948) 286–290.
30. G. N. Smith and C. Stocker, *Arch. Biochem. Biophys.,* **21** (1949) 95–100.
31. P. Bernfeld, *Methods Enzymol.,* **1** (1955) 149–153.
32. J. F. Robyt and W. J. Whelan, *Anal. Biochem.,* **45** (1972) 510–516.
33. I.D. Fleming and H. F. Pegler, *Analyst,* **88** (1963) 967–970.
34. B. L. Horecker and P. Z. Smyrniotis, *Methods Enzymol.,* **1** (1955) 323–327; R. D. DeMoss, *Methods Enzymol.,* **1** (1955) 328–334; H. Schrachter, *Methods Enzymol.,* **41** (1975) 4.
35. E. Fischer, *Chem. Ber.,* **26** (1893) 2400–2407; **28** (1895) 1145–1151.
36. R. U. Lemieux, A. A. Pavia, J. C. Martin, and K. A. Watanabe, *Can. J. Chem.,* **47** (1969) 4427–4436.
37. R. U. Lemieux, *Pure Appl. Chem.,* **25** (1971) 527–535.
38. R. U. Lemieux and A. R. Morgan, *Can. J. Chem.,* **43** (1965) 2205–2212.

3.9 References for Further Study

"Polarimetry, saccharimetry, and the sugars," F. J. Bates, *Circular of the National Bureau of Standards C440,* U.S. Government Printing Office, Washington, D.C. (1942).

"The formation of furan compounds from hexoses," F. H. Newth, *Adv. Carbohydr. Chem.,* **6** (1951).

"The saccharinic acids," J. C. Sowden, *Adv. Carbohydr. Chem.,* **12** (1957).

"The Lobry DeBruyn–Alberda Van Ekenstein transformation,"J. C. Speck, Jr., *Adv. Carbohydr. Chem.,* **13** (1958).

"Mutarotation of sugars in solution. Part I. History, basic kinetics, and composition of sugar solutions," W. W. Pigman and H. S. Isbell, *Adv. Carbohydr. Chem.,* **23** (1968).

"Mutarotation of sugars in solution. Part II. Catalytic processes, isotope effects, reaction mechanisms, and biochemical aspects," H. S. Isbell and W. W. Pigman, *Adv. Carbohydr. Chem.*, **24** (1969).

"Mutarotation and actions of acids and bases," W. W. Pigman and E. F. L. J. Anet, in *The Carbohydrates,* Vol. 1A, Chap. 4, (W. W. Pigman and D. Horton, eds.) pp. 165–192, Academic, New York (1972).

"Dehydration reactions of carbohydrates," M. S. Feather and J. F. Harris, *Adv. Carbohydr. Chem.*, **28** (1973).

"The composition of reducing sugars in solution," S. J. Angyal, *Adv. Carbohydr. Chem.*, **42** (1984).

"The composition of reducing sugars in solution: current aspects," S. J. Angyal, *Adv. Carbohydr. Chem.*, **49** (1991).

Chapter 4

Modifications

Abbreviations Used in Chapter 4

Ac	acetyl
Ac$_2$O	acetic anhydride
AcOH	acetic acid
Bn	benzyl
Bu	butyl
Bz	benzoyl
DAST	diethylaminosulfur trifluoride
DMF	dimethylformamide
DMSO	dimethylsulfoxide
HMPA	hexamethyl phosphoramide
Me	methyl
Ph	phenyl
Ph$_3$P	triphenylphosphine
Pv	pivaloyl (trimethyl acetyl)
Ra Ni	Raney nickel
Tf	trifyl (trifluoromethyl sulfonyl)
Tf$_2$O	trifylic anhydride (trifluoromethylsulfonyl anhydride)
THF	tetrahydrofuran
Tr	trityl (triphenyl methyl)
Ts	tosyl (*p*-toluene sulfonyl)
TsOH	tosic acid

4.1 Formation of Carboxylic Acid Esters

Carbohydrates are polyhydroxy aldehydes or ketones. Therefore, one of their primary chemical properties is that of a polyalcohol. It is natural then that one of the first types of derivatives to be formed was esters. The formation of a completely acetylated carbohydrate can be obtained by the reaction with acetic anhydride in the presence of either a basic or an acidic catalyst. When reducing carbohydrates react with acetic anhydride in the presence of sodium acetate at 4°C, the β-anomer is favored (reactions 4.1 and 4.2).

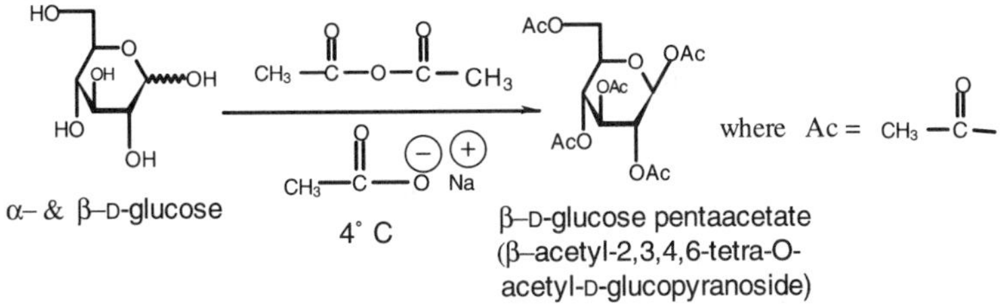

α– & β–D-glucose   4° C   β–D-glucose pentaacetate
(β–acetyl-2,3,4,6-tetra-O-
acetyl-D-glucopyranoside)

(4.1)

where Ac = CH₃—C—

maltose   4° C   β–maltose octaacetate
(β–acetyl-2,3,6,2',3',4',6'-
hepta-O-acetyl-maltose)

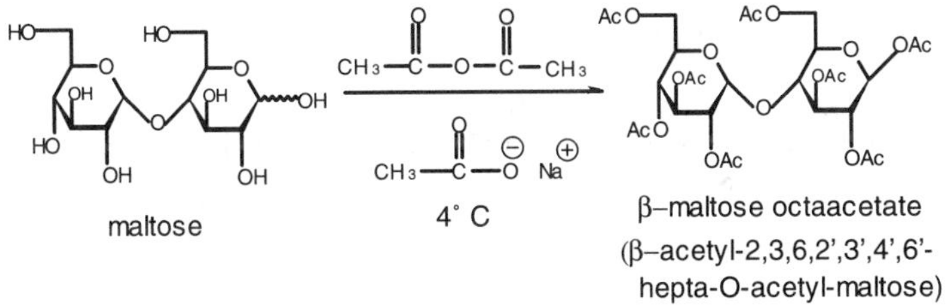

(4.2)

Nonreducing carbohydrates react to give acetylation of all of the hydroxyl groups as shown by the acetylation of sucrose (reaction 4.3).

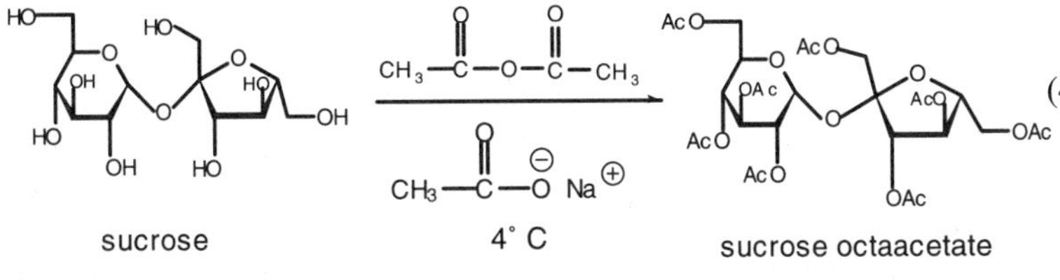

sucrose   4° C   sucrose octaacetate

(4.3)

Starting with a mixture of α- and β-D-glucose, reaction of acetic anhydride in pyridine at 0°C results in a mixture of α- and β-D-glucopyranoside pentaacetate. Likewise, starting with pure α- or pure β-D-glucopyranose, the reaction in pyridine at 0°C retains the configuration of the anomeric carbon to give α-D-glucose pentaacetate (reaction 4.4) and β-D-glucose pentaacetate (reaction 4.5), respectively.

Starting with a mixture of α- and β-D-glucose, reaction in acetic anhydride with an acid catalyst, such as zinc chloride, at 24°C favors the formation of α-D-glucose pentaacetate (reaction 4.6).

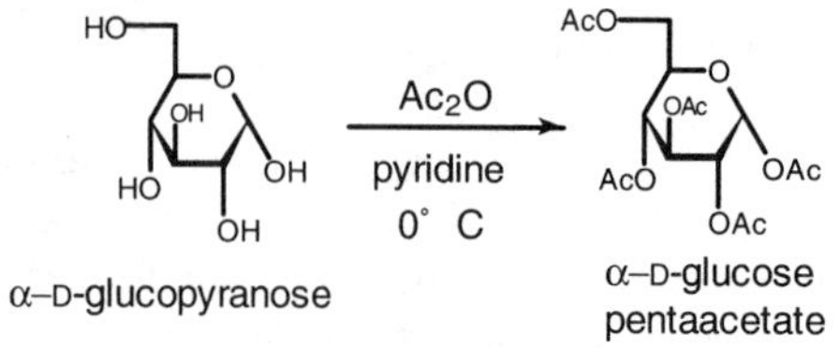
HO
O
OH
HO
OH
OH
α–D-glucopyranose
Ac2O
pyridine
0° C
AcO
O
OAc
AcO
OAc
OAc
α–D-glucose
pentaacetate

(4.4)

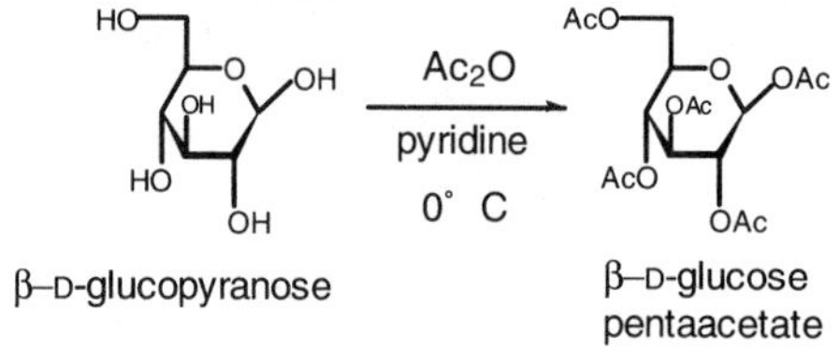
HO
O
OH
OH
HO
OH
β–D-glucopyranose
Ac2O
pyridine
0° C
AcO
O
OAc
OAc
AcO
OAc
β–D-glucose
pentaacetate

(4.5)

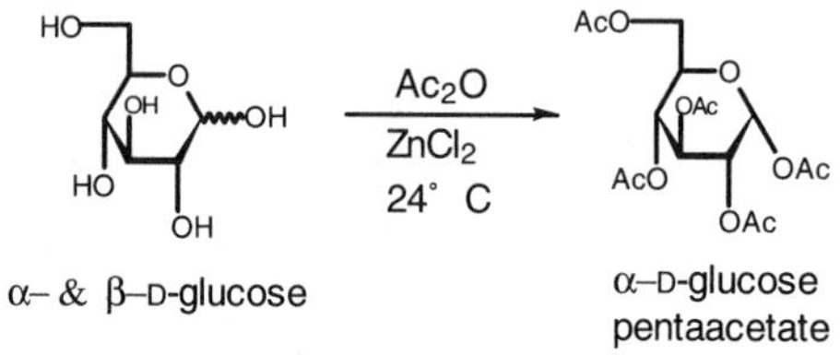
HO
O
OH
OH
HO
OH
α– & β–D-glucose
Ac2O
ZnCl2
24° C
AcO
O
OAc
AcO
OAc
OAc
α–D-glucose
pentaacetate

(4.6)

Another common carbohydrate ester is benzoylate that is formed when benzoyl chloride is added to a pyridine solution of the carbohydrate (reaction 4.7).

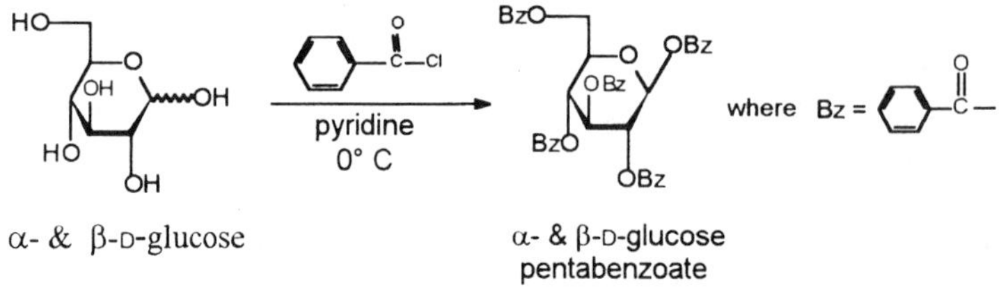
HO
O
OH
OH
HO
OH
α- & β-D-glucose
O
C—Cl
pyridine
0° C
BzO
O
OBz
OBz
BzO
OBz
where Bz =
O
C—
α- & β-D-glucose
pentabenzoate

(4.7)

The formation of acetate or benzoate esters produces crystalline carbohydrate derivatives that were much sought by the organic chemists of the late nineteenth century. Because of their facile formation and easy removal under basic conditions (see discussion later), acetate and benzoate esters can be used to protect carbohydrate alcohol groups from undergoing reaction in synthetic procedures. Furthermore, the acetate derivatives are often volatile, especially the alditol acetates, permitting their use in gas chromatographic analyses of carbohydrates.

Besides acetates and benzoates many other types of carboxylic acid esters can be formed, such as chloroacetates, trifluoroacetates, pivalates, carbonates, and thiocarbonates, by reaction with chloroacetic anhydride, trifluoroacetic anhydride, pivaloyl chloride (2,2-dimethyl propanoyl chloride), phosgene or methylchloroformate, and carbon disulfide or carbonyl sulfide, respectively. Each of the ester derivatives imparts specific chemical properties to the carbohydrates so they can be used selectively in synthetic schemes. It was found that the sulfonic acid esters could be selectively displaced by nucleophiles to produce modification or substitution of the particular carbon atom to which they were attached (see the discussion in section 4.2).

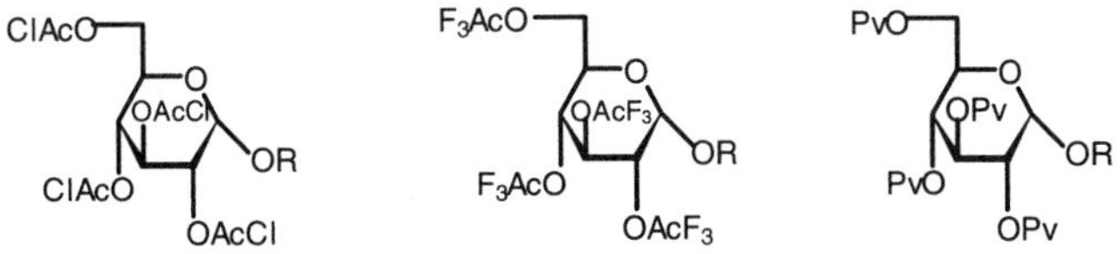

2,3,4,6-tetra-O-chloro-acetyl-α–D-glucopyrano-side

2,3,4,6-tetra-O-trifluoro-acetyl-α-D-glucopyrano-side

2,3,4,6-tetra-O-pivaloyl-α-D-glucopyranoside

where

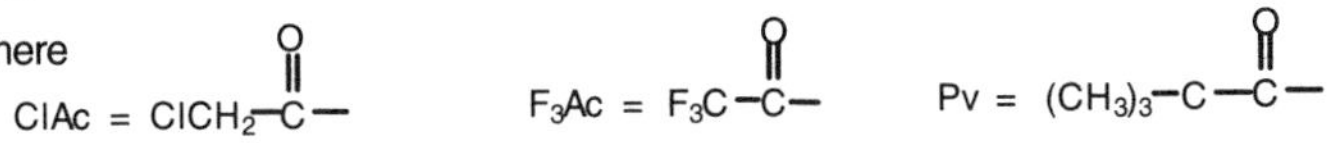

$ClAc = ClCH_2-\overset{\overset{O}{\|}}{C}-$

$F_3Ac = F_3C-\overset{\overset{O}{\|}}{C}-$

$Pv = (CH_3)_3-C-\overset{\overset{O}{\|}}{C}-$

The chloroacetates and trifluoroacetates are more labile to base and therefore easier to remove than are the acetates or benzoates; the chloroacetate and trifluoroacetate can be readily and selectively removed in synthetic schemes. Phosgene is highly reactive and bifunctional and gives intramolecular cross-linked carbonates or cyclic carbonates. The bifunctional reactions can be prevented when synthesizing carbonates by using methylchloroformate as the reactant.

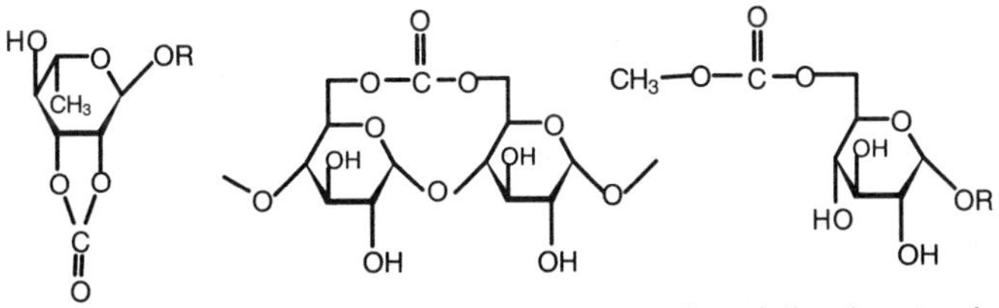

2,3– carbonate α-L-rhamnopyranoside

6,6'-intramolecular carbonate of maltose unit in amylose

6-methyl carbonate of α-D-glucopyranoside

Carbonates are also formed when carbohydrates are reacted with cyanogen bromide, followed by reaction with a second molecule containing a reactive amine or alcohol group. These reactions are used to couple various ligands such as enzymes, antibodies, affinity ligands, and so forth to an insoluble carbohydrate carrier for use in affinity chromatography or for the immobilization of enzymes (see section 7.8 in Chapter 7).

Carbohydrate esters of almost any kind of carboxylic acid can be readily synthesized by using an equimolar mixture of trifluoroacetic acid anhydride and the desired carboxylic acid (reaction 4.8). The very mild conditions of this acylating technique gives acyl esters of acid-labile glycosides in high yields.

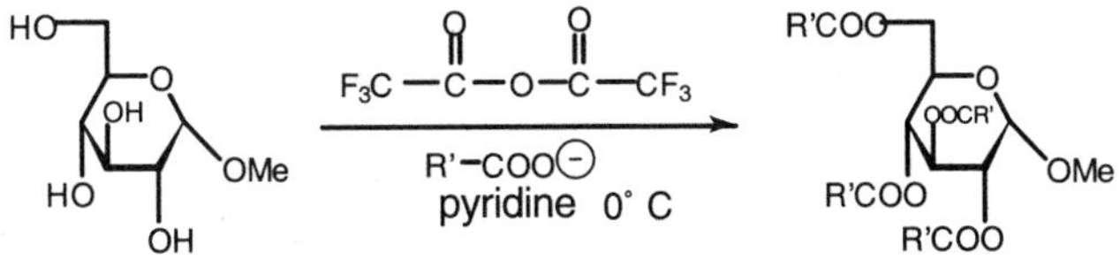

(4.8)

The acyl groups of carbohydrate esters are stable to mild acidic conditions but are removed with relatively mild basic conditions. A common method used to remove carbohydrate ester groups is to dissolve the ester in methanol that contains 0.025 M sodium methoxide at 4°C for 20 hr. The reaction, which is called the *Zemplén reaction,* after its discoverer, is smooth and gives the free carbohydrate in high yields [1] (reaction 4.9).

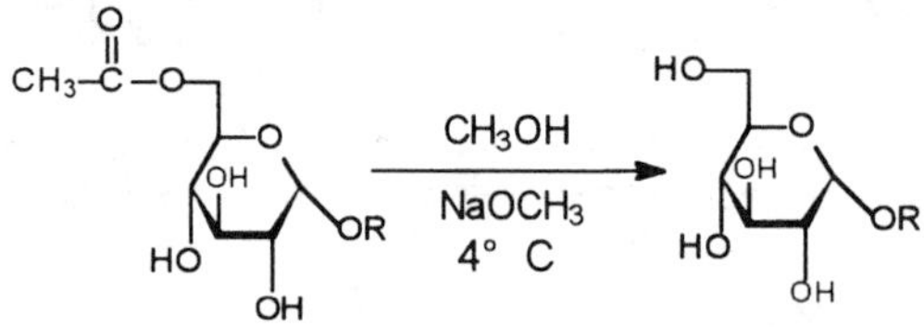

$$(4.9)$$

Zemplén Reaction to Remove Ester Groups

Under the mild basic conditions, the acetate group shows a tendency to migrate to a neighboring free hydroxyl group, particularly if it is a primary hydroxyl group (reaction 4.10).

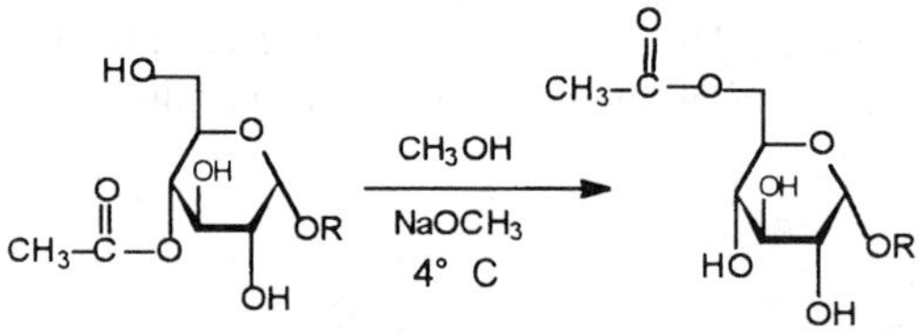

$$(4.10)$$

Acetyl migration in the Zemplén Reaction

A recent application of completely or nearly completely esterified carbohydrates is the formation of fat substitutes such as *Olestra* (a sucrose product from Procter and Gamble), which is obtained by transesterification of seven or eight fatty acids from soybean or cottonseed edible oils.

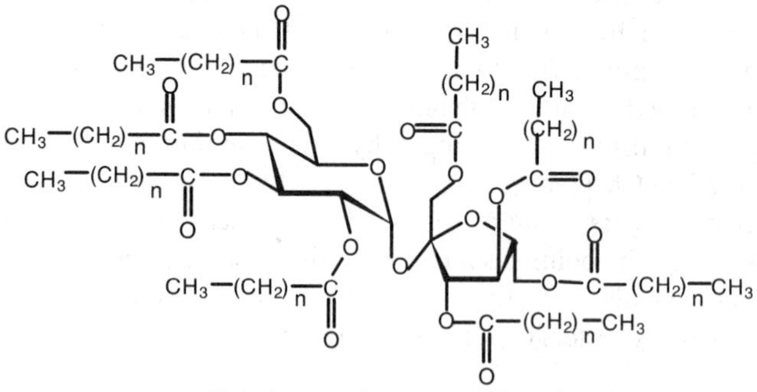

Octa fatty acyl sucrose (Olestra)
where n = 12-24 (in increments of 2)

Borate ions react with carbohydrates in aqueous solutions to give complexes (borate esters) that are fairly specific for *cis*-1,2-diols [2]. These complexes have been used in separation methods of carbohydrates that essentially differ from each other by the *cis-trans* configuration of their alcohol groups [3a]. The formation of borate complexes has been used to shift the equilibrium from an aldose to a ketose, as in the conversion of D-glucose to D-fructose [3b] and the conversion of lactose to lactulose (β-D-galactopyranosyl-(1 → 4)-D-fructose) [4].

Many other types of esters have been prepared or observed for carbohydrates. Nitrate esters of carbohydrates are widely known and used as explosives; for example, cellulose nitrate (gun powder) is used as a propellant for firearms (see Chapter 7), and nitroglycerine (dynamite) is obtained by nitrating the sugar alcohol, glycerol. These esters are formed by reaction of the carbohydrate with nitric acid, catalyzed by sulfuric acid. Sulfate esters occur naturally in a variety of polysaccharides found in both plants and animals (see section 6.3 in Chapter 6). Chemical synthesis of sulfate esters can be accomplished by using chlorosulfonic acid in pyridine. It has already been mentioned in Chapter 2 that phosphate esters play an important biological role as the agents by which energy of the sun is saved and used as a source of energy in biological systems. Carbohydrate phosphate esters play a direct and important role in the biochemical metabolism that involves both synthesis and degradation (see Chapters 10 and 11).

4.2 Formation of Sulfonic Acid Esters

Sulfonic acid esters are more versatile than carboxylic acid esters due to their stability toward acids and bases and the way they are cleaved. Cleavage of carboxylic acid esters occurs by carbonyl-oxygen fission (reaction 4.11), while cleavage of sulfonic acid esters occurs by alkyl-oxygen fission (reaction 4.12). This latter chemical property makes them very useful in the synthesis of many carbohydrate derivatives and analogues in which the nucleophile is substituted onto the carbon atom of the carbohydrate.

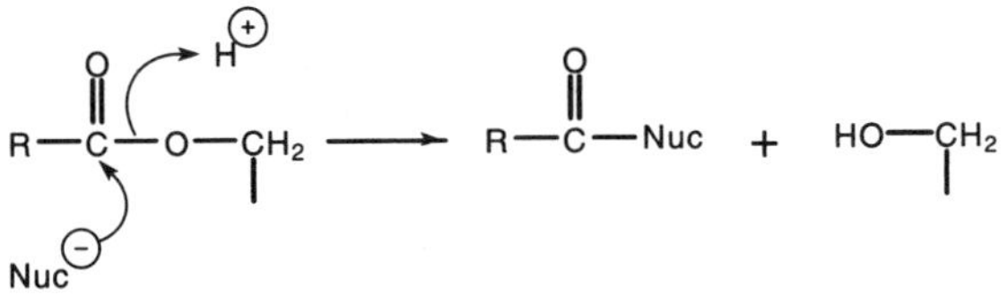

$$\text{(4.11)}$$

nucleophilic cleavage of carbohydrate carboxylic acid esters

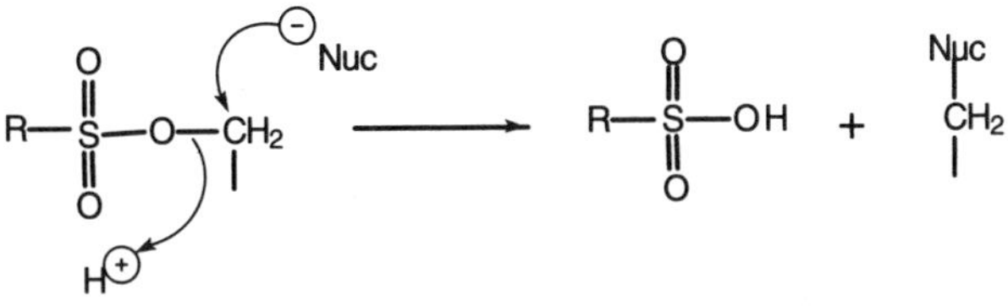

$$\text{(4.12)}$$

nucleophilic cleavage of carbohydrate sulfonic acid esters

The esters are formed by reaction of the carbohydrate dissolved in pyridine with a sulfonyl chloride at 20°C or below. Various kinds of sulfonyl chlorides have been employed for different purposes. Two of the most common are methane sulfonyl chloride (mesyl chloride, CH_3–SO_2–Cl) and p-toluene sulfonyl chloride (tosyl chloride). Both reagents can esterify all of the free hydroxyl groups if a sufficient amount of reactant is used and the temperature is kept at 20°C or below [5]. Because of the higher reactivity of the primary hydroxyl group of carbohydrates, tosyl chloride selectively forms sulfonate esters with primary alcohols by using 1.1 mole of reactant to 1.0 mole of primary alcohol group at 20°C [6] (reaction 4.13). Another sulfonic acid, tripsyl chloride (2,4,6-triisopropyl benzene sulfonyl chloride), also has a high degree of selectivity for primary alcohol groups due to the steric hindrance provided by the isopropyl groups substituted onto the phenyl ring.

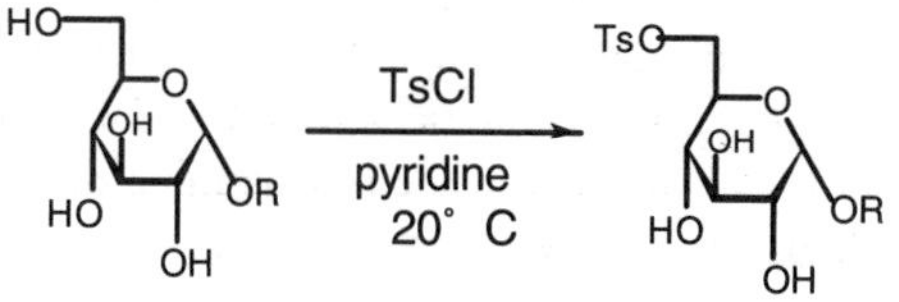

(4.13)

Specific tosylation of a C-6 hydroxyl group

where Ts = CH_3—⟨ ⟩—SO_2—

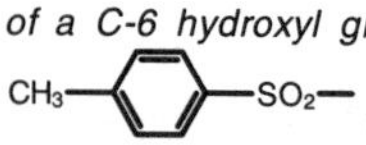

Many nucleophiles can be used to displace the tosyl or tripsyl group to give substituted carbohydrates, for example, halides (Cl^-, Br^-, and I^-), azide (N_3^-, thioacetate (CH_3COS^-), and so forth (reactions shown in 4.14).

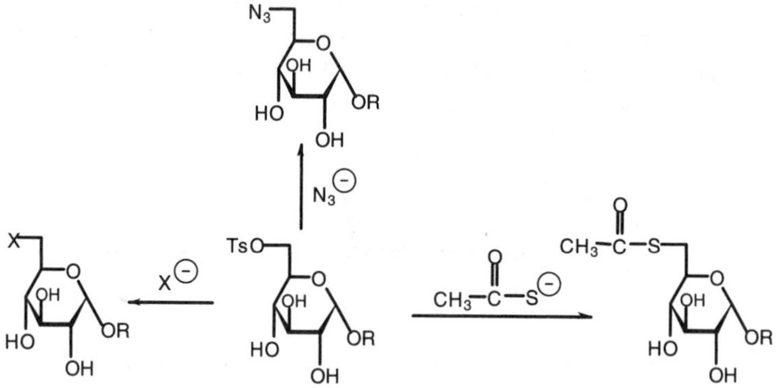

(4.14)

Displacement of tosyl group by various nucleophiles

For nucleophilic displacement of a secondary hydroxyl group, a particular free hydroxyl group is esterified with trifyl anhydride (trifluoromethane sulfonyl anhydride) to give the trifyl ester. The displacement by a nucleophile takes place with inversion of configuration. An example would be the formation of methyl-4-azido-α-D-glucopyranoside from the readily obtained methyl-2,3,6-tri-O-benzoyl-4-trifyl-α-D-galactopyranoside [7] (reaction 4.15).

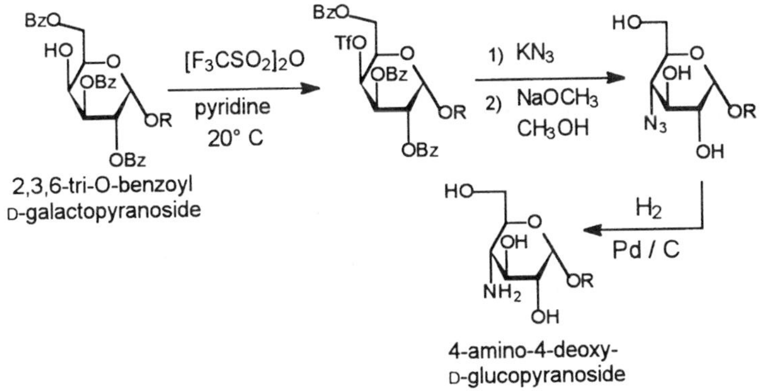

$$(4.15)$$

2,3,6-tri-O-benzoyl
D-galactopyranoside

4-amino-4-deoxy-
D-glucopyranoside

The azide can be readily reduced to give amino sugars.

4.3 Formation of Ethers

There are four types of carbohydrate ethers that are commonly formed. Each has a specific function in structural or synthetic carbohydrate chemistry.

a. Methyl Ethers

Methyl ethers have played an important role in determining the position(s) of glycosidic linkage(s) in oligo- and polysaccharides. The method was pioneered by Haworth [8] in the early part of the twentieth century by using repeated methylations with dimethylsulfate. A more modern procedure was developed by Hakomori [9] that gives complete methylation of the free hydroxyl groups in a single reaction. The Hakomori reagent is prepared by treating dimethyl sulfoxide with sodium hydride to give methyl sulfinyl carbanion (CH_3–SO–CH_2^-). This reagent is strongly basic and dissolves most oligo- and polysaccharides by removing protons from the free hydroxyl groups. Methyl iodide is added and the iodide is displaced by the alkoxide ions formed from the free hydroxyl groups, giving an O-methyl ether. The methylated saccharide is then acid hydrolyzed, and the O-methylated monosaccharides are separated and identified, providing information about the positions of the glycosidic bonds in the original carbohydrate. For example, if 2,3,4-tri-O-methyl-D-glucose and 2,3,6-tri-O-methyl-D-galactose are formed, the saccharide has a glycosidic linkage to the C-6 position of D-glucose and to the C-4 position of D-galactose. See section 12.4 in Chapter 12 for more details about the use of methylation in linkage analysis.

b. Trityl Ethers

Trityl (triphenyl methyl) ethers are formed by reaction of the carbohydrate, usually dissolved in pyridine, with trityl chloride (triphenylchloromethane). This reagent shows a high degree of regioselectivity for reaction with primary hydroxyl groups in the presence of secondary hydroxyl groups, due to the high degree of steric hindrance of the triphenylmethyl structure [10,11] (reaction 4.16). The trityl group is readily removed by treatment with mild acid at low temperature [10,12]. Trityl groups, thus, play an important role in the reversible protection of primary alcohols in carbohydrates.

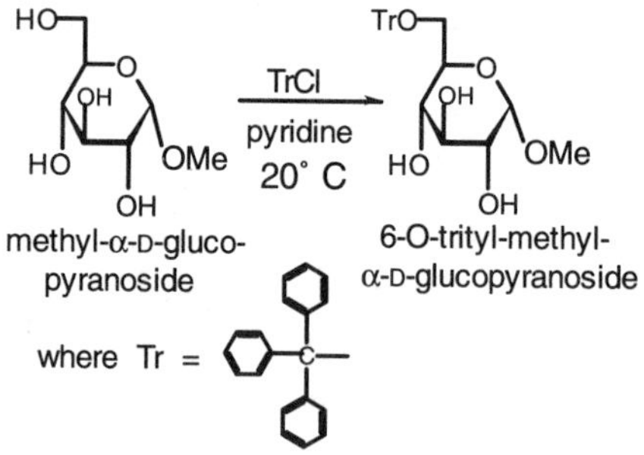

methyl-α-D-gluco-
pyranoside

6-O-trityl-methyl-
α-D-glucopyranoside

where Tr =

(4.16)

c. Benzyl Ethers

Benzyl ethers of carbohydrates are formed with both primary and secondary hydroxyl groups by reaction with benzyl chloride or bromide in a strong alkaline solution of the carbohydrate. The carbohydrate is dissolved directly in the benzyl halide, containing 4.5 M potassium hydroxide, and heated to 90–100°C for several hours [13] (reaction 4.17). Sometimes the carbohydrate is dissolved in dioxane or dimethylformamide (DMF), containing 4.5 M potassium hydroxide and benzyl halide, and heated for several hours. Difficult benzylations are accomplished by dissolving the carbohydrate in DMF that was treated with sodium hydride, similar to the Hakomori reagent, followed by the addition of benzyl halide [14].

The advantage of using benzyl protecting groups over acyl protecting groups is that they are very stable over a broad range of both acidic and basic conditions. If base sensitive groups such as esters are present, benzylation can be accomplished by using trichloroacetimidate in the presence of an acid catalyst [14] (reaction 4.18). Under neutral conditions, benzylation is accomplished by using silver triflate as a catalyst [15] (reaction 4.19).

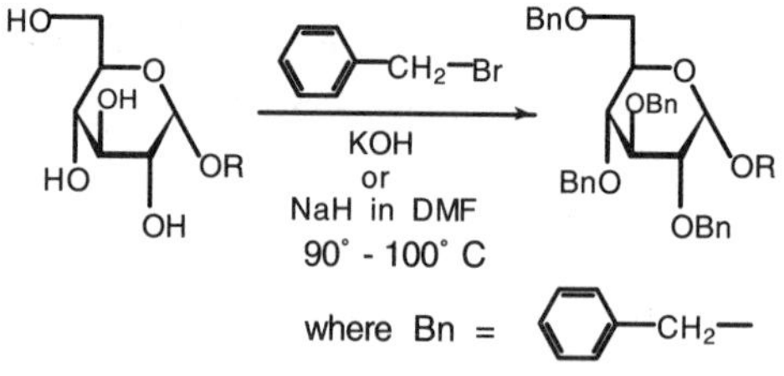

where Bn = $-CH_2-$

(4.17)

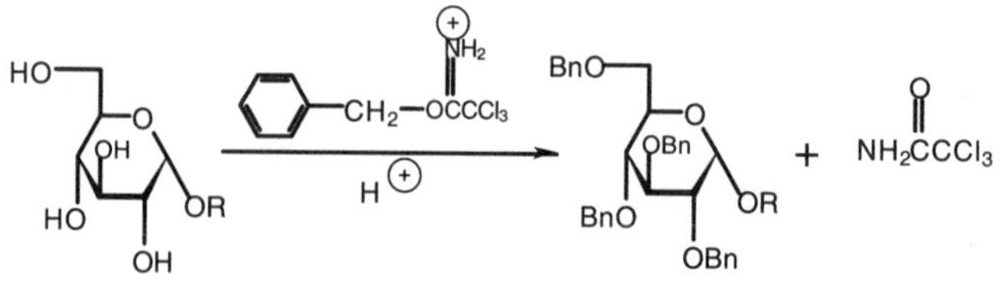

(4.18)

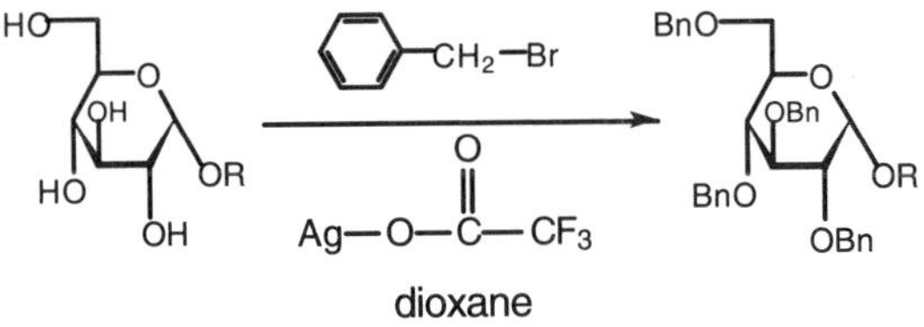

$$(4.19)$$

The benzyl groups can be removed by catalytic hydrogenolysis, using Pd/C catalyst [16]. A very simple, specific, and mild alternative method of removal uses ferric chloride in methylene chloride for 15–30 min at 20°C, and results in > 70% yields [17]. Methyl ethers and esters are not removed by this procedure.

d. Trialkylsilyl Ethers

Silyl ethers are formed with carbohydrate hydroxyl groups by reaction with chlorotrimethylsilane in pyridine with hexamethyldizilazane. Trimethylsilyl ethers are stable in neutral and basic condition but are hydrolyzed in aqueous acidic conditions [18]. The major uses of the silyl ethers is the formation of highly volatile carbohydrate derivatives that can be used in gas chromatography and gas chromatography/mass spectrometric analysis [19,20].

4.4 Formation of Acetals and Ketals

Two carbohydrate alcohol groups can react with ketones to produce ketals, and with aldehydes to produce acetals. By far the most common reaction with a ketone is the reaction with acetone that results: isopropylidene derivatives, and the most common reaction with an aldehyde is the reaction with benzaldehyde that results in a benzylidene derivative. The isopropylidene and benzylidene groups can be readily removed by acid hydrolysis. Acetone and benzaldehyde react with some specificity and they can be used to protect certain arrangements of carbohydrate alcohol groups. In general, acetone reacts with cis-diols to give five-membered cyclic structures, and benzaldehyde reacts with 1,3-cis-diols to give six-membered cyclic structures. The products favor the most thermodynamically stable structure, and the difference observed in the specificity of the reaction of acetone and benzaldehyde involves the affect of the methyl and phenyl substituents. The formation of a six-membered ring by benzaldehyde gives a chair conformation that can place the phenyl substituent in an equatorial position, whereas the formation of a six-membered isopropylidene ring would give chair conformations in which one of the two methyl groups would be axial in either of the two possible chair conformations, thus not providing any selectivity. The formation of a five-membered ring avoids this placement of a methyl in the axial position and hence is favored.

One of the best-known isopropylidene derivatives is diacetone glucose (1,2:5,6-di-O-isopropylidene D-glucofuranoside). It is prepared by the reaction of D-glucose in anhydrous acetone at 20°C with sulfuric acid or zinc chloride as catalysts [21] (reaction 4.20). The reaction of acetone has a preference for a cis-vicinal diol.

When D-glucose forms the furanose ring, it has two such diols at positions 1 and 2, and 5 and 6 (reaction 4.20). This then is the acetone derivative of D-glucose that is formed. Diacetone D-glucose is easily crystallized from cyclohexane [21].

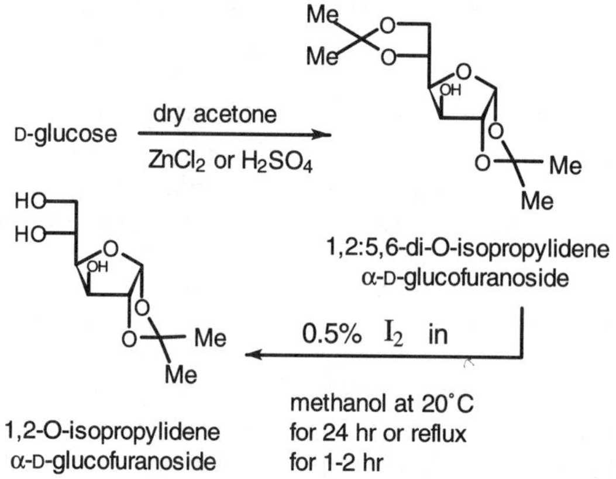

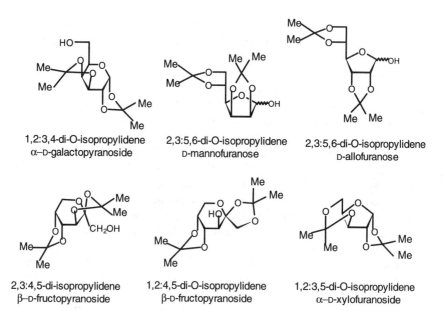

$$D\text{-glucose} \xrightarrow[\text{ZnCl}_2 \text{ or H}_2\text{SO}_4]{\text{dry acetone}}$$

1,2:5,6-di-O-isopropylidene
α-D-glucofuranoside   (4.20)

0.5% I$_2$ in

1,2-O-isopropylidene
α-D-glucofuranoside

methanol at 20°C
for 24 hr or reflux
for 1-2 hr

D-Galactose gives 1,2:3,4-di-O-isopropylidene D-galactopyranoside; D-mannose gives 2,3:5,6-di-O-isopropylidene D-mannofuranose; D-fructose gives two isopropylidenes, 2,3:4,5-di-O-isopropylidene D-fructopyranoside and 1,2:4,5-di-O-isopropylidene D-fructopyranoside; D-allose gives 2,3:5,6-di-O-isopropylidene D-allofuranose; and D-xylose gives 1,2:3,5,-di-O-isopryplidene D-xylofuranose [22].

1,2:3,4-di-O-isopropylidene
α–D-galactopyranoside

2,3:5,6-di-O-isopropylidene
D-mannofuranose

2,3:5,6-di-O-isopropylidene
D-allofuranose

2,3:4,5-di-O-isopropylidene
β–D-fructopyranoside

1,2:4,5-di-O-isopropylidene
β-D-fructopyranoside

1,2:3,5-di-O-isopropylidene
α–D-xylofuranoside

The isopropylidene groups can be removed by aqueous acid. When there are two isopropylidene groups, one of the groups is usually more easily removed than the other, producing a monoisopropylidene derivative. For 1,2:5,6-di-O-isopropylidene D-glucofuranoside, the isopropylidene group at the 5,6-position is hydrolyzed over 40 times faster than the isopropylidene group at position 1,2, resulting to the 1,2-O-isopropylidene derivative [23]. The 1,2-O-isopropylidene group is hydrolyzed over 500 times faster than the glycosidic bond of glycosides or disaccharides [24]. Another very smooth and facile method for the selective removal of the 5,6-isopropylidene group of diacetone glucose is the reaction with 0.5% (w/v) I_2 in methanol for 24 hr at 20°C, or 1–2 hr at reflux temperature [25] (reaction 4.20).

Diacetone D-glucose provides a D-glucose derivative that can be specifically modified at C-3. 1,2-O-Isopropylidene D-glucofuranoside also provides a derivative that can be selectively modified at C-6 or C-5 (see modifications below in sections 4.7 and 4.9).

The specificity for the formation of isopropylidene derivatives is shown by the reactions of the alditols. D-Glucitol reacts with acetone and dry HCl to give 1,2:3,4:5,6-tri-O-isopropylidene D-glucitol (reaction 4.21), and with zinc chloride as the catalyst, it provides a mixture of 1,2-O-isopropylidene D-glucitol and 1,2:5,6-di-O-isopropylidene D-glucitol [26] (reaction 4.22). Reaction of the tri-O-isopropylidene derivative with aqueous acid gives 3,4-O-isopropylidene D-glucitol in which the two end isopropylidene groups are removed [27]. Reaction of D-mannitol with acetone and zinc chloride gives 1,2:5,6-di-O-isopropylidene-D-mannitol [28] (reaction 4.23). Use of boric acid with sulfuric acid as the catalysts, gives 1,2:3,4:5,6-tri-O-isopropylidene D-mannitol [29] (reaction 4.24). Reaction of sucrose with 2,2-dimethoxypropane in DMF with tosic acid as the catalyst gives 4,6-O-isopropylidene derivative in 15% yield and 1'2:4,6-di-O-isopropylidene sucrose in 55% yield [30] (reaction 4.25).

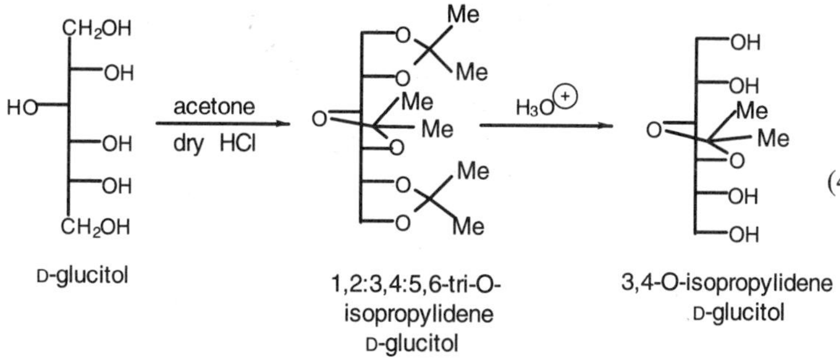

D-glucitol   1,2:3,4:5,6-tri-O-isopropylidene D-glucitol   3,4-O-isopropylidene D-glucitol   (4.21)

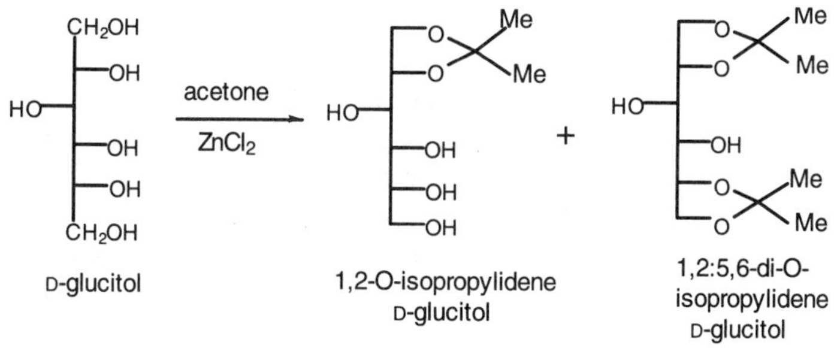

D-glucitol   1,2-O-isopropylidene D-glucitol   1,2:5,6-di-O-isopropylidene D-glucitol   (4.22)

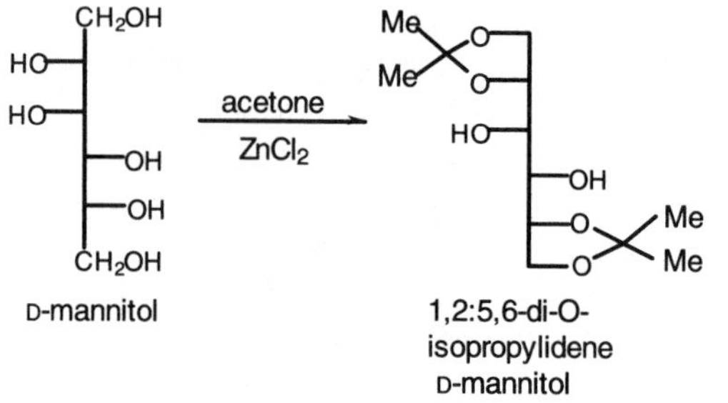
acetone
ZnCl2
D-mannitol
1,2:5,6-di-O-
isopropylidene
D-mannitol

(4.23)

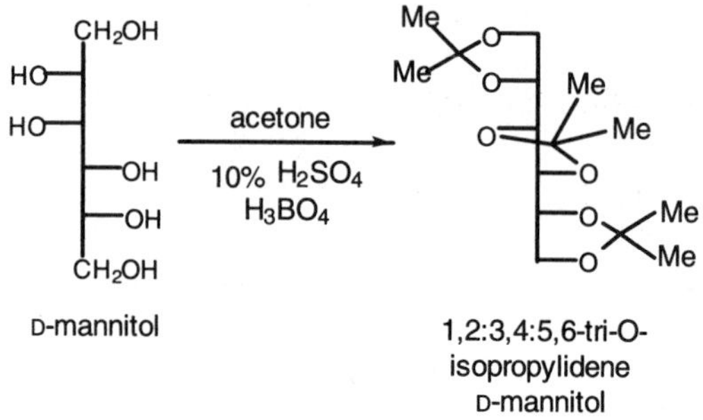
acetone
10% H2SO4
H3BO4
D-mannitol
1,2:3,4:5,6-tri-O-
isopropylidene
D-mannitol

(4.24)

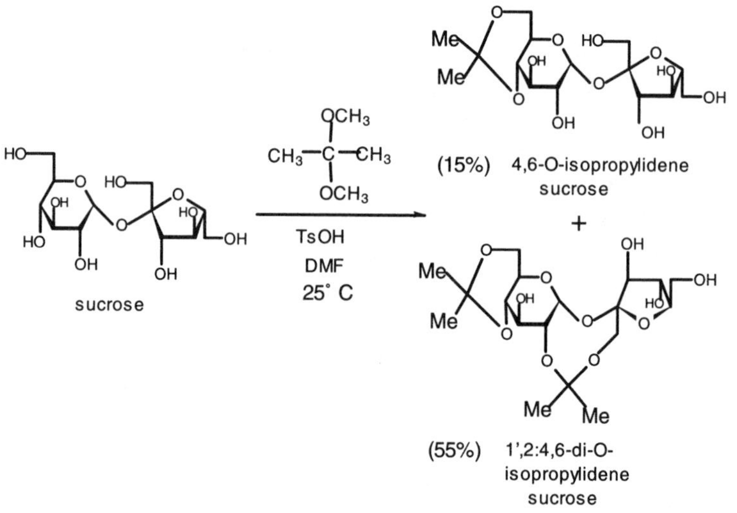
sucrose
TsOH
DMF
25° C
(15%) 4,6-O-isopropylidene
sucrose
+
(55%) 1',2:4,6-di-O-
isopropylidene
sucrose

(4.25)

Reaction of methyl-α-D-glucopyranoside with benzaldehyde, catalyzed by tosic acid in DMF, gives methyl-4,6-*O*-benzilidene α-D-glucopyranoside (reaction 4.26). The 4,6-*O*-benzilidene sucrose has been synthesized in 35% yield using α,α-dibromotoluene [31] or 1,1-dimethyoxybenzaldehyde [32] (reaction 4.27). The former has the advantage of not requiring an acid catalyst and is run in pyridine, and the latter requires tosic acid as a catalyst and is run in DMF.

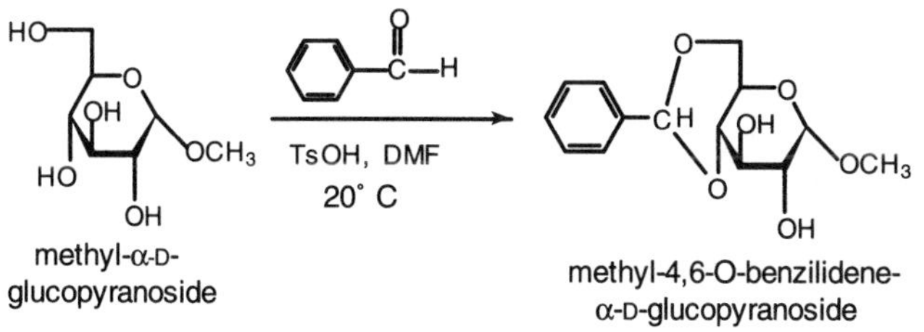

(4.26)

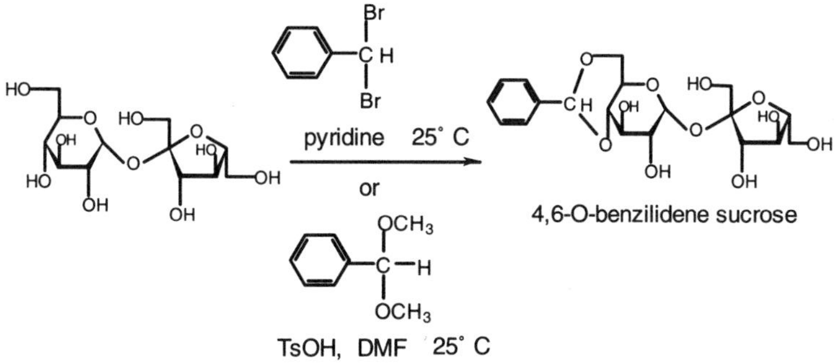

methyl-α-D-
glucopyranoside

methyl-4,6-O-benzilidene-
α-D-glucopyranoside

(4.27)

4,6-O-benzilidene sucrose

As with the reaction with acetone, the reaction of the alditols with dimethoxy-acetal benzaldehyde illustrates the specificity of forming benzilidenes [28]. D-Glucitol reacts to give 2,4-O-benzilidene D-glucitol as the major product and 1,3:2,4-di-O-benzilidene D-glucitol as a minor product (reaction 4.28). D-Mannitol reacts to give a single product, 1,3:4,6-di-O-benzilidene D-mannitol (reaction 4.29); D-xylitol reacts to give 2,4-O-benzilidene D-xylitol (reaction 4.30); and galactitol gives 1,3:4,6-di-O-benzilidene galactitol [28] (reaction 4.31).

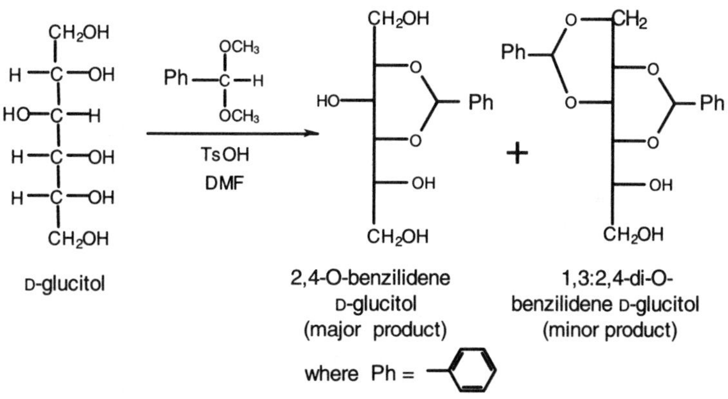

(4.28)

D-glucitol

2,4-O-benzilidene
D-glucitol
(major product)

1,3:2,4-di-O-
benzilidene D-glucitol
(minor product)

where Ph =

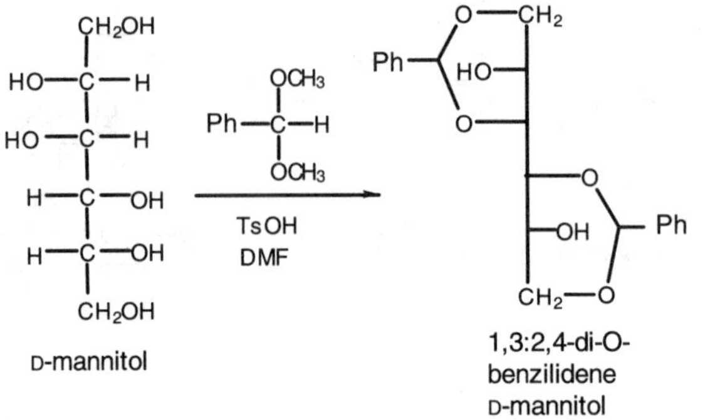

(4.29)

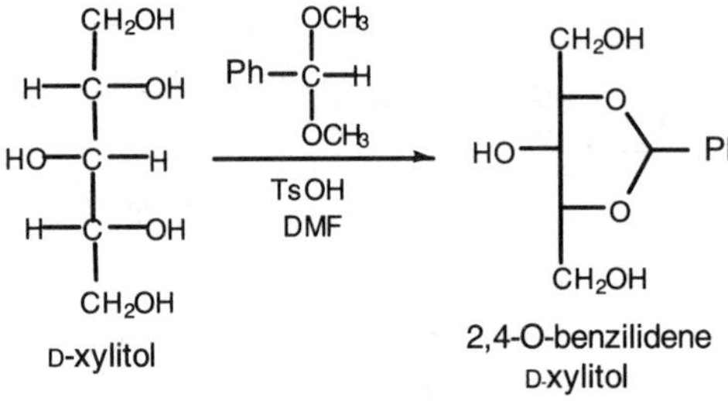

D-mannitol

1,3:2,4-di-O-benzilidene
D-mannitol

(4.30)

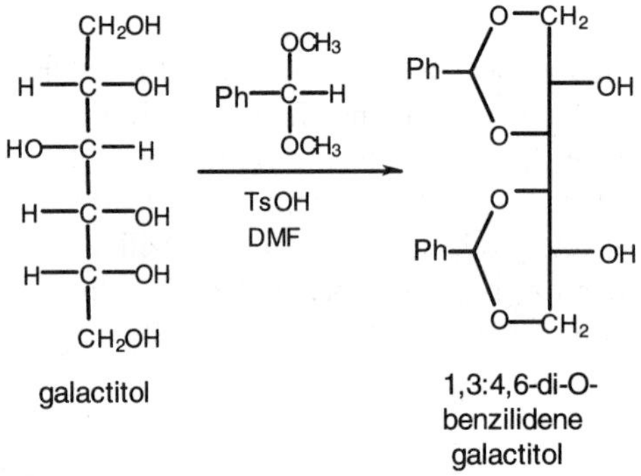

D-xylitol

2,4-O-benzilidene
D-xylitol

(4.31)

galactitol

1,3:4,6-di-O-benzilidene
galactitol

The benzilidene group also can be removed after treatment for 24 hr at 20°C in 0.1% (w/v) I_2 in methanol, or 1–2 hr at reflux temperature [25]. The 4,6-benzilidene group is asymmetrically attached to the pyranose ring and can be asymmetrically cleaved. When the 4,6-benzilidene is treated with sodium cyanoborohydride and hydrogen chloride gas, in tetrahydrofuran, specific cleavage of the C-4-O bond occurs to give the 6-O-benzyl derivative and a free C-4-OH [33,34] (reaction 4.32). Reaction with trimethylamine/borane and aluminum chloride in THF results in the same cleavage, but when the solvent is changed to toluene, the opposite ring opening is obtained, which produces a free C-6-OH and the 4-O-benzyl derivative [35] (reactions 4.33 and 4.34). When the reductant is lithium aluminum hydride and aluminum chloride, a similar ring opening is obtained, producing a free C-6-OH [36]. These differences in the selectivity of the benzilidene ring opening can be very useful in the development of protective group strategies.

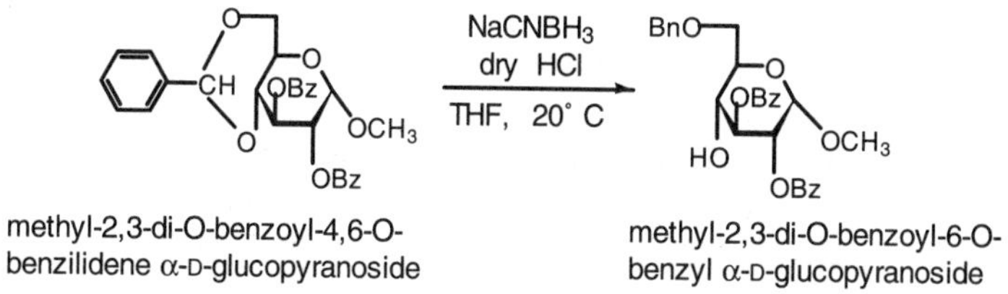
NaCNBH3
dry HCl
THF, 20° C
methyl-2,3-di-O-benzoyl-4,6-O-benzilidene α-D-glucopyranoside
methyl-2,3-di-O-benzoyl-6-O-benzyl α-D-glucopyranoside

(4.32)

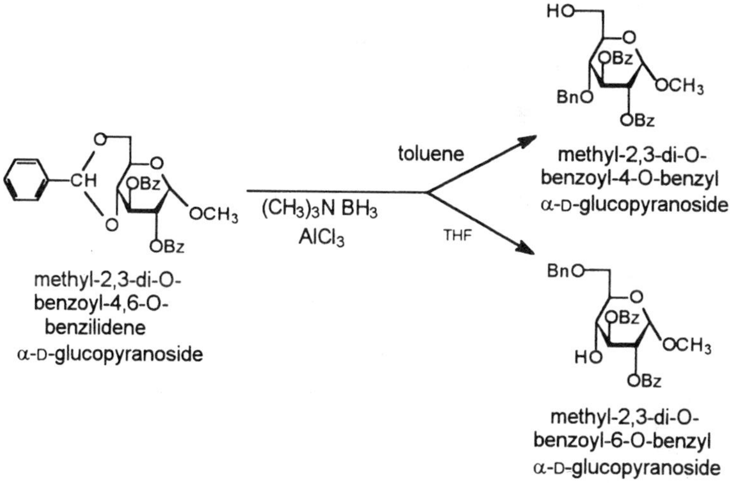
toluene
(CH3)3N BH3
AlCl3
THF
methyl-2,3-di-O-benzoyl-4,6-O-benzilidene α-D-glucopyranoside
methyl-2,3-di-O-benzoyl-4-O-benzyl α-D-glucopyranoside
methyl-2,3-di-O-benzoyl-6-O-benzyl α-D-glucopyranoside

(4.33)

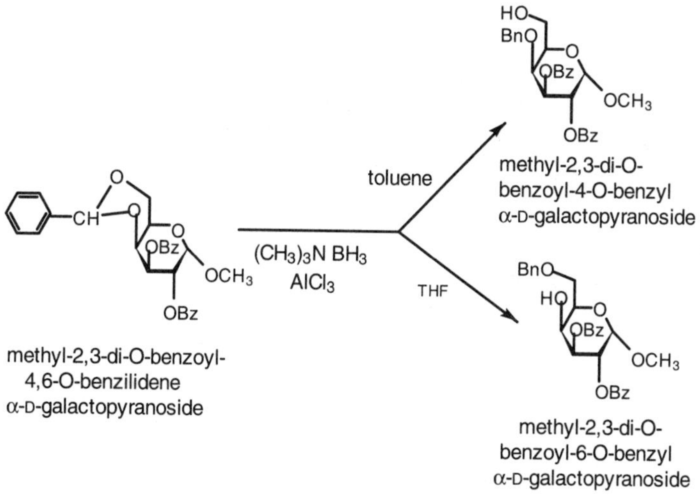
toluene
(CH3)3N BH3
AlCl3
THF
methyl-2,3-di-O-benzoyl-4,6-O-benzilidene α-D-galactopyranoside
methyl-2,3-di-O-benzoyl-4-O-benzyl α-D-galactopyranoside
methyl-2,3-di-O-benzoyl-6-O-benzyl α-D-galactopyranoside

(4.34)

4.5 Modifications at C-1

a. Reduction of Aldehyde and Ketone Carbonyls

The aldo or keto groups of carbohydrates can be readily reduced in aqueous solution at pH values between 6 and 10 with sodium borohydride to give sugar alcohols (alditols, reaction 4.35). The rate of the reduction can be accelerated by heating at 50°C. The reduction of ketoses results in two epimeric alditols (reaction 4.36). Use of tritiated sodium borohydride gives a tritium atom attached to C-1.

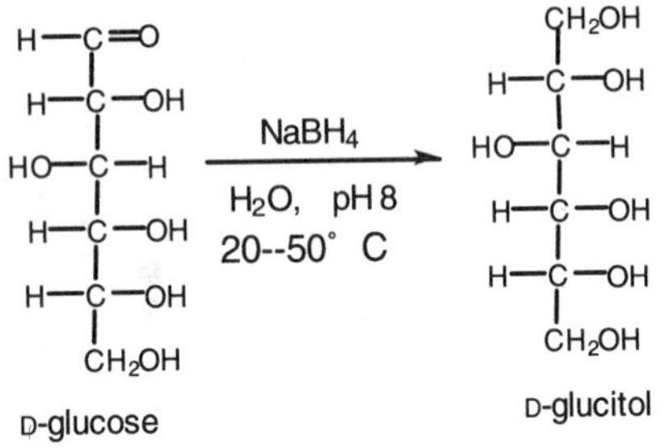

(4.35)

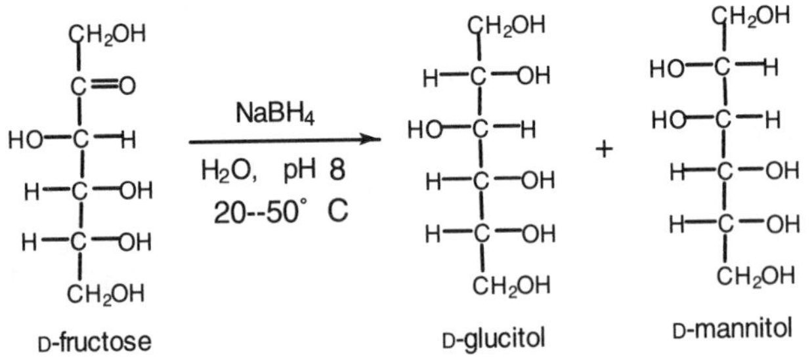

(4.36)

D-Glucose can be specifically reduced in the presence of other sugars by the enzyme, D-*sorbitol dehydrogenase,* using the reduced form of the coenzyme nicotinamide adenine dinucleotide (NADH). The reaction is reversible, and D-glucitol (D-sorbitol) can be specifically oxidized to D-glucose using the oxidized form of the coenzyme, NAD$^+$.

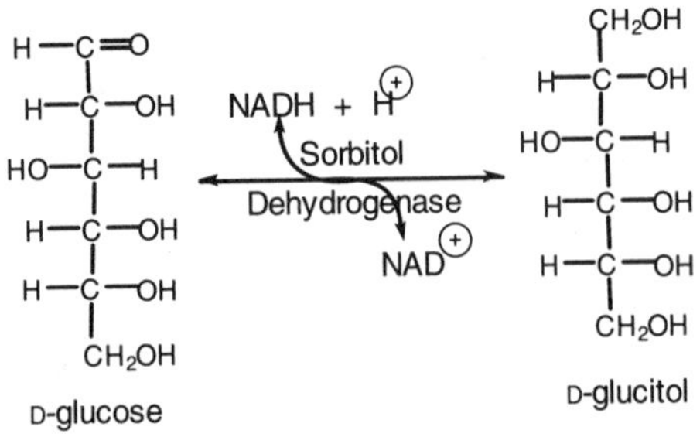

(4.37)

Sorbitol dehydrogenase is obtained commercially from sheep liver. It occurs naturally in humans in two specific locations, sperm and the eye. Its function in the eye is to keep the concentration of D-sorbitol low; D-sorbitol is a factor in the formation of cataracts. It is, however, the reaction of sorbitol dehydrogenase in the reverse reaction with the high concentrations of D-glucose resulting from an uncontrolled diabetic condition that leads to diabetic cataracts.

b. Reduction of Thioacetals and the Protection of C-1

Dithioacetals of reducing sugars are obtained by reaction with a dithiol, such as ethane dithiol, and zinc chloride [37] (reaction 4.38). The acetylated dithioacetal can be desulfurized by cartalytic reduction using Raney nickel to reduce the carbonyl group to a hydrocarbon [38] (reaction 4.38). The 1,1-dithioacetal can be used to reversibly protect the aldehyde group, since it can be readily removed by acid hydrolysis.

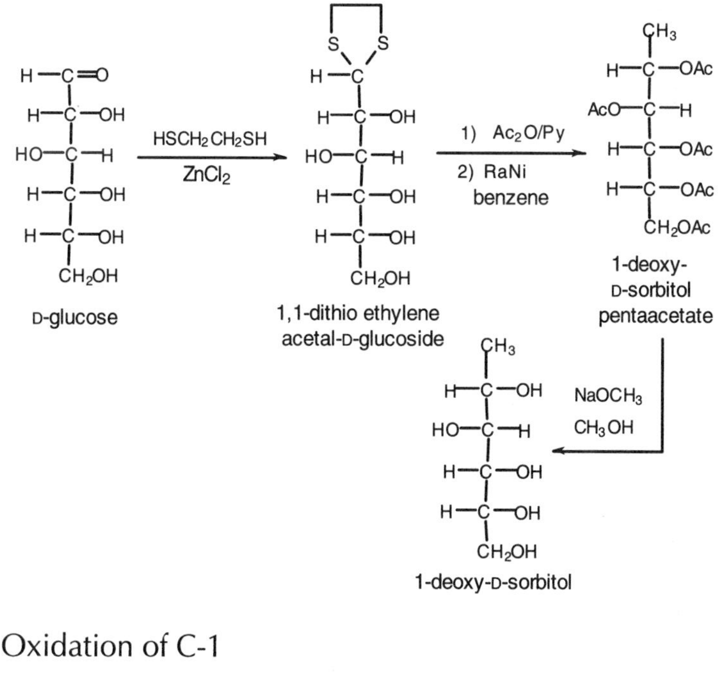

(4.38)

c. Oxidation of C-1

The use of halogens to oxidize carbohydrates was an early reaction used in the development of carbohydrate chemistry in the nineteenth century. The aldehyde carbon is oxidized to give an aldonic acid.

$$RCHO + Br_2 + H_2O \rightarrow RCOOH + 2\,HBr$$

$$RCHO + HOBr \rightarrow RCOOH + HBr$$

The β-anomer of D-glucopyranose is specifically oxidized by bromine-water at pH 5–6 to give the 1,5-D-gluconolactone [39], which is sometimes called δ-gluconolactone. As the reaction progresses, the concentration of acid increases and the lactone ring is opened to give D-gluconic acid. The reaction is inhibited by the formation of HBr and can be greatly accelerated by the use of a slightly acidic buffer containing calcium or barium carbonate and will provide exclusively the lactone [39,40]. 1,5-D-Gluconolactone is commercially prepared by using bromine-water [39].

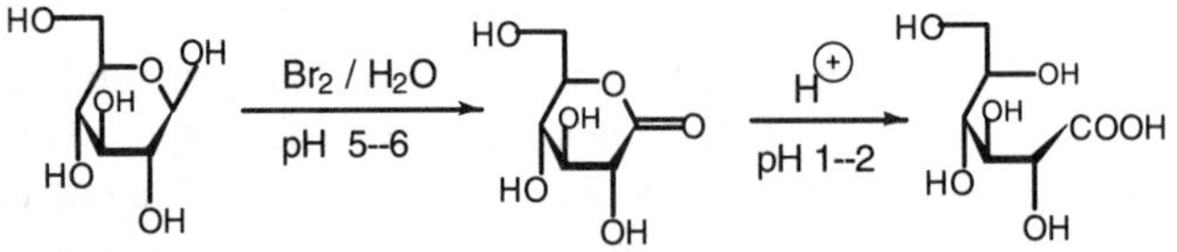

β-D-gluco   1,5-D-gluconolactone   D-gluconic
pyranose   (δ-D-gluconolactone)   acid

(4.39)

The use of an alkaline solution of iodine gives sodium D-gluconate (reaction 4.40).

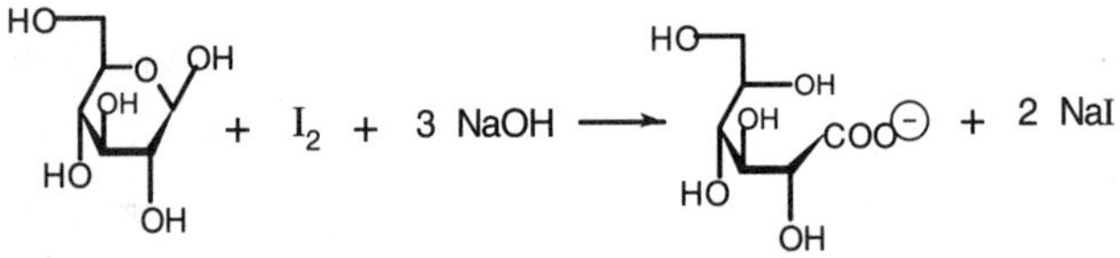

β-D-gluco   sodium
pyranose   D-gluconate

(4.40)

As already discussed in Chapter 3, section 3.5, the enzyme *glucose oxidase* will specifically oxidize the β-anomer of D-glucopyranose to give δ-gluconolactone, and this specific oxidation is used as an analytical procedure for the quantitative determination of D-glucose in the presence of other kinds of carbohydrates.

d. Chain Elongation

i. Fischer-Kiliani Synthesis

It already has been mentioned (Chapter 2) that the cyanohydrin reaction of adding hydrogen cyanide to carbohydrates played an important role in the elucidation of the configurational structures of the carbohydrates. The reaction increases the length of the carbohydrate chain by one carbon but, unfortunately, forms two epimeric products that must be separated. Sometimes one of the two products is produced in much larger amounts than the other, facilitating the separation, but sometimes also providing the unwanted product. The nitrile products can be converted into the aldonic acids by hydrolysis with sodium hydroxide. The aldonic acid can be converted into the lactone by treatment with sulfuric acid. The lactone can then be reduced by sodium borohydride to the hemiacetal (reaction 4.41).

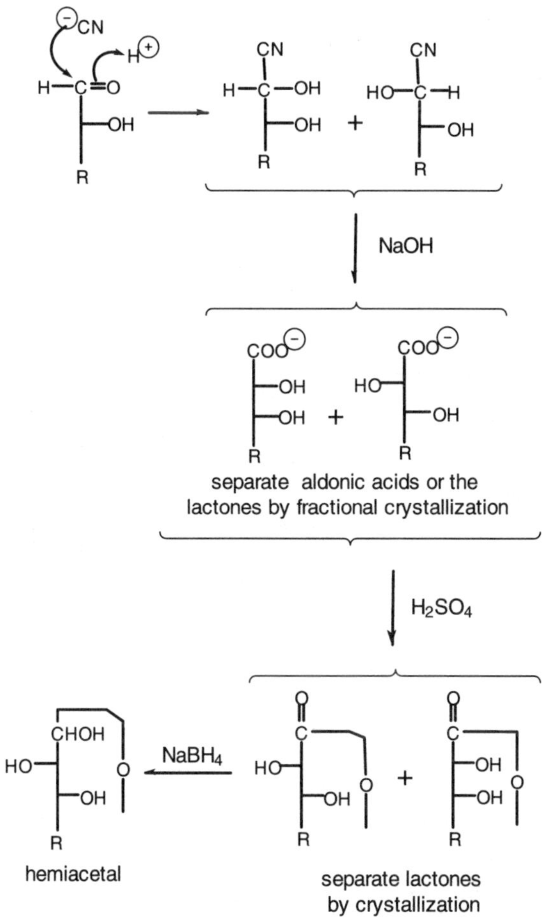

(4.41)

Today, the Fischer-Kiliani synthesis is used primarily to synthesize higher-carbon sugars and to prepare 1-^{14}C-labeled carbohydrates.

ii. The Nitroalkane Synthesis and the Nef Reaction

Another method for lengthening the carbohydrate chain is the addition of a nitroalkane to the aldehyde group. The nitroalkane is treated with sodium methoxide to form a carbanion that adds to the aldehyde group to give a 1-nitro-1-deoxyalditol. The alditol can be converted into an aldehyde by treatment with sodium hydroxide, followed by acid; this is the *Nef reaction* [41] (reaction 4.42).

L-Glucose can be synthesized from L-arabinose by the Fischer-Kiliani synthesis, but the synthesis is long and tedious [42] due in part to the difficulty in separating L-gluconic acid from the preponderant L-mannonic acid. In contrast, the synthesis of L-glucose by the addition of nitromethane to L-arabinose results in the formation of two epimeric 1-nitroalcohols that are readily separated by fractional crystallization [43]. This results in the synthesis of L-glucose (reaction

4.42) in relatively pure and large amounts from L-arabinose, one of the few L-carbohydrates that occurs naturally. L-Arabinose is widely distributed in plants in complex carbohydrate gums, hemicelluloses, and pectin.

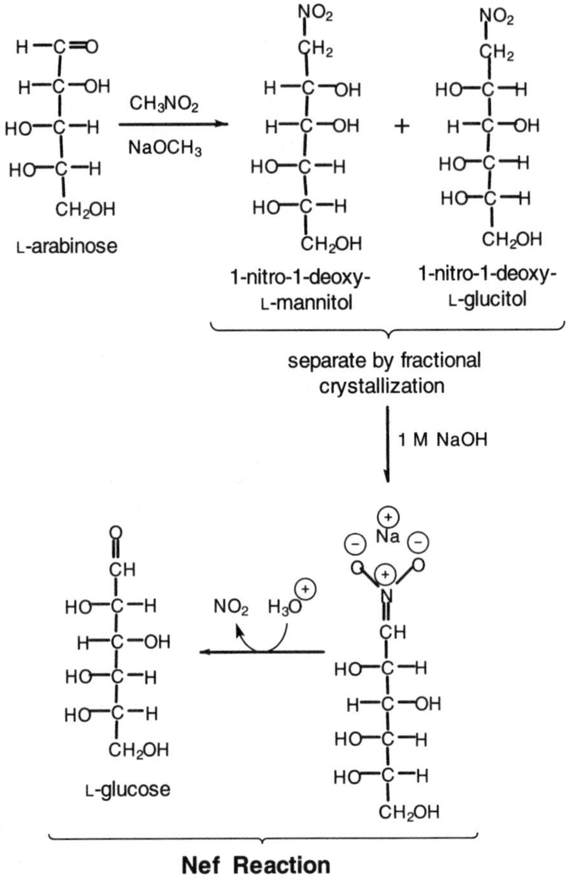

CH3NO2
NaOCH3
L-arabinose
1-nitro-1-deoxy-L-mannitol
1-nitro-1-deoxy-L-glucitol
separate by fractional crystallization
1 M NaOH
NO2 H3O
L-glucose
Nef Reaction

(4.42)

iii. Synthesis of Ketoses with 2-Nitroethanol (Two-Carbon Addition)

The Nef reaction of the 2-nitroalditol that results from the condensation of 2-nitroethanol with a carbohydrate produces a ketose with two additional carbon atoms (reaction 4.43).

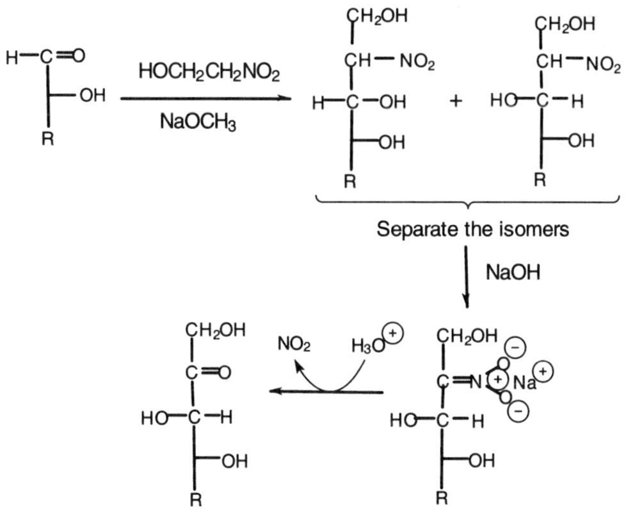

(4.43)

iv. Synthesis of 2-Deoxy Carbohydrates with Methyl Cyanide (Addition of Two Carbons)

A combination of the techniques of the nitroalkane and the cyanohydrin reaction, using methyl cyanide, gives 2-deoxy carbohydrates containing two additional carbon atoms. The addition takes place in a manner similar to the nitroalkane condensation. A carbanion is produced by the treatment of methyl cyanide with sodium methoxide, and this adds to the aldehyde group of the carbohydrate. Two epimeric nitriles are produced that are converted into a carbohydrate by using the reactions of the Fischer-Kiliani synthesis (reaction 4.44).

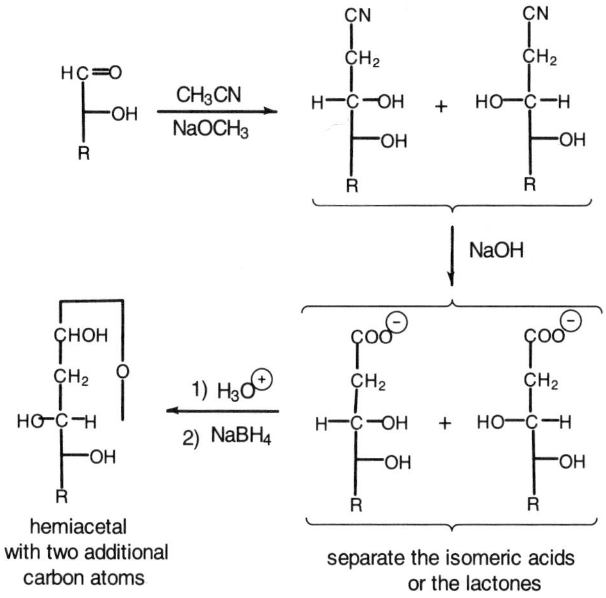

(4.44)

hemiacetal
with two additional
carbon atoms

separate the isomeric acids
or the lactones

The principal uses of these methods are the synthesis of higher unnatural carbohydrates and the synthesis of specifically labeled (^{13}C or ^{14}C) carbohydrates using $^*CN^-$, $^*CH_3-NO_2$, $HO-^*CH_2-^*CH_2-NO_2$, $HO-^*CH_2-CH_2-NO_2$, $HO-CH_2-^*CH_2-NO_2$, $^*CH_3-^*CN$, $^*CH_3-CN$, or CH_3-^*CN, where * indicates the possible positions of the label.

e. Chain-Length Reduction

i. Ruff Degradation

The Ruff degradation [44] was also used in the early studies of the configurational structure of carbohydrates. It starts with a soluble salt of the aldonic acid of the sugar whose chain is to be reduced. The aldonic acid is oxidized by hydrogen peroxide and ferric ion to give the α-keto acid, which upon warming easily undergoes decarboxylation to give a carbohydrate chain with one less carbon (reaction 4.45). The details of the method are given in volume 1 of *Methods in Carbohydrate Chemistry* [45].

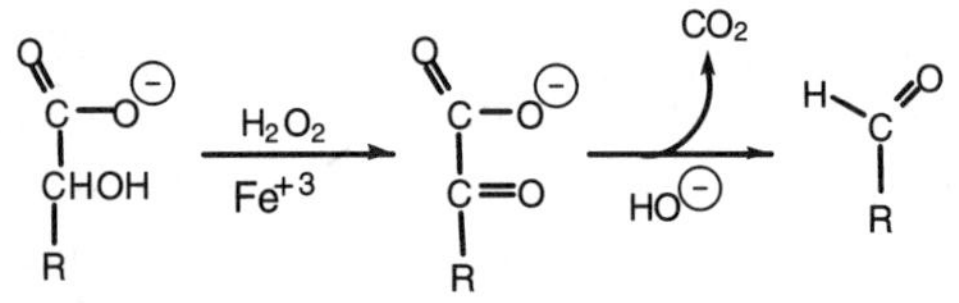

$$ \tag{4.45} $$

ii. Wohl Degradation

The Wohl degradation also transforms one aldose into another with one less carbon. The first step in the degradation is the formation of the oxime, which is acetylated by acetic anhydride and sodium acetate to give O-acetyl nitrile (reaction 4.46). For the next step in the degradation, Zemplén [1] provided an improvement in the procedure by using sodium methoxide to simultaneously remove the cyanide and acetyl groups in one step. A modified Wohl degradation was introduced in 1950 [46] in which the oxime is reacted with 2,4-dinitrofluorobenzene that, on treatment with sodium methoxide, removes hydrogen cyanide and 2,4-dinitrophenolate, and produces a carbohydrate with one less carbon (reaction 4.47).

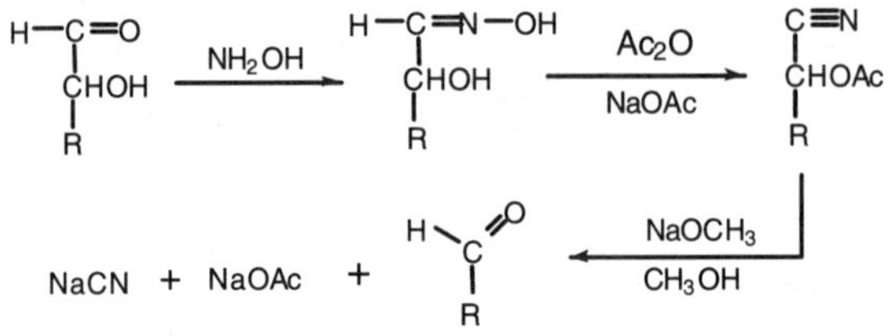

$$ \tag{4.46} $$

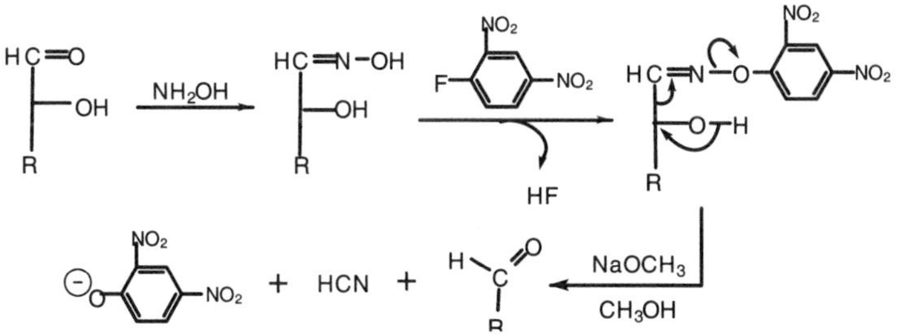

$$(4.47)$$

f. Substitution at C-1, the Reducing Carbon

The principal compound that has been used to make substitutions at the reducing carbon (C-1) is the peracetylated carbohydrate from which the α-1-bromide can be made. For example, D-glucose pentaacetate undergoes reaction with HBr in glacial acetic acid to give 1-bromo-2,3,4,6-tetra-O-acetyl-α-D-glucopyranoside (bromoacetyl glucose) [47] (reaction 4.48). The 1-bromide can then be displaced by a number of nucleophiles to give substitution.

Formation of 2,3,4,6-tetra-O-acetyl-D-glucopyranose is obtained by treating bromoacetyl glucose in acetone containing silver carbonate [48,49] (reaction 4.48).

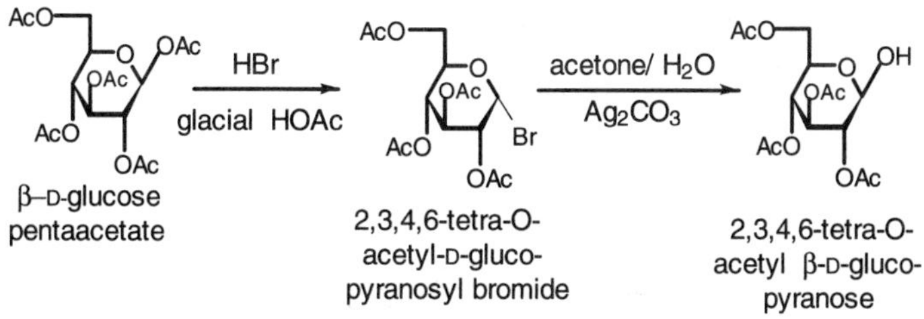

$$(4.48)$$

The reaction of bromoacetyl glucose with lithium aluminum hydride in diethyl ether gives the substitution of hydrogen for the hemiacetal hydroxyl at C-1, and the formation of the nonreducing, 1,5-anhydro-D-glucitol [50,51] (reaction 4.49). The excess $LiAlH_4$ is removed by the addition of ethyl acetate, which is reduced to ethanol.

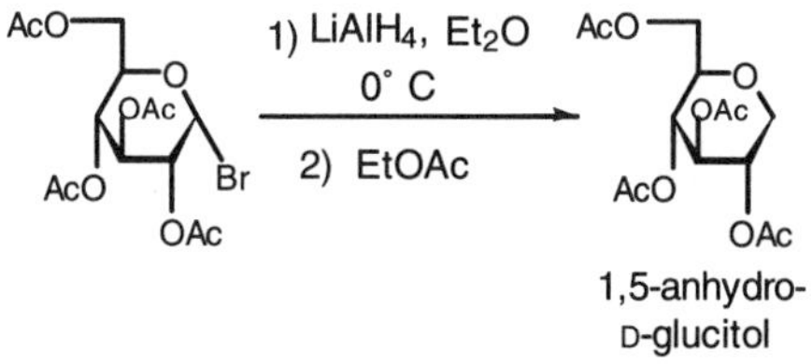

$$(4.49)$$

The 1-phosphate can be prepared by reaction of bromoacetyl glucose with silver phosphate in benzene or toluene. The anomeric form of the resulting product depends on the type of silver salt used. When monosilver phosphate is the nucleophile, the β-1-phosphate is obtained [52] (reaction 4.50). Use of trisilver phosphate [53] gave a 10% yield of the α-1-phosphate (reaction 4.51), the biological form of α-glucose-1-phosphate; and the use of diphenyl silver phosphate, followed by catalytic hydrogenation [54], gives a 37% yield of α-1-phosphate (reaction 4.52).

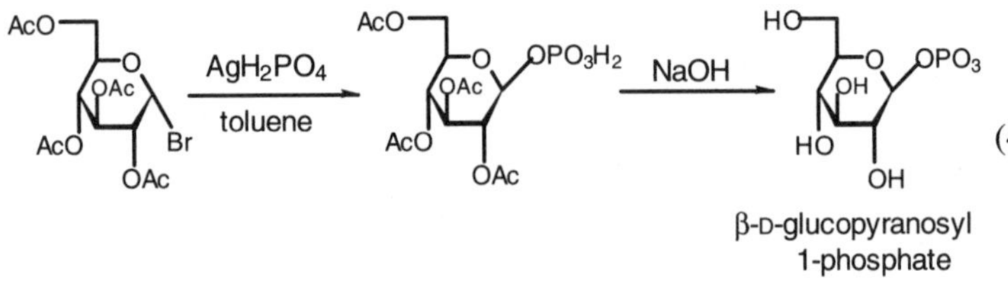

$$\beta\text{-D-glucopyranosyl}$$
$$\text{1-phosphate}$$

(4.50)

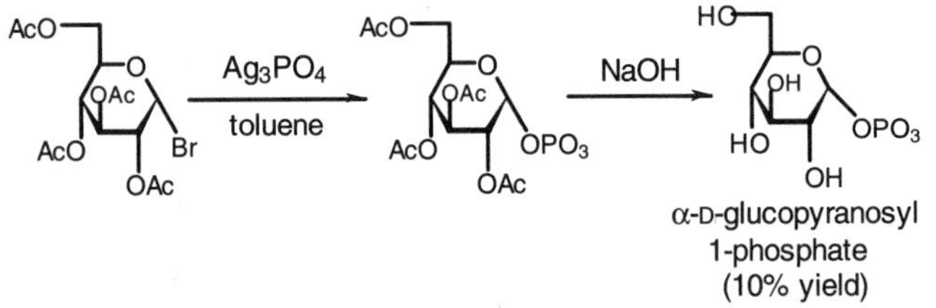

$$\alpha\text{-D-glucopyranosyl}$$
$$\text{1-phosphate}$$
$$\text{(10\% yield)}$$

(4.51)

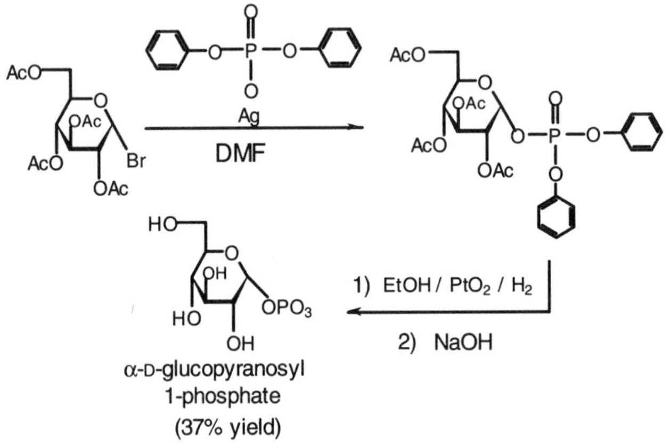

$$\alpha\text{-D-glucopyranosyl}$$
$$\text{1-phosphate}$$
$$\text{(37\% yield)}$$

(4.52)

The reaction of bromoacetyl glucose and bromoacetyl galactose with sodium azide in DMF gives 68 and 75% yields, respectively, of the 1-β-azide [55] (reaction 4.53). The azide can be easily reduced by catalytic hydrogenation to give 1-glycosylamines.

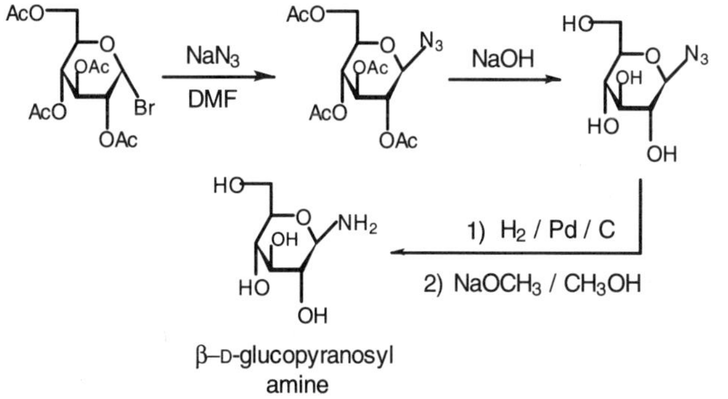

(4.53)

β–D-glucopyranosyl
amine

The direct substitution of peracetylated mono- and disaccharides by trimethyl-silyl azide, in the presence of a Lewis acid, produces the β-azide without the formation of the 1-bromide intermediate [56] (reaction 4.54).

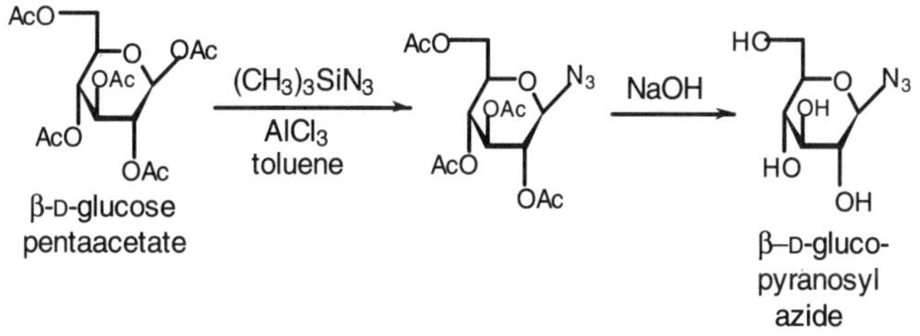

(4.54)

β-D-glucose
pentaacetate

β–D-gluco-
pyranosyl
azide

The β-glycopyranosyl amines can also be formed directly by the reaction of ammonium chloride and ammonia in methanol [57,58].

The α-1,2-*cis*-glycosyl azides are synthesized by reaction of 3,4,6-tri-*O*-acetyl-2-*O*-trichloroacetyl-β-D-glucopyranosyl chloride (for the synthesis of this compound, see reaction 4.57) with sodium azide in hexamethylphosphoramide (HMPA) at 20°C [59] (reaction 4.55).

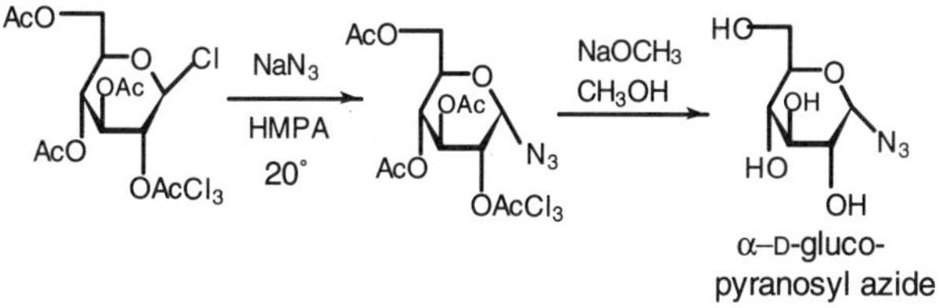

(4.55)

α–D-gluco-
pyranosyl azide

Reaction of bromoacetyl glucose with trimethylammonium bromide [60,61] or with alkali [62] forms 1,6-anhydro-D-glucopyranoside, commonly called levoglucosan (reaction 4.56).

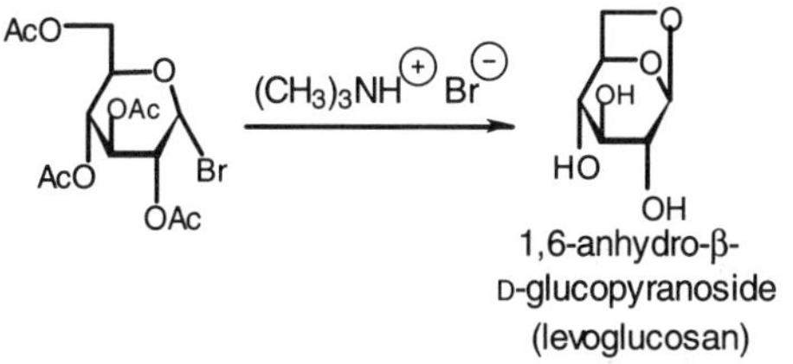

(4.56)

1,6-anhydro-β-
D-glucopyranoside
(levoglucosan)

One of the best-known carbohydrates of the anhydro class is 1,2-anhydro-3,4,6-tri-O-acetyl-α-D-glucopyranoside, commonly called Brigl's anhydride. It is prepared from β-D-glucopyranoside pentaacetate by reaction with phosphorous pentachloride in carbon tetrachloride to give 3,4,6-tri-O-acetyl-2-O-trichloro-acetyl-β-D-glucopyranosyl chloride from which 1,2-anhydro-tri-O-acetyl-α-D-glucopyranoside is obtained by further treatment with dry ammonia in benzene [63,64] (reaction 4.57).

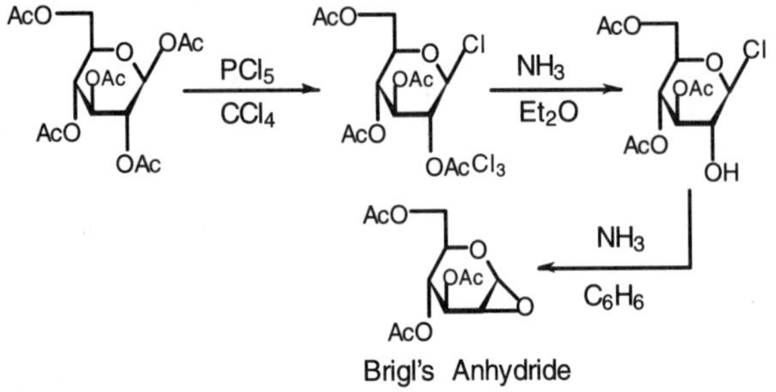

(4.57)

Brigl's Anhydride

The 1,2-anhydride is very reactive and has been used to synthesize β-glyco-sidic linkages [63,65,66]. At elevated temperatures (80–100°C) α-glycosidic linkages are formed [64,65]. β-Glycosides are also synthesized from bromoacetyl glucose by the Koenigs-Knorr reaction (see section g.i.).

The reactivities of the glycosyl halides are in the following decreasing order: I > Br > Cl > F. Glycosyl fluorides are much more stable and easier to handle than other glycosyl halides. They can be isolated as free sugars and are relatively stable in aqueous solutions, in contrast to the glycosyl chlorides or bromides that cannot be isolated as the free sugars and are rapidly hydrolyzed by water. The 1-fluoro de-rivatives are important compounds as glycosyl donors in carbohydrate synthesis [67–69], and they also have use as substrates for the study of enzyme mechanisms and as glycosyl donors in the enzymatic synthesis of saccharides [70].

The β-glycosyl fluorides are prepared by reaction of peracylated glycosyl chloride or bromide with silver fluoride [71] or silver tetrafluoroborate [72] (re-action 4.58).

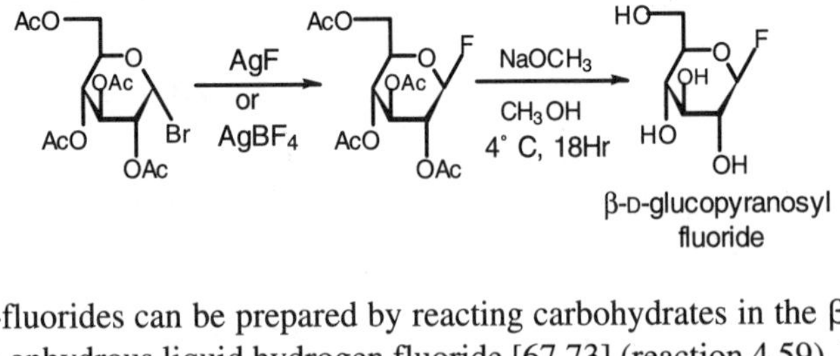

(4.58)

β-D-glucopyranosyl
fluoride

The α-fluorides can be prepared by reacting carbohydrates in the β-anomeric form with anhydrous liquid hydrogen fluoride [67,73] (reaction 4.59). Liquid hy-drogen fluoride is a difficult material to handle, and the procedure is rather severe. A much milder and highly specific procedure for the formation of α-1-fluorides uses pyridinium poly (hydrogen fluoride) [74,75] (reaction 4.60).

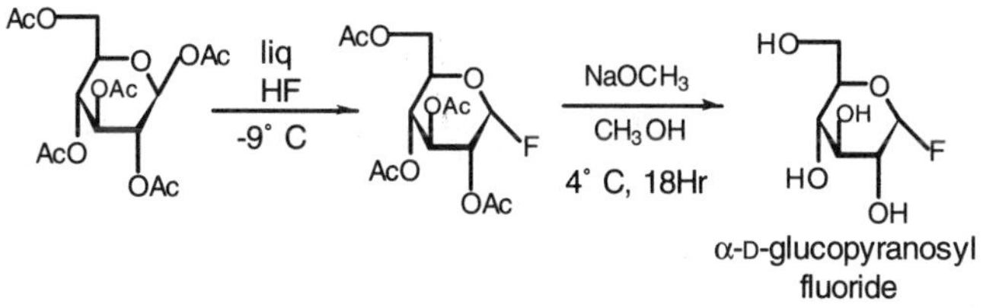

$$(4.59)$$

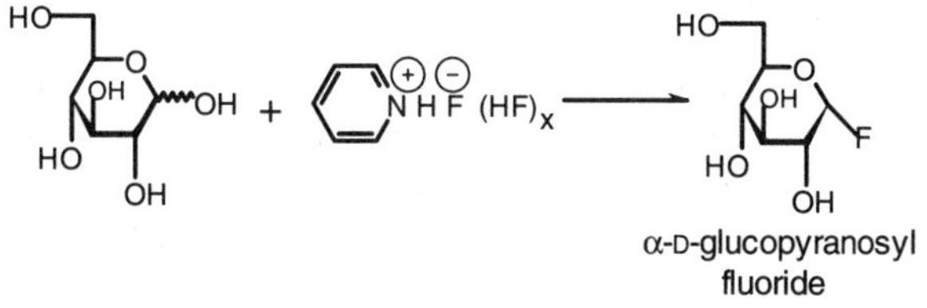

α-D-glucopyranosyl
fluoride

$$(4.60)$$

α-D-glucopyranosyl
fluoride

g. Formation of Glycosides

Several methods have been reported for the synthesis of glycosides, that is, sugars that have noncarbohydrate groups attached to the reducing carbon as an acetal. We have already discussed the Fischer method (Chapter 2) for the synthesis of α-glycosides formed between a carbohydrate and an alkyl alcohol. This method is not successful when aryl groups are involved.

i. Koenigs-Knorr Method

Koenigs and Knorr [66] prepared a number of β-glycosides by reacting 2,3,4,6-tetra-O-acetyl-α-D-glucopyranosyl bromide with alkyl alcohols in pyridine, with silver carbonate as a catalyst (reaction 4.61). This general reaction became known as the *Koenigs-Knorr method*. The reaction forms the β-anomer and can be used for preparing both alkyl and aryl glycosides. The method has been widely used for the synthesis of a number of complex β-glycosides [76]. The glycosyl bromides react faster than the chloride. Glycosyl iodides are extremely reactive and unstable. Glycosyl fluorides are unreactive under the conditions of the Koenigs-Knorr reaction.

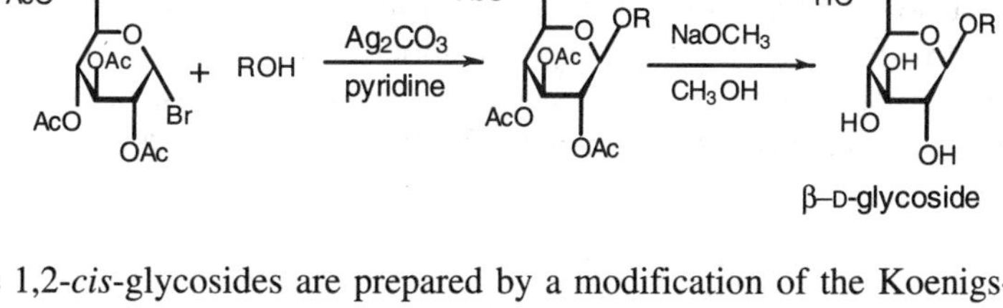

$$(4.61)$$

β-D-glycoside

The 1,2-*cis*-glycosides are prepared by a modification of the Koenigs-Knorr reaction. Reaction of the peracetylated carbohydrate with phosphorous pentachloride in carbon tetrachloride gives 3,4,6-tri-O-acetyl-2-O-trichloroacetyl-β-D-glucopyranosyl chloride [77] (reaction 4.57), which reacts with alkyl and aryl nucleophiles to form the α-glycopyranoside (reaction 4.62).

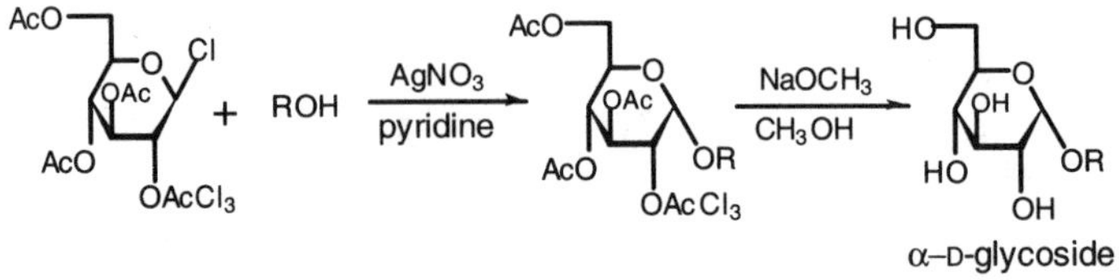

(4.62)

α–D-glycoside

ii. Trichloroacetimidate Method

Carbohydrate trichloroacetimidates have become very useful derivatives for the activation of the carbohydrate moiety. The carbohydrate can be readily transferred to form many types of glycosides [78]. Glycosides are readily formed by reaction of the hemiacetal hydroxyl with trichloroacetonitrile under base catalysis conditions. The β-anomer is rapidly and preferentially formed first as a favored kinetic product. Under base catalyzed conditions, however, it is slowly anomerized to give the α-anomer, which is the more stable thermodynamic product, due to the anomeric effect (reaction 4.63). Thus, either the α- or the β-anomer can be isolated in pure form and in high yield, depending on the length of time that the reaction is allowed to proceed.

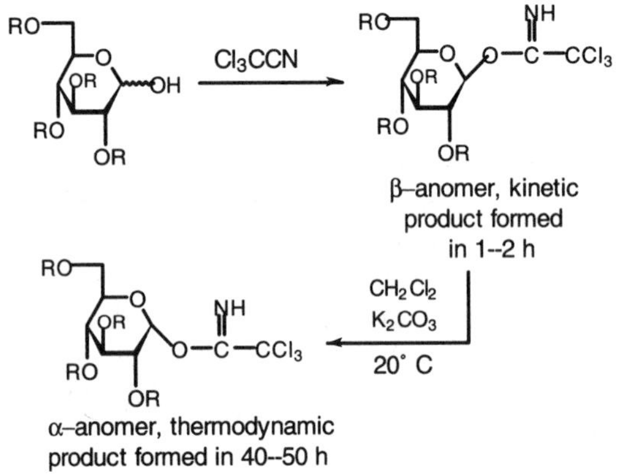

β–anomer, kinetic
product formed
in 1–2 h

(4.63)

α–anomer, thermodynamic
product formed in 40–50 h

The trichloroacetimidates are readily displaced to give substitution at C-1. They react with hydrogen halides to give chlorides, bromides, and fluorides, with retention of configurations (reaction 4.64). Carboxylic acids, being relatively weak acids, react to give inversion of configuration (reaction 4.65).

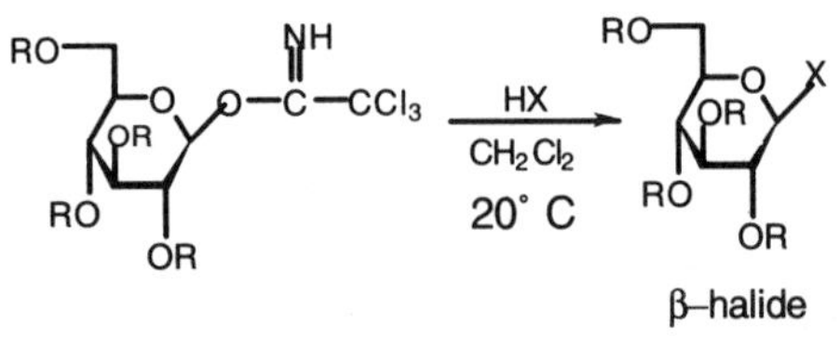

(4.64)

β–halide

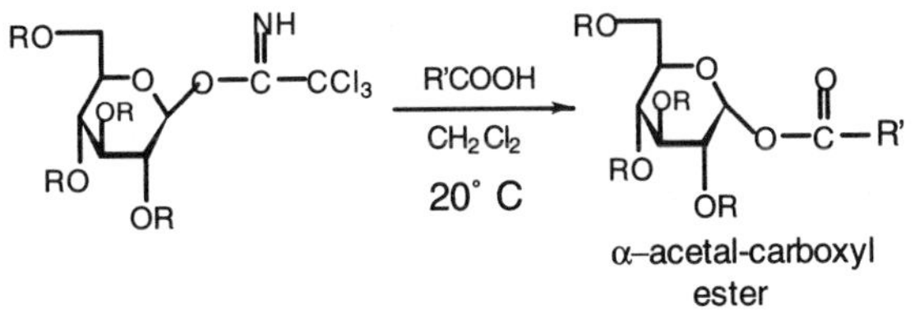

$$(4.65)$$

α–acetal-carboxyl
ester

Mono- and diphosphate esters displace the trichloroacetimidate and form the glycosyl derivative. This is useful in the synthesis of glycolipids [78]. Acid-catalyzed α-anomerization can convert the β-anomer into the α-anomer (reaction 4.66).

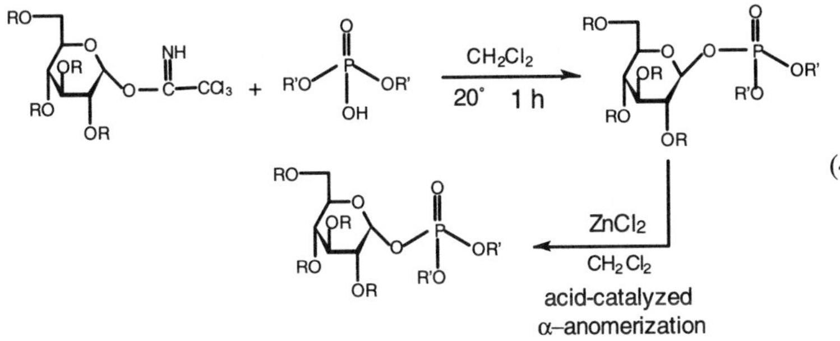

$$(4.66)$$

acid-catalyzed
α–anomerization

Many glycosides that were difficult to form or were formed only in low yields using the Koenigs-Knorr method can often be formed in acceptable yields using trichloroacetimidates [78]. The method has found uses in the chemical synthesis of O-linked glycopeptides [79]. Several different kinds of glycosyl donors can be formed into trichloroacetimidates, such as D-glucopyranose, D-galactopyranose, N-acetyl D-glucosamine, N-acetyl D-galactosamine, N-acetyl D-mannosamine, D- and L-fucose, D- and L-rhamnose, N-acetyl lactosamine, N-acetyl D-muramic acid, and chitobiose as well as complex oligosaccharides [78]. The trichloroacetimidate method, thus, has wide utility in that either α- or β-anomers can be formed, depending on the conditions, in pure form and in high yields. The further application of diastereocontrolled glycosyl transfer widens the scope of synthetic reactions and the consequent formation many complex saccharides, glycolipids, and glycopeptides.

h. Formation of Glycosidic Linkages between Monosaccharide Residues

The anhydrides have been used to synthesize higher saccharides. Several disaccharides have been synthesized by reacting Brigl's anhydride with 2,3,4,6-tetra-O-acetyl-β-D-glucopyranose in benzene at 90°C for 37 hr to give α,β-trehalose heptaacetate [80,81] (reaction 4.67).

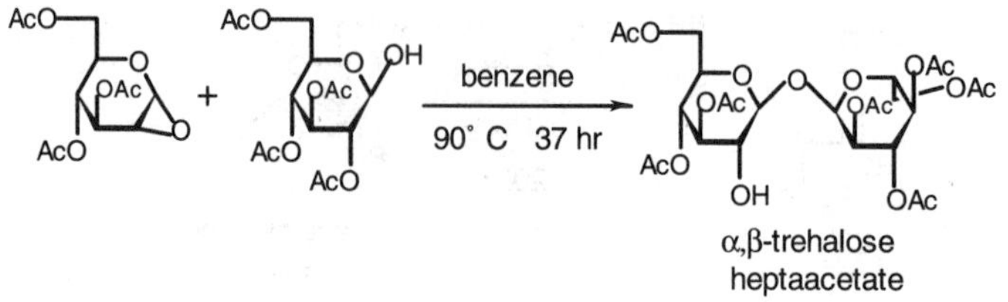

$$(4.67)$$

α,β-trehalose
heptaacetate

β-Maltose heptaacetate was synthesized using Brigl's anhydride with 1,2,3,6-tetra-O-acetyl-β-D-glucopyranoside at 90°C in benzene for 20 hr [82] (reaction 4.68).

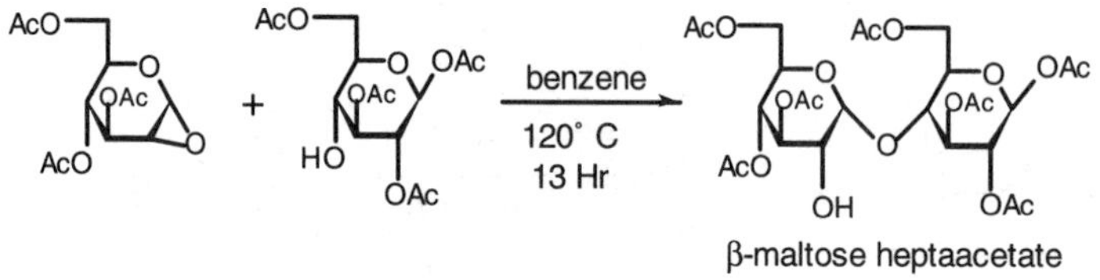

$$(4.68)$$

β-maltose heptaacetate

Sucrose was first chemically synthesized using Brigl's anhydride and 1,3,4,6-tetra-O-acetyl-β-D-fructofuranose (the latter prepared by reacting inulin triacetate with HBr in glacial acetic acid [83] (reaction 4.69).

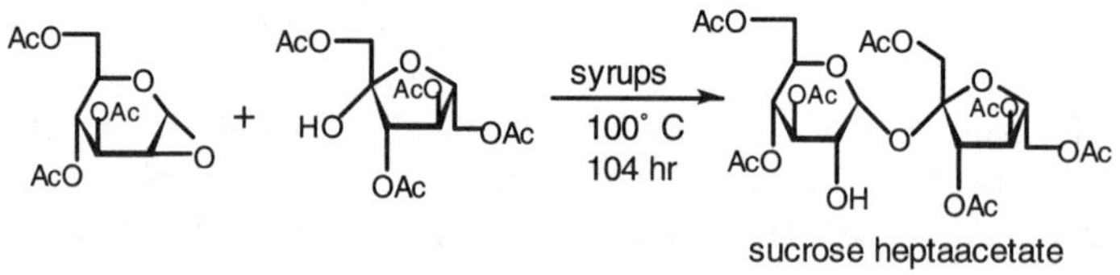

$$(4.69)$$

sucrose heptaacetate

β,β-Trehalose octaacetate was prepared by reaction of bromoacetyl glucose with 2,3,4,6-tetra-O-acetyl-β-D-glucopyranose in chloroform, with silver carbonate and iodine as catalysts [48] (reaction 4.70). β,β-Trehalose also has been synthesized by the reaction of 2,3,4,6-tetra-O-acetyl-β-D-glucopyranose with 2,3,4,6-tetra-O-acetyl α-D-glucopyranosyl trichloroacetimidate [84]. With these syntheses, the two nonnaturally occurring trehaloses (α,β- and β,β-) were chemically synthesized.

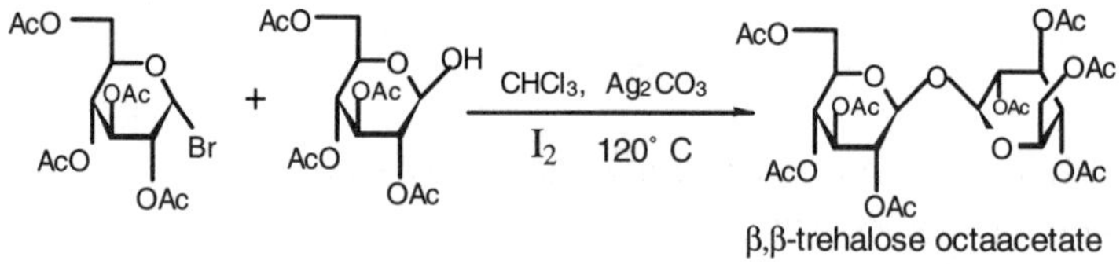

$$(4.70)$$

β,β-trehalose octaacetate

Glycosidic linkages containing sulfur substituted for oxygen have been synthesized. The 1-α-thio-D-glucoside is generated by reaction of the β-chloride with tetra-butylammonium thioacetate (reaction 4.71). This is then reacted with methyl-2,3,6-tri-O-benzoyl-4-O-trifyl-α-D-galactopyranoside (reaction 4.72, see section 4.8 for the preparation of the C-4 OH D-galactopyranoside) in HMPA to give 4^1-thiomaltose [85] (reaction 4.73).

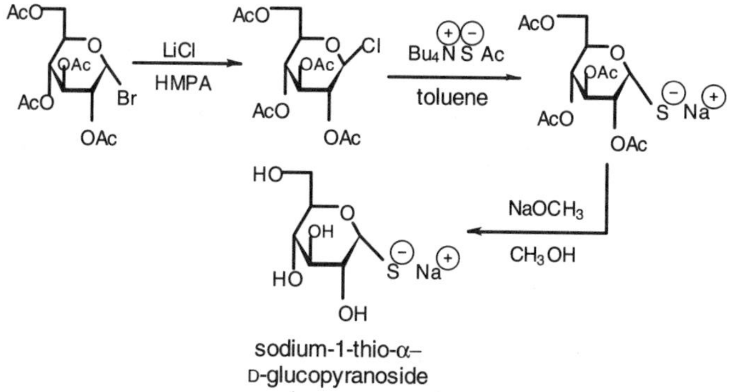

sodium-1-thio-α–
D-glucopyranoside

(4.71)

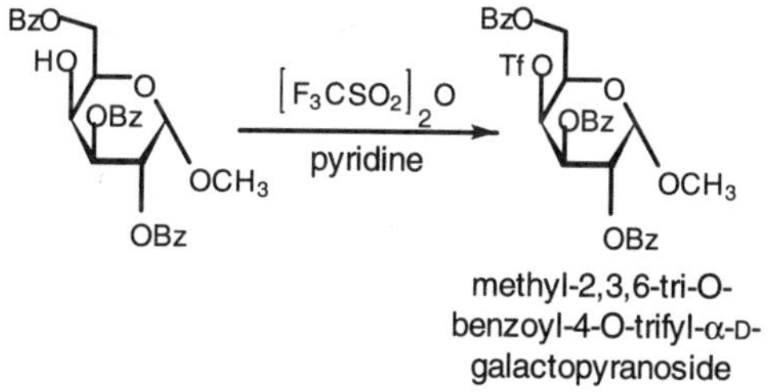

methyl-2,3,6-tri-O-
benzoyl-4-O-trifyl-α-D-
galactopyranoside

(4.72)

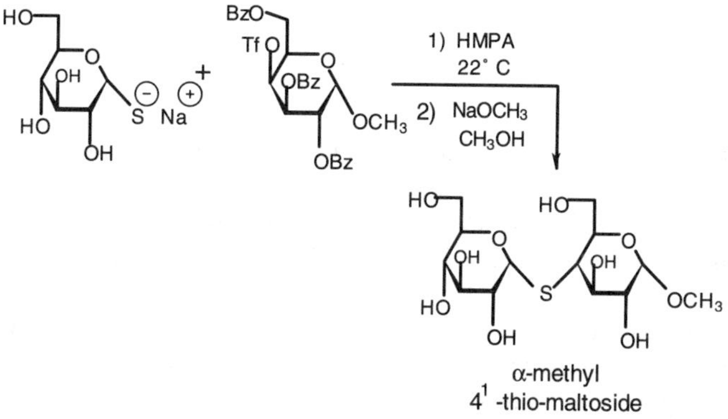

α-methyl
4^1 -thio-maltoside

(4.73)

6^2-Thiopanose, a trisaccharide, has been synthesized in a similar manner by reacting 6^2-deoxy-6^2-iodomaltose heptaacetate with the 1-α-thio-D-glucopyranoside [86] (reaction 4.74).

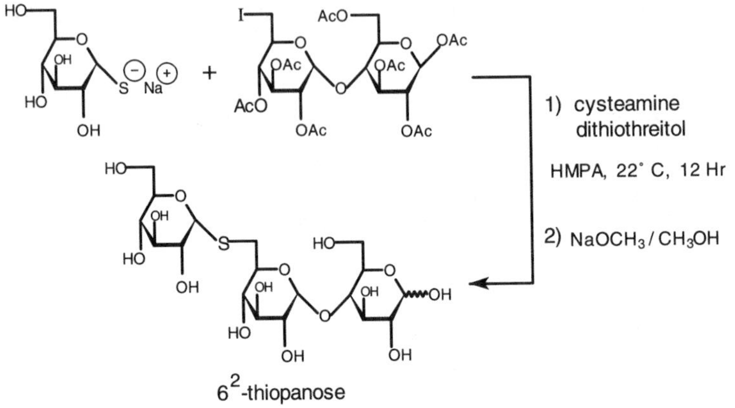

$$(4.74)$$

6²-thiopanose

4.6 Modifications at C-2

The first reported carbohydrate modification at C-2 was the formation of D-glucal (D-glucose with a carbon-carbon double bond between C-1 and C-2) from bromoacetyl glucose [87]. This led to the synthesis of 2-deoxy-D-glucose on the addition of water to the double bond [88] (reaction 4.75). The optimization and preparative details for the formation of 2-deoxy-D-glucose have been reported [89].

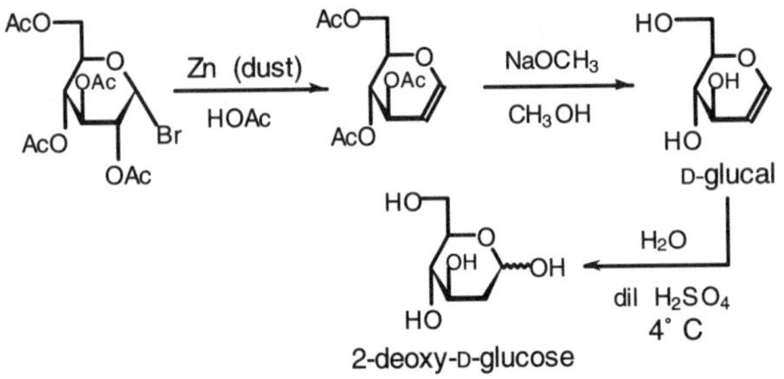

$$(4.75)$$

2-deoxy-D-glucose

Glycals have been formed from many other types of carbohydrates, for example, from D-xylose, L-arabinose, L-fucose, L-rhamnose, lactose, maltose, and cellobiose, by using similar reactions [90,91]. Many different reagents can be added to the double bond, giving modifications at C-2. For example, hydrogen can reduce the double bond to form 2-deoxy-1,5-anhydro-D-alditol (reaction 4.76); hydrogen halides can add to the double bond to give 2-deoxy-1-α-halides that can be used to synthesize 2-deoxyglycosides (reaction 4.77); halides (e.g., bromine) can add to the double bond to form 2-bromo-2-deoxy-1-aldopyranosyl bromide [90] (reaction 4.78) that can give 2-bromo-2-deoxy-β-D-glycosides; hydrogen fluoride can be added from pyridinium poly(hydrogen fluoride) to produce 2-deoxy-α-D-aldopyranosyl fluoride (reaction 4.79); fluorine can be added to C-2 by the reaction with fluoroxytrifluoromethane [91] (reaction 4.80). Boiling of the tri-O-acetyl glycal with water will result in the elimination of the C-3 acetyl group and the isomerization of the double bond to give 2,3-unsaturated sugar [90] (reaction 4.81) that can be further modified by addition to the double bond; addition of hydrogen gives 2,3-dideoxy-D-aldose (reaction 4.81); addition of hydrogen bromide gives 2-deoxy-3-bromo-3-deoxy-D-glucose (reaction 4.82); and the addition of halide (e.g., bromine) gives 2,3-dibromo-2,3-dideoxy-D-aldose (reaction 4.83).

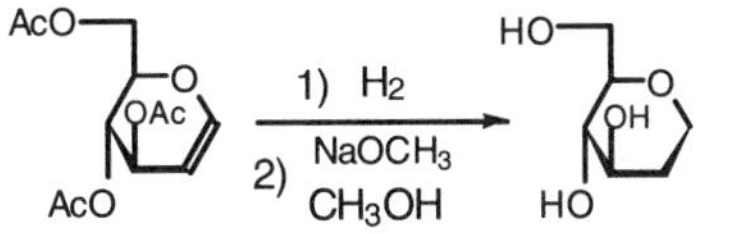

(4.76)

2-deoxy-1,5-
anhydro-
D-glucitol

(4.77)

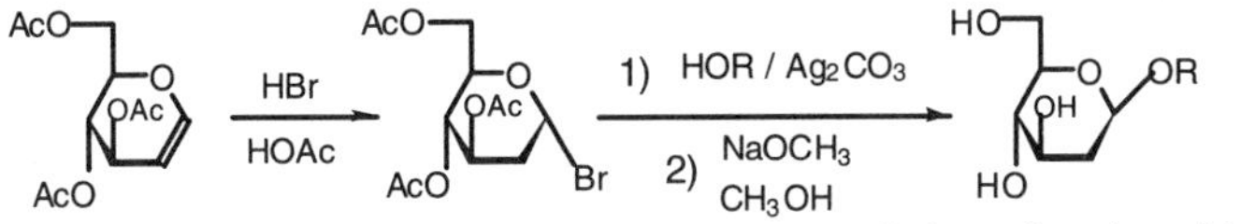

2-deoxy-β-D-glycoside
or oligosaccharide

(4.78)

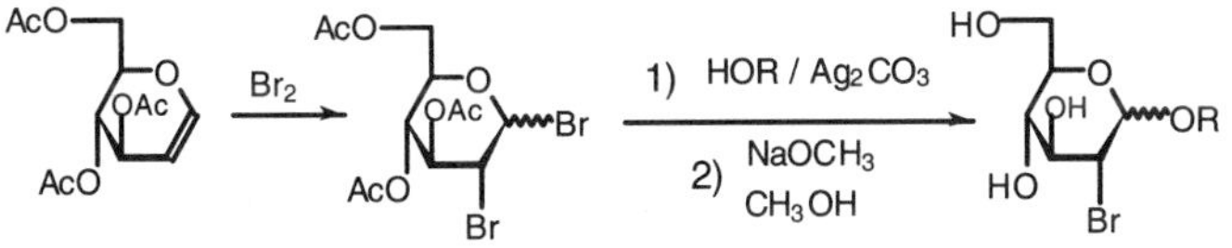

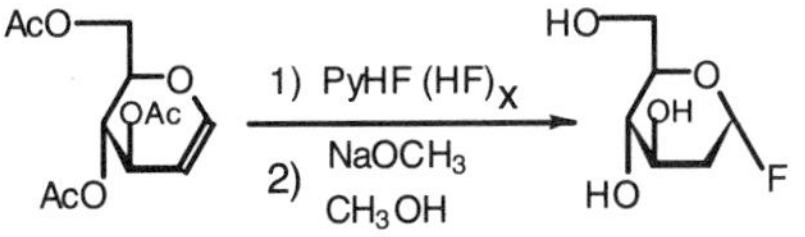

2-bromo-2-deoxy-
D-glycoside or
oligosaccharide

(4.79)

2-deoxy-α-D-
glucopyranosyl
fluoride

(26%)   (34%)

(4.80)

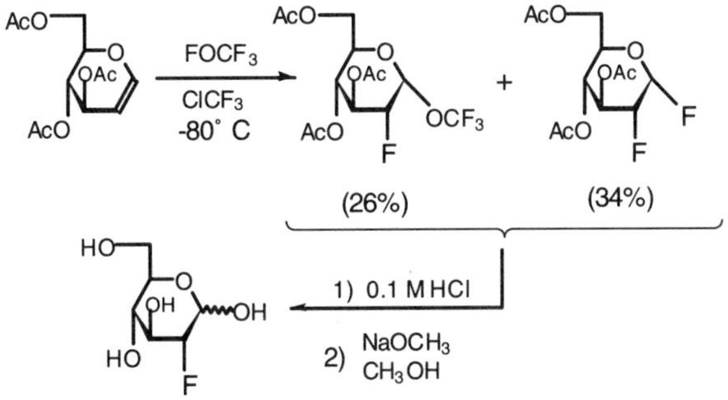

2-deoxy-2-fluoro-
D-glucose

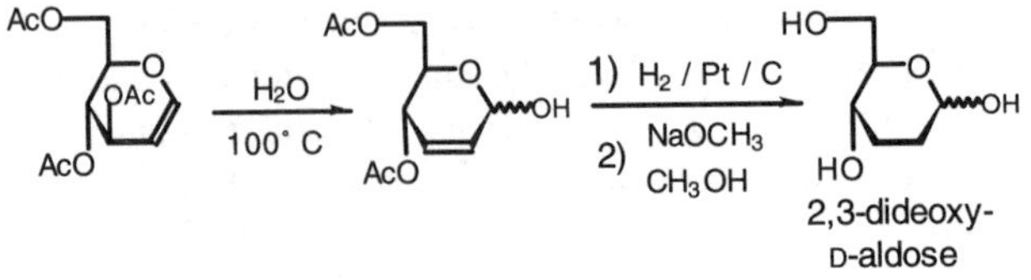

$$(4.81)$$

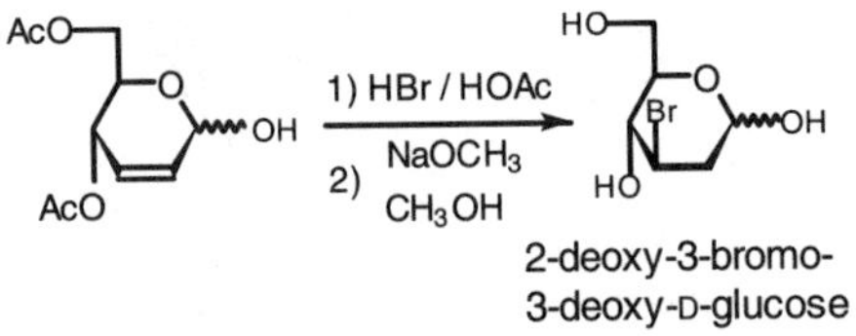

2,3-dideoxy-
D-aldose

$$(4.82)$$

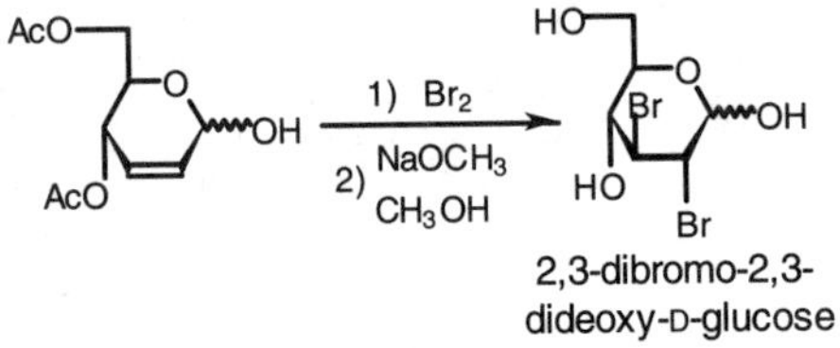

2-deoxy-3-bromo-
3-deoxy-D-glucose

$$(4.83)$$

2,3-dibromo-2,3-
dideoxy-D-glucose

The 2-deoxy-α-1-bromide (reaction 4.77) can be used to synthesize 2-deoxy-β-D-glycosides (reaction 4.77) or oligosaccharides, the latter with a 2-deoxy group in the nonreducing residue, as shown in reaction 4.84, to give 2^2-deoxycellobiose.[1]

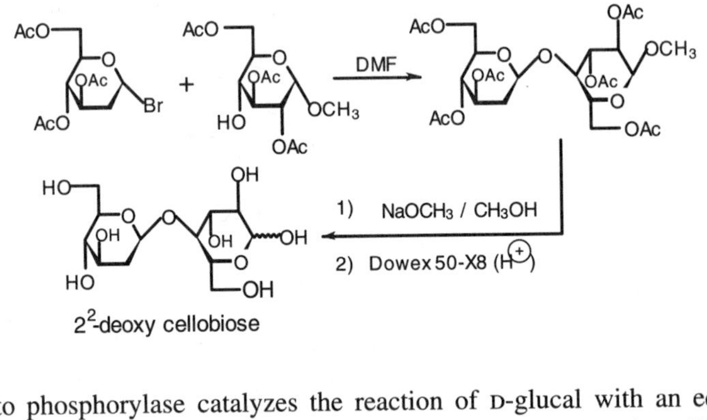

$$(4.84)$$

2^2-deoxy cellobiose

Potato phosphorylase catalyzes the reaction of D-glucal with an equimolar amount of phosphate to give 2-deoxy-α-D-glucopyranosyl phosphate [92] (reaction 4.85). Reaction of potato phosphorylase with D-glucal, maltotetraose, and catalytic amounts of phosphate gives 2-deoxymaltooligosaccharide [92] (reaction 4.86). When this latter reaction is followed by a reaction with sweet potato β-amylase, $2^{1,2}$-dideoxymaltose is obtained [93] (reaction 4.87).

---

[1]See Appendix A for the numbering and naming of substituted oligosaccharides.

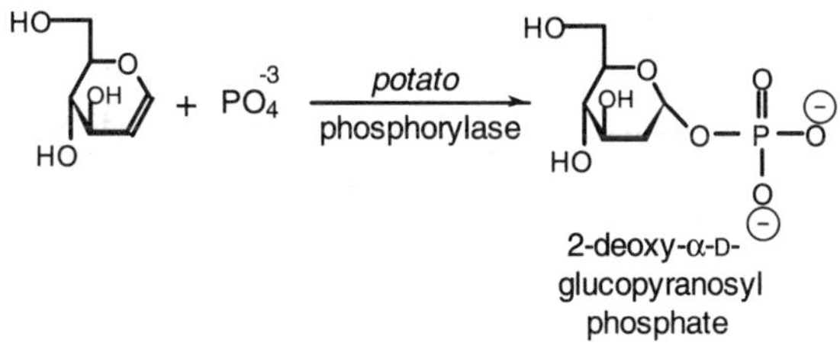
HO
HO
OH
O
+ PO_4^{-3}
potato
phosphorylase
HO
HO
OH
O
O
P
O
O
2-deoxy-α-D-
glucopyranosyl
phosphate

(4.85)

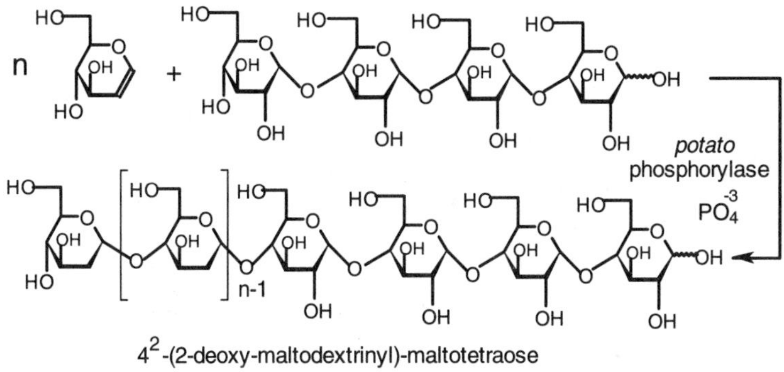
n
HO
HO
OH
O
+
OH
potato
phosphorylase
PO_4^{-3}
n-1
4^2-(2-deoxy-maltodextrinyl)-maltotetraose

(4.86)

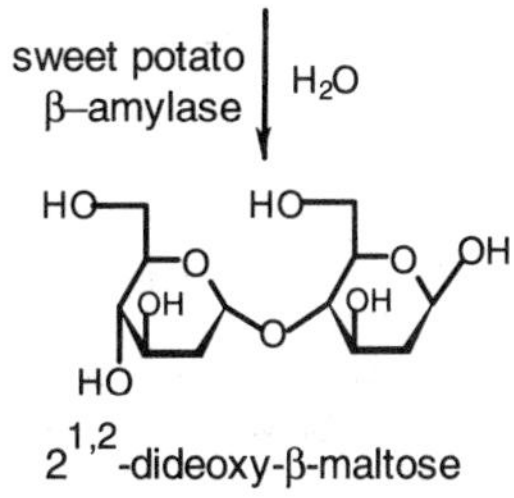
sweet potato
β–amylase
H_2O
HO
OH
O
$2^{1,2}$-dideoxy-β-maltose

(4.87)

The C-2 hydroxyl group of D-glucose or D-allose can be obtained in a free form for reaction by using diacetone glucose or diacetone allose. The 5,6-isopropylidene group is removed and the three free hydroxyl groups at C-3, C-5, and C-6 are benzoylated. The 1,2-isopropylidene group is removed and the α-1-methyl glycoside is formed (reaction 4.88). The C-2 hydroxyl group of the D-glucose can then be modified by a double displacement of a trifyl group, as illustrated by the reaction with azide form 2-azido-2-deoxy-D-glucose (reaction 4.89). The azide can be catalytically reduced to give 2-amino-2-deoxy-D-glucose, or reductively acetylated by reaction with thioacetic acid [94] to give 2-acetamido-2-deoxy-D-glucose (reaction 4.89). The trifyl group can be reduced with sodium borohydride in acetonitrile [95] to give 2-deoxy-D-glucose (reaction 4.90). The free C-2 hydroxyl carbon can be oxidized to give the 2-keto compound (reaction 4.91).

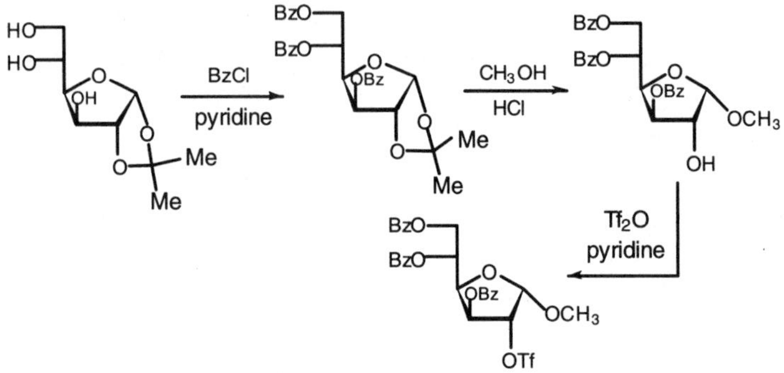

(4.88)

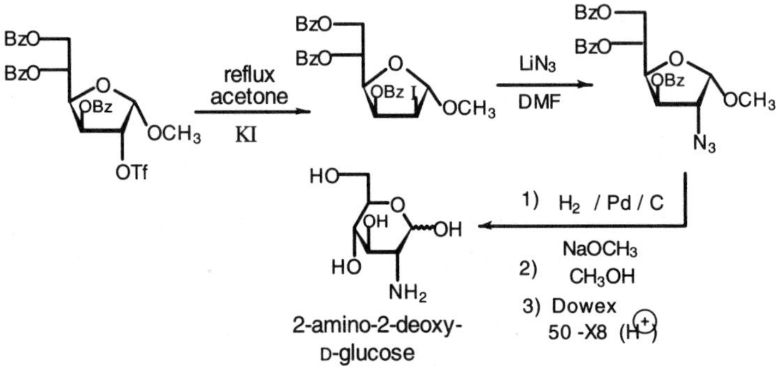

(4.89)

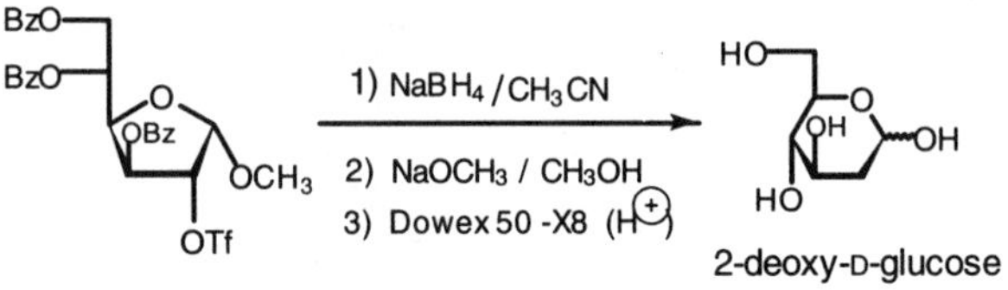

(4.90)

(4.91)

Ribonucleosides can be reduced at C-2 to give 2-deoxyribonucleosides if the C-3 and C-5 hydroxyl groups are protected by the formation of 1,1,3,3-tetraiso-propylidisiloxane (3′,5′-*O*-TPDS) derivative [96]. The 2-*O*-triflate is formed by reaction with triflic anhydride, which can be reduced with sodium borohydide in acetonitrile, followed by removal of the protecting group by tetra-*n*-butyl-ammonium fluoride, to give overall yields of 57 to 88% of 2′-deoxyribonucleo-side [97] (reaction 4.92).

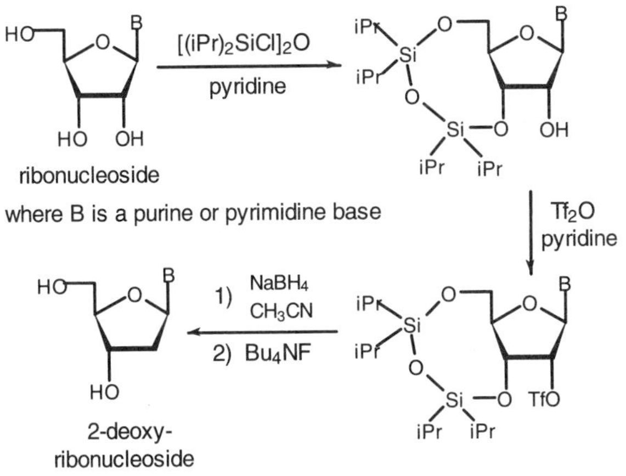

(4.92)

4.7 Modifications at C-3

The 2′,3′-dideoxynucleotides have found important uses as antiviral and anti-cancer drugs [97] and as reagents in determining the sequence of DNA [98]. They are incorporated into the ends of a growing DNA chain by DNA polymerase, where they terminate chain synthesis. Several different methods have been used to synthesize the 2′,3′-dideoxynucleosides (nucleotides) by mesylating (reaction with methylsulfonyl chloride) the 3′-hydroxyl group, followed by displacement with iodide and catalytic hydrogenation to form the 3′-O-tosyl derivative, and eliminate it with sodium methoxide to produce the 2′,3′-unsaturated sugar, which is catalytically hydrogenated [99].

An easier and more direct method begins with the protection of the C-5′ hydroxyl group by the formation of the trityl ether of either the ribonucleoside or the 2′-deoxyribonucleoside, followed by the formation of the 2′,3′-di-O-triflate or the 3′-O-trifylate, respectively, that can be reduced with sodium borohydride in acetonitrile [95]. The removal of the trityl group gives the 2′,3′-dideoxyribonucleoside (reaction 4.93).

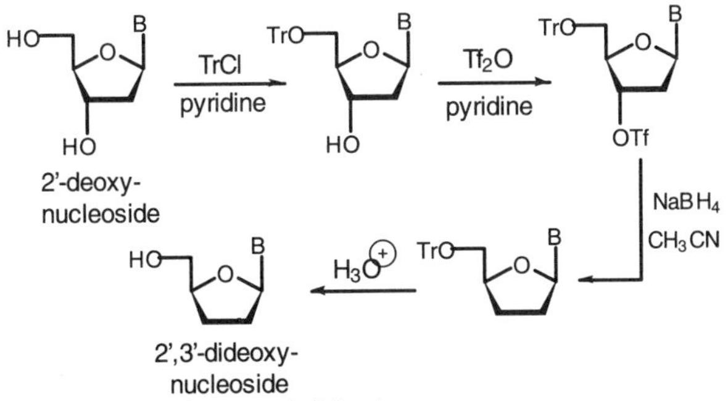

(4.93)

where B is a purine or pyrimidine base,
Tr is trityl (triphenylmethyl), and Tf is trifyl (trifluoromethylsulfonyl)

Replacement of the C-3′ hydroxyl group of D-ribonucleosides (or ribonu-
cleotides) with azido or amino groups forms 3′-azidothymidine (AZT) or 3′-
amino-3′-deoxythymidine, both of which are used in the treatment of AIDS and
cancer. These compounds are synthesized by the mesylation of 1-(2′-deoxy-5′-O-
trityl-β-D-lyxofuranosyl)-thymidine [100]. The mesyl group is displaced by reaction with lithium azide in DMF at 100°C. The azide can then be catalytically hydrogenated to give the 3′-amino analog (reaction 4.94).

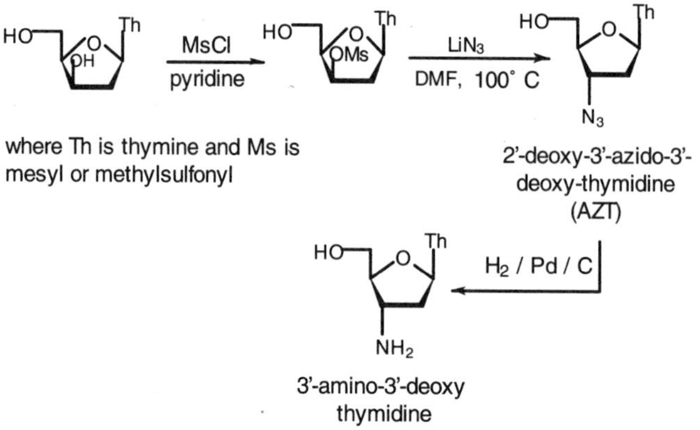

where Th is thymine and Ms is
mesyl or methylsulfonyl

2′-deoxy-3′-azido-3′-
deoxy-thymidine
(AZT)   (4.94)

3′-amino-3′-deoxy
thymidine

The modification of C-3 of D-glucose involves 1,2:5,6-diacetone glucose or
1,2:5,6-diacetone allose. These isopropylidene derivatives leave the C-3 hydroxyl
group free for reaction and modification. Reactions of the C-3 hydroxyl group of
1,2:5,6-diacetone allose usually go more smoothly and provide higher yields than
reactions at the C-3 hydroxyl group of diacetone glucose, which is sterically hin-
dered. Derivatives of the C-3 hydroxyl group of D-allose that can undergo S_N2
type displacements produce inversion of the configuration and the formation of a
modified D-glucose. This is fortunate because D-glucose analogs and derivatives
are of primary interest for biochemical studies.

Diacetone allose can be formed from diacetone glucose by first oxidizing the
C-3 hydroxyl carbon with acetic anhydride in dimethylsulfoxide [101], followed
by reduction with sodium borohydride (reaction 4.95). The reduction forms exclusively the allose configuration.

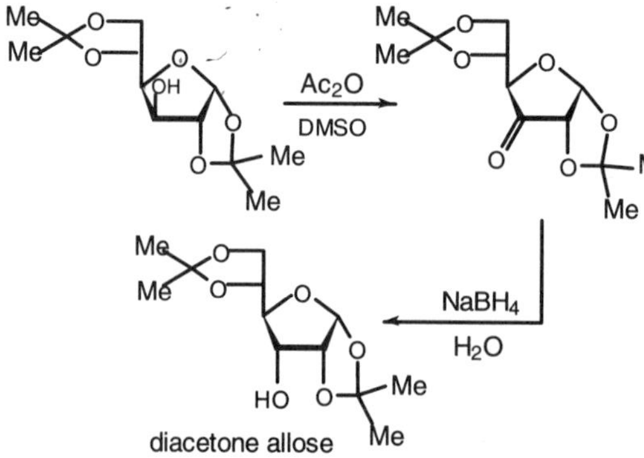

diacetone allose   (4.95)

Diacetone allose can be chlorinated at C-3 by reaction with triphenylphosphine and carbon tetrachloride, a common chlorinating reaction of alcohols [102]. The reaction gives inversion of the configuration and the formation of 3-chloro-3-deoxy-D-glucose after the removal of the isopropylidene groups (reaction 4.96).

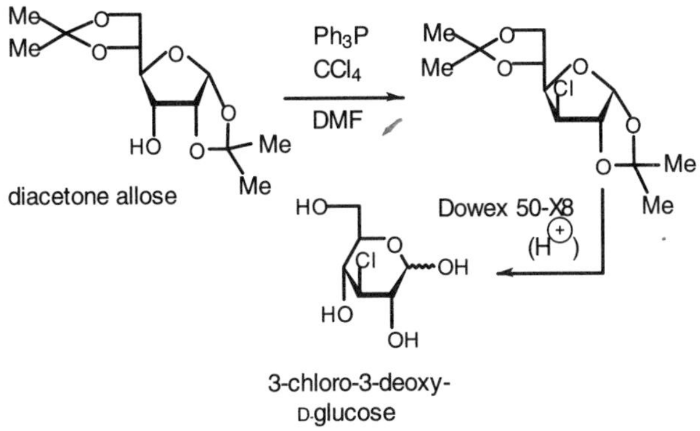

diacetone allose

3-chloro-3-deoxy-
D.glucose

(4.96)

Reaction of diacetone glucose with triphenylphosphine and carbontetrachloride results in rearrrangement, with chlorination at C-6 and not at C-3 [102] (reaction 4.97).

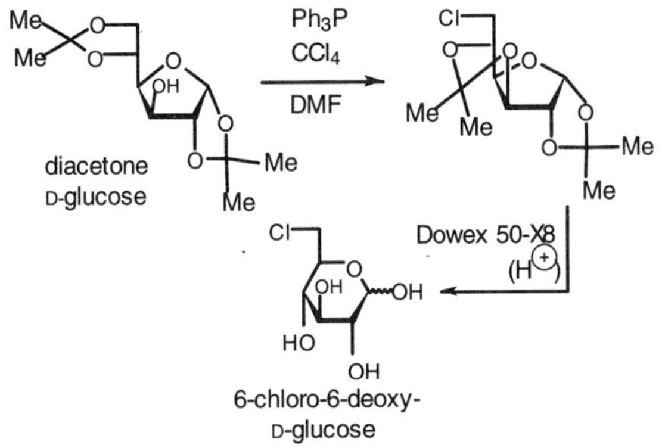

diacetone
D-glucose

6-chloro-6-deoxy-
D-glucose

(4.97)

Reaction of diacetone allose with triflic anhydride gives the 3-O-trifyl derivative, which can be reduced with sodium borohydride, followed by the removal of the isopropylidene groups, to give 3-deoxy-D-glucose (reaction 4.98).

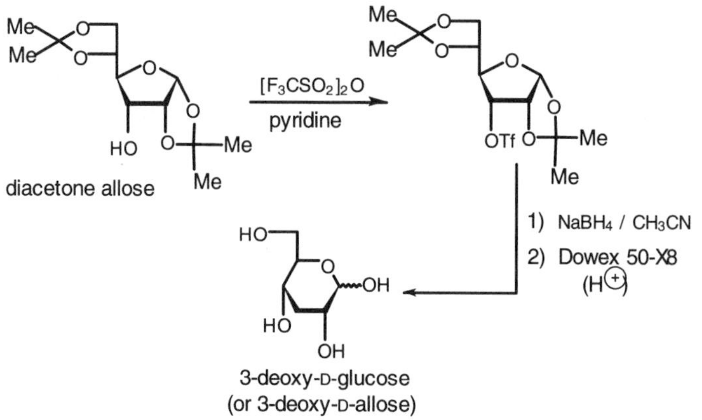

diacetone allose

[F₃CSO₂]₂O / pyridine

(4.98)

1) NaBH₄ / CH₃CN
2) Dowex 50-X8 (H⊕)

3-deoxy-D-glucose
(or 3-deoxy-D-allose)

Reaction of diacetone glucose in a similar manner provides some 3-deoxy-D-glucose, but it also gives an elimination reaction and the formation of 3,4-D-glucosene [95] (reaction 4.99)

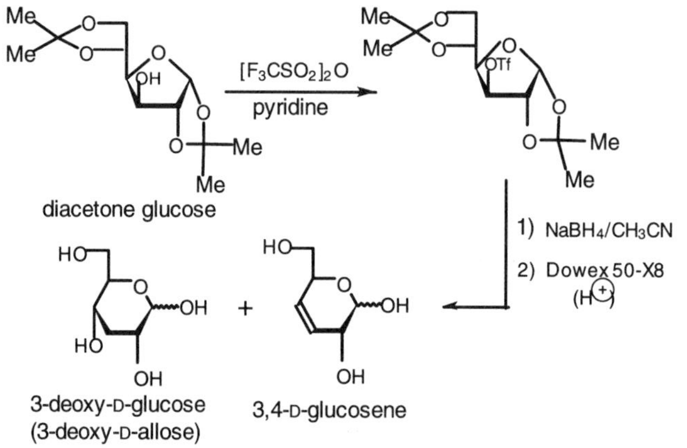

diacetone glucose

[F₃CSO₂]₂O / pyridine

(4.99)

1) NaBH₄/CH₃CN
2) Dowex 50-X8 (H⊕)

3-deoxy-D-glucose   3,4-D-glucosene
(3-deoxy-D-allose)

Primary and secondary hydroxyl groups can be fluorinated by reaction with diethylaminosulfur trifluoride (DAST) [103]. The reaction of asymmetric secondary alcohols produces inversion of the configuration. Thus, the reaction of the protected diacetone allose with DAST, followed by removal of the protective groups, gives 3-deoxy-3-fluoro-D-glucose [104] (reaction 4.100).

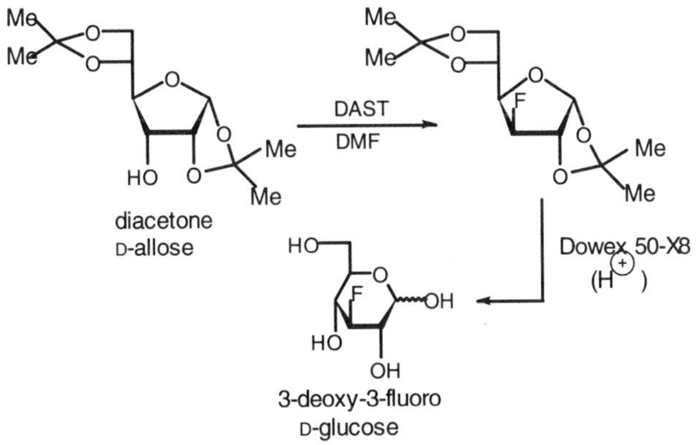

diacetone
D-allose

(4.100)

Dowex 50-X8
(H⊕)

3-deoxy-3-fluoro
D-glucose

Although the reaction of DAST with carbonyl compounds gives the *gem*-di-fluoride, the reaction of 1,2:5,6-di-*O*-isopropylidene-3-keto-α-D-glucofuranoside (3-keto-diacetone glucose) gave decomposition, and the 3-*gem*-difluoride was not formed [105].

3-*O*-Benzoyl-1′,2:4,6-di-*O*-isopropylidene sucrose can be prepared by reaction with acetone and benzoyl chloride [106]. The other free hydroxyl groups of sucrose can be further protected by reaction with *tert*-butylchlorodimethyl silane in DMF to give 3′,4′,6′-tri-*O*-*tert*-butyldimethyl silyl-1′,2:4,6-di-*O*-isopropylidene sucrose. The 3-*O*-benzoyl group can be selectively removed by reaction with lithium aluminum hydride to give the free C-3 hydroxyl group [107] (reaction 4.101).

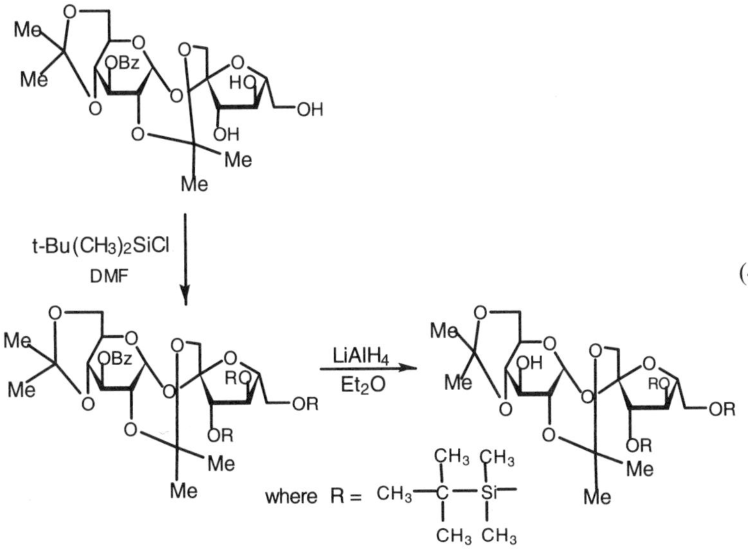

t-Bu(CH₃)₂SiCl
DMF

(4.101)

LiAlH₄
Et₂O

where R = $CH_3-\overset{\underset{\displaystyle CH_3}{|}}{\underset{\underset{\displaystyle CH_3}{|}}{C}}-\overset{\underset{\displaystyle CH_3}{|}}{\underset{\underset{\displaystyle CH_3}{|}}{Si}}-$

The free C-3 hydroxyl group can be derivitized to 3-*O*-(methylthio)thiocar-bonyl sucrose. This derivative can be homolytically reduced with tributyltin hydride to give 3-deoxysucrose. The free C-3 hydroxyl carbon can be oxidized with pyridinium chromate to form 3-ketosucrose that can be reduced with sodium borohydride to give allosucrose (reaction 4.102). The C-3 hydroxyl group of the protected allosucrose can be reacted with triflic anhydride to give the 3-O-trifyl derivative, which, in turn, can be reduced with sodium borohydride to form 3-deoxysucrose (reaction 4.103). The C-3 hydroxyl group of the protected allosucrose can also be fluorinated with DAST to give 3-deoxy-3-fluorosucrose [107] (reaction 4.104).

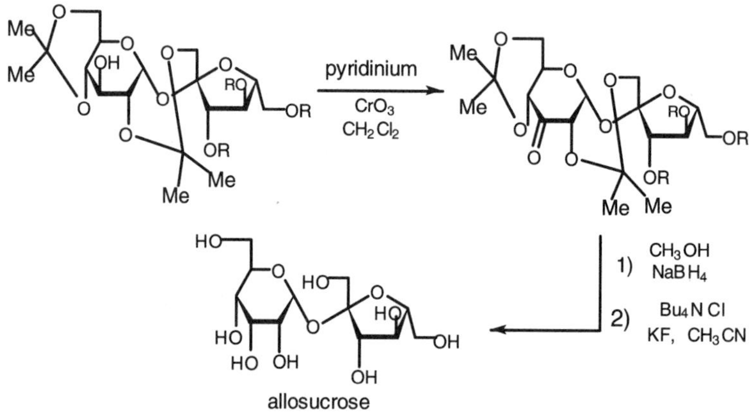

(4.102)

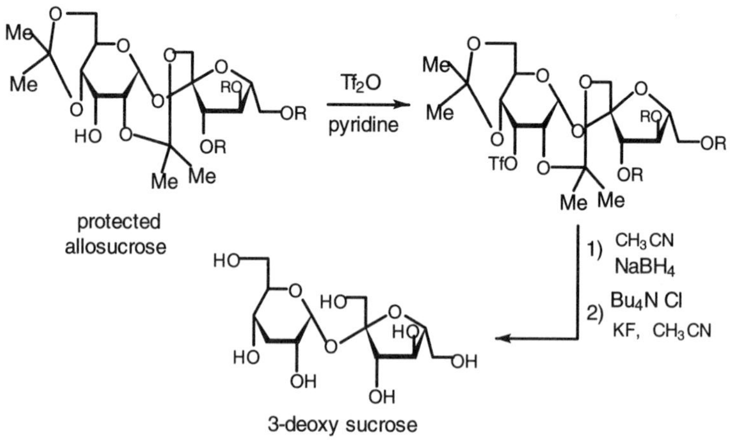

allosucrose

protected
allosucrose

(4.103)

3-deoxy sucrose

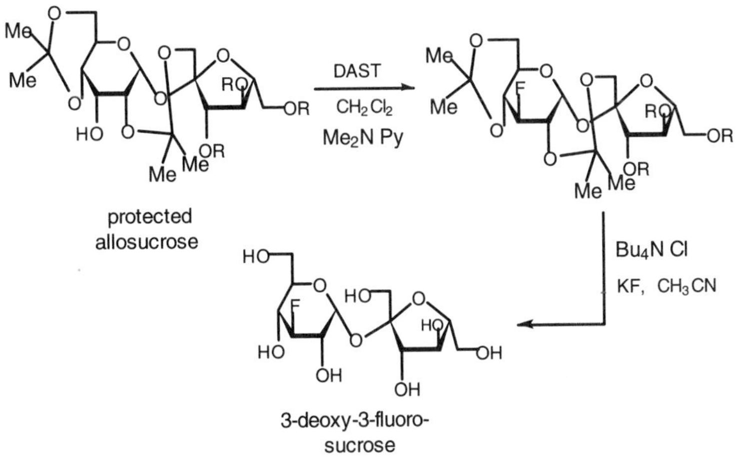

protected
allosucrose

(4.104)

3-deoxy-3-fluoro-
sucrose

A bacterial enzyme has been reported that specifically oxidizes the C-3 hydroxyl carbon of several mono-, di-, and trisaccharides to produce 3-keto sugars [108]. The enzyme, *glycoside-3-dehydrogenase,* is elaborated by *Agrobacterium tumefaciens.* It oxidizes a wide variety of monosaccharides (D-glucose, methyl-α-D-glucopyranoside, methyl-β-D-glucopyranoside, α-D-glucopyranosyl-1-phosphate, 2-amino-2-deoxy-D-glucose, 2-deoxy-D-glucose, D-galactose, and D-mannose) and a variety of di- and trisaccharides (cellobiose, cellobionate, maltose, maltobionate, sucrose, α,α-trehalose, leucrose, lactose, lactulose, melibiose, melezitose, and raffinose). With reducing sugars and onic acids, the C-3 hydroxyl group of the terminal nonreducing residue is oxidized [109]. With α,α-trehalose, only one of the two D-glucose residues is oxidized, and with sucrose, only the D-glucose residue is oxidized (reaction 4.105).

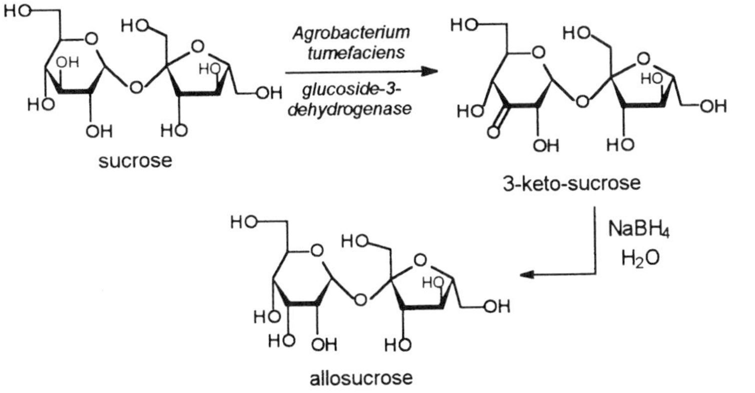

sucrose

3-keto-sucrose   (4.105)

allosucrose

The various 3-keto analogs also can be used to give regioselective modifications. Reduction of 3-ketosucrose with sodium borohydride exclusively gives allosucrose [110] (reaction 4.105). Reaction of the keto group with hydroxylamine and its derivatives, with allyl and benzyl groups for oximes. Reductive amination of the oxime gives 3-amino-3-deoxyallosucrose [110] (reaction 4.106).

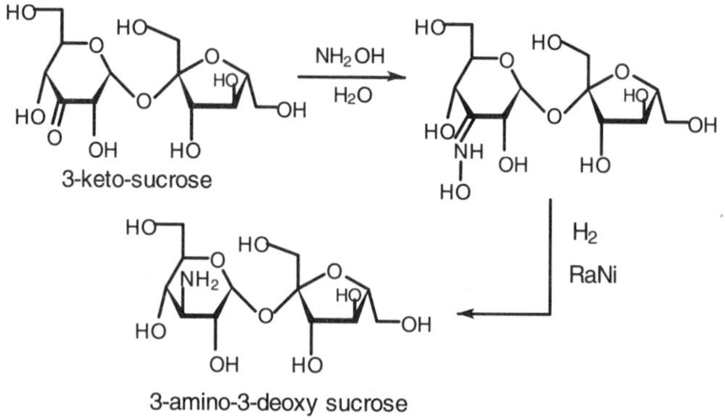
NH2OH
H2O
3-keto-sucrose
H2
RaNi
3-amino-3-deoxy sucrose

(4.106)

The 3-amino group can be easily *N*-acylated to give acrylamide or fatty acid esters [110]. Silyation of 3-ketosucrose at the free hydroxyl groups with chlorotrimethyl silane provides a derivative that can be modified by Grignard-type reactions introducing allyl or alkyl chains onto the carbohydrate ring [110] (reaction 4.107).

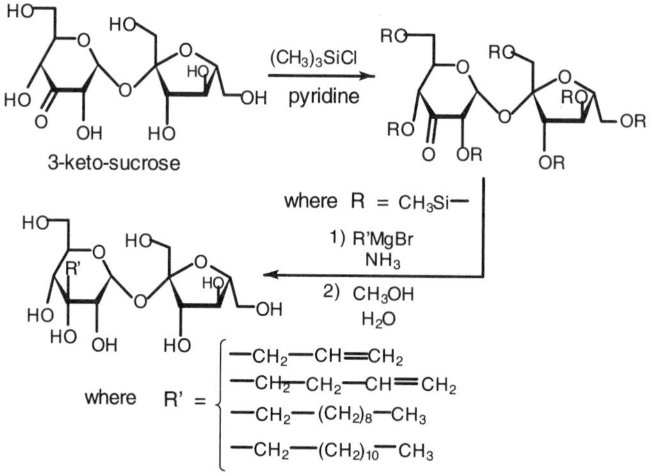
(CH3)3SiCl
pyridine
3-keto-sucrose
where R = CH3Si—
1) R'MgBr
NH3
2) CH3OH
H2O
where R' =
—CH2—CH═CH2
—CH2—CH2—CH═CH2
—CH2—(CH2)8—CH3
—CH2—(CH2)10—CH3

(4.107)

When 3-ketosucrose is treated with alkali, a β-elimination of D-fructose occurs, producing the endiolone [110] (reaction 4.108).

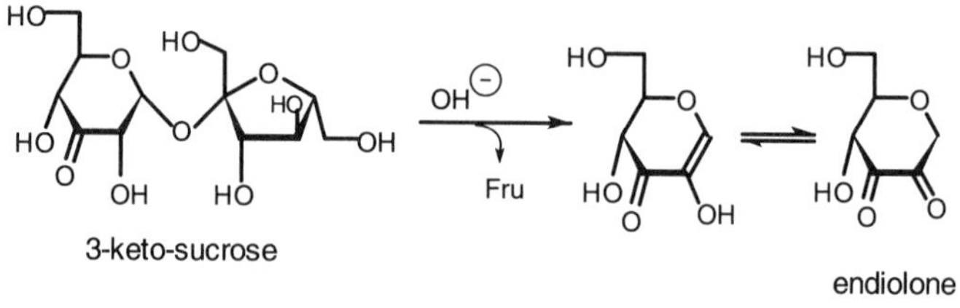
3-keto-sucrose
Fru
endiolone

(4.108)

4.8 Modifications at C-4

The C-4 hydroxyl group of D-glucose, D-mannose, and D-galactose can be obtained in a free form for modification by selective benzoylation [7]. Reaction of methyl α-D-galactopyranoside with 4.2 molar equivalents of benzoyl chloride in pyridine at −30°C gave a 65% yield of methyl 2,3,6-tri-O-benzoyl-α-D-galactopyranoside. This particular specificity was rationalized by the fact that axial hydroxyl groups undergo acylation less rapidly that equatorial hydroxyl groups. Reaction of methyl α-D-mannopyranoside with 3.1 molar equivalents of benzoyl chloride gave a 56% yield of methyl 2,3,6-tri-O-benzoyl D-mannopyranoside. A similar reaction of methyl α-D-glucopyranoside with 3.1 molar equivalents of benzoyl chloride resulted in two tribenzoates, methyl 2,3,6-tri-O-benzoyl α-D-glucopyranoside as the major (67%) product, and methyl 2,4,6-tri-O-benzoyl α-D-glucopyranoside as a minor (28%) product.

Methyl-4-deoxy α-D-glucopyranoside can be obtained by reaction of the 2,3,6-tribenzoate with triflic anhydride, reduction with sodium borohydride in acetonitrile, and removal of the benzoyl groups (reaction 4.109).

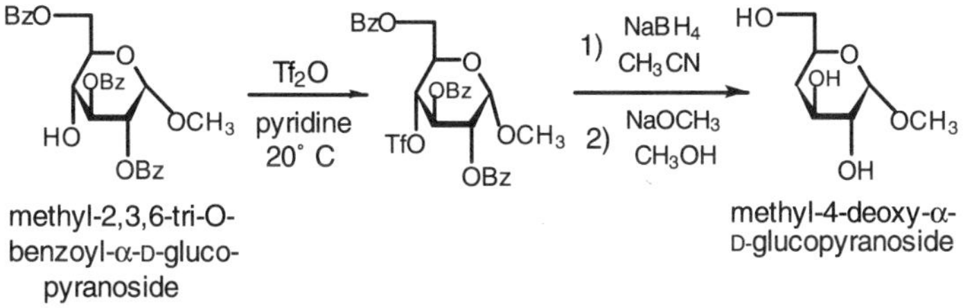

$$(4.109)$$

methyl-2,3,6-tri-O-
benzoyl-α-D-gluco-
pyranoside

methyl-4-deoxy-α-
D-glucopyranoside

In a similar manner, the 4-O-trifyl group of methyl 2,3,6-tri-O-benzoyl-4-O-trifyl α-D-galactopyranoside can be displaced by various nucleophiles such as azide or thioacetate to give 4-amino-4-deoxy D-glucose (reaction 4.110) or 4-deoxy-4-thio D-glucose (reaction 4.111).

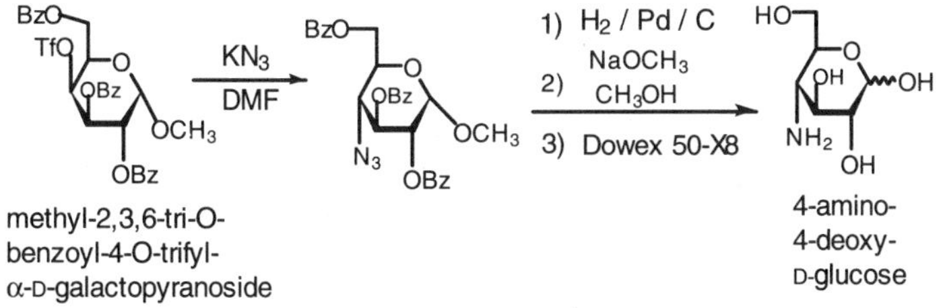

$$(4.110)$$

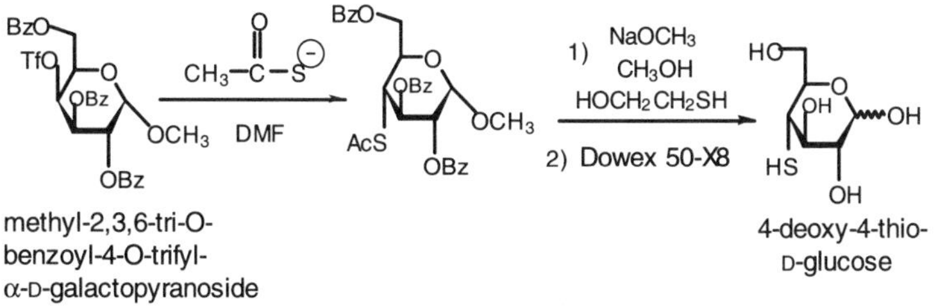

methyl-2,3,6-tri-O-
benzoyl-4-O-trifyl-
α-D-galactopyranoside

4-amino-
4-deoxy-
D-glucose

$$(4.111)$$

methyl-2,3,6-tri-O-
benzoyl-4-O-trifyl-
α-D-galactopyranoside

4-deoxy-4-thio-
D-glucose

The methyl 2,3,6-tri-O-benzoyl-α-D-galactopyranoside can be fluorinated with DAST to give 4-deoxy-4-fluoro-D-glucose (reaction 4.112).

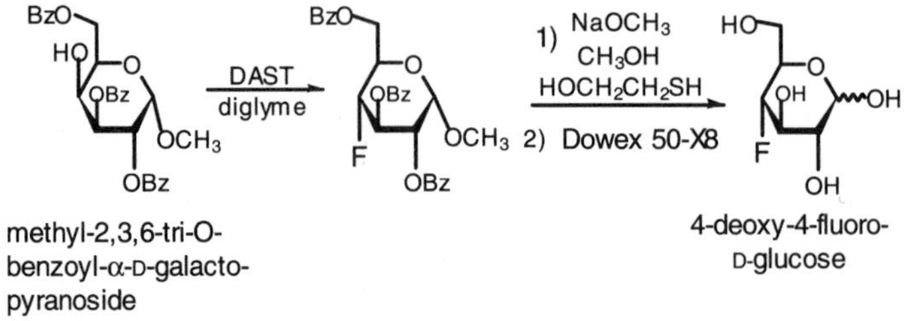

methyl-2,3,6-tri-O-
benzoyl-α-D-galacto-
pyranoside

4-deoxy-4-fluoro-
D-glucose

(4.112)

The same type of reactions can be conducted with methyl 2,3,6-tri-O-benzoyl-4-O-trifyl α-D-glucopyranoside to give modified D-galactoses. The free 4-hydroxyl-carbon can be oxidized with pyridinium chromate to produce the 4-keto derivative, which, in turn, can be fluorinated with DAST to provide the 4-*gem*-difluoride (reaction 4.113).

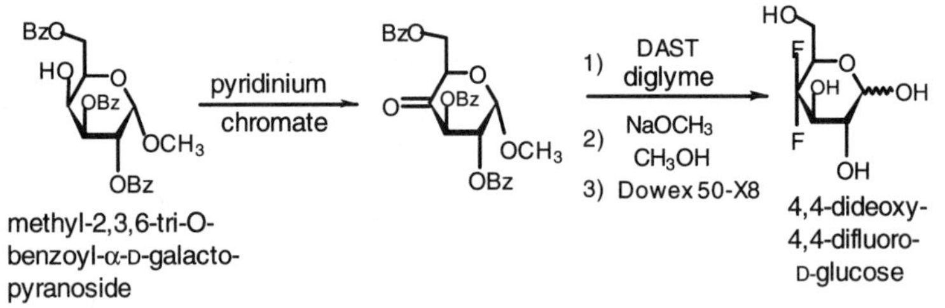

methyl-2,3,6-tri-O-
benzoyl-α-D-galacto-
pyranoside

4,4-dideoxy-
4,4-difluoro-
D-glucose

(4.113)

The modification of the C-4 position of sucrose can be accomplished by first forming the 4,6-benzilidene derivative (see reaction 4.27), followed by benzoylation. The benzilidene group can then be selectively cleaved to form the free C-4 hydroxyl group by reaction with trimethylamine borane and aluminum chloride in THF [35] (see reaction 4.33). The 4-O-trifyl group can be formed and displaced by iodide. The iodide can then be displaced by a nucleophile such as azide, followed by reduction to form 4-amino-4-deoxysucrose (reaction 4.114). If the trifyl group is directly displaced by azide, the reduced product is 4-amino-4-deoxygalactosucrose (reaction 4.115).

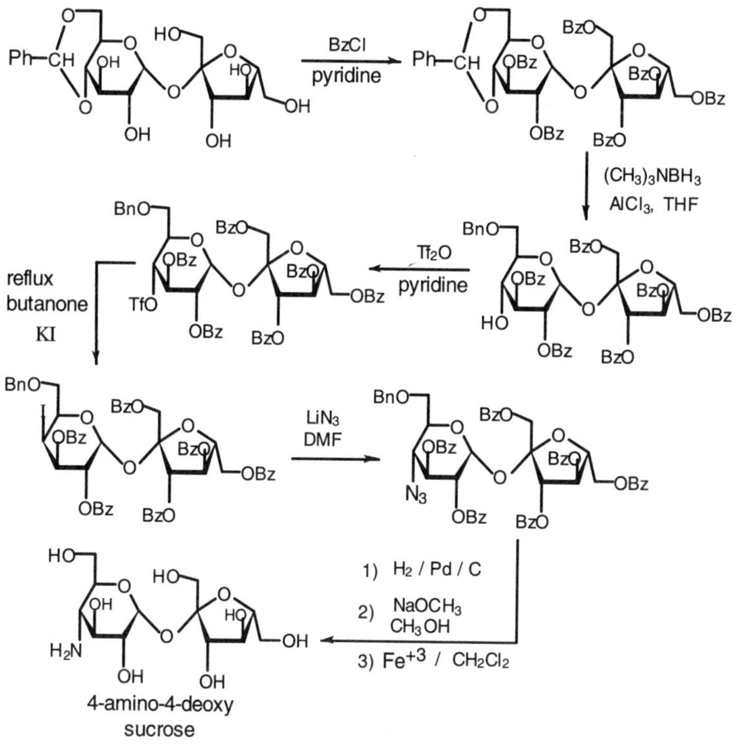

(4.114)

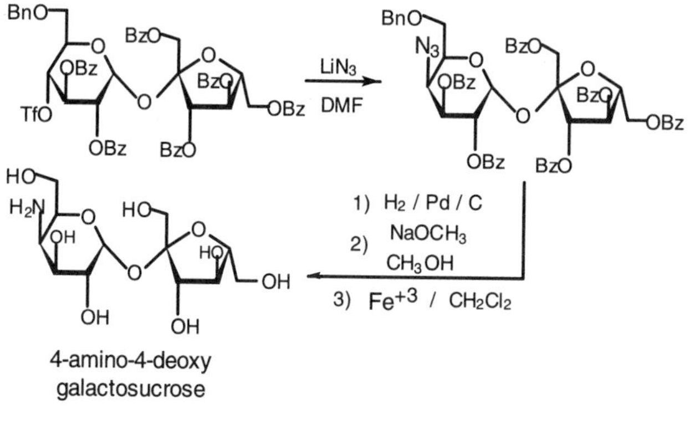

4-amino-4-deoxy
sucrose

(4.115)

4-amino-4-deoxy
galactosucrose

The C-4′ position of maltose can be modified in a similar way. Maltose, being a reducing sugar, usually has to have C-1 protected. C-1 is often protected by forming the 1,2-O-isopropylidene derivative. It has been obtained in a 33% yield and represented one of four isopropylidene compounds that were formed [111]. These compounds had to be separated on a silica-gel column. The 1,2-isopropylidene maltose is then reacted with 1,1-dimethoxy benzaldehyde to give the 4′,5′-benzilidene derivative (reaction 4.116). The other free hydroxyl groups are protected by benzoylation, and the C-4′-O bond is selectively cleaved to produce the free C-4′

hydroxyl group; it then can be reacted with trifylic anhydride, iodide, azide, and catalytically reduced to give 4^2-amino-4^2-deoxymaltose (reaction 4.116).

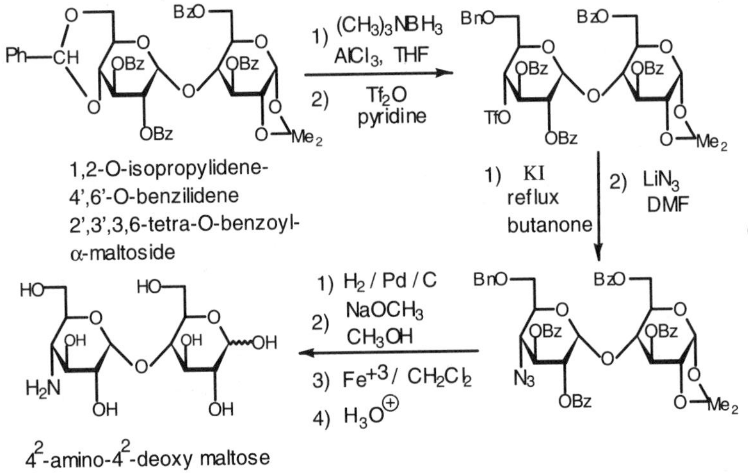

1) (CH3)3NBH3 AlCl3, THF
2) Tf2O pyridine
1,2-O-isopropylidene-
4',6'-O-benzilidene
2',3',3,6-tetra-O-benzoyl-
α-maltoside
1) KI reflux butanone
2) LiN3 DMF
1) H2 / Pd / C
2) NaOCH3 CH3OH
3) Fe+3/ CH2Cl2
4) H3O⊕
4²-amino-4²-deoxy maltose

(4.116)

The 4^2-amino-4^2-deoxyisomaltose analog can be synthesized a little more easily by starting with melibiose [α-D-galactopyranosyl-(1 → 6)-D-glucose]. The reducing C-1 group is protected by a 1,2-isopropylidene group; the 4′,6′-*O*-benzilidene group is formed, and the other hydroxyl groups are benzoylated. The benzilidene group is selectively cleaved to create the free C-4′-hydroxyl group that is trifylated and directly displaced by azide to give 4^2-amino-4^2-deoxyisomaltose (reaction 4.117).

4^2-Amino-4^2-deoxycellobiose can be synthesized in a similar manner starting with lactose [β-D-galactopyranosyl-(1 → 4)-D-glucose].

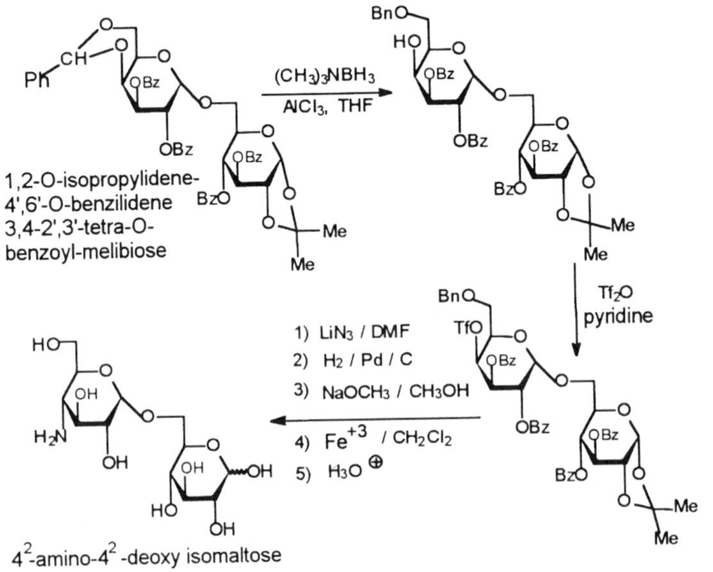

(CH3)3NBH3 AlCl3, THF
1,2-O-isopropylidene-
4',6'-O-benzilidene
3,4-2',3'-tetra-O-
benzoyl-melibiose
Tf2O pyridine
1) LiN3 / DMF
2) H2 / Pd / C
3) NaOCH3 / CH3OH
4) Fe+3 / CH2Cl2
5) H3O ⊕
4²-amino-4² -deoxy isomaltose

(4.117)

4.9 Modifications at C-5 and Substitution for the Ring Oxygen

The substitution at C-5 for D-glucose starts with 1,2:5,6-di-O-isopropylidene-3-O-benzoyl-α-D-glucofuranoside (diacetone glucose) and the selective removal of the 5,6-isopropylidene group by treatment with iodine in methanol [25] (see reaction 4.20). C-6 is tritylated and the free C-5 hydroxyl is triflylated. The C-5-O-trifyl group can be reduced with sodium borohydride in acetonitrile to give 5-deoxy-D-glucose, after removal of the protecting groups (reaction 4.118).

The C-5-O-trifyl group can be displaced by iodide that in turn can be displaced by azide, with overall retention of the configuration (reaction 4.119) to give 5-amino-5-deoxy-D-glucose and the synthesis of nojirimycin [112]. Nojirimycin readily looses water, forming the imine that can be reduced with sodium borohydride to give the stable product, dihydronojirimycin (reaction 4.119).

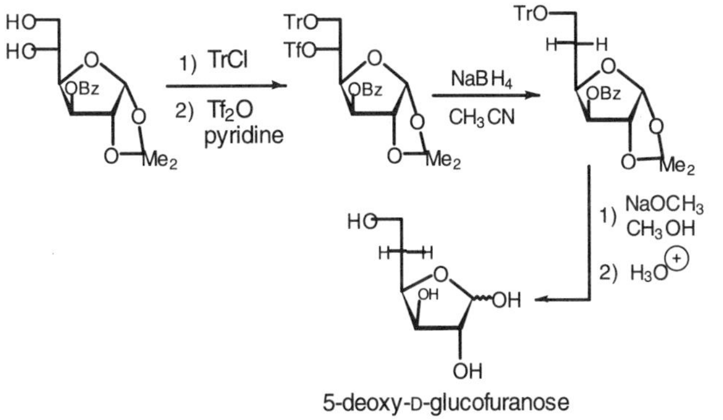

5-deoxy-D-glucofuranose   (4.118)

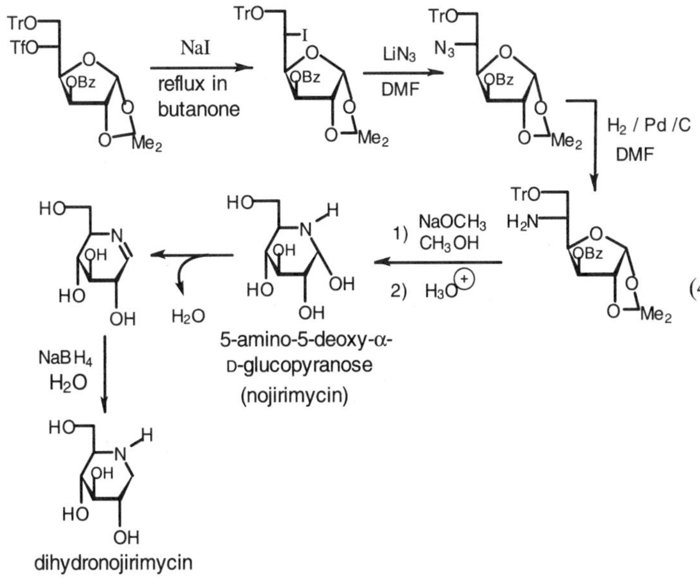

5-amino-5-deoxy-α-
D-glucopyranose
(nojirimycin)

dihydronojirimycin   (4.119)

Starting with the 1,2-*O*-isopropylidene-α-D-glucofuranoside, the C-6 hydroxyl can be tritylated and the C-5 hydroxyl carbon oxidized with acetic anhydride in DMSO to give a 75% yield of the 5-keto compound. The oxime of the keto compound can be formed and reduced with lithium aluminum hydride in diethyl ether [113]. Two C-5 diastereoisomers are formed in a ratio of 9:1, with the D-isomer predominating (reaction 4.120).

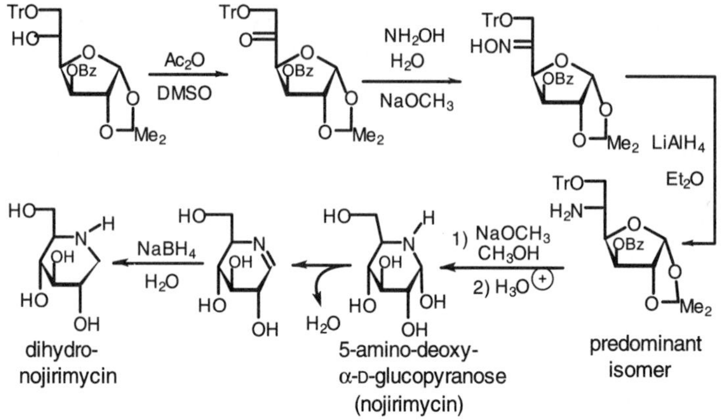

(4.120)

dihydro-
nojirimycin

5-amino-deoxy-
α-D-glucopyranose
(nojirimycin)

predominant
isomer

Nojirimycin is a naturally occurring antibiotic that is elaborated by several strains of *Streptomyces*. It is relatively unstable and must be isolated as the stable bisulfite adduct. It can also be stabilized by reduction with sodium borohydride to give dihydronojirimycin which retains its full antibiotic activity.

The iodo group of 1,2-isopropylidene-5-iodo-6-*O*-trityl α-D-glucofuranoside can also be displaced by thioacetate to form the 5-acetylthio derivative [114]. After the removal of the protecting groups, 5-deoxy-5-thio-D-glucopyranose is obtained (reaction 4.121).

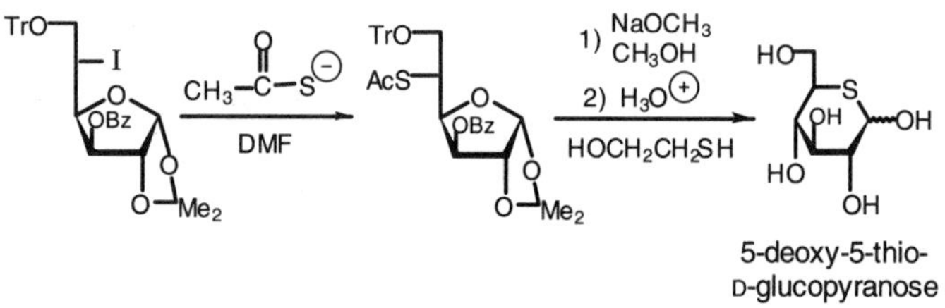

(4.121)

5-deoxy-5-thio-
D-glucopyranose

5-Thio-D-xylose has been prepared in a similar manner [112]. 1,2-*O*-Isopropylidene-5-*O*-tosyl α-D-xylofuranoside was reacted with the sodium salt of benzylthio alcohol, followed by reaction with sodium amide and acid (reaction 4.122).

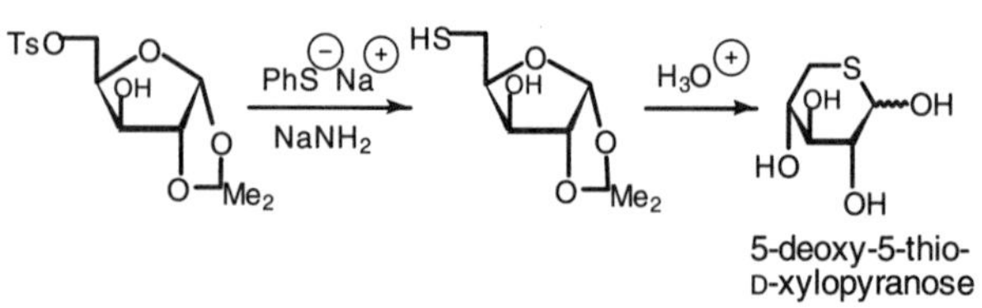

(4.122)

5-deoxy-5-thio-
D-xylopyranose

5-Deoxy-5-thio-D-glucose is a close analog of D-glucose, with sulfur replacing the ring oxygen. It is believed to interfere with the cellular transport of D-glucose. It is reported to reversibly interfere with spermatogenesis and to be an effective anticancer agent [115].

The substitution of the phosphorous atom at C-5 followed by introduction into the pyranose ring has been accomplished for D-xylose [116]. The 1,2-*O*-isopropylidene α-D-xylofuranoside is tosylated at C-5 and methylated at C-3. The tosyl group at C-5 is displaced to give the 5-iodo derivative, which is displaced with triethyl phosphite, followed by reduction with lithium aluminium hydride to give the C-5 phosphine analog (reaction 4.123).

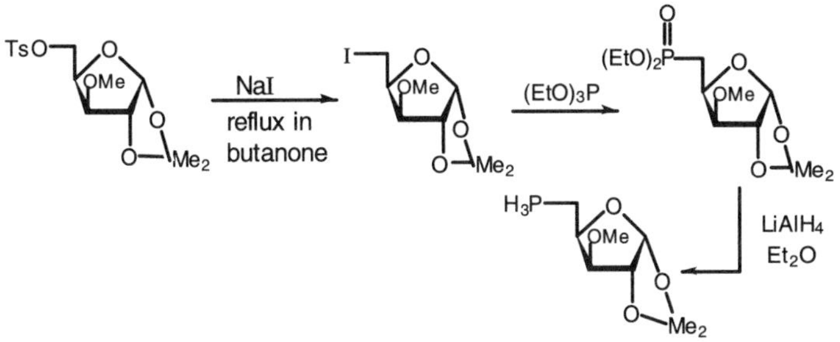

(4.123)

Treatment of the phosphine analog with acid results in the placement of a phosphorous atom in the pyranose ring. This compound is readily oxidized in air to give the phosphinyl analog which can be further oxidized with hydrogen peroxide to give the phosphinic acid derivative of D-xylopyranose (reaction 4.124).

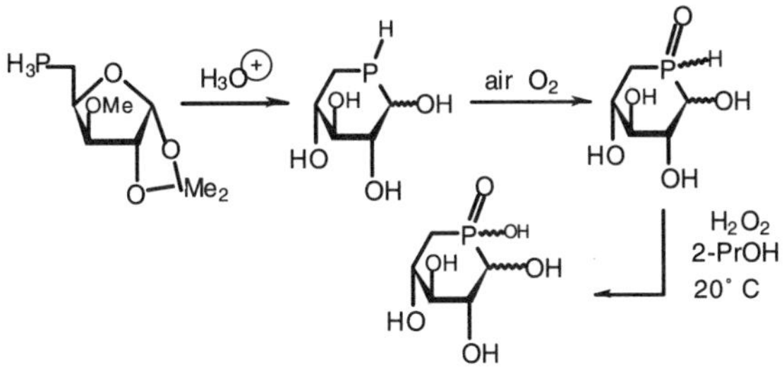

(4.124)

Analogous reactions can be carried out with 1,2-*O*-isopropylidene-3-*O*-methyl-5-iodo-6-*O*-trityl α-L-idofuranoside to form the phosphinic acid derivative of D-glucopyranose (reaction 4.125).

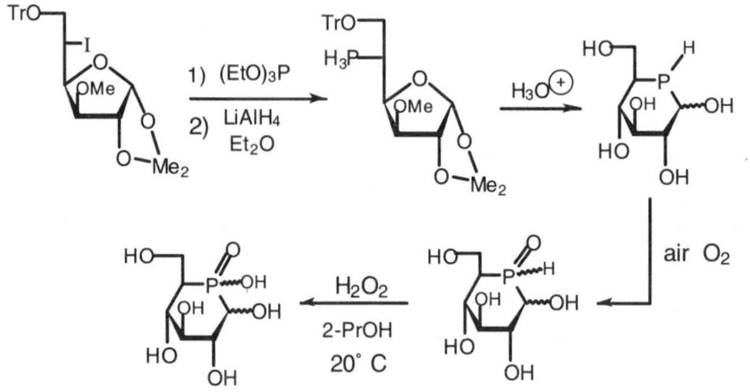

$$(4.125)$$

4.10 Modifications of C-6 in Hexopyranoses

Strategies similar to what we have seen for the secondary alcohol groups are used to chemically modify or substitute for the primary alcohol groups of carbohydrates. However, the greater reactivity of primary alcohol groups and the use of bulky modifying reagents allow a greater degree of flexibility and variety of modifying reactions that can be used to selectively derivatize and substitute the primary alcohols in the presence of secondary alcohols without the need for their protection.

For example, the bulky trityl chloride selectively reacts with primary alcohol groups (see section 4.3b). The secondary alcohol groups can then be protected by acetylation or benzoylation, and the trityl group selectively removed by mild acid hydrolysis (see reaction 4.16). The reaction of tosyl chloride has also been used to selectively derivatize the primary alcohols [6]. The tosyl group can then be displaced by various nucleophiles such as halides or azide (reaction 4.14).

The substitution of a primary hydroxyl group by a halide, can be accomplished directly by reaction with triphenylphosphine and carbontetrachloride or carbontetrabromide [117]. The direct substitution of iodine can be obtained by reaction of triphenylphosphine, iodine, and imidazole [118]. These reactions are illustrated by the substitution of the primary hydroxyl groups of C-6 and C-6' of sucrose (reactions 4.126).

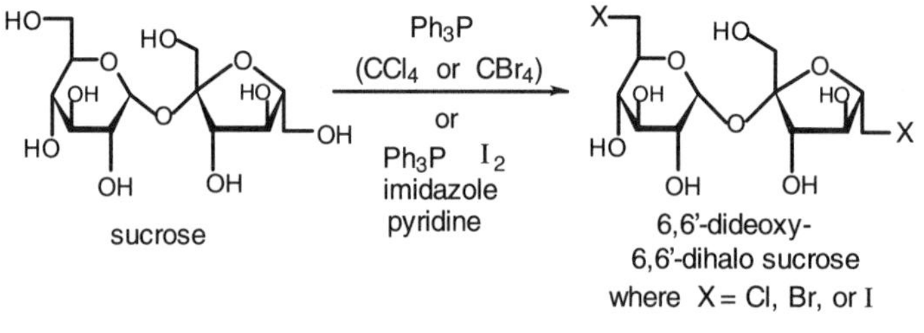

$$(4.126)$$

The primary hydroxyl group at the C-1' position on the fructose ring of sucrose is much less reactive than the hydroxyl groups at the C-6 and C-6' positions, due to intramolecular hydrogen bonds formed between the C-1′ hydroxyl group with the ring oxygen of the glucose unit and/or the C-2 hydroxyl of the glucose unit [119].

Methyl 6-deoxy-6-fluoro-α-D-glucopyranoside can be prepared by tritylating the C-6 hydroxyl group, benzoylating the C-2,-3, and -4 hydroxyl groups, removing the trityl group, and fluorinating with DAST. A similar scheme can be used to prepare 6,6′-dideoxy-6,6′-difluorosucrose [120]. The C-6 and C-6′ hydroxyl groups are tritylated, the remaining free hydroxyl groups are benzoylated, the trityl groups are removed, and the C-6 and C-6′ hydroxyl groups are substituted by fluorine after reaction with DAST (reaction 4.127).

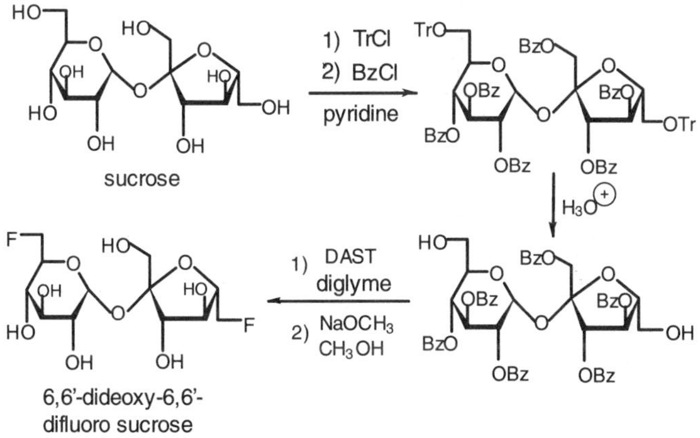

(4.127)

6,6′-Dideoxysucrose can be prepared by specifically iodinating the sucrose with triphenylphosphine, iodine, and imidazole [119], followed by catalytic hydrogenation (reaction 4.128).

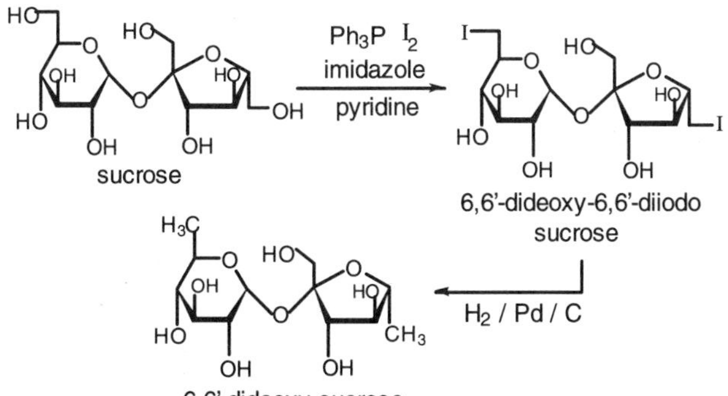

(4.128)

The C-6 primary alcohol group of sucrose can be selected by the formation of 4,6-O-benzilidene sucrose [32] (reaction 4.27). The free hydroxyl groups are then benzoylated and the 4,6-O-benzilidene group is selectively cleaved by reaction with trimethylamine/borane and aluminum chloride in toluene [34] to produce the

free C-6 hydroxyl (see reactions 4.33 and 4.129). The C-6 hydroxyl can then be
iodinated with triphenylphosphine, iodine, and imidazole [118]. This is followed
by catalytic hydrogenation to give 6-deoxysucrose (reaction 4.130), or the iodo
group can be displaced by a nucleophile such as thioacetate to give 6-deoxy-6-
thiosucrose (reaction 4.131).

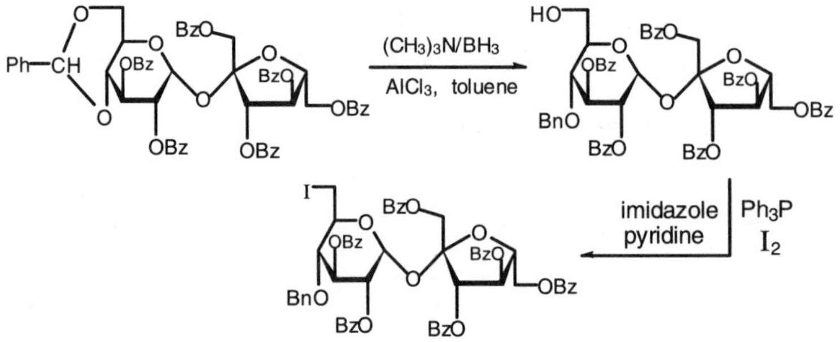

(4.129)

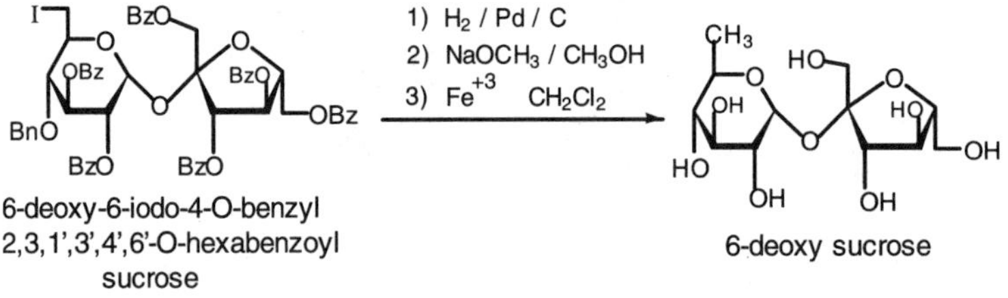

(4.130)

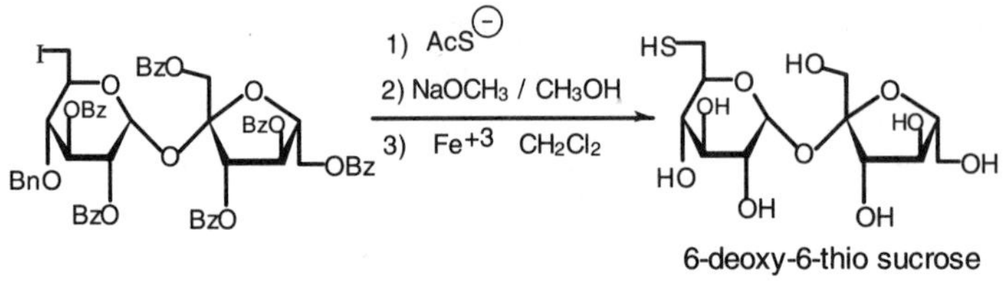

6-deoxy-6-iodo-4-O-benzyl
2,3,1',3',4',6'-O-hexabenzoyl
sucrose

6-deoxy sucrose

(4.131)

6-deoxy-6-thio sucrose

When 2,3,1',3',4',6'-hexa-O-benzoyl-4-O-benzyl sucrose (see reaction 4.129)
is reacted with DAST, 6-deoxy-6-fluorosucrose can be obtained (reaction 4.132).

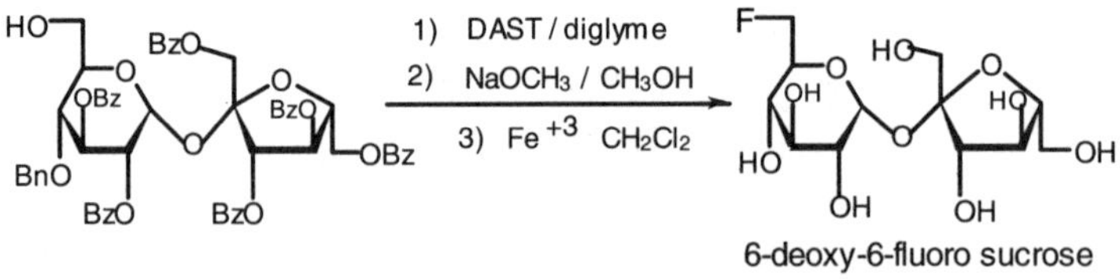

(4.132)

6-deoxy-6-fluoro sucrose

Another scheme for selectively modifying the C-6 hydroxyl group of sucrose
involves the removal of the benzilidene group from 2,3,1',3',4',6'-hexa-O-
benzoyl-4,6-O-benzilidene sucrose by refluxing with iodine in methanol [25].
The resulting hexa-O-benzoyl sucrose can be tritylated at C-6 and benzoylated at

C-4. The trityl group can then be removed by hydrolysis, and the free C-6 hydroxyl group modified, for example, by fluorinating with DAST to give 6-deoxy-6-fluorosucrose [121] (reaction 4.133), or halogenating with triphenylphosphine.

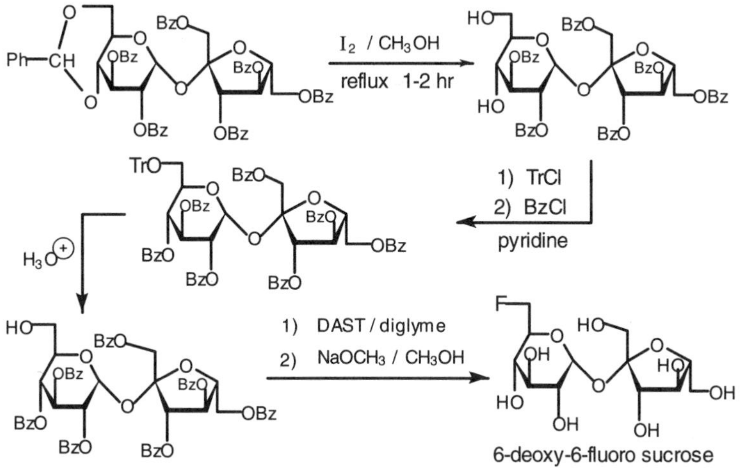

(4.133)

6-deoxy-6-fluoro sucrose

Starting with the hexa-O-benzoyl sucrose, both C-4 and C-6 can be reduced. The C-6 hydroxyl group of the hexa-O-benzoyl sucrose can be replaced with iodine by reaction with triphenylphosphine, iodine, and imidazole. The iodinated derivative can then be reacted with sulfuryl chloride to give chlorination of C-4, with inversion of the configuration, forming 2,3,1′,3′,4′,6′-hexa-O-benzoyl-4-chloro-4,6-deoxy-6-iodo-galactosucrose [122]. This derivative can be reduced at C-4 and C-6 with tributyltin hydride in the presence of a radical initiator to give 4,6-dideoxysucrose (reaction 4.134).

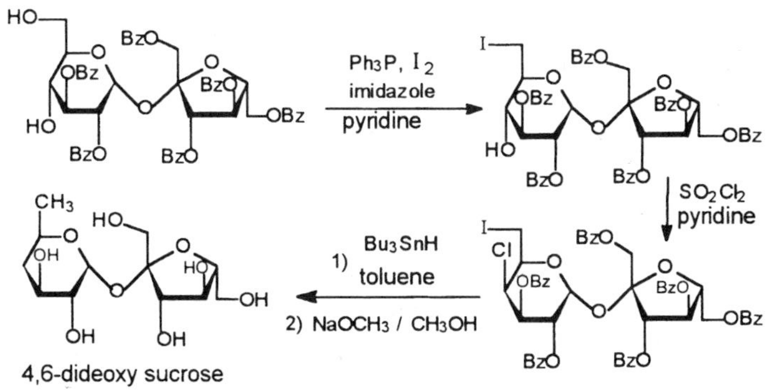

(4.134)

4,6-dideoxy sucrose

Sucrose can be selectively oxidized at C-6 by reacting 2,3,4,1′,3′,4′,6′-hepta-O-benzoyl sucrose with pyridinium chromate to convert C-6 into an aldehyde. The aldehyde group can be reduced with sodium borohydride to create specifically labeled 6-[^{3}H]sucrose (reaction 4.135).

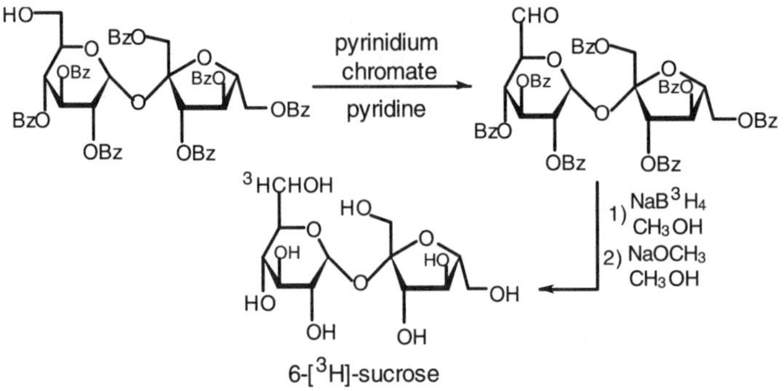

$$(4.135)$$

6-[^{3}H]-sucrose

By making the tosyl derivative of 2,3,4,1′,3′,4′,6′-hepta-O-benzoyl sucrose, followed by displacement with azide, reduction, and removal of the protecting groups, 6-amino-6-deoxysucrose is obtained (reaction 4.136). Alternatively, the iodo group of 2,3,4,1′,3′,4′,6′-hexa-O-benzoyl-6-deoxy-6-iodosucrose can be replaced by azide, reduced, and the protecting groups removed to give 6-amino-6-deoxysucrose.

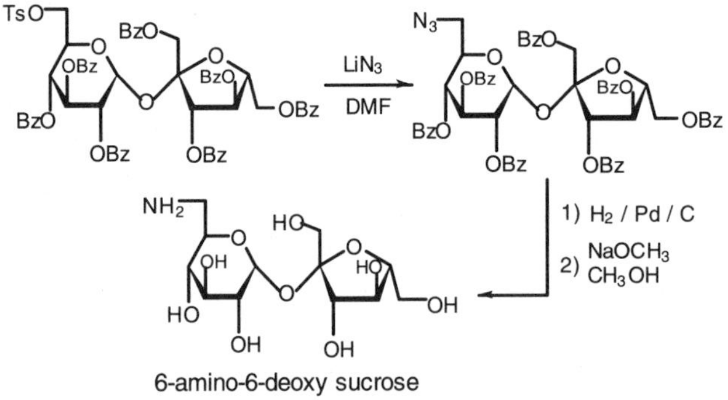

$$(4.136)$$

6-amino-6-deoxy sucrose

Sucrose can be specifically tosylated at positions 6 and 6′ by a new type of two-phase reaction in which the sucrose is dissolved in an aqueous, alkaline solution, and tosyl chloride dissolved in toluene is slowly added to the sucrose solution [123]. Reaction takes place at the interface between the two immiscible solvents to give 6,6′-di-O-tosyl sucrose at 95–99% yields in the organic phase (reaction 4.137). The 6,6′-di-O-tosyl sucrose can then be used to make many modifications at C-6 and C-6′ by nucleophilic displacement of the tosyl groups. As an example, 3,6:3′,6′-dianhydrosucrose is formed when dissolved in methanol with catalytic amounts of sodium methoxide (reaction 4.138).

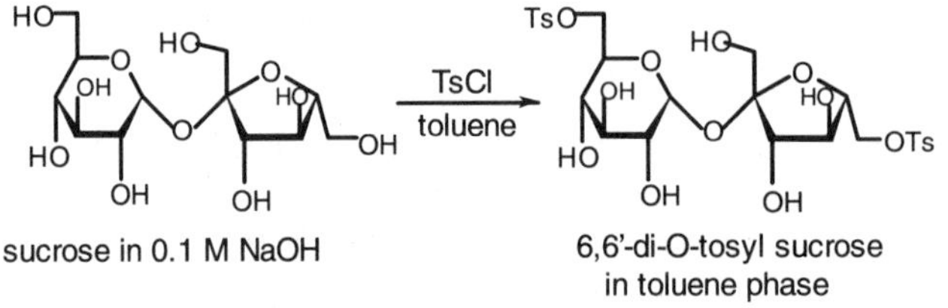

$$(4.137)$$

sucrose in 0.1 M NaOH

6,6′-di-O-tosyl sucrose
in toluene phase

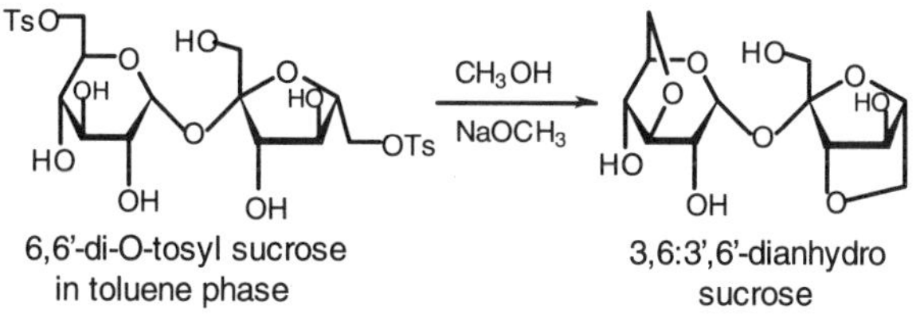

6,6'-di-O-tosyl sucrose
in toluene phase

3,6:3',6'-dianhydro
sucrose

(4.138)

The primary alcohol groups of maltose (C-6 and C-6') and α,α-trehalose can be directly modified by tosylation in pyridine, followed by displacement of the tosyl groups with azide, and catalytic reduction to give $6^1,6^2$-diamino-$6^1,6^2$-dideoxy-maltose and 6,6'-diamino-6,6'-dideoxy-α,α-trehalose, respectively [124].

The C-6' hydroxy carbon (the primary alcohol group on the nonreducing glucose residue) of maltose can be selectively modified by monotosylating cyclomaltohexaose or maloheptaose (see Chapter 8) [125]. The tosyl group can be displaced by nucleophiles such as azide to give 6-monoazidocyclomaltodextrin. This derivative can then be hydrolyzed by *Aspergillus oryzae* α-amylase to give 6^2-amino-6^2-deoxymaltose and maltose [126]. The maltose can be removed by fermentation with yeast to give the aminomaltose derivative [127].

The C-6 hydroxyl carbon (the primary alcohol group on the reducing-end glucose residue) of maltose can be specifically modified by first blocking the nonreducing-end C-4' and C-6' hydroxyl groups by forming the 4',6'-O-benzilidene derivative, followed by tosylation of the C-6 primary alcohol group on the reducing-end glucose residue. The 6-O-tosyl group can then be displaced by nucleophiles such as azide, followed by catalytic hydrogenation to form 6^1-amino-6^1-deoxymaltose (reaction 4.139).

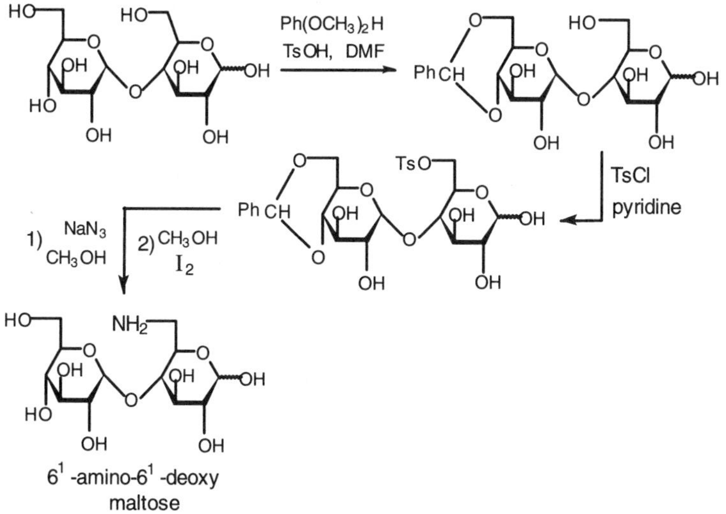

6^1-amino-6^1-deoxy
maltose

(4.139)

Primary alcohol groups can be mildly oxidized to aldehydes by reacting the tosyl derivative with DMSO in collidine at 135°C for 1–2 hr [128]. This was specifically applied to the oxidation of the primary alcohol group of cyclomaltoheptaose. The use of a bifunctional tosyl, diphenyl-4,4′-disulfonyl, gave a cyclo-maltoheptaose dialdehyde [128].

Primary alcohol groups can also be mildly and selectively oxidized to carboxyl groups by reaction with 2,2,6,6-tetramethyl-1-piperidine oxoammonium ion (TEMPO) in the presence of hypochlorite and bromide [129,130]. The specificity for the oxidation of primary alcohols in the presence of secondary alcohols in car-bohydrates occurs because of the bulky nature of the TEMPO reagent, similar to the specificity obtained with the bulky trityl chloride. The mechanism for the ox-idation of primary alcohols with TEMPO is given in reaction 4.140.

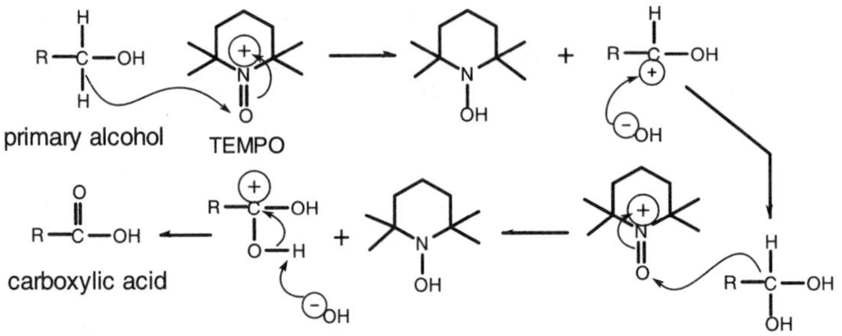

(4.140)

Reaction of 6-deoxy-6-iodo or 6-bromo-6-deoxy carbohydrates with a hard base such as silver fluoride gives β-elimination of the halide and the formation of a 5,6-ene for hexopyranosides [131] (reaction 4.141). Catalytic reduction of the unsaturated bond results in inversion of the configuration at C-5 and the formation of 6-deoxy-L-sugars [132] (reaction 4.141). Reduction of the unsaturated bond with hydrogen and Raney nickel, however, retains the configuration [132] (reac-tion 4.141).

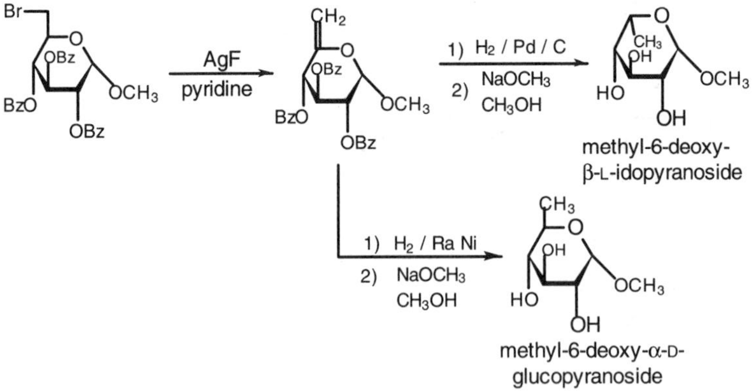

(4.141)

(4.142)

Carboxyl groups of uronic acid can be reduced to primary alcohol groups by reaction of carbodiimide and sodium borohydride [133] (reaction 4.143)

$$(4.143)$$

4.11 Summary of the Strategies Presented for the Chemical Modification of Carbohydrates

The primary principle in the specific chemical modification of carbohydrate hydroxyl carbons involves the exclusive selection of the hydroxyl group to be modified in the free state. This is usually attained by selectively reacting (protecting) the other hydroxyl groups. Acetyl and benzoyl esters are most commonly used as protecting groups because they can be formed in high yields and are easily and quantitatively removed. Cyclic isopropylidene ketals and cyclic benzilidene acetals are also used as protecting groups. Acetone or 2,2-dimethoxypropane reacts with adjacent *cis*-hydroxyl groups to give five-membered isopropylidene rings, and benzaldehyde, or 1,1-dimethoxybenzaldehyde reacts with alternate *cis*-hydroxyl groups to give six-membered benzilidene rings. Isopropylidene groups can often be selectively removed with acid or iodine in methanol, producing two free hydroxyl groups. Benzilidene rings also can be removed with iodine in methanol to provide two free hydroxyl groups, or they can be selectively cleaved with trimethylamine/borane and aluminum chloride to give either a free secondary hydroxyl group or a free primary hydroxyl group, depending on the solvent. Benzyl ethers are used as protecting groups when acid conditions are to be avoided and when alkaline conditions are to be used in the modifying reaction. Trityl (triphenylmethyl) ethers are used to selectively protect primary alcohols. The trityl groups can be readily removed by acid hydrolysis.

The various types of carbohydrate hydroxyl groups have variable reactivities. The hemiacetal hydroxyl is the most reactive; the next reactive hydroxyl is the primary hydroxyl group. The C-2 hydroxyl is usually the most reactive of the secondary alcohols. Equatorial hydroxyl groups are more reactive than axial hydroxyl groups. Strategies are developed for the selective protection of hydroxyl groups by taking advantage of their variable reactivities and by the use of different reagents that themselves have specificities for different kinds of hydroxyl groups.

After selective protection and, in some instances, selective deprotection or selective cleavage, the resulting hydroxyl group(s) that are free can be derivatized to form sulfonyl esters. The most common are tosyl (*p*-toluene sulfonyl) or trifyl (trifluoromethyl sulfonyl) esters. The sulfonyl ester groups can be selectively displaced, by nucleophilic attack, onto the sugar carbon atom, resulting in substitution by the nucleophile. The presence of the three fluorine atoms on the trifyl group makes it particularly labile toward nucleophilic displacement. Conditions can be obtained in which primary alcohols can be specifically tosylated in the presence of secondary hydroxyl groups.

Various kinds of reducing agents are used to reduce carbohydrate groups. Sodium borohydride is a general reducing reagent that can be easily handled and used in aqueous solutions to reduce aldehydes, ketones, and lactones. Trifyl esters are readily reduced by sodium borohydride in acetonitrile to give deoxy sugars. Other nucleophiles such as azide can displace either tosyl or trifyl groups. The resulting azido derivative can be readily reduced by catalytic hydrogenation to give amino sugars, or reacted with thioacetic acid to form acetamido sugars. For more rigorous reductions, lithium aluminum hydride or tributyl tin hydride can be used in nonprotic solvents.

The thioacetate nucleophile can displace sulfonyl esters to produce acetyl thioesters from which the acetyl group can be removed to form thio sugars. The substitution of amino, sulfhydryl, or ethyl phosphite groups at C-5 of either pentoses or hexoses results in the substitution of nitrogen, sulfur, and phosphorus atoms, respectively, for the oxygen atom in the pyranose ring.

The substitution of halogens onto sugars plays an important role in the production of synthetic intermediates that are used for further modification. The peracetylated reducing-end bromides or chlorides give activated glycosyl units that can be used to form glycosides and oligosaccharides. Both α- and β-1-trichloroacetimidates can be formed and readily displaced to give either α- or β-glycosides. The 1-fluorides, which are more stable in water than the other halides, can be synthesized from the free reducing sugars by reaction with pyridinium poly(hydrogen fluoride). Secondary bromides, chlorides, and iodides can be obtained by nucleophilic displacement of sulfonyl esters to give inversion of configuration. Primary bromides and chlorides can be selectively obtained by reaction with triphenylphosphine and carbontetrabromide or carbontetrachloride. Primary iodides are selectively formed by reaction of triphenylphosphine, iodine, and imidazole. Fluorine atoms can be substituted for hydroxyl groups by reaction of the free hydroxyl groups with diethylaminosulfur trifluoride (DAST). Reaction of DAST with secondary hydroxyl groups results in fluorination with inversion of the configuration. DAST will also react with many carbonyl groups to give *gem*-difluorides.

Primary and secondary hydroxyl groups can be mildly oxidized to carbonyl groups (aldehydes or ketones) by reaction with pyridinium dichromate. Primary tosyl groups can also be oxidized to aldehydes by reaction with DMSO in collidine. Primary hydroxyl groups can be mildly and selectively oxidized, in the presence of secondary alcohols, to carboxyl groups by reaction with 2,2,6,6-tetramethyl-1-piperidine oxoammonium ion (TEMPO) to form uronic acids. Uronic acid carboxyl groups can be reduced to primary alcohols by reaction with carbodiimide and sodium borohydride.

The combination of these various strategies and reactions can, thus, be used to chemically modify carbohydrate groups and produce new carbohydrates that have altered chemical and biological activities.

4.12 Literature Cited

1. G. Zemplén, *Chem. Ber.,* **59** (1926) 1258–1261.
2. J. Böeseken, *Adv. Carbohydr. Chem.,* **4** (1949) 189–202.
3a. J. F. Robyt, *Carbohydr. Res.,* **40** (1975) 373–374.
3b. H. Pelmore and M. C. R. Symons, *Carbohydr. Res.,* **155** (1986) 206–215.
4. K. B. Hicks, D. L. Raupp, and P. W. Smith, *J. Agric. Food Chem.,* **32** (1984) 288–292.
5. R. S. Tipson, *Adv. Carbohydr. Chem.,* **8** (1953) 108–215.
6. F. D. Cramer, *Methods Carbohydr. Chem.,* **2** (1963) 244–245.
7. J. M. Williams and A. C. Richardson, *Tetrahedron,* **23** (1967) 1369–1378.
8. W. N. Haworth, *J. Chem. Soc.,* **107** (1915) 8–14.
9. S. Hakomori, *J. Biochem. (Tokyo),* **55** (1964) 205–208.
10. B. Helfrich, *Adv. Carbohydr. Chem.,* **3** (1948) 79–111.
11. L. Hough, K. Mufti, and R. Khan, *Carbohydr. Res.,* **21** (1972) 144–147.
12. R. Khan, *Carbohydr. Res.,* **22** (1972) 441–445.
13. M. Bessodes, D. Komiotis, and K. Antonakis, *Tetrahedron Lett.,* **27** (1986) 579–580.
14. G. Zemplén, L. Csürös, and S. Angyal, *Chem. Ber.,* **70** (1937) 1848–1853.
15. T. Iversen and D. R. Bundle, *J. Chem. Soc. Chem. Commun.,* (1981) 1240–1242.
16. J. M. Berry and L. D. Hall, *Carbohydr. Res.,* **47** (1976).
17. C. M. McCloskey, *Adv. Carbohydr. Chem.,* **12** (1957) 148–149.
18. M. H. Park, R. Takeda, and K. Nakanishi, *Tetrahedron Lett.,* **28** (1987) 3823–3824.
19. C. C. Sweeley, R. Bentley, M. Makita, and W. W. Wells, *J. Am. Chem. Soc.,* **85** (1963) 2497–2500.
20. G. C. S. Dutton, *Adv. Carbohydr. Chem. Biochem.,* **28** (1973) 23–54.
21. O. T. Schmidt, *Methods Carbohydr. Chem.,* **2** (1963) 319–325.
22. A. N. DeBelder, *Adv. Carbohydr. Chem.,* **20** (1965) 219–301.
23. H. W. Coles, L. D. Goodhue, and R. M. Hixon, *J. Am. Chem. Soc.,* **51** (1929) 523–525.
24. K. Freudenberg, W. Durr, and H. von Hochstetter, *Chem. Ber.,* **61** (1928) 1735–1738.
25. W. A. Szarek, A. Zamojski, K. N. Tiwari, E. R. Ison, *Tetrahedron Lett.,* **27** (1986) 3827–3830.
26. B. C. Pressman, L. Anderson, H. A. Lardy, *J. Am. Chem. Soc.,* **72** (1950) 240–243.
27. T. G. Bonner, E. J. Bourne, R. F. J. Cole, and D. Lewis, *Carbohydr. Res.,* **21** (1972) 29–37.
28. S. A. Barker and E. J. Bourne, *Adv. Carbohydr. Chem.,* **7** (1951) 137–207.
29. E. Baer, *J. Am. Chem. Soc.,* **67** (1945) 338–339.
30. R. Khan and K. S. Mufti, *Carbohydr. Res.,* **43** (1975) 247–253.
31. R. Khan, *Carbohydr. Res.,* **32** (1974) 375–379.
32. R. Khan, K. S. Mufti, and M. R. Jenner, *Carbohydr. Res.,* **65** (1978) 109–113.
33. P. J. Garegg and H. Hultberg, *Carbohydr. Res.,* **93** (1981) C10–C11.
34. P. J. Garegg, H. Hultberg, and S. Wallin, *Carbohydr. Res.,* **108** (1982) 97–101.
35. M. Ek, P. J. Garegg, H. Hultberg, and S. Oscarson, *J. Carbohydr. Chem.,* **2** (1983) 305–311.
36. J. Gelas, *Adv. Carbohydr. Chem. Biochem.,* **39** (1981) 71–102.
37. M. L. Wolfrom and A. Thompson, *J. Am. Chem. Soc.,* **56** (1934) 880–883.
38. M. L. Wolfrom and J. V. Karabinos, *J. Am. Chem. Soc.,* **66** (1944) 909–913.

39. H. S. Isbell and W. W. Pigman, *J. Res. Natl. Bur. Stand.,* **10** (1933) 337–356.

40. J. W. Green, *Adv. Carbohydr. Chem.,* **3** (1948) 151–152.

41. J. U. Nef, *Ann. Chem.,* **280** (1894) 263–267.

42. E. Fischer, *Chem. Ber.,* **23** (1890) 2611–2615.

43. J. C. Sowden and H. O. L. Fischer, *J. Am. Chem. Soc.,* **69** (1947) 1963–1966.

44. O. Ruff, *Chem. Ber.,* **31** (1898) 1573–1577; O. Ruff and G. Ollendorf, *Chem. Ber.,* **33** (1900) 1798–1802; O. Ruff, *Chem. Ber.,* **34** (1901) 1362–1365.

45. H. G. Fletcher, Jr., *Methods Carbohydr. Chem.,* **1** (1962) 77–79; R. L. Whistler and J. N. BeMiller, *Methods Carbohydr. Chem.,* **1** (1962) 79–80; G. N. Richards, *Methods Carbohydr. Chem.,* **1** (1962) 180–182.

46. F. Weygand and R. Lowenfeld, *Chem. Ber.,* **83** (1950) 559–567.

47. B. Helferich and K. Weis, *Chem. Ber.,* **89** (1956) 314–318.

48. C. M. McCloskey, R. E. Pyle, and G. H. Coleman, *J. Am. Chem. Soc.,* **66** (1944) 349–350.

49. P. Z. Allen, *Methods Carbohydr. Chem.,* **1** (1962) 372–374.

50. R. K. Ness, H. G. Fletcher, Jr., and C. S. Hudson, *J. Am. Chem. Soc.,* **72** (1950) 4547–4548.

51. H. G. Fletcher, Jr., *Methods Carbohydr. Chem.,* **2** (1963) 197–198.

52. F. J. Reither, *J. Am. Chem. Soc.,* **67** (1945) 1056–1057.

53. H. W. Kosterlitz, *Biochem. J.,* **33** (1939) 1087–1089.

54. T. Posternak, *J. Am. Chem. Soc.,* **72** (1950) 4824–4825.

55. W. Pfleiderer and E. Bühler, *Chem. Ber.,* **89** (1966) 3022–3026.

56. H. Paulsen, Z. Györgydeák, and M. Friedmann, *Chem. Ber.,* **107** (1974) 1568–1571.

57. H. S. Isbell and H. Frush, *J. Org. Chem.,* **23** (1958) 1309–1319.

58. D. E. Walker and B. Axelrod, *Arch. Biochem. Biophys.,* **195** (1979) 392–395.

59. Z. Györgydeák and L. Szilagyi, *Ann. Chem.,* **499** (1977) 1987–1989.

60. F. Micheel, *Chem. Ber.,* **62** (1920) 687–690.

61. P. Karrer and A. P. Smirnoff, *Helv. Chim. Acta,* **4** (1921) 817–822.

62. G. Zemplén, R. Bognár, and G. Pongor, *Acta Chim. Acad. Sci. Hung.,* **19** (1956) 285–291.

63. P. Brigl, Z. *Physiol. Chem.,* **116** (1921) 1–5; **122** (1922) 245–250.

64. R. U. Lemieux and J. Howard, *Methods Carbohydr. Chem.,* **2** (1963) 400–402.

65. W. J. Hickinbottom, *J. Chem. Soc.,* (1928) 3140–3142.

66. W. Koenigs and E. Knorr, *Chem. Ber.,* **34** (1901) 957–981.

67. F. Micheel and A. Klemer, *Adv. Carbohydr. Chem.,* **16** (1961) 85–103.

68. T. Mukaiyama, Y. Murai, and S. Shoda, *Chem. Lett.,* (1981) 431–434; T. Mukaiyama, Y. Hashimoto, and S. Shoda, *Chem. Lett.,* (1983) 935–938.

69. S. Hashimoto, M. Hayashi, and R. Noyori, *Tetrahedron Lett.,* **25** (1984) 1379–1382.

70. J. E. G. Barnett, W. T. S. Jarvis, and K. A. Munday, *Biochem. J.,* **105** (1967) 669–672; D. S. Genghof and E. J. Hehre, *Proc. Soc. Exp. Biol. Med.,* **140** (1972) 1298–1301; G. Okada and E. J. Hehre, *Carbohydr. Res.,* **26** (1973) 240–243; M. Ariki and T. Fukui, *J. Biochem. (Tokyo),* **78** (1975) 1197–1201; C. D. Poulter and H. C. Rilling, *Acc. Chem. Res.,* **11** (1978) 307–337; B. Y. Tao, P. J. Reilly, and J. F. Robyt, *Biochim. Biophys. Acta,* **995** (1989) 214–220.

71. L. D. Hall, J. F. Manville, and N. S. Bhacca, *Can. J. Chem.,* **47** (1969) 1–8.

72. K. Igarashi, T. Honma, J. Irisawa, *Carbohydr. Res.,* **11** (1969) 577–587.

73. D. H. Brauns, *J. Am. Chem. Soc.,* **45** (1923) 833–835; S. Kitahata, C. F. Brewer, D. S. Genghof, T. Sawai, and E. J. Hehre, *J. Biol. Chem.,* **256** (1981) 6017–6026.

74. M. Hayashi, S.-I. Hashimoto, and R. Noyori, *Chem. Lett.*, (1984) 1747–1750.

75. W. A. Szarek, G. Grynkiewicz, B. Doboszewshi, and G. W. Hay, *Chem. Lett.*, (1984) 1751–1754.

76. K. Igarashi, *Adv. Carbohydr. Chem. Biochem.*, **34** (1977) 243–283.

77. W. J. Hickinbottom, *J. Chem. Soc.*, (1929) 1676–1687.

78. R. R. Schmidt and W. Kinzy, *Adv. Carbohydr. Chem. Biochem.*, **50** (1994) 21–120.

79. H. Kunz, *Angew. Chem. Int. Ed. Engl.*, **26** (1987) 294–308.

80. W. N. Haworth and W. J. Hickinbottom, *J. Chem. Soc.*, (1931) 2847–2849.

81. R. U. Lemieux and H. F. Bauer, *Can. J. Chem.*, **32** (1954) 340–344.

82. R. U. Lemieux, *Can. J. Chem.*, **31** (1953) 949–951.

83. R. U. Lemieux and G. Huber, *J. Am. Chem. Soc.*, **75** (1953) 4118–4119; **78** (1956) 4117–4118.

84. S. J. Cook, R. Khan, and J. M. Brown, *J. Carbohydr. Chem.*, **3** (1984) 343–348.

85. M. Blanc-Muesser, J. Defaye, and H. Driguez, *Carbohydr. Res.*, **67** (1978) 305–328.

86. S. Cottaz, H. Driguez, and B. Svensson, *Carbohydr. Res.*, **228** (1992) 299–305.

87. E. Fischer, *Chem. Ber.*, **47** (1914) 196–200.

88. M. Bergmann, H. Schotte, and W. Lechinsky, *Chem. Ber.*, **55** (1922) 158–160.

89. W. G. Overend, M. Stacey, and J. Stanec, *J. Chem. Soc.*, (1949) 2841–2845.

90. B. Helferich, *Adv. Carbohydr. Chem.*, **7** (1952) 210–245.

91. J. Adamson, A. B. Foster, L. D. Hall, R. N. Johnson, and R. H. Hesse, *Carbohydr. Res.*, **15** (1970) 351–359.

92. B. Evers, P. Mischnick, and J. Thiem, *Carbohydr. Res.*, **262** (1994) 335–341.

93. B. Evers, M. Petricek, and J. Thiem, *Carbohydr. Res.*, **300** (1997) 153–159.

94. T. Rosen, I. M. Lico, and D. T. W. Chu, *J. Org. Chem.*, **53** (1988) 1580–1582; T. Suami, Y. Fukuda, J. Yamamoto, Y. Saito, M. Ito, and S. Ohba, *J. Carbohydr. Chem.*, **1** (1982) 9–15.

95. E.-P. Barrette and L. Goodman, *J. Org. Chem.*, **49** (1984) 176–178.

96. M. J. Robins, J. S. Wilson, and F. Hansske, *J. Am. Chem. Soc.*, **105** (1983) 4059–5065.

97. P. Herdewiju, J. Balzarini, E. DeClercq, R. Pauwels, M. Baba, S. Broden, and H. Vanderhaeghe, *J. Med. Chem.*, **30** (1987) 1270–1280.

98. F. Sanger, S. Nicklen, and A. R. Coulson, *Proc. Natl. Acad. Sci. USA*, **74** (1977) 5463–5467.

99. A. F. Russell and J. G. Moffatt, *Biochemistry*, **8** (1969) 4889–4896; K. Geider, *Eur. J. Biochem.*, **27** (1974) 555–563.

100. J. P. Horowitz, J. Chua, and M. Noel, *J. Org. Chem.*, **29** (1964) 2076–2078.

101. R. F. Butterworth and S. Hanessian, *Synthesis*, (1971) 70–88.

102. C. R. Haylock, L. D. Melton, K. N. Slessor, and A. S. Tracey, *Carbohydr. Res.*, **16** (1971) 375–382.

103. W. J. Middleton, *J. Org. Chem.*, **40** (1975) 574–578; T. J. Tweson and M. J. Welsh, *J. Org. Chem.*, 43 (1978) 1090–1094; M. Sharma and W. Korytnyk, *Tetrahedron Lett.*, (1977) 573–576.

104. S. G. Withers, D. J. MacLennan, I. P. Street, *Carbohydr. Res.*, **154** (1986) 127–144.

105. R. A. Sharma, I. Kavai, Y. L. Fu, and M. Bobek, *Tetrahedron Lett.*, (1977) 3433–3436.

106. T. P. Binder and J. F. Robyt, *Carbohydr. Res.*, **132** (1984) 173–177.

107. T. P. Binder and J. F. Robyt, *Carbohydr. Res.*, **147** (1986) 149–154.

108. M. J. Bernaerts, J. Furnelle, and J. De Ley, *Biochim. Biophys. Acta*, **69** (1963) 322–330.

109. J. Van Beeumen and J. De Ley, *Eur. J. Biochem.,* **6** (1968) 331–343.

110. M. Pietsch, M. Walter, and K. Buchholz, *Carbohydr. Res.,* **254** (1994) 183–194.

111. Y. Ueno, K. Hori, R. Yamauchi, M. Kiso, A. Hasegawa, and K. Kato, *Carbohydr. Res.,* **89** (1981) 271–278.

112. R. L. Whistler and R. E. Gramera, *J. Org. Chem.,* **29** (1964) 2609–2610.

113. S. Inouye, T. Tsuroka, T. Ito, and T. Niida, *Tetrahedron,* **24** (1968) 2125–2144.

114. R. L. Whistler, M. S. Feather, and D. L. Ingles, *J. Am. Chem. Soc.,* **84** (1962) 122–124.

115. R. L. Whistler and W. C. Lake, *Biochem. J.,* **130** (1972) 919–925; R. L. Whistler, *Science* **186** (1974) 431–433.

116. H. Yamamoto and S. Inokawa, *Adv. Carbohydr. Chem.,* **42** (1984) 140–142.

117. A. K. M. Anisuzzaman and R. L. Whistler, *Carbohydr. Res.,* **61** (1978) 511–518.

118. P. J. Garegg and B. Samuelsson, *J. Chem. Soc. Chem. Commun.,* (1979) 978–980.

119. R. Khan, *Pure Appl. Chem.,* **56** (1984) 833–844; F. W. Lichtenthaler, S. Immel, and U. Kreis, in *Carbohydrates as Raw Materials,* pp. 1–32 (F. W. Lichtenthaler, ed.) VCH Weinheim, Germany (1991).

120. J. N. Zikopoulos, S. H. Eklund, and J. F. Robyt, *Carbohydr. Res.,* **104** (1982) 245–251.

121. S. H. Eklund and J. F. Robyt, *Carbohydr. Res.,* **177** (1988) 253–258.

122. A. Tanriseven and J. F. Robyt, *Carbohydr. Res.,* **186** (1989) 87–94.

123. R. Mukerjea and J. F. Robyt, unpublished results.

124. S. Umezawa, T. Tsuchiya, S. Nakada, and K. Tatsuta, *Bull. Chem. Soc. Jpn.,* **40** (1967) 395–401.

125. L. D. Melton and K. N. Slessor, *Carbohydr. Res.,* **18** (1971) 29–37.

126. L. D. Melton and K. N. Slessor, *Canad. J. Chem.,* **51** (1973) 327–332.

127. K. Kitaoka and J. F. Robyt, unpublished results.

128. J. Yoon, S. Hong, K. A. Martin, and A. W. Czarnik, *J. Org. Chem.,* **60** (1995) 2792–2795.

129. A. E. T. De Nooy, A. C. Besemer, and H. Van Bekkum, *Carbohydr. Res.,* **269** (1995) 89–98.

130. P. S. Chang and J. F. Robyt, *J. Carbohydr. Chem.,* **15** (1996) 819–830.

131. B. Helferich and E. Himmen, *Chem. Ber.,* **61** (1928) 1825–1831; **62** (1929) 2136–2141.

132. T. M. Cheung, D. Horton, and W. Weckerle, *Carbohydr. Res.,* **58** (1977) 139–151.

133. R. L. Taylor and H. E. Conrad, *Biochemistry,* **11** (1972) 1383–1387.

4.13 References for Further Study

"Trityl ethers of carbohydrates," B. Helferich, *Adv. Carbohydr. Chem.,* **3** (1948).

"Applications in the carbohydrate field of reductive desulfurization by Raney nickel," H. G. Fletcher, Jr. and N. K. Richtmyer, *Adv. Carbohydr. Chem.,* **5** (1950).

"The nitromethane and 2-nitroethanol syntheses," J. C. Sowden, *Adv. Carbohydr. Chem.,* **6** (1951).

"Acetals and ketals of the tetritols, pentitols, and hexitols," S. A. Barker, *Adv. Carbohydr. Chem.,* **7** (1952).

"The glycals," B. Helferich, *Adv. Carbohydr. Chem.,* **7** (1952).

"Relative reactivities of hydroxyl groups of carbohydrates," J. M. Sugihara, *Adv. Carbohydr. Chem.,* **8** (1953).

"Sulfonic esters of carbohydrates," R. S. Tipson, *Adv. Carbohydr. Chem.*, **8** (1953).

"Glycosylamines," G. P. Ellis and J. Honeyman, *Adv. Carbohydr. Chem.*, **10** (1955).

"The glycosyl halides and their derivatives," L. J. Haynes and F. H. Newth, *Adv. Carbohydr. Chem.*, 10 (1955).

"Benzyl ethers of sugars," C. M. McCloskey, *Adv. Carbohydr. Chem.*, **12** (1957).

"The carbonates and thiocarbonates," L. Hough, J. E. Priddle, and R. S. Theobald, *Adv. Carbohydr. Chem.*, **15** (1960).

"Applications of trifluoroacetic anhydride in carbohydrate chemistry," T. G. Bonner, *Adv. Carbohydr. Chem.*, **16** (1961).

"Glycosyl fluorides and azides," F. Micheel and A. Klemer, *Adv. Carbohydr. Chem.*, **16** (1961).

"Developments in the chemistry of thio sugars," D. Horton and D. H. Hutson, *Adv. Carbohydr. Chem.*, **18** (1963).

"Unsaturated sugars," R. J. Ferrier, *Adv. Carbohydr. Chem.*, **20** (1965).

"Halogenated carbohydrates," J. E. G. Barnett, *Adv. Carbohydr. Chem.*, **22** (1967).

"Sulfonic esters of carbohydrates. Part I," D. H. Ball and F. W. Parrish, *Adv. Carbohydr. Chem.*, **23** (1968).

"Cyclic monosaccharides having nitrogen or sulfur in the ring," H. Paulsen and K. Todt, *Adv. Carbohydr. Chem.*, **23** (1968).

"Deoxyhalogeno sugars," W. A. Szarek, *Adv. Carbohydr. Chem. Biochem.*, **32** (1973).

"Dithioacetals of sugars," J. D. Wander and D. Horton, *Adv. Carbohydr. Chem. Biochem.*, **32** (1976).

"Relative reactivities of hydroxyl groups in carbohydrates," A. H. Haines, *Adv. Carbohydr. Chem. Biochem.*, **33** (1976).

"The chemistry of sucrose," R. Khan, *Adv. Carbohydr. Chem. Biochem.*, **33** (1976).

"Cyclic acetals of the aldoses and aldosides," A. N. DeBelder, *Adv. Carbohydr. Chem. Biochem.*, **34** (1977).

"The Koenigs-Knorr reaction," K. Igarashi, *Adv. Carbohydr. Chem. Biochem.*, **34** (1977).

"Fluorinated carbohydrates," A. A. E. Penglis, *Adv. Carbohydr. Chem. Biochem.*, **38** (1981).

"Sugar analogs having phosphorus in the hemiacetal ring," H. Yamamoto and S. Inokawa, *Adv. Carbohydr. Chem. Biochem.*, **42** (1984).

"Chemistry and developments of fluorinated carbohydrates," T. Tsuchiya, *Adv. Carbohydr. Chem. Biochem.*, **48** (1990).

"Anomeric-oxygen activation for glycoside synthesis: the trichloroacetimidate method," R. R. Schmidt and W. Kinzy, *Adv. Carbohydr. Chem. Biochem.*, **50** (1994).

"Synthetic methods for carbohydrates," in *ACS Symposium Series 39* (H. S. El Khadem, ed.) American Chemical Society, Washington, D.C. (1976).

Carbohydrate Chemistry: Monosaccharides and Their Oligomers, H. S. El Khadem, Academic, New York (1988).

"Trends in synthetic carbohydrate chemistry," in *ACS Symposium Series 386* (D. Horton, L. D. Hawkins, and G. J. McGarvey, eds.) American Chemical Society, Washington, D.C. (1989).

Monosaccharides: Their Chemistry and Their Roles in Natural Products, P. M. Collins and R. J. Ferrier, John Wiley & Sons, New York (1995).

Chapter 5

Sweetness

5.1 The Sweet Taste of Sugars and the Development of the Sweet-Taste Hypothesis

Cane sugar was known to early humans as a sweet and pleasant-tasting material. The most common physical property ascribed to carbohydrates is sweetness. Not all carbohydrates, however, are sweet, and different carbohydrates have different degrees of sweetness. After it was determined that carbohydrates differed from each other by the stereochemical arrangement of their chiral hydroxyl groups, it was also recognized that they had different degrees of sweetness. D-Glucose is sweet, D-fructose and D-xylose are much sweeter, D-galactose is devoid of sweetness, and D-mannose is bitter. Among the disaccharides, sucrose is very sweet, maltose is mildly sweet, lactose is only very slightly sweet, and gentiobiose is bitter. These observations suggest that the differences in sweetness of the carbohydrates are due to differences in the stereochemical arrangements of their chiral hydroxyl groups. Table 5.1 lists the relative sweetness of a number of carbohydrates.

The relative degrees of sweetness of different carbohydrates and other compounds can be compared by using a taste panel. Using different concentrations of sucrose and other carbohydrates, the members of the taste panel are preselected to ensure that they are sufficiently sensitive to sweetness and are able to rank a set of samples consistently in a certain order. The usual practice is to determine the relative sweetness of substances by making comparisons with a standard such as sucrose. A 1 M or a 10% (w/v) solution of the standard is assigned a value of 1.00, or sometimes 100, and other substances are compared relative to the standard on an equivalent weight basis. For example, a substance that is 100 times sweeter than sucrose would give the same degree of sweet taste as 1 M sucrose when the concentration of the substance was 0.01 M, and a 1 M solution of a substance that was 1/2 as sweet as sucrose

Table 5.1 Sweetness of Sugars[a]

Sugar	Relative sweetness[b]
β-D-Xylopyranose	180
β-D-Fructopyranose	180
Sucrose	100
β-D-Glucopyranose	82
α-D-Glucopyranose	74
α-D-Galactose	32
β-D-Galactose	21
α-D-Mannose	32
β-D-Mannose	bitter
β-Lactose	32
α-Lactose	16
β-Maltose	32
Raffinose	1
D-Glucitol	50
Maltitol	63
4-O-β-D-Galactopyranosyl-D-glucitol (lactitol)	34
4-O-β-D-Glucopyranosyl-D-glucitol (cellobitol)	11

[a]From R. S. Shallenberger and T. E. Acree, "Chemical structure of compounds and their sweet and bitter taste," in *Handbook of Sensory Physiology IV. Chemical Senses 2. Taste* (L. M. Beidler, ed.) Springer-Verlag, Berlin (1971) and C.-K. Lee, *Food Chem.*, **2** (1977) 95–105.
[b]Measured by a taste panel on an equal weight basis in solution.

would give the same degree of sweet taste as a 0.5 M solution of sucrose. It would then be said that the first substance had a sweet value of 100 relative to sucrose, and the second substance had a sweet value of 0.5 relative to sucrose.

One of the first explanations to be put forward for the cause of the sweetness of carbohydrates involved the presence of several hydroxyl groups whose specific stereochemistry elicited a sweet taste [1]. But it was also recognized that certain compounds other than carbohydrates were sweet, sometimes more so than carbohydrates. Compounds such as chloroform, L-alanine, saccharin, 2-amino-4-nitrobenzene, and 6-chloro-D-tryptophan are sweet, and most of these do not have any hydroxyl groups or at most have only one.

By considering the chiral hydroxyl nature of carbohydrates, Shallenberger, et al. [2–4] proposed that the chemical structure responsible for the sweet taste was a glycol group. By taking into consideration the structures of both carbohydrate and noncarbohydrate sweet compounds, Shallenberger and Acree [5,6] proposed a more definitive hypothesis involving a structural relationship that was common to all sweet-tasting compounds. Their hypothesis stated that there were two electronegative atoms, A and B, that were separated by a distance greater than 2.5 Å but less than 4 Å. They further proposed that one of the atoms was polarized by having a covalently attached hydrogen atom. A and B were usually an oxygen atom or a nitrogen atom, but in some compounds they were a chlorine atom or a carbon atom. One of the requisite groups was AH, a hydrogen bond donor, and the other group was B, a hydrogen bond acceptor. It was further proposed that the AH and B groups in sweet sugars were vicinal pairs of hydroxyl groups that were *cis* or *gauche,* and in sugars that were not sweet, the hydroxyl pairs were *trans* or *eclipsed.* This latter part of the hypothesis was consistent with a distance of 2.5–4.0 Å between the hydroxyl groups.

Shallenberger and Acree further postulated that the sweet taste was elicited when the AH and B groups of the sweet compound interacted with complementary proton-accepting and proton-donating groups on the sweet-tasting protein receptor (see Fig. 5.1). Shallenberger and Acree intuitively assigned the AH group of a reducing carbohydrate to the hemiacetal hydroxyl, and the B group to the C-2 hydroxyl. For β-D-fructopyranose, they assigned the AH group to the C-2 hydroxyl (hemiketal hydroxyl), and the B group to the C-1 primary hydroxyl. Lindley and Birch [7] confirmed this assignment by replacing the hydroxyl group at C-2 of D-fructopyranose with a hydrogen, and by forming a glycoside with the C-2 hemiketal hydroxyl, creating compounds that were considerably less sweet than β-D-fructopyranose.

The examination of several classes of intensely sweet compounds, such as 2-substituted 5-nitroanilines [8], cyclamate, saccharin, and the α-amino acids [9], indicated that in addition to the AH and B groups there was also the existence of a hydrophobic group, X, that produced an enhancement of the sweetness of the compound. Kier [10] postulated that AH, B, and X made up a triangular *glucophore* in which the X group was separated from AH by 3.5 Å and from the B group by 5.5 Å (see Fig. 5.2). The presence of this type of glucophore was then assigned to groups on many noncarbohydrate, sweet-tasting compounds (see Fig. 5.3) and eventually to groups on carbohydrates (see Figs. 5.4 and 5.5).

In most cases, the D-α-amino acids are sweet, and the L-enantiomers are bitter or not sweet (see Table 5.2). This is in contrast with the D- and L-enantiomers of the sugars which both have the same degree of sweetness [11]. This difference was explained by Shallenberger et al. [11] as being due to the fact that carbohydrates have several chiral centers that can act as the AH, B, and X groups of a glucophore, whereas the α-amino acids have the AH, B, and X groups all attached to a single, chiral center. The triangular glucophores of D- and L-glucose can both form complementary binding with the three-site sweet-taste protein receptor, but only the D-α-amino acids have the correct geometry to form complementary com-

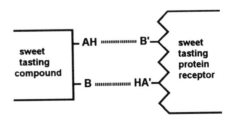

Figure 5.1. Shallenberger and Acree hypothesis for the interaction of the *AH* and *B* units of a sweet compound, with complementary groups on the sweet-tasting protein receptor.

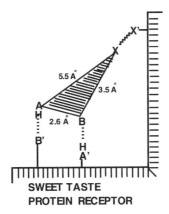

Figure 5.2. Binding of the Shallenberger-Kier AH-B-X glucophore with the complementary AH'-B'-X' groups at the sweet-tasting protein receptor site.

plexes with the three-site sweet-taste protein receptor; the L-α-amino acid can form complexes only with two of the three sites (see Fig. 5.4) and hence do not elicit a sweet taste.

The relatively low degrees of sweetness of carbohydrates when compared with the sweetness of some noncarbohydrate compounds such as cyclamates, saccharin, certain aminoacids, and so on (see Fig. 5.3) can be explained by the relatively weak hydrophobic character of the C-6 hydroxymethyl group found in many pyranoses. However, the presence of a hydrophobic sweetness intensifier can explain why certain carbohydrates are much sweeter than other carbohydrates; for example, D-fructose and D-xylose are much sweeter than D-glucose and sucrose (see Table 5.1). Both D-fructopyranose and D-xylopyranose (the predominant forms of D-fructose and D-xylose in solution and in the crystalline state) have methylene groups that are not substituted by a hydroxyl group, and hence are more hydrophobic and produce a sweeter taste (see Fig. 5.5).

The presence of an unsubstituted methylene group in producing or enhancing sweetness is further supported by the sweetness of the 1,5-anhydro-D-hexitols: 1,5-anhydro-D-glucitol, 1,5-anhydro-D-mannitol, and 1,5-anhydro-D-galactitol [12]. Of the three 1,5-anhydro-D-hexitols, the parent sugars, β-D-mannose and β-D-galactose, are completely devoid of sweetness. In the case of the 1,5-anhydro-D-hexitols, however, the glucophore is shifted to accommodate the C-1 methylene group as the hydrophobic sweet intensifier (see Fig. 5.5).

Birch and Lee [13] reported that 6-deoxy-D-glucose (D-quinivose), 6-deoxy-D-mannose (D-rhamnose), and 6-deoxy-L-galactose (L-fucose) are all sweet. As might have been anticipated, the replacement of the hydroxyl group of the parent sugar with a hydrogen at C-6 increases the hydrophobic character of these sugars and produces an enhancement of their sweetness over the sweetness of the parent sugar.

Further support of the importance of the hydrophobic methylene group in producing sweetness is obtained when D-fructose is heated in solution. The sweetness

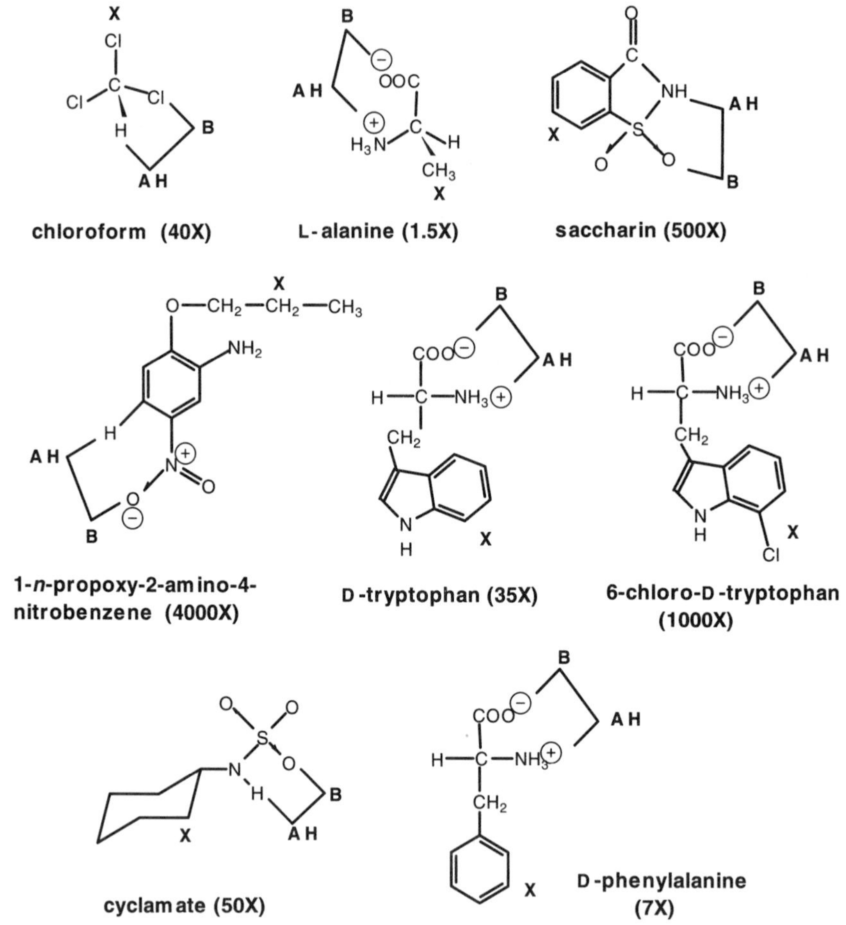

Figure 5.3. Sweet tasting molecules showing the *AH-B-X* groups postulated by Shallenberger-Kier hypothesis. The numbers in parenthesis are the degrees of sweetness relative to sucrose.

is significantly decreased [14,15] because of a shift in the mutarotational equilibrium, which results in a decreased amount of the β-D-fructopyranose form and a consequent decrease in the presence of the methylene group in the glucophore.

5.2 Naturally Occurring Sweet Glycosides

There are a few, intensely sweet, naturally occurring bisglycosides. Stevioside and rebaudoside A are found in the leaves of the Paraguayan shrub *Stevia rebaudiana* Bertoni (yerba dulce). Both have steviol as the aglycone [16] to which there are two glycosyl groups attached, a β-D-glucopyranosyl attached to the carboxyl

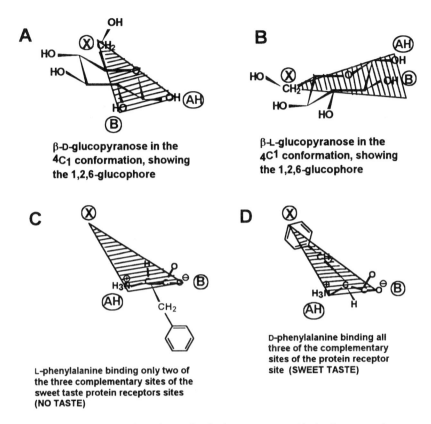

A

β-D-glucopyranose in the
$4C_1$ conformation, showing
the 1,2,6-glucophore

B

β-L-glucopyranose in the
$4C_1$ conformation, showing
the 1,2,6-glucophore

C

L-phenylalanine binding only two of
the three complementary sites of the
sweet taste protein receptors sites
(NO TASTE)

D

D-phenylalanine binding all
three of the complementary
sites of the protein receptor
site (SWEET TASTE)

Figure 5.4. Proposed binding of D- and L-β-glucopyranose with the three complementary binding sites of the sweet-taste protein receptor, and the binding of D-phenylalanine with the three sites, but the binding of L-phenylalanine with only two of the sites.

group by an acetal-ester linkage, and a β-linked sophorosyl group for stevioside [17,18] (see Fig. 5.6A); and a 3^2-O-β-D-glucopyranosyl-β-sophorosyl group for rebaudoside A [19] (see Fig. 5.6B). Stevioside is reported to be about 280 times sweeter than sucrose, with a mild licorice aftertaste, and rebaudoside A is 190 times sweeter than sucrose [20]. The β-D-glucopyranosyl group is reported to be essential for sweetness since steviol sophoroside is not sweet.

Osladin, a sterol bisglycoside, is found in the fern *Polypodium vulgare* [21]. It has a 2-O-α-L-rhamnopyranosyl-β-D-glucopyranosyl (neohesperidosyl) group at one end and an α-L-rhamnopyranosyl group at the other end (see Fig. 5.6C). It has been reported to be 3,000 times sweeter than sucrose.

Glycyrrhizin is a glycoside with a uronic acid disaccharide, 2-O-β-D-glucopyranosyl uronic acid-α-D-glucopyranosyl uronic acid, attached to a polyterpenoid aglycone (see Fig. 5.6D). It is found in licorice root and is 50 times sweeter than sucrose, with a licorice taste [22]. Glycyrrhizin has medical and dental applications: it has been used to treat ulcers; it also has been used as an antiinflammatory

A

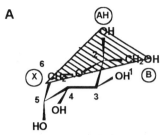

β-D-fructopyranose $_4$C^1 conformation
showing the 1,2,6-glucophore with the
methylene group at C-6

B

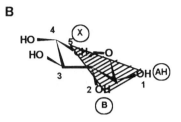

β-D-xylopyranose^{4}C$_1$ conformation
showing the 1,2,5-glucophore with
the methylene group at C-5

C

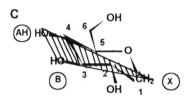

1,5-anhydro-D-glucitol, showing the
1,3,4-glucophore with the methylene
group at C-1

D

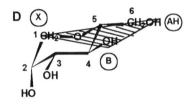

1,5-anhydro-D-mannitol, showing
the 1,4,6-glucophore with the
methylene group at C-1

Figure 5.5. Proposed *AH-B-X* glucophores for **A,** β-D-fructopyranose; **B,** β-D-xylopyranose; **C,** 1,5-anhydro-D-glucitol; **D,** 1,5-anhydro-D-mannitol, showing the involvement of
a hydrophobic methylene group sweetness intensifier.

agent; it inhibits carbohydrate metabolism of dental bacteria; and it apparently in-
terferes in the adherence of dental bacteria to smooth surfaces, thereby helping to
prevent dental plaque formation and tooth decay. Glycyrrhizin also affects tooth
enamel by raising fluoride uptake and reducing its solubility [23].

5.3 Synthesis of Supersweet Sucroses

In the 1970s Hough and associates were carrying out a series of chemical modifications of sucrose. In one series of experiments they were replacing the hydroxyl
groups with chlorine atoms by reaction with sulfuryl chloride in pyridine/chloro-
form at low temperatures. The reaction initially formed chlorosulfate esters with
four hydroxyl groups at positions 4,6,1', and 6'. These esters each undergo nucleophilic displacement by a chloride ion to give 6'-chlorosucrose, 6,6'-dichlorosu-
crose, 4,6,6'-trichloro-galactosucrose, and 4,6,1',6'-tetrachlorogalactosucrose
[24–27] (see Fig. 5.7A). During the course of these modifications, it was apparently discovered by accident that the 4,6,1',6'-tetrachloro-4,6,1',6'-tetradeoxy-

Table 5.2. Sweetness of Enantiomeric α-Amino Acids Relative to Sucrose

Amino acid	Sweet taste[a]	
	L-Isomer	D-Isomer
Tryptophan	Bitter	Sweet (35× suc.)
Phenylalanine	Bitter	Sweet (7× suc.)
Histidine	Bitter	Sweet (7× suc.)
Tyrosine	Bitter	Sweet (6× suc.)
Leucine	Bitter	Sweet (4× suc.)
Alanine	Sweet (1.5× suc.)	Bitter
Glycine	Sweet (1.5× suc.)[b]	
Arginine	NS[c]	Slightly sweet
Aspartic acid	NS	NS
Isoleucine	NS	NS
Lysine	NS	NS
Proline	NS	NS
Serine	NS	NS
Threonine	NS	NS
Valine	NS	NS

[a]Data from ref. [8].
[b]Glycine is not chiral and exists in only one form.
[c]NS, not sweet.

galactosucrose had a very sweet taste, comparable to some of the intensely sweet noncarbohydrate compounds [27].

The 6,6'-dichloro-6,6'-dideoxysucrose can be synthesized in high yields (>70%) by reaction of sucrose with triphenylphosphine in carbontetrachloride and pyridine [28] (see reaction 4.123 in Chapter 4). It, however, is not sweet at all. The 4,6,1',6'-tetrachloro-4,6,1',6'-tetradeoxygalactosucrose was first synthesized by preparing the 6,1',6'-tri-O-(methylphenylsulfonyl) sucrose, followed by reaction with lithium chloride and chlorination of the C-4 hydroxyl with sulfuryl chloride [27] (see Fig. 5.7B). This tetrachlorogalactosucrose was found to be 200 times sweeter than sucrose.

The synthesis of 4,1',6'-trichloro-4,1',6'-trideoxygalactosucrose in which the C-6 hydroxyl group is not substituted proved to be the sweetest of the chlorodeoxysucroses. It was prepared by first forming 6,1',6'-tri-O-trityl sucrose, followed by acetylation and the removal of the trityl groups with dilute acid. In the removal of the trityl group, the acetyl group on C-4 migrates to C-6, leaving C1', C-4, and C-6' free. These hydroxyls are then chlorinated with sulfuryl chloride, followed by deacetylation [29] (see Fig. 5.7C). This chlorodeoxysucrose turns out to be 2,000 times sweeter than sucrose [30] and has been given the trivial name, *sucralose.*

An even sweeter chlorodeoxysucrose was later synthesized by the reaction of sucralose with triphenylphosphine and diethyl azodicarboxylate to give the 3',4'-lyxo epoxide, followed by stereospecific opening of the epoxide with chloride to

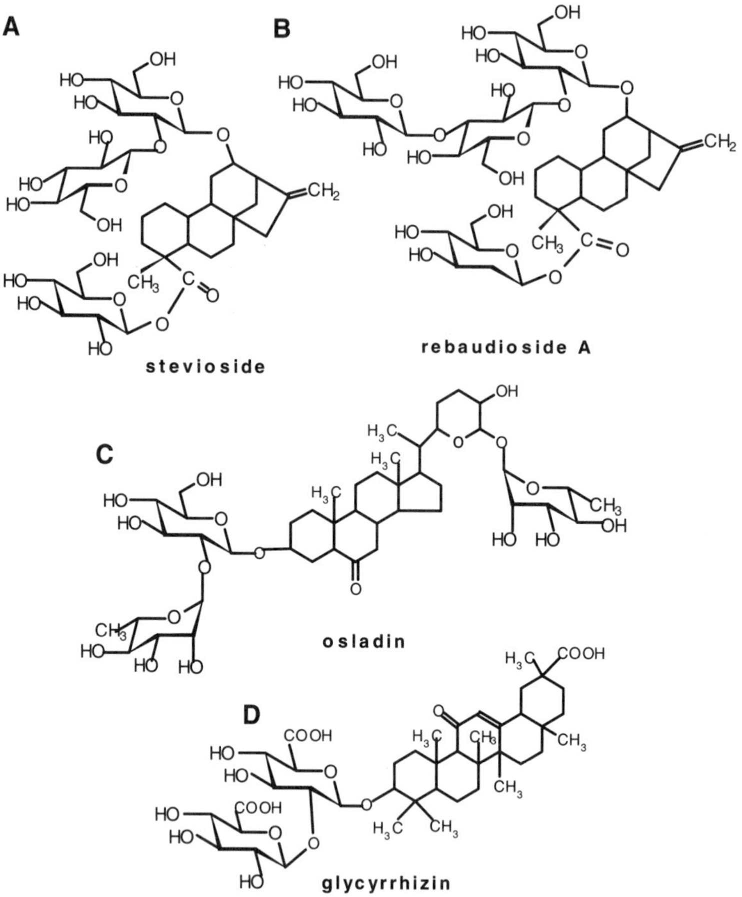

Figure 5.6. Structures of naturally occurring sweet glycosides.

give 4,1′,4′,6′-tetrachloro-4,1′,4′,6′-tetradeoxygalactosucrose [31], a product that was reported to be 2,200 times sweeter than sucrose [32].

When bromide is substituted for chloride, the bromo analogue, 4′-bromo-4,1′,6′-trichloro-4,1′,4′,6′-tetradeoxygalactosucrose, is formed (see Fig. 5.7D). It is reported to be 3,000 times sweeter than sucrose. Substitution of iodine at C-4′ in sucralose also forms an intensely sweet deoxyhalogalactosucrose, 4,1′,6′-trichloro-4′-iodo-4,1′,4′,6′-deoxygalactosucrose, that is reported to be 7,000 times sweeter than sucrose. The all-bromo analogue, 4,1′,4′,6′-tetrabromo-4,1′,4′,6′-tetra-deoxygalactosucrose is the most intensely sweet deoxyhalogalac-tosucrose reported, being 7,500 times sweeter than sucrose [32]. Table 5.3 gives the relative sweetness for several deoxyhalosucroses.

Hough and Khan [30] concluded that the 1′-chloro (halo) substituent on the fructose moiety was critical for an intense sweet taste. Addition of chlorine to the

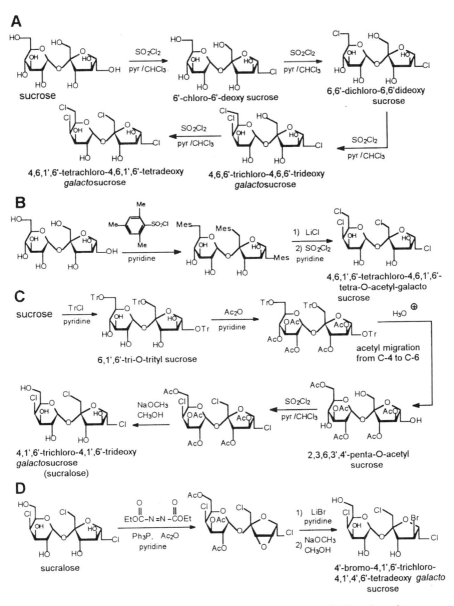

Figure 5.7. Synthesis of intensely sweet chlorodeoxy sucroses: **A,** Reaction of sucrose with sulfuryl chloride in pyridine/chloroform at 0;dgC; **B,** synthesis of 4,6,1',6'-tetra-chloro-4,6,1',6'-tetradeoxygalactosucrose; **C,** synthesis of 4,1',6'-trichloro-4,1',6'-tri-deoxygalactosucrose (sucralose); **D,** synthesis of 4'-bromo-4,1',4',6'-tetradeoxygalacto-sucrose.

Table 5.3 Sweetness of Deoxyhalo Derivatives of Sucrose

Sugar	Relative sweetness
Sucrose[a]	1
6-Chloro-6-deoxysucrose	NS[b]
6,6'-Dichloro-6,6'-dideoxysucrose	NS (bitter)
4-Chloro-4-deoxygalactosucrose	5
6'-Chloro-6'-deoxysucrose	20
4,6,1',6'-Tetrachloro-4,6,1',6'-tetradeoxysucrose	100
4,6,1',6'-Tetrachloro-4,6,1',6'-tetradeoxygalactosucrose	200
1'-6'-Dichloro-1',6'-dideoxysucrose	500
4,1'-Dichloro-4,1'-dideoxysucrose	600
4,1'-6'-Trichloro-4,1',6'-trideoxygalactosucrose (sucralose)	2000
1',4',6'-Tribromo-4-chloro-4,1',4',6'-tetradeoxygalactosucrose	30
1',4',6'-Trichloro-4,1',4',6'-tetradeoxy-4-fluorogalactosucrose	200
4,1',6'-Trichloro-4,1',4',6'-tetradeoxy-4'-fluorogalactosucrose	1000
4,1',6'-Trichloro-4,1',4',6'-tetradeoxy-4'-iodogalactosucrose	7000
4,1',4',6'-Tetrabromo-4,1',4',6'-tetradeoxygalactosucrose	7500

[a]Data for the first 11 compounds are from L. Hough and R. Khan, *Trends Biol. Sci.,* **3** (1978) 61–63, and the data for the remaining compounds are from G. Jackson et al., Br. Pat. 8,211,427A (1982).
[b]NS, not sweet

C-6' and C-4, with inversion, provided an additional increase in the sweetness. Addition of chlorine to the C-6 position was deleterious to the sweet-tasting character. The substitution at C-4, with inversion, produced compounds that were sweeter than those substituted at C-4, with retention of the configuration. Inversion of the C-4 hydroxyl to give galactosucrose, however, formed a compound that was devoid of sweetness. Hence, the substitution of a chlorine atom, with inversion, significantly contributed to the sweetness.

Hough and Khan [30] attempted to explain the increased sweetness of their chlorosucrose derivatives by using the Shallengerber/Kier triangular glucophore hypothesis. They postulated that sucrose and the chlorosucroses had two glucophores: a 1',2,4-glucophore, with the chloro group at C-4 being the hydrophobic group; and a 1',2,6'-glucophore, with the hydrophobic group at C-6' (see Fig. 5.8).

Using computer-aided models of the three-dimensional molecular conformation of sucrose and sucralose, Lichtenthaler and Immel [33] obtained a major breakthrough in the understanding of how sweet compounds might bind to the sweet-tasting protein receptor and elicit a sweet-taste response. Their calculations and modeling indicate that there is not a single hydrophobic binding point in the sucrose molecule and the sweet-taste hydrophobic receptor protein. They found that there is an extended hydrophobic region on sucrose that encompasses the entire carbon back side of the fructose moiety. The AH and B groups were proposed to be the C-2 and C-3 hydroxyl groups of the glucose moiety. The C-2 hydroxyl is intramolecularly hydrogen bonded by the C-1' hydroxyl group, since it is the predominant conformer of sucrose in solution. This intramolecular hydrogen bond

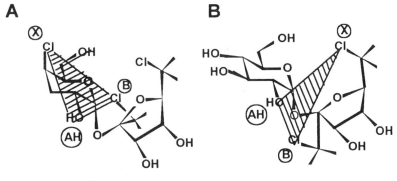

A

The 1',2,4-glucophore of 1',4,6'- trichloro *galacto*sucrose

B

The 1',2'6'-glucophore of 1',6'- dichloro sucrose

Figure 5.8. Two glucophores of chlorosucroses proposed by Hough and Khan [29].

increases the electropositive character of the C-2-OH, suggesting that it is the AH group involved as the hydrogen bond donor to the protein B-group receptor (see Fig. 5.9A).

The much sweeter compound 1',6'-dichlorosucrose (500 times sweeter than sucrose) has the hydrophobic chlorine atoms at both C-1' and C-6' interacting with the hydrophobic region of the sweet-taste receptor, thereby enhancing the sweetness. Sucralose, 4,1',6'-trichloro-4,1',6'-trideoxygalactosucrose, places the C-4 chlorine such that it too interacts with the hydrophobic pocket, resulting in a further increase in the sweetness to 2,000 times that of sucrose (see Fig. 5.9B). This latter hydrophobic interaction is suggested by the fact that galactosucrose, with a hydroxyl group at C-4 instead of the chlorine, is not hydrophobic and not sweet.

The substitution of chlorine at C-6 and at C-6 and C-6' in sucrose creates sucrose analogues that are not sweet. The substitution of chlorine at C-6 of sucralose forms an analogue that has its sweetness diminished by an order of magnitude. A possible explanation of these observations is that in the strong hydrophobic interaction of the C-6 chlorine with the hydrophobic pocket of the receptor, the requisite hydrogen bonds with the AH and B groups of the receptor are destroyed or greatly weakened in the case of 6-chloro-6-deoxysucrose and 6,6'-dichloro-6,6'-dideoxysucrose (see Fig. 5.9C), and significantly diminished in 4,6,1',6,'-trichloro-4,6,1',6'-tetradeoxysucrose to significantly decrease the sweetness (see Fig. 5.9D).

These hypotheses seem to better explain the enhanced sweetness of deoxy-halogen-substituted sucroses than the Hough and Khan hypothesis of two glucophores within the substituted sucrose molecule. The general concept of a sweet molecule forming a complex with the sweet-tasting protein at the receptor site to elicit a sweet taste, however, is appealing in that it follows similar kinds of protein-ligand interactions to create biological responses, such as in enzyme-

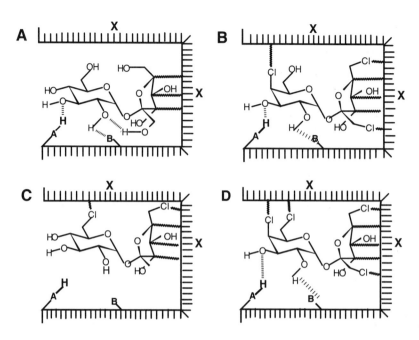

Figure 5.9. Proposed binding of sucrose and chlorodeoxy-substituted sucroses with the sweet-taste receptor. **A** Binding of sucrose, showing the AH and B hydrogen bonding and the carbon backside hydrophobic binding of the fructose moiety; **B** binding of sucralose, showing the increased hydrophobic interaction of 4,1′,6′-chloro groups with the hydrophobic receptor site; **C** binding of 6,6′-dichloro-6,6′-dideoxysucrose with the hydrophobic site and the loss of the hydrogen bonds with the AH and B groups of the receptor due to the hydrophobic binding of the 6-chloro group, pulling the molecule away from the AH and B groups; **D**, binding of 4,6,1′,6′-tetrachloro-4,6,1′,6′-tetradeoxygalacto-sucrose, showing the increased lengths of the hydrogen bonds with AH and B due to the hydrophobic interaction of the 6-chloro group to give diminution of the sweetness.

substrate complex formation, antigen-antibody complex formation, and hormone-receptor complex formation.

The Shallenberger-Kier AH-B-X sweetness hypothesis is probably too simple to explain all of the observations of sweet-tasting compounds, but the hypothesis has proved very useful in developing an understanding of the structural relationship with a sweet-taste response. The extrapolation of the concept of a single hydrophobic site to a broader area of hydrophobic interaction as proposed by Lichtenthaler and Immel greatly expands the scope of the hypothesis and gives a better insight into why certain substitutions in carbohydrates result in an intensification of sweetness and why other substitutions result in a diminution in sweetness. As more information is obtained, the hypotheses will undoubtedly be further refined to give us an even more complete understanding of the causes of sweetness and its intensification.

5.4 Literature Cited

1. R. W. Moncrieff, *The Chemical Senses,* 3rd edn., L. Hill, London (1967).
2. R. S. Shallenberger, *J. Food Sci.,* **28** (1963) 584–586.
3. R. S. Shallenberger, *Agric. Sci. Rev.,* **2** (1964) 11–20.
4. R. S. Shallenberger, T. E. Acree, and W. E. Guild, *J. Food Sci.,* **30** (1965) 560–563.
5. R. S. Shallenberger and T. E. Acree, *Nature,* **216** (1967) 480–482.
6. R. S. Shallenberger and T. E. Acree, *J. Agric. Food Chem.,* **17** (1969) 701–703.
7. M. G. Lindley and G. G. Birch, *J. Sci. Food Agric.,* **26** (1975) 117–124.
8. J. J. Blanksma and D. Hoegen, *Recl. Trav. Chim. Pays-Bas,* **65** (1946) 333–337.
9. J. Solms, *J. Agric. Food Chem.,* **17** (1969) 686–688.
10. L. B. Kier, *J. Pharm. Sci.,* **61** (1972) 1394–1397.
11. R. S. Shallenberger, T. E. Acree, and C. Y. Lee, *Nature,* **221** (1969) 555–556.
12. G. G. Birch, C. K. Lee, and M. G. Lindley, *Stärke,* **27** (1975) 51–56.
13. G. G. Birch and C. K. Lee, *J. Food Sci.,* **39** (1974) 947–949.
14. Y. Tsuzuki and J. Yamazaki, *Biochem. Z.,* **323** (1953) 525–531.
15. C. K. Lee, S. E. Mattai, and G. G. Birch, *J. Food Sci.,* **40** (1975) 390–393.
16. E. Mosettig and W. R. Nes, *J. Org. Chem.,* **20** (1955) 884–899.
17. H. B. Wood, Jr., R. Allerton, H. W. Diehl, and H. G. Fletcher, Jr., *J. Org. Chem.* **20** (1955) 875–883.
18. E. Vis and H. G. Fletcher, Jr., *J. Am. Chem. Soc.,* **78** (1956) 4709–4710.
19. H. Kohda, R. Kasai, K. Yamasaki, K. Murakami, and O. Tanaka, *Phytochemistry,* **15** (1976) 981–983.
20. H. Schultz and J. J. Pilgrim, *Food Res.,* **22** (1957) 206–213.
21. J. Jibza, L. Dolejs, V. Herout, and F. Sorm, *Tetrahedron Lett.,* (1971) 1329–1332.
22. B. Crammer and R. Ikan, *Chem. Soc. Rev.,* **6** (1977) 431–465.
23. M. N. Sela and D. Steinberg, in *Progress in Sweeteners,* pp. 71–96 (T. H. Grenby, ed.) Elsevier Applied Science, Amsterdam (1989).
24. J. M. Ballard, L. Hough, A. C. Richardson, and P. H. Fairclough, *J. Chem. Soc. Perkin Trans. I,* (1973) 1524–1528.
25. L. Hough, S. P. Phadnis, and E. Tarelli, *Carbohydr. Res.,* **44** (1975) 37–44.
26. H. Parolis, *Carbohydr. Res.,* **48** (1976) 132–135.
27. L. Hough and S. P. Phadnis, *Nature,* **263** (1976) 800–801.
28. A. K. Anisuzzaman and R. L. Whistler, *Carbohydr. Res.,* **61** (1978) 511–518.
29. P. H. Fairclough, L. Hough, and A. C. Richardson, *Carbohydr. Res.,* **40** (1975) 285–298; R. Khan and R. S. Mufti, U.K. Pat. 2,079,749 (1980).
30. L. Hough and R. Khan, *Trends Biol. Sci.,* **3** (1978) 61–63.
31. C.-K. Lee, *Carbohydr. Res.,* **162** (1978) 53–63.
32. G. Jackson, M. R. Jenner, R. A. Khan, C.-K. Lee, K. S. Mufti, G. D. Patel, and E. B. Rathbone, Br. Pat. 8,211,427 A (1982).
33. F. W. Lichtenthaler and S. Immel, *Int. Sugar J.,* **97(1153)** (1995) 13–22.

5.5 References for Further Study

"Intensification of sweetness," L. Hough and R. Khan, *Trends Biol. Sci.,* **3** (1978) 61–63.
Carbohydrate Sweeteners in Foods and Nutrition, P. Koivistoinen and L. Hyvönen, eds., Academic, New York (1980).

"The chemistry and biochemistry of the sweetness of sugars," C.-K. Lee, *Adv. Carbohydr. Chem. Biochem.,* **45** (1987) 199–352.

"Enhancement of the sweetness of sucrose by conversion into chlorodeoxy derivatives," L. Hough and R. Khan, in *Progress in Sweeteners,* pp. 97–120, (T. H. Grenby, ed.) Elsevier Applied Science, London (1989).

"Sucralose: unveiling its properties and applications," M. R. Jenner, in *Progress in Sweeteners,* pp. 121–142 (T. H. Grenby, ed.) Elsevier Applied Science, London (1989).

"Old roots—new branches: evolution of the structural representation of sucrose," F. W. Lichtenthaler, S. Immel, and U. Kreis, in *Carbohydrates as Organic Raw Materials,* pp. 1–33 (F. W. Lichtenthaler, ed.) VCH, Weinheim, Germany (1990).

"Applications of the chemistry of sucrose," L. Hough, in *Carbohydrates as Organic Raw Materials,* pp. 33–58 (F. W. Lichtenthaler, ed.) VCH, Weinheim, Germany (1990).

"The sweetness of sugar," I. Ramirez, in *Sugar: A User's Guide to Sucrose,* pp. 71–81 (N. L. Pennington and C. W. Baker, eds.) Van Nostrand Reinhold, New York (1990).

Chapter 6

Polysaccharides I

Structure and Function

6.1 Introduction: Structure and Classification of Polysaccharides

Polysaccharides, as the name implies, are composed of many monosaccharide units that are joined one to the other by an acetal linkage to give a long chain. The acetal linkage between the monosaccharide units is formed by the reaction of the hemiacetal hydroxyl group of one unit with an alcohol group of another unit, splitting out water to give a *glycosidic bond.* The bond can have either the α- or the β-configuration and can be to any one of the alcohol groups at C-2, C-3, C-4, or C-6 of, for example, D-glucose (see, Fig. 6.1). The chain can be either linear or branched. The branches can be single monosaccharide units, chains of two or more monosaccharide units of uniform number, or chains of a variable number of monosaccharide units.

Polysaccharides can be divided into two classes: (1) *homopolysaccharides* consisting of only one kind of monosaccharide and (2) *heteropolysaccharides* consisting of two or more kinds of monosaccharide units. In the latter type, the arrangement or sequence of the monosaccharide units is usually in a definite, repeating pattern, rather than random.

Homopolysaccharides can be further divided by the type(s) of glycosidic linkages joining the monosaccharide units. The linkages can be *homolinkages* with either an α- or a β-configuration to a single position (exclusive of any branch linkages). That is, the polysaccharide would have a single kind of monosaccharide linked by one type of bond $\alpha\text{-}1 \rightarrow 3$, $\alpha\text{-}1 \rightarrow 4$, and so on, or $\beta\text{-}1 \rightarrow 3$, $\beta\text{-}1 \rightarrow 4$, and so on. Homopolysaccharides can have heterolinkages with a mixture of α- and β-configurations and/or a mixture of positions. That is, homopolysaccharides with heterolinkages would have a single monosaccharide unit linked $\alpha\text{-}1 \rightarrow 4$ and $\beta\text{-}1 \rightarrow 4$, or $\alpha\text{-}1 \rightarrow 4$ and $\alpha\text{-}1 \rightarrow 6$, or $\alpha\text{-}1 \rightarrow 4$ and $\beta\text{-}1 \rightarrow 6$, and so forth. As with heteropolysaccharides, heterolinked polysaccharides usually have a definite

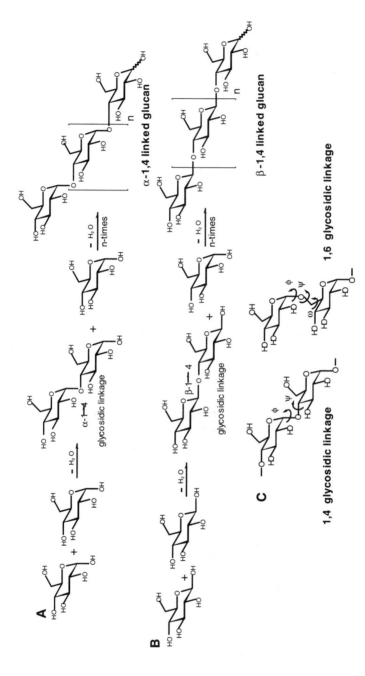

Figure 6.1. The glycosidic linkage: **A**, formation of α-1 → 4 glycosidic linkage; **B**, formation of β-1 → 4 glycosidic linkage; **C**, rotations around the 1 → 4 and 1 → 6 glycosidic linkages.

pattern for the arrangement of the linkages. Thus, a homopolysaccharide could be heterolinked with alternating α-1 $\rightarrow$ 4 and β-1 $\rightarrow$ 4 linkages, or heterolinked with alternating α-1 $\rightarrow$ 6 and α-1 $\rightarrow$ 3 linkages, or it might have a repeating sequence of two α-1 $\rightarrow$ 4 linkages and one α-1 $\rightarrow$ 6 linkage. Thus, many different combinations of α- and β-configurations and linkage positions are possible to give homopolysaccharides with a great diversity of structure.

Heteropolysaccharides can have this same kind of linkage diversity, but now the different kinds of linkages can be associated with one or more of the different kinds of monosaccharide units, giving an almost infinite degree of diversity of structure to heteropolysaccharides. Polysaccharides can, thus, not only have different sequences of monosaccharide units but they can also have different sequences of glycosidic linkages. Superimposed on this diversity of monosaccharide and linkage sequence, there are different kinds of branching, as mentioned above. This allows the possibility of a very high degree of diversity for polysaccharides and their structure-function relationships.

The size of polysaccharide molecules can vary. They most often occur as polydisperse molecules that have a range of 100 to 100,000 monosaccharide units, giving a range of molecular weights of 16,000–16,000,000 daltons (Da). There are a number of methods used to determine the molecular weight of a polysaccharide preparation. The simplest, most commonly used, and most correct molecular weight is the *number-average molecular weight* that is based on methods of counting the number of molecules in a given weight of polysaccharide. The methods that have been used for counting are reducing value measurement, osmometry, and cryoscopy. Another type of molecular weight is the *weight-average molecular weight* that is obtained using light-scattering techniques. The ratio of the number-average molecular weight to the weight-average molecular weight (M_n/M_w) is an indication of the degree of polydispersity. A monodisperse preparation, that is, a preparation in which all of the molecules have the same number of monosaccharide units and hence the same molecular weight, the ratio would be 1.0. In the instance of a preparation with an exponential or most probable distribution of molecular weights, the ratio would be 0.5. The number of monosaccharide units in a polysaccharide molecule is termed the *degree of polymerization* or d.p.

The conformation of the individual monosaccharide residues in a polysaccharide is relatively fixed. The residues joined together by the glycosidic linkage, however, can rotate around the bonds of the linkage to give different chain conformations. The glycosidic linkage to a secondary alcohol group has two bonds around which the monosaccharide residues can rotate, and the glycosidic linkage to a primary alcohol group has three bonds around which the monosaccharide residues can rotate (see Fig. 6.1C). The most commonly observed secondary and tertiary structures of polysaccharides are the single and double helical structures.

The different kinds of primary structures that result in secondary and tertiary structures give different kinds of properties such as water solubility, aggregation and crystallization, viscosity, gellation, digestibility, and biological recognition. Polysaccharides are recognized to have a variety of biological functions, some of which are (1) storage of the chemical energy obtained from the sun in the process of photosynthesis; (2) structural material for the cell walls of plants and microorganisms and the exoskeletons of insects and other arthropods; (3) protection of organisms, especially microorganisms, from changes in the environment such as changes in temperature, pH, and concentration of oxygen (polysaccharides absorb oxygen on their surfaces and prevent its passage, thereby producing an anaerobic environment); (4) adaptation and fixation of organisms to a specific environmental niche; (5) protection of organisms from invasion by other organisms and viruses, and protection against unwanted destruction by the immunological process; (6) alteration of biological environments to produce a desired condition such as the prevention of blood coagulation or the prevention of drying; (7) the action as a lubricant in the movement of muscles and joints; (8) structure of skin, cartilage, and cornea; and (9) biological recognition involved in infection and immunity, cell-cell interaction, and receptor binding and response.

There are literally hundreds of different kinds of polysaccharides found in nature, especially when we consider microbial polysaccharides. It is therefore impossible to describe the structure and function of all of the known polysaccharides. We have focused on those that are best known and studied, those that have agricultural, biological, and commercial importance, those that have interesting properties and applications, and in some cases those that have an unusual structure. By doing this, we cover a broad range of polysaccharide structure, function, and application.

6.2 Plant Polysaccharides

a. Starch

i. Occurrence and Structure

Starch is a polysaccharide composed exclusively of D-glucose and is one of the three most abundant organic compounds found on the earth, cellulose and murein being the other two abundant compounds. Starch is found in the leaves of higher green plants and in the stems, seeds, roots, and tubers of many higher plants, where it serves to store the chemical energy obtained from the light energy of the sun in the process of photosynthesis. It is also found in green unicellular algae. Starch is biosynthesized in cytoplasmic organelles called *plastids.* Starch is also found in *amyloplasts,* which are plastids that specialize in the storage of starch [1]. The starch in amyloplasts exist as water-insoluble granules. The size, shape, and morphology of the starch granules are characteristic of the particular botanical source [2]. Figure 6.2 is a composite of scanning electron micrographs of starch granules from different botanical sources, illustrating the differences in size, shape, and morphology.

The starch found in amyloplasts forms an important source of nutritional energy in the human diet. A high proportion of the world's food energy is in the form of starch that is obtained from cultivated plants (see Chapter 1).

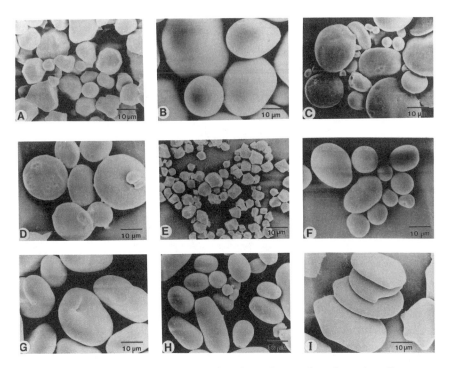

Figure 6.2. Scanning electron micrographs of starch granules: **A,** maize; **B,** potato; **C,** wheat; **D,** rye; **E,** rice; **F,** lentil bean; **G,** green pea; **H,** avocado; **I,** shoti. From ref. [2], reprinted by permission of the author and publisher.

ii. Chemical Composition of Starch in the Granules

Most starches are composed of two types of polysaccharides: *amylose,* a mixture of linear polysaccharides of D-glucose units linked α-1 → 4 to each other (see Fig. 6.3A), and *amylopectin,* a mixture of branched polysaccharides of D-glucose units linked α-1 → 4, with 5% α-1 → 6 branch linkages (see Fig. 6.3B). From the percentage of branch linkages, we can calculate that the branch chains have an average of 20 D-glucose units (1 ÷ 0.05). Both amylose and amylopectin have a distribution of sizes with different average numbers (d.p.) of glucose residues. The average number of glucose residues for amylose can vary from 250 to 5,000, and the average number of glucose residues for amylopectin can vary from 10,000 to 100,000.

The amylose component is present in the granules of many starches in amounts of 15–30% by weight, and the amylopectin component is present in corresponding amounts of 85–70% by weight. There are exceptions, such as the waxy varieties of maize, barley, rice, and sorghum that are composed entirely of amylopectin. There also are high-amylose varieties such as amylomaize-5 which is 50% amylose by weight, and amylomaize-7, which is 70% amylose by weight. Recently, National Starch and Chemical Co. (Bridgewater, NJ, USA) have obtained a maize starch variety that is 100% amylose [3].

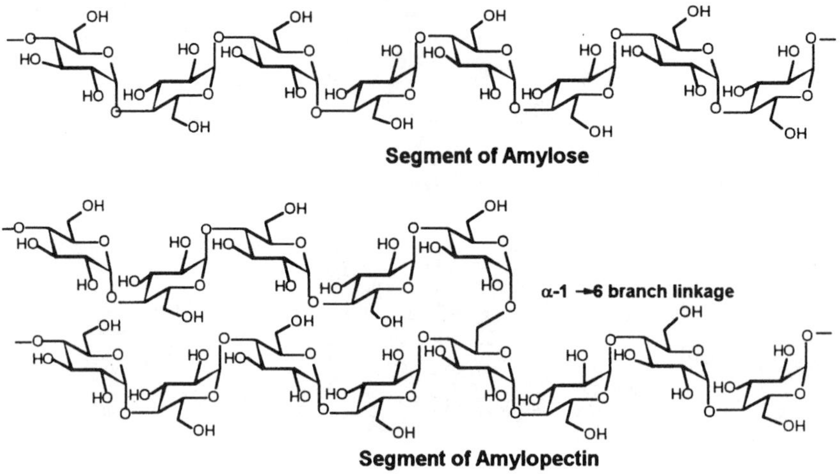

Segment of Amylose

Segment of Amylopectin

Figure 6.3. Structures of the starch components: **A,** structural formula for amylose; **B,** structural formula for a segment of amylopectin.

iii. Fractionation of Starch into its Components

Extraction of starch granules with warm water gave a water-soluble fraction that was the linear amylose component [4–6]. It turned out, however, that this water-extractable component was only a fraction of the amylose in the granule. It was discovered that the amylose and amylopectin components could be more completely separated by first solubilizing the granules by heating in boiling water, followed by the addition of a mixture of pentyl alcohols or 1-butanol to the hot starch solution. On cooling, an amylose-butanol or pentanol complex precipitated [7–9]. This was removed by centrifugation, dissolved in water and further purified by a second precipitation. The amylose can be obtained as a dry powder by treating the complex with dry acetone five or six times, followed by a final treatment with ethanol, and drying under vacuo at 40°C. The amylopectin component remains in the solution after the removal of the amylose complex and can be precipitated by the addition of two volumes of ethanol. The precipitated amylopectin can be treated with acetone and ethanol, which is similar to the treatment of the amylose-butanol complex, to produce a dry powder.

It was further shown that several other organic compounds such as nitroalkanes [10], thymol [11], *t*-butyl alcohol, and tetrachloroethane [12,13] will also form a precipitable complex with amylose.

iv. Structure and Properties of the Amylose Component

Amylose is a linear molecule containing three kinds of D-glucose residues: (1) one reducing-end residue with a free hemiacetal hydroxyl group, substituted only at C-4; (2) many D-glucose residues substituted at C-1 and C-4; and (3) one nonreducing D-glucose residue substituted only at C-1 (see Fig. 6.3A). Because of the hot water extraction experiment, amylose was erroneously considered a water-soluble material. The precipitated linear amylose molecule is essentially water in-soluble. It can be dissolved (10 g/100 mL) in 0.1 M sodium hydroxide or in di-methylsulfoxide (DMSO). When the alkaline solution is diluted with water and neutralized, or the DMSO solution is diluted with water so that the amylose con-centration is 1 mg/mL, the amylose will remain in solution indefinitely. At higher concentrations of 2, 5, or 10 mg/mL, the amylose will precipitate from solution. Starch chemists call this precipitation *retrogradation*. When amylose is in an aqueous solution, it has a random coil structure with a variable amount of single helical structure composed of six, seven, or eight glucose residues per turn of the helix [14–16]. The helices are constantly undergoing transitions between helical structure and random coil structure (see Fig. 6.4A). When amylose undergoes ret-rogradation, the molecules associate together to form double helices that further associate to give the precipitate.

When a solution of triiodide is added to the amylose solution, a deep blue color with an absorbance maximum of 645 nm is formed. The iodine-iodide forms a com-plex with the amylose by inducing a regular helix of six glucose residues per turn of the helix. The iodine-iodide is complexed in the hydrophobic interior of the helix [17] (see Fig. 6.4B). The color of the iodine-iodide complex is dependent on the number of glucose residues in the chain [18]. A chain of less than 10 glucose residues does not give a color. As the number of glucose residues increases, there is a pro-gressive change in color from red, to red-purple, purple, and blue. The absorbance maximum changes from 490 nm (d.p. 15) to 645 nm (d.p. 366 and higher) [18].

Organic molecules such as 1-butanol form a similar complex with amylose in which the 1-butanol molecules are complexed in the hydrophobic interior of the helix. These complexes have crystalline properties and produce X-ray diffraction patterns called a *V-pattern* [19] (see Fig. 6.6). Electron microscopy and electron diffraction studies have indicated that the complex is a folded helical chain with a lamellar structure [20–24]. The helical chain folds every 100 Å, giving an an-tiparallel structure in which the helices are 13 Å in diameter (see Fig. 6.4C).

v. Structure and Properties of the Amylopectin Component

Amylopectin is a much larger molecule than amylose. It may be considered to consist of several amylose chains, one linked to another by α-1 $\rightarrow$ 6 branch link-ages. Amylopectin has been determined to have 5% branch linkages. If the mole-cule contains 10,000 D-glucose residues, there would be 500 branch linkages per molecule. This would give a complex "tree" or "bush" structure (see Fig. 6.5A). Meyer and Bernfeld [25] proposed such a structure in 1940. Based on the study of the structure of branched oligosaccharides obtained from α-amylase and acid hy-drolyses, French [26] proposed a modified tree structure called a cluster or race-mose structure. A similar structure was proposed by Nikuni [27] (see Fig. 6.5B). By studying the distribution of chain lengths (number of glucose residues per chain) using an isoamylase debranching enzyme and column chromatography, Hizukuri [28] defined the cluster structure more accurately, as shown in Fig. 6.5C.

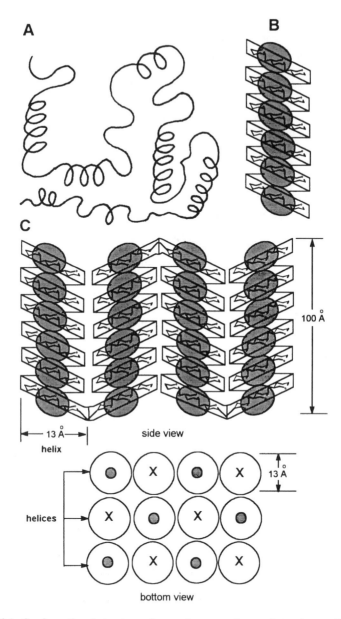

Figure 6.4. Conformational structures for amylose complexes: **A,** random coil helix of amylose in solution; **B,** amylose iodine-iodide complex; **C,** lamellar structure for organic molecule–amylose complex.

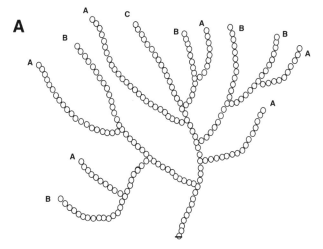

A

Meyer-Bernfeld "tree" structure for amylopectin [5]

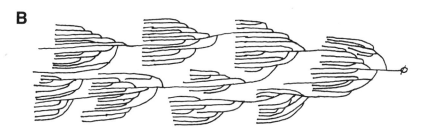

B

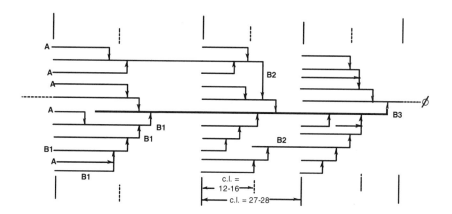

C CLUSTER (RACEMOSE) MODEL FOR AMYLOPECTIN STRUCTURE

Figure 6.5. Fine structure of amylopectin: **A,** Meyer-Bernfeld "tree" structure for amylopectin, adapted from ref. [25]; **B,** French/Nikuni cluster structure for amylopectin, from refs. [26,27] with permission from the publisher; **C,** detailed cluster structure for amylopectin, reprinted from ref. [28], with kind permission from Elsevier Science Ltd, The Boulevard, Langford Lane, Kidlington OX5 1GB, UK and the author.

Amylopectin has four kinds of D-glucose residues: (1) one reducing-end D-glucose residue; (2) many D-glucose residues substituted at C-1 and C-4; (3) 5% of the D-glucose residues substituted at C-1, C-4, and C-6 from the branched glucose residue; and (4) 5% of the D-glucose residues from the nonreducing ends of the branch chains.

The branch chains consist of three types: (1) B2-chains that are relatively long and join the individual clusters together; (2) B1-chains that have one or more chains attached to them by branch linkages; and (3) A-chains that are branch chains without any other chains attached to them. Amylopectin as a whole gives a red-wine color with iodine-iodide reagent, reflecting the relatively short average chain length (d.p. 20–25) of the individual chains in the amylopectin molecule.

vi. Structure of the Starch Granule

The common starches are readily identifiable by light microscopy by observing their size, shape, and the position of the hilum (the center of the granule) [29]. Tuber starches are generally large and ellipsoidal. Some are spherical or hemispherical. Cereal starches are smaller and pancake-shaped or polyhedral (see Fig. 6.2).

As we have previously stated, most starch granules are composed of amylose and amylopectin in a ratio of 1:4 or 1:3. Tuber starches in addition contain covalently linked phosphate; potato starch has 0.06%, and shoti starch has 0.18% covalent phosphate [30a]. The cereal starches are devoid of phosphate but contain 1–5% lipid [30b], which is believed to be complexed with the amylose component in a helical structure.

Native starch granules give distinct X-ray diffraction patterns of three types [30c,30d]: an A-pattern characteristic of cereal starches such as maize, wheat, and rice starches, a B-pattern characteristic of tuber, fruit, and stem starches such as potato, banana, and sago starches; and a C-pattern that is obtained from some starches and is intermediate between the A- and B-patterns and may be due to mixtures of A-type and B-type granules (see Fig. 6.6). X-ray diffraction indicates that the starch granules have crystalline properties. However, they are only partly crystalline and vary in their degree of crystallinity [31]. Waxy starches that are 100% amylopectin showed a crystallinity of 40%, and high-amylose starches showed a crystallinity of only 15%.

This difference in crystallinity in these different kinds of starches suggest that it is the amylopectin component that is primarily involved in the crystalline regions of the starch granule, and the amylose component is in the amorphous regions of the granule. There are indications that the α-1 $\rightarrow$ 6 branch linkages of amylopectin facilitate crystallite formation. This was suggested by the leaching of amylose from the granule with warm water, leaving behind the amylopectin with its crystallinity largely intact. This is also suggested by the relative preferential reaction of the amylose component with chemical reagents [32]. Further, acid or α-amylase digestion of starch granules produced resistant crystalline concentric shells that appeared as stacks of lamellae [33–37].

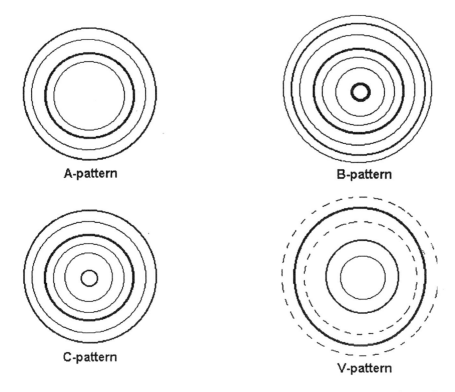

A-pattern   B-pattern

C-pattern

V-pattern

Figure 6.6. X-ray powder patterns for starch granules and amylose complexes, from ref. [30d], reprinted by permission of the author and publisher.

From a consideration of the X-ray diffraction patterns of acid- and enzyme-treated starch granules, and the construction of space-filling models, Kainuma and French [38] postulated that the chains in amylopectin formed parallel double helices (see Fig. 6.7). These double helices provided the best interpretation for the structure of the crystalline regions of the starch granule that were resistant to acid and enzyme hydrolysis. The α-1 $\rightarrow$ 6 branch linkages of amylopectin are ideal points for promoting double helices. Model structures for amylopectin showed that parallel arrays of short branches were also suited for forming double helices and crystallites [39]. Proof of the existence of both double and single helices in the crystalline regions of the starch granule was obtained by X-ray diffraction and ^{13}C-NMR studies [40]. ^{13}C-NMR of four types of starch granules indicated double helical contents of 38–53% [40].

The nonreducing ends of the starch chains have been shown to be directed toward the granule surface [41]. French postulated [26] that the starch granule consisted of crystalline regions that were interspersed with amorphous regions. He further postulated that the crystalline regions consist of double helical chains of amylopectin that are ordered by association with each other, and the amorphous

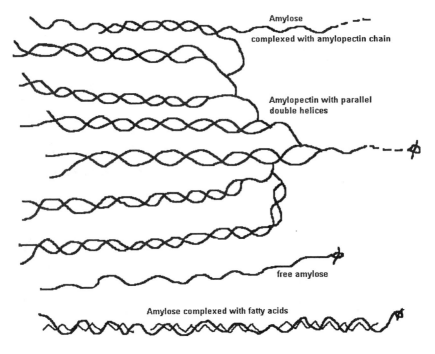

Figure 6.7. Tertiary structures of the component molecules in the starch granule.

regions consist of starch chains that are nonassociated and are hydrated to form a gel. The amorphous regions are less dense, more open, and much more susceptible to acid and enzymatic hydrolysis than the crystalline regions.

Even though starch granules appear by light microscopy to be smooth and dense, scanning electron microscopy shows that some starch granules, particularly the cereal starches, have pores [42–44]. Starch granules actually contain significant amounts of space between the crystalline regions and the starch chains so that both small molecules, such as salts, oligosaccharides, and amino acids [45,46], and large molecules, such as enzymes [35,36,42,43,47], can penetrate into the granule.

b. Cellulose

i. Source, Structure, and Properties

Cellulose has been said to be the most abundant organic compound on the Earth. It is the major structural component of the cell wall of higher plants. We have already mentioned that it occurs in a very pure form in the cotton boll. It is also a major component of flax (80%), jute (60–70%), and wood (40–50%). Pure cellulose is also formed by some bacteria such as *Acetobacter xylinum* and related species

[48a,48b]. Grasses such as papyrus and bamboo are also important sources. Cellulosic pulps can be obtained from many agricultural by-products such as sugarcane, sorghum bagasse, corn stalks, and straws of rye, wheat, oats, and rice [48c].

Celluloses from all sources are high molecular weight linear polysaccharides of D-glucopyranose units linked β-1 → 4 (see Fig. 6.8A). No evidence has been found for branching. The conformation of the β-1 → 4-linked structure gives chains that have every other glucose residue rotated 180°, providing a high propensity to form intermolecular hydrogen bonds (see Fig. 6.8B). This results in large aggregates of parallel cellulose chains that have crystalline properties and give an X-ray diffraction pattern [49].

The tertiary structure of parallel-running intermolecular hydrogen-bonded cellulose chains further associate by hydrogen bonds and van der Waals forces to produce three-dimensional microfibrils. The microfibrils give an X-ray diffraction pattern that indicates a regular, repeating crystalline structure interspersed by less-ordered paracrystalline regions. A model for the microfibrils was proposed by Hess et al. [50] and is presented in Fig. 6.8C.

The β-1 → 4 glycosidic linkage produces a structure that has low water solubility. Even a small cellodextrin chain with only six glucose residues, cellohexaose, has a much lower (<0.005 times) water solubility than its α-1 → 4-linked maltodextrin counterpart, maltohexaose. The further secondary and tertiary structural effect of cellulose chains to associate and form fibers makes it very water insoluble and impermeable to water. This highly associated, water-insoluble aggregiate provides a strong matrix for the cell walls of plants, keeping the cellular components in and the environmental components out.

ii. Isolation of Cellulose

Cellulose can be isolated from raw cotton by cutting the cotton linters into short lengths and extracting them in a Soxhlet with chloroform for 18 hr, followed by treatment with 95% ethanol for another 18 hr. This removes the waxes. The fiber is then boiled for 8 hr with 1% (w/v) sodium hydroxide (2 g cotton/100 mL) in a nitrogen atmosphere. This removes the small amount of pectin. The cotton is then washed with water until neutral, and then sequentially treated with acetone, ethanol, and ether, followed by drying under vacuo at 20°C [51]. The resulting product is called *alpha cellulose*.

Wood is very often thought of as a source of cellulose. It is the starting material for the manufacture of most paper. Wood actually is a complex material composed of cellulose (40–50%), hemicellulose (25–60%, see the following section), pectin (1–4%), and lignin (20–25%). The isolation of pure, undegraded cellulose from wood is difficult, requiring several steps. The wood from coarse sawdust or wood chips is made into a 40- to 60-mesh powder. For coniferous woods that contain resins, the powder is extracted in a Soxhlet with a solution of benzene and ethanol (2:1, v/v) for 18 hr, followed by treatment with 95% ethanol for 8 hr and ether for 4 hr. Taking about 50 g, the air-dried material is suspended in 800 mL of

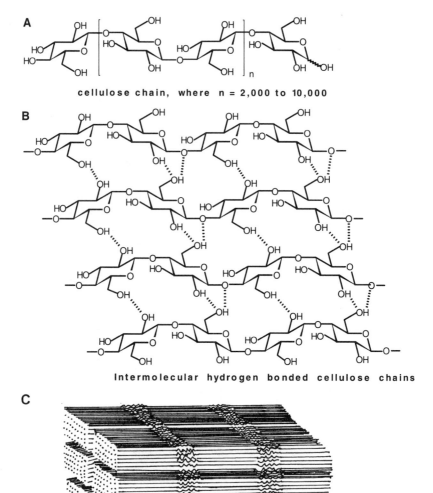

A

cellulose chain, where n = 2,000 to 10,000

B

Intermolecular hydrogen bonded cellulose chains

C

crystalline   paracrystalline

Cellulose microfibrils

Figure 6.8. Structure of cellulose: **A,** Structural formula for cellulose; **B,** intermolecular hydrogen bonded cellulose chains; **C,** structure of cellulose microfibrils, from ref. [50] and reprinted by permission of the authors and publisher.

hot water containing 3 mL of glacial acetic acid and 7.5 g of sodium chlorite. The flask is covered with an inverted beaker and heated to 70°C on a steam bath for 1 hr; a second addition of acetic acid and sodium chlorite is made, and the mixture heated for another hour. This last step is repeated for a third time. The solid material is usually white and is filtered and washed with water and air dried. This product is called *holocellulose*. Cellulose is prepared from holocellulose by placing 25 g into 850 mL of boiling water under a nitrogen atmosphere for 1 hr, followed by filtering; 500 mL of cold 4% sodium hydroxide is added for 4 hr; 500 mL of 18% sodium hydroxide is added, and the temperature is brought to 95°C for 4 hr, followed by the filtration of the solid; the solid is dispersed in 500 mL of 24% potassium hydroxide and allowed to stand for 2 hr at 20°C. The resulting solid is washed with 4% acetic acid, followed by several washings with water until the washings are neutral. Up to the acetic acid wash, all of the treatments are done under an atmosphere of nitrogen to prevent oxidation [52]. The chlorous acid treatment removes the lignin, and the alkaline treatments remove the hemicelluloses.

Cellulose is insoluble in most solvents. It can be dissolved in solutions of zinc chloride [53] or by treating with Cadoxen, a solution of 5% (w/v) of cadmium oxide in 28% (v/v) of ethylenediamine in water [54–56a]. It can be dissolved in Hakamori Reagent for methylation analysis [56b].

There are several polysaccharides that have structures similar to cellulose, with β-1 → 4 linked monosaccharide units. They are the hemicelluloses, chitin, murein, algin, and xanthan, and related bacterial polysaccharides (see sections to follow).

c. Hemicelluloses

As discussed in the previous section, cell walls of higher plants consist of cellulose, hemicelluloses, and lignin. The three substances are physically entangled and held together by secondary forces such as hydrogen bonding and van der Waals forces. In separating the components of the plant cell wall, the noncarbohydrate component, lignin, is separated first by treating the plant material with chlorous acid at 70–75°C. The resulting material is known as holocellulose. The hemicelluloses are then extracted from the holocellulose with alkaline solutions of 2–18% (w/v) sodium hydroxide.

The hemicelluloses are a heterogeneous group of polysaccharides that vary from plant to plant and from one plant part to another. There are four basic types of hemicellulose polysaccharides: D-xyloglucans, composed of D-xylopyranose attached to a cellulose chain; D-xylans, composed of D-xylose; D-mannans, composed of D-mannose; and D-galactans, composed of D-galactose. These polysaccharides are similar to cellulose in having their main chains linked β-1 → 4. Most of the hemicelluloses are, however, heteropolysaccharides with one to three monosaccharides units linked to the main monosaccharide chains.

D-Xyloglucan is an extensively studied hemicellulose [57,58]. The structure consists of β-1 → 4-linked D-glucopyranose residues with single D-xylopyranose residues linked α-1 → 6 to the glucan chain [59,60] (see Fig. 6.9A). There are some xyloglucans from rape, nasturtium, and *Tamarindus indica* seeds that also have D-galactopyranosyl units linked β-1 → 2 to the D-xylopyranosyl residues [59,61,62]. Some D-xyloglucans also contain L-fucopyranose units linked β-1 → 2 to the D-galactopyranose units [70].

D-Xylan is composed of D-xylopyranose linked β-1 → 4 with various kinds of branch units [63]. The most common branch chain is 4-*O*-methyl-D-glucopyranosyl uronic acid units linked α-1 → 2 [64–66] (see Fig. 6.9B). Substitution of the D-glycopyranosyl uronic acid units α-1 → 3 onto the xylan main chain has also been observed [63]. Another branch side chain that is commonly found in D-xylans is L-arabinofuranosyl units linked α-1 → 3 [63] (see Fig. 6.9B). The uronic acids and the L-arabinose units can occur on the same D-xylan chain or on separate chains. In addition, the D-xylopyranose units are often *O*-acetylated to the extent of 3–17% [67].

The main chain of D-xylan is very much like that of cellulose except that each sugar residue lacks a primary alcohol group. The absence of the primary alcohol group reduces the formation of intermolecular hydrogen bonds and the formation of microfibrils. Further, the substitution of uronic acids onto the xylan chain makes them acidic polysaccharides and much more water soluble than cellulose chains.

Hemicellulose composed exclusively of D-mannose is found in palm seed endosperm [68]. In particular, D-mannan is found in tagua palm seed, where it is called "vegetable ivory" or "ivory nut mannan". It is β-1 → 4 linked and forms microfibrils similar to cellulose [65]. Another type of D-mannan is D-galactomannan which has D-galactopyranose units linked α-1 → 6 to the β-D-mannan chain [69,70] (see Fig. 6.9C). This D-galactomannan has been obtained from seed pods of the locust bean and has been used in food products. In particular, a D-galactomannan, known as *guar gum,* is obtained from the seeds of the legume *Cyamopsin tetragonolobus* which is an annual plant grown primarily in the arid and semiarid regions of India [71]. The D-galactopyranose unit is linked, on the average, to every second D-mannose unit in a random manner in groups of ones, twos, and threes [72]. The polysaccharide is water soluble, with a high molecular weight of $1–2 \times 10^6$ Da [73]. A similar polysaccharide is known as *tara gum.* The differences between locust bean, guar, and tara gums are in the increasing ratios of D-galactopyranosyl to D-mannopyranosyl units. All three of the gums have relatively high water solubility and form highly viscous solutions. They are used as food thickeners and stabilizers in dessert gels, processed seafoods, ice creams, and frozen yogurts [71].

The major D-galactan hemicellulose is an arabinogalactan obtained from softwoods such as larch, black spruce, pine, Douglas fir, cedar, and juniper [72]. Western larch is high in arabinogalactan. The larch arabinogalactans have main chains of D-galactopyranosyl units linked β-1 → 3 with branch chains of

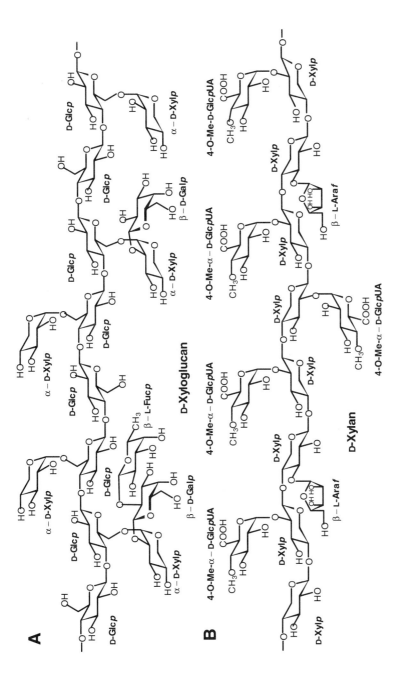

Figure 6.9. Structures of the major hemicelluloses: **A**, D-xyloglucan; **B**, D-xylan.

(continued)

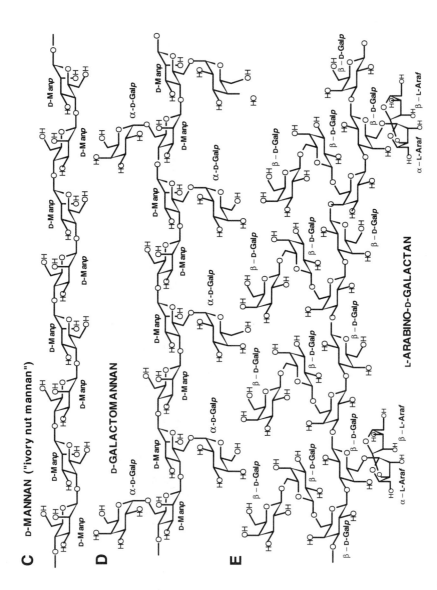

C D-MANNAN ("ivory nut mannan")

D D-GALACTOMANNAN

E L-ARABINO-D-GALACTAN

Figure 6.9. *(continued)* **C**, D-mannan; **D**, D-galactan; **E**, L-arabino-D-galactan.

β-1 $\rightarrow$ 6-linked disaccharides of β-D-galactopyranosyl-(1 $\rightarrow$ 6)-D-galactopyranose and, to a lesser degree, an α-1 $\rightarrow$ 6 linked β-L-arabinofuranosyl (1 $\rightarrow$ 3)-L-arabinofuranose disaccharide [74] (see Fig. 6.9D).

d. Pectins

Pectins are polysaccharides that occur in all plants, primarily in the cell wall, in low amounts of 1–5%. They are, however, particularly prevalent in fruits, where the amounts are much higher. For example, apple pulp contains 10–15% (w/w) pectin, and orange and lemon rinds contain 20–30% (w/w). Pectins act as an intercellular cementing material that gives body to fruits and helps them keep their shape. When fruit becomes overripe, the pectin is broken down into its constituent monosaccharide sugars. As a result, the fruit becomes soft and loses it firmness.

Commercial pectins are isolated almost exclusively from either citrus peels or apple pulp that is a by-product of the preparation of juices for human consumption. The waste material from the production of sugar (sucrose) from sugar beets also contains substantial amounts of pectin, although it is not of the same quality as citrus- or apple-derived pectins for gel formation due to the relatively high degree of O-acetylation of the monosaccharide units and a relatively low molecular weight [75]. Pectins are used commercially in foods as thickeners, emulsifiers, and gels and are the gelling agent in the production of jellies.

Pectin is isolated from citrus peel or apple pulp by extraction with an acidic aqueous solution of pH 1–3 at 50–90°C for 1–2 hr. The resulting suspension is filtered and the filtrate is concentrated. The pectin is precipitated from the concentrate by the addition of two volumes of 2-propanol (isopropyl alcohol) or ethanol.

All pectins are polysaccharides composed of D-galactopyranosyl uronic acid units linked α-1 $\rightarrow$ 4. A relatively large number of the carboxyl groups of the uronic acids exist as methyl esters. Some pectins also have 2-O-acetyl or 3-O-acetyl groups on the D-galactopyranosyl uronic acid units. A typical average molecular size is 100,000 Da. There are also a small number (1 in 25) of α-L-rhamnopyranosyl units attached to the C-2 position of the D-galactopyranosyl uronic acid units. See Fig. 6.10 for the structure of a pectin segment. Pectins are reported with a range of average molecular weights of 20,000–400,000.

Pectins are divided into two categories: (1) *pectinic acids* which are polygalacturonic acids that are partly esterified with methanol, and (2) *pectic acids* which are polygalacturonic acids with no or only a negligible amount of methyl ester. The pectinic acids can be subdivided into two types: *LM-pectins* (low methyl pectins) are those that contain less than 50% methyl esters, and *HM-pectins* (high methyl pectins) are those that contain greater than 50% methyl esters.

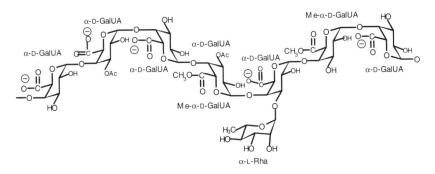

Figure 6.10. Structure of a segment of pectin.

One of the most prominent characteristics of pectins is their ability to form gels. Various physical factors, such as temperature, pH, concentration of the pectin, concentration of solutes (soluble solids), and concentration of divalent cations affect the formation and strength of pectin gels. In addition, various structural features of the pectins themselves affect the formation of gels, such as the molecular weight, degree of esterification, degree of amidation (usually produced by treating the pectin with ammonia), presence of acetyl groups in the galacturonan chain, and presence of α-L-rhamnose units. The structure of the pectin and its physical properties are interdependent in that a structural feature can influence differences in physical properties.

Pectins will form gels at low concentrations (0.3–0.7% w/v). The gels are formed by first heating solutions of the pectins and allowing them to cool. The temperatures at which the gels form is the gellation temperature. LM-pectins gel almost instantaneously, whereas HM-pectins gel after a time lag. Once formed, the HM-pectins cannot be melted, but the LM-pectins are thermoreversible and can be melted and reformed repeatedly.

The rate of gelation is directly proportional to the degree of esterification of the carboxyl group [75]. Rapid-gelling pectins have 70–75% methyl ester; medium-gelling pectins have 65–70% methyl ester; and slow-gelling pectins have 55–65% methyl ester. Relatively low levels of O-acetylation inhibit gelling. The inhibition becomes absolute when one out of every eight D-galactouronic acid units is acetylated [76]. Gels made from pectins with high molecular weights form stronger gels than gels made with pectins with lower molecular weights [77].

The formation of pectin gels is directly dependent on the concentration of other soluble solids (e.g., sugars). HM-pectins will gel only when the concentration of soluble solids is higher than 55% w/v. Increasing the soluble solids also increases the gelation temperature and the strength of the gel. LM-pectins, on the other hand, will gel when the concentration of the soluble solids is zero, if there is present about 3% w/v of a divalent cation such as Ca^{+2} [78]. The concentration of pectin required for the formation of gels is inversely proportional to the concentration of the soluble solids. At fixed solute concentrations, an increase in the concentration of pectin produces stronger gels.

The pH of the pectin solution is important. A pH of 3 is typical for the formation of pectin gels. HM-pectins generally will not form gels above pH 3.5; LM-pectins will not form gels above pH 6.5. A decrease in the pH from 3.0 to 2.0 increases the strength of both HM- and LM-pectin gels [75].

To form a gel, the pectin molecule must have several areas along its linear structure that will form complexes with several other areas along the structure of other pectin molecules. These areas are called *junction zones* and must be of a limited size so as to form a gel and not form a precipitate. The substitution of α-L-rhamnopyranosyl units at the 2-*O*-position of the D-galacturonan chain produces discontinuities in the chain and thereby limits the size of the junction zones [79]. X-ray diffraction studies have shown that the sodium salt of pectin exists as a tight helix, with three D-galacturonic acid residues per turn of the helix [80]. It was further shown that calcium pectate, pectic acid, and pectinic acid all had this same helical structure, but that the helices were arranged relative to each other in different crystalline arrays [81]. It was suggested that pectinic acid helices packed as parallel chains in which there was a columnar stacking of the methyl ester groups, providing cylindrical hydrophobic areas that were parallel to the helix axis. This produces an aggregation of the chains that were stabilized by hydrophobic bonding of the methyl groups. The structure is also stabilized by hydrogen bonding between hydroxyl groups of residues on different chains [82]. The calcium pectate has calcium ions complexed with carboxylate and hydroxyl groups from two different helical chains running in antiparallel directions [81], giving a cross-linked, three-dimensional gel matrix. An alternative model is the so-called egg box model that has been put forward for the formation of calcium alginate gels (see section g.i. in this chapter).

e. Exudate Gums

Exudate gums are polysaccharides produced by plants to seal wounds in their bark. The plant exudes an aqueous solution of polysaccharide around a break in the bark, covering the injury. The water evaporates, leaving a polysaccharide that hardens and prevents infection and desiccation of the plant.

One such material is *gum arabic* produced by *Acacia* trees that grow primarily in arid regions of the African continent. Gum arabic has been used by humans in various applications for at least 5,000 years. Today, the majority of the commercial production is from the Republics of Sudan, Nigeria, and Senegal. The thorny trees grow to a maximum height of 7–8 meters. Besides being used for the production of the gum, the trees are used to prevent encroachment of the desert. The gum is produced by cutting a 5-cm by 60–cm section of the bark from the tree during the dry season. After 1–2 months, the gum dries into a transparent mass that can be colorless, yellow, pale pink, or amber in color. The masses are harvested, dissolved in water, filtered to remove nonsoluble debris, and spray dried to a fine powder that is used in foods and pharmaceuticals.

The structure of gum arabic is relatively complex, with a main chain of D-galactopyranosyl units linked β-1 → 3 and β-1 → 6, along with D-glucopyra-nosyl uronic acid units linked β-1 → 6. The nonreducing ends are terminated with α-L-rhamnopyranosyl units. The main chains carry three types of branch trisac-charides that have α-L-rhamnopyranose, β-D-glucuronic acid, β-D-galactopyra-nose, and α-L-arabinofuranose units with 1 → 3, 1 → 4, and 1 → 6 glycosidic linkages [83].

Gum arabic has a high water solubility, up to 50% (w/v), but has a relatively low viscosity. Concentrated solutions have the consistency of a gel.

Another of the commercial exudate gums is *gum ghatti* or *Indian gum.* It is ob-tained from a large tree grown in the deciduous forests of India and Sri Lanka. Gum ghatti also has a complex structure containing L-arabinose, D-galactose, D-man-nose, D-glucuronic acid, and D-xylose units in the molar ratio of 10:6:2:2:1 [84].

A third exudate gum that is used commercially is *gum tragacanth,* which was also known and used in the third century B.C.E. It is obtained primarily from trees grown in Iran, Syria, and Turkey. It is a water-soluble gum with a complex, highly branched, high molecular weight arabinogalactan structure with D-xylopyranose and α-L-fucopyranose branch units [85]. The gum produces solutions of high vis-cosities at low concentrations and forms gels at concentrations of 2–4% (w/v). It is used as an emulsifying agent.

All three of these gums have applications in confectionery, cosmetics, textiles, paper coatings, water paints, pastes and polishes, ice creams, and salad dressings. They are used in jellies, fruit gums, gum drops, and cough drops. The high water solubility and relatively low viscosity of gum arabic is useful in bakery products, where it stabilizes and emulsifies flavors at high temperatures. Gum arabic has also been used to stabilize beer foams.

The stabilizing properties of these are useful in cosmetics, where they increase the viscosity, assisting in the application of lotions and creams to the skin, giving a protective coating and a smooth feeling. The gums are used as binding agents to give "body" to cosmetics. Gum ghatti is used to stabilize oily substances and to emulsify waxes to form liquid and paste waxes.

Gum tragacanth is used as a suspending agent for many water-insoluble prod-ucts, where it prevents the settling out of these materials. It is used as a base for jelly lubricants, tooth pastes, medicinal oil emulsions of steroids and fat-soluble vitamins, and insect repellents.

The high viscosity and acid stability of gum tragacanth is useful in foods in sauces, spices, and condiments containing vinegar as an ingredient. It also has been used in ice cream, where it prevents crystallization of lactose at low temper-atures and gives a smooth texture.

Because of the rising costs of collecting the gums by hand, and the relatively low yields per tree, the exudate gums are being replaced by polysaccharides ob-tained by chemically modifying starch, cellulose, and alginate, and by the use of xanthan obtained from a bacterial fermentation.

f. Fructans

There are essentially two kinds of fructans, inulin and levan. Both are found in plants and are produced extracellularly by certain species of bacteria.

Inulin is a polysaccharide containing D-fructofuranosyl units linked β-2 → 1 (see Fig. 6.11A) [86]. Inulins are found in the roots and tubers of the family of plants known as *Compositae,* including aster, dandelions, dahlias, cosmos, burdock, goldenrod, chicory, lettuce, and Jerusalem artichokes. Other sources are from the *Liliacae* family, including lily bulbs, onion, hyacinth, choursysanthemum, and tulips. Inulins are also produced by certain species of algae [87]. Inulins serve as a reserve polysaccharide in the plant, replacing starch or in addition to starch. Inulins occur as relatively low molecular weight polysaccharides with only 20–30 D-fructofuranose units and have molecular weights of 3,000–5,000 Da. They have D-glucopyranosyl units linked to the C-2 position of the terminal reducing-end D-fructofuranose unit, giving a nonreducing sucrose unit at the end of the chain [86].

Levans are polysaccharides containing D-fructofuranosyl units linked β-2 → 6 (see Fig. 6.11B). They are found primarily in grasses [87]. Levans have β-2 → 1 branch linkages of one to four D-fructofuranose units linked onto the main chains. They are of higher molecular weight than the inulins, having 100–200 D-fructofuranose units. None of the fructans have the high molecular weights found in starch, cellulose, and the bacterial polysaccharides. Further, the furanosyl structure is considerably more acid labile than the pyranosyl structures found in most other polysaccharides.

It has been suggested that a rich source of inulin, such as Jerusalem artichokes, might be a source of D-fructose for use as a sweetening agent. However,

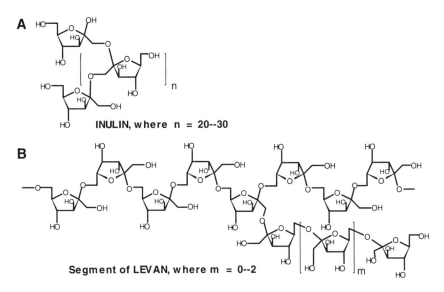

Figure 6.11. Structures of fructans: **A,** structure of inulin; **B,** structure of levan.

the commercial production of high-fructose corn syrups by the enzymatic conversion of glucose derived from corn starch has proved to be a more feasible source of fructose.

g. Seaweed Polysaccharides

i. Algin (Alginic Acid)

Brown seaweeds contain the polysaccharide algin or alginic acid in amounts of 18–40% (w/v) of the plant material. The majority of the alginic acid is located between the cells, where it comprises 80–85% of the total, with the remaining in the cell wall [88]. In the native state, the polysaccharide exists primarily as the Ca^{+2} salt complex [89]. One of the major species of seaweed that contains alginic acid is the giant kelp, *Macrocystis pyrifera,* that grows to lengths of 2,500 m and contains 20–30% algin. It grows along the California coast of the United States, the northwestern and southwestern coasts of South America, and the southeastern coasts of Australia and New Zealand [90]. *M. pyrifera* grows in relatively calm waters in large, dense beds. The plant is a perennial that can essentially be harvested on a continuing basis. Its rapid growth permits up to four cuttings per year. At the time of harvesting, mature beds consisting of a dense mat of fronds float on the surface. Cutting and harvesting of this mat allows light to penetrate into the water and reach the immature fronds, thus stimulating their growth. Blades under the boat cut the kelp about 1 m below the surface, and the cut kelp is conveyed into a hold [91].

Another brown seaweed with several species that produce algin is from the genus *Laminaria. L. hyperborea, L. digitata,* and *L. japonica* grow along the U.S. north Atlantic coast, the Canadian coasts of Nova Scotia and New Foundland, the eastern coast of Japan, the western coast of Ireland, the eastern coast of England, the western coasts of Norway and Europe, and the northwestern coasts of Africa [90]. Algins are also synthesized by two kinds of bacteria, *Pseudomonas aeruginosa* [92] and *Azotobacter vinelandii* [93].

The polysaccharide can exist in a number of chemical forms. It is composed of uronic acids. If the acid groups are in the acid form (-COOH), the polysaccharide is called *alginic acid,* which is water insoluble. If the acid groups are in the carboxylate form (-COO⁻) it is as the alginate or sodium salt (-COONa) or sodium alginate, which is water soluble. If the sodium ions are replaced by a divalent metal ion such as calcium, barium, or strontium, the polysaccharide is crosslinked by the metal ions to form gels.

In the isolation of algin, the fronds or stipes of seaweed are washed several times with water to remove impurities. They are then immersed in 0.5% hydrochloric acid for 24 hr. The acid is decanted, and the process is repeated two or more times. The fronds are then washed with water until the acid is removed. Then they are covered with a 2% (w/v) solution of sodium carbonate for 15–18 hr [94]. On the addition of the alkali, the fronds swell and loose their shape. The mixture is frequently stirred and becomes viscous. The solution is diluted so that it can be easily filtered through coarse filter paper. The filtrate is warmed to 60°C and filtered through a thin layer of activated charcoal. After cooling, the alginic acid is precipitated by the addition of hydrochloric acid [94] or by the addition of two volumes of ethanol [95]. The precipitate is then treated with 95% ethanol for 12–18 hr, filtered, and air dried. The insoluble alginic acid can be converted to the soluble sodium alginate by treatment with 5% sodium carbonate solution to give a solution of sodium alginate, which is spray dried and obtained as a powder.

Alginic acid or sodium alginate is a linear polysaccharide composed of two monosaccharide units, β-D-mannopyranosyl uronic acid and α-L-gulopyranosyl uronic acid, each linked $1 \rightarrow 4$ [96] (see Fig. 6.12A). The amounts of D-mannuronic acid and L-guluronic acid are in the ratio of 2:1 for most algins, with the exception of the algin from *L. hyperborea* in which the ratio is 1:2.3 [97]. The ratios can vary with the algal species, the type of tissue, and the age of the plant. *Ascophyllum nodosum* receptacles contain only D-mannuronic acid residues. The bacterial algins are also rich in D-mannuronic acid, but the ratio can be made to vary from 0.4 to 2.4, depending on the species of bacterium and the culturing conditions [98].

Algin is a copolymer containing blocks of the two types of residues [99] and segments that alternate in a fairly regular manner [100] (see Fig. 6.12A). X-ray diffraction studies have indicated that the D-mannuronic acid units are in the C1 ($^4C_1$) conformation, and the L-guluronic acid residues are in the 1C ($_4C^1$) conformation [101]. The combination of these conformational features is unusual and gives unique properties to the alginates.

One of the most unusual properties that the alginates have is their ability to instantly form gels when they come into contact with solutions of divalent metal ions such as calcium, barium, strontium, cadmium, copper, cobalt, lead, nickel, or zinc. Calcium is the most frequently used cation to form gels [102].

The affinity for divalent cations and the formation of gels are related mainly to the amount of L-guluronic acid present in the alginate [103]. There is evidence that two axially linked L-guluronic acid residues form binding sites for the calcium ions, and that sequences of L-guluronic acid between two or more alginate molecules creates a gel network of calcium cross-linking the alginate molecules [104]. This structure has been termed an egg box because the calcium ions are similar to eggs fitting into positions in a box made by the alginate chains, as shown in Fig. 6.12B. The chains making up the box can be formed in three dimensions, resulting in the so-called nesting of the boxes. The complex with the calcium ion is formed by two carboxylate groups and two C-2 hydroxyl groups from four L-guluronic acid residues [105,106], as shown in Fig. 6.12B.

The calcium alginate gels contain 2–3% alginate and 97% water. The biological role that calcium alginate may play in seaweeds is as a protective agent against desiccation of the plant cells at low tide. Because of its surface activity and relatively high viscosity at low concentrations, sodium alginate can serve as an excellent stabilizer of emulsions and suspensions and can act as a bulking agent in food preparations. This, coupled with its lack of toxicity, have made alginate useful in many commercial products such as ice cream, cheese, salad dressings, bar-

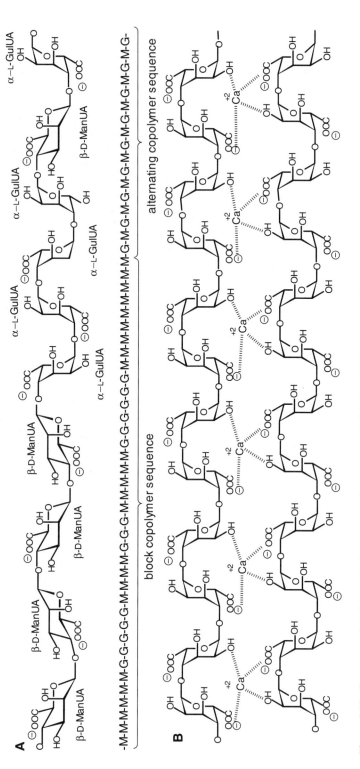

Figure 6.12. Structures of alginic acid: **A**, structure of a segment of alginic acid; **B**, structure of the calcium alginate gel by the complexing of calcium ions with α-ʟ-guluronate blocks—the so-called egg box structure.

becue sauces, frozen foods, fruit pie fillings, syrups, chocolate milk, pharmaceuticals, water paints, films, and paper coatings [91].

Calcium alginate also has a number of important uses. The "pimento" in stuffed olives is mostly calcium alginate with some pimento flavoring and red dye. It has been found that real pimento is much too unstable and readily decomposes. Calcium alginate can be spun into fibers that can be woven into cotton or wool cloth. Treatment of the cloth with sodium carbonate solution removes the calcium alginate, leaving behind a very light-weight cloth. Dry calcium alginate can be pressed into tablets with drugs. When the tablet becomes wet, the alginate swells, dispersing the drug and hastening its dissolution. Calcium alginate is used in some cases as a wound dressing. It reduces the trauma of wound healing and can shorten the healing time. Calcium alginate gels also find uses in the forming of molds and impressions that are used in forensic science, art, dentistry, and medicine.

One of the most important uses of calcium alginate gels is in the entrapment and encapsulation that they provide for a variety of substances such as whole cells [107], enzymes [108a], and hormones [109]. Such materials can be used in many biotechnological and biomedical applications [110].

The relative ease of their preparation makes calcium alginate gels particularly useful in these applications. A 2–5% (w/v) solution of sodium alginate is prepared. Cells or a concentrated solution of enzyme are added to the alginate solution. The sodium alginate-cell suspension or enzyme solution is then added dropwise from a pipette or syringe with a diameter of 0.22–1.0 mm to a solution of 20–100 mM calcium chloride. The alginate solution, on coming into contact with the calcium ions, instantly forms beads, entrapping the cells or enzymes. Various hardening agents, such as sodium alginate itself [108b], ethyleneimine [108c], or chitosan [108d], can be added to the calcium alginate beads. Alternative procedures for forming gels involve the forcing of the sodium alginate solutions through a series of pores into a calcium chloride solution or through a slit into a calcium chloride solution to make threads, sheets, and membranes. Films can very easily be cast by spreading the sodium alginate solutions onto glass plates, with or without a glycerol plasticizer, and allowing the alginate to dry. The dried films can be treated with a calcium solution to make the calcium alginate films.

ii. Carrageenans

Carrageenans constitute the primary structural polysaccharides of red seaweeds of the Rodophyta of which *Eucheuma cottonii, Eucheuma spinosum, Chondrus crispus,* and *Iridaea laminarioides* are prominent forms. These seaweeds attach themselves to rocks and are found as intertidal growths along the north Atlantic coasts of Europe and North America and the western Pacific coasts of Korea and Japan [111]. There have been some efforts to "farm" these seaweeds by culturing the algae in tanks that are continuously fed nutrients. The carrageenans are isolated from washed, freshly harvested alga. The alga are dried, milled, and extracted with acetone in a Soxhlet. The material is suspended in 75°C water for 20 min, and then sodium chlorite is added to 4% (w/v) with a sufficient amount of hydrochloric acid to make the pH 3.5, to produce chlorous acid. After 60 min, the mixture is neutralized by the addition of sodium acetate and allowed to stand about 1 hr; the solution is heated to 60°C and centrifuged while hot. The clear centrifugate is dialyzed for 24 hr against 1% (w/v) sodium acetate; 20% (w/v) potassium chloride is slowly added to the dialyzate, with vigorous stirring, to form a gelatinous precipitate, which is collected by centrifugation. The precipitate is dissolved in 1% sodium acetate, dialyzed against sodium acetate solution, and precipitated with potassium chloride. The process of dissolving, dialyzing, and precipitating is repeated two more times, with the last dialysis against distilled water. The carrageenans are precipitated from the dialyzate by the addition of 3 volumes of ethanol or 2-propanol, followed by vacuo drying at 40°C [112].

The carrageenans are a family of polysaccharides composed of three linear D-galactans. Two of them, κ-carrageenan and ι-carrageenan are composed of alternating sequences of β-1 → 4 D-galactopyranose and α-1 → 3 3,6-anhydro-D-galactopyranose. They differ from each other by the number and position of O-sulfate groups. κ-carrageenan has the sulfate group at C-4 of the D-galactopyranose unit, and ι-carrageenan has a sulfate group at C-4 of the D-galactopyranose unit and at C-2 on the 3,6-anhydro-D-galactopyranose unit. The D-galactopyranose unit is in the C1 ($^4C_1$) conformation, and the 3,6-anhydro-D-galactopyranose is in the 1C ($_4C^1$) conformation [113]. See Fig. 6.13 for the structures of κ- and ι-carrageenan. The two types of conformations give a high degree of flexibility to the polysaccharide chain, as all of the glycosidic bonds are equatorial [113]. The increased flexibility permits the polysaccharide chain to assume an ordered secondary helical structure.

κ- and ι-Carrageenan can exist as parallel, double helical tertiary structures, both as solids and in solution [114]. The sulfate groups are located on the outside of the double helix, with their negative charges countered best by potassium ions. Sodium and lithium ions inhibit the formation of the double helix. The formation of double helices by potassium ion induces the formation of gels by forming intermolecular aggregates between double helices [115,116].

The third carrageenan, λ-carrageenan, has a structure similar to κ- and ι-carrageenan with the exceptions that the 3,6-anhydro-D-galactopyranose unit is replaced by 2,6-di-O-sulfato-α-D-galactopyranose, and the D-galactopyranose unit is sulfated at C-2 (see Fig. 6.13). In contrast to the other two carrageenans, both of the D-galactopyranose residues in the repeating disaccharide unit are in the C1 ($^4C_1$) conformation. λ-Carrageenan has the highest negative charge, having three sulfate groups per repeating disaccharide unit in contrast with κ-carrageenan, which has one sulfate group per repeating disaccharide unit, and ι-carrageenan, which has two sulfate groups per repeating disaccharide unit (see Fig. 6.13).

The strength of the gels is dependent on the molecular size of the chains, the concentration of the carrageenan, and the concentration of potassium ion [111]. κ- and ι-Carrageenan will form gels with both K^+ and Ca^{2+}. κ-Carrageenan forms the strongest gels [111] and ι-carrageenan forms the best gels with Ca^{+2}, although they are soft gels [111].

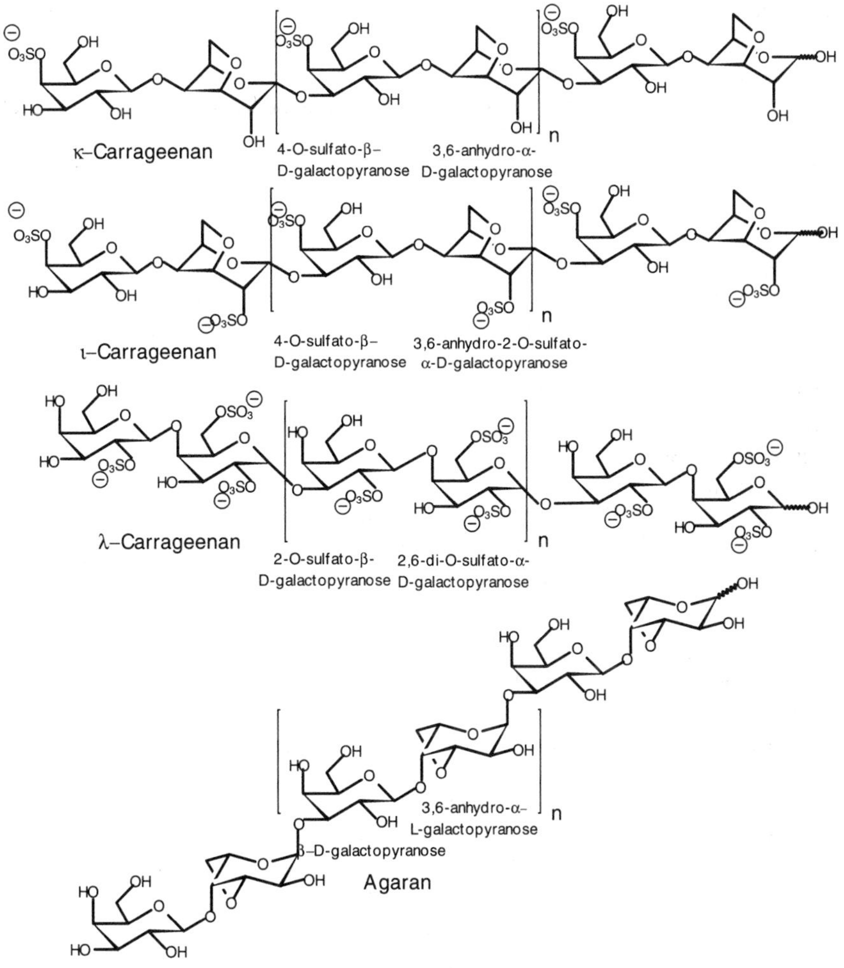

Figure 6.13. Structures of carrageenans and argaran.

iii. Agar

Agar is a mixture of two polysaccharides, agaran and agaropectin [117]. They are obtained from the red-purple seaweeds that occur in the class of Rhodophyceae. The principal seaweeds that produce agar are from the species *Gelidium* that grow in the waters along the coasts of Japan, Mexico, Chile, South Africa, Portugal, Spain, and New Zealand. Agar was apparently discovered in Japan in the middle of the seventeenth century when a Japanese innkeeper threw some seaweed jelly on the ground on a cold night, expecting it to disappear into the ground when the sun hit it in the morning. But, after several days of freezing and thawing, a mass remained that could be dissolved in boiling water and, on cooling, formed a hard gel. The seaweed plants are harvested by hand by divers who tear them from their sites of growth under the water.

Agar is obtained from the plants by heating them in a dilute (0.007 M) solution of hydrochloric acid at 80°C for 8–10 hr. Calcium hypochlorite and sodium bisulfite are added for an additional 5 hr to remove the lignin and bleach the material. The hot mixture is filtered through a coarse filter, and the liquor containing the agar at a concentration of 0.5–1.0% (w/v) is cleared by sedimentation. The clear solution is allowed to solidify by lowering the temperature. The solid is cut into pieces, washed, dissolved in hot water, and then allowed to solidify. This is repeated several times to remove the agaropectin fraction that does not form a gel.

Agaran has a structure similar to κ- and ι-carrageenan. The difference being that the second residue in the repeating disaccharide unit has the L-configuration instead of the D-configuration. Further, agaran has very few if any O-sulfate groups, and approximately 20% of the D-galactopyranosyl units have a 2-O-methyl group. See Fig. 6.13 for a comparison of the structures of the carrageenans and argaran.

Because the 3,6-anhydrogalactopyranose unit has the L-configuration, it exists in the 1C conformation. This is similar to the 3,6-anhydro-D-galactopyranose units of the κ- and ι-carrageenans and of the L-guluronic acid units of alginic acid.

Agaran is one of the best-known gel-forming polysaccharides. It forms thermoreversible gels, melting and dissolving and solidifying between 45 and 90°C. Solutions remain liquid down to about 50°C, where the agaran begins to solidify. Solidification is usually complete between 40 and 50°C. The strength of the gel is dependent on the concentration of the agaran, with gels forming at concentrations as low as 0.1% (w/v). Very firm gels are formed at a concentration of 3%.

Anyone who has taken a course in microbiology and/or worked with microorganisms knows that agar (agaran) is a valuable material for the culturing of microorganisms on a solid surface [118]. It was first used for this purpose in 1881 by Robert Koch, a German bacteriologist. It is still the most commonly used solid support for the culturing of microorganisms. Agaran is relatively inert biologically, because very few organisms are able to break it down into its component sugars and metabolize these sugars. It also has found use as a material for the preparation of impressions used in forensic chemistry, medicine, dentistry, and art. Because of its relatively inert, stable, and nontoxic nature, agaran gels are used as a solid support for the electrophoretic separation of proteins and oligonucleotides. It, thus, is widely used in the separation and analysis of the purity of proteins and in the analysis of DNA sequences. It has also found use in size-exclusion chromatography, especially in the separation of very large macromolecules by size, with exclusion limits of 500,000–50,000,000 Da. These chromatographic gels are marketed by Bio-Rad Laboratories (Richmond, CA, USA) as their Bio-Gel A series.

Agar is not digestible by humans and is nontoxic. Thus, it has found some uses in foods as an emulsifying and stabilizing agent, especially in foods that need to be heated to relatively high temperatures. The Japanese prepare a "sweet" using agar. Sugar, spices, and food coloring are added to the hot agar, after which it is allowed to solidify. The agar sweet is especially used by the Japanese when drinking green tea.

Agar is used in treating acute diarrhea. It is administered as a solid or as a thick slurry and provides a nonirritating bulking material that absorbs water and swells in the bowel, giving form to the stool. It is also used as an ingredient in slow-release capsules, in suppositories, and as a carrier in topical ointments.

h. Dietary Fibers

Dietary fibers are polysaccharides from natural foods found primarily in plant cell walls, and include cellulose, hemicelluloses, and pectins. They are β-glucans that are not digestible by humans, and thereby provide noncaloric bulk in the diet. Dietary fibers also include polysaccharides such as pectins, alginates, exudate gums, carrageenans, and agar that are added to foods and are not digestible.

6.3 Animal Polysaccharides

a. Glycogen

Glycogen is a polysaccharide that serves as a form of reserve storage of chemical energy in animals. It is thought of primarily as a mammalian polysaccharide, although it is also found in fish, insects, bacteria, fungi, protozoa, and yeasts. In mammals, it occurs in all tissues, but it is particularly prevalent in liver and skeletal muscle. The latter contains the largest amount of glycogen in the human body, and the former has the highest concentration.

Among fish, glycogen is particularly prevalent in the muscle of shellfish. It is synthesized by several species of bacteria. The blue-green algae, which are photosynthesizing bacteria and not eukaryotic algae, synthesize glycogen as the reserve storage polysaccharide, rather than starch. Glycogen is also synthesized by a number of nonphotosynthesizing bacteria such as *Escherichia coli* and *Neisseria perflava*.

Glycogen was first obtained from mammalian liver in the middle of the nineteenth century. The liver is cut into small pieces and 1 g of minced liver is added to 2 mL of 50% (w/v) sodium hydroxide and heated in a boiling water bath for 3 hr. After cooling, centrifugation, and neutralization, the glycogen is precipitated by the addition of 4 volumes of ethanol to the clear extract [119a,119b]. It was later shown that the strong alkaline conditions produce degradation of the glycogen, and that the molecular weight of glycogen that was extracted with cold, dilute trichloroacetic acid was about 10 times the molecular weight that was obtained by hot alkaline extraction [120].

In the trichloroacetic acid procedure, the minced liver is treated with 10% (w/v) TCA at 2°C for 3 min, the suspension is filtered through muslin, and the glycogen is precipitated by the addition of 4 volumes of ethanol to the filtrate [121].

An aqueous extraction procedure was developed by taking frozen liver or muscle and grinding it to a powder under liquid nitrogen, and then homogenizing the powder at 4°C with 6 volumes of 45% phenol for 4 min. The mixture is stirred for 1 hr at 20°C, and the milky upper phase is centrifuged. The glycogen mixed with RNA is then precipitated by the addition of 3 volumes of ethanol. The precipitate is dissolved in water and dialyzed. The RNA is removed by treatment with ribonuclease, followed by precipitation of the glycogen with 4 volumes of ethanol [122].

Methylation analysis of glycogen showed that it was an α-1 → 4-linked glucan with 10% α-1 → 6 branch linkages [123]. It is a highly water-soluble, ramified glucan that has structural properties similar to amylopectin. Glycogen, however, is more highly branched than amylopectin (10% vs. 5%) and has an average chain length of 10–12 vs. an average chain length of 20–23 for amylopectin [124]. Another major difference is that amylopectin occurs as a partial crystalline material in a starch granule, and glycogen is amorphous and does not occur in a large crystalline granule.

Glycogen, like amylopectin, is a polydisperse molecule ranging in size from 1×10^6 to 2×10^9 Da, with the largest amounts at the lower molecular weights [125]. In liver, glycogen is in the cytoplasm associated with the endoplasmic reticulum [126]. Electron microscopy has shown that glycogen does occur as an aggregate of spherical particles of approximately 25 nm [127]. These aggregates are called β-particles and can be isolated as separate entities from muscle. Even larger aggregates were observed in which there were approximately 100 β-particles associated to give what is called an α-glycogen particle. The exact nature of what holds these particles together is not known. It has been postulated that the particles are held together by some kind of labile covalent bonds, since the particles are not dissociated by exposure to hydrogen-bond breaking reagents such as urea or guanidine · HCl, but are dissociated by dilute acid [128].

The compact structure, low osmotic pressure, and ready enzymatic availability of about 40% of the glucose residues of glycogen make it an efficient material as a reserve carbohydrate energy source. Glycogen serves two distinct physiological roles in mammals. Liver glycogen is used to maintain a constant level of blood glucose. Skeletal muscle, brain, and heart muscle glycogens supply glucose as an immediately available source of chemical energy for the physiological function of these tissues. Liver possesses the enzymes to provide free glucose for the blood; muscle, brain, and heart tissues lack the enzyme, glucose phosphatase, that converts glucose-6-phosphate into free glucose in the liver. Instead, the glucose-6-phosphate is rapidly converted into ATP in muscle, brain, and heart for energy (see Chapter 11). The brain normally uses about 100 g of glucose from glycogen per day [129]. Liver glycogen is capable of providing 100–150 mg of glucose per min to the blood over a sustained period of 12 hr if necessary [129].

b. Glycosaminoglycans

Glycosaminoglycans comprise a group of polysaccharides found primarily in mammalian tissues. They are composed of repeating disaccharide units of uronic acids and 2-acetamido-2-deoxy sugars. The linkages are at positions 3 and 4 and are primarily β, but there are two that have α-linkages. These polysaccharides are most often attached to a protein backbone, forming what is called a *proteoglycan* [130]. Proteoglycans are macromolecular complexes with a protein central core to which several polysaccharide chains are covalently attached. The polysaccharides and proteoglycans serve a variety of functions in mammalian tissues.

i. Hyaluronic Acid

This glycosaminoglycan consists of a linear chain of the repeating disaccharide units of [→ 4)-β-D-glucopyranosyl uronic acid-(1 → 3)-*N*-acetyl-2-amino-2-deoxy-β-D-glucopyranosyl-(1 →] (see Fig. 6.14A). It is composed of 500–50,000 monosaccharide residues per molecule [130].

Hyaluronic acid is widely distributed in mammalian cells and tissues. It is found in synovial fluid that lubricates the joints, in the vitreous humor of the eye, and in connective tissues such as the umbilical cord, the dermis, and the arterial wall. It also occurs as a capsular material around certain bacteria, usually pathogenic, such as *Streptococci* [130].

ii. Chondroitin sulfate

Chondroitin sulfate consists of a repeating disaccharide of [→ 1)-β-D-glucopyranosyl uronic acid-(1 → 3)-*N*-acetyl-2-amino-2-deoxy-β-D-galactopyranosyl-(1 →]. It differs from hyaluronic acid by having D-galactosamine instead of D-glucosamine in the second residue of the repeating disaccharide unit. It also has *O*-sulfate groups attached to the monosaccharide residues. There are two kinds of chondroitin sulfates, with the sulfate group esterified to the hydroxyl group at either C-4 or C-6 of the *N*-acetyl-2-amino-2-deoxy-D-galactopyranose unit (see Fig. 6.14B). These two sulfated polysaccharides occur separately or in mixtures, depending on the tissue [130].

The chondroitin sulfates are major components of cartilage. Proteochrondroitin sulfates are found in the cornea of the eye, the aorta, skin, and lung tissue, where they are located between fibrous protein molecules and provide a soft, pliable texture. Hyaluronic acid and chondroitin sulfate represent heteropolysaccharides with two monosaccharides units with heterolinkages of β-1 → 3 and β-1 → 4.

iii. Dermatan Sulfate

Dermatan sulfate may be derived from chrondroitin-4-sulfate by the action of a C-5 epimerase that inverts the carboxylate group of the D-glucuronic acid residue.

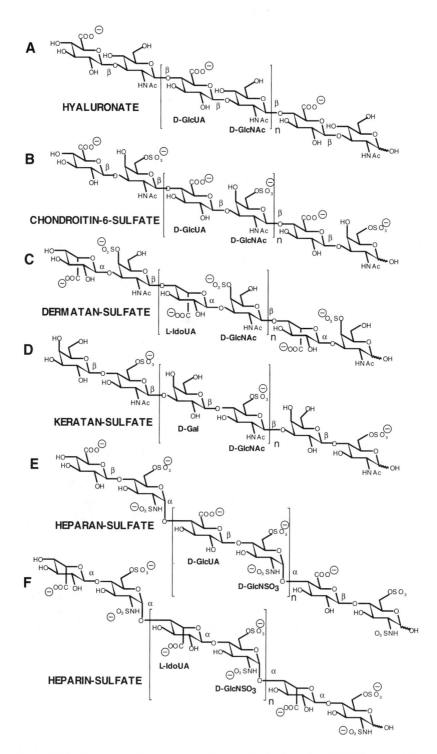

Figure 6.14. Structures of glycosaminoglycans: **A,** hyaluronic acid; **B,** chondroitin-6-sulfate; **C,** dermatan sulfate; **D,** keratan sulfate; **E,** heparan sulfate; **F,** heparin sulfate.

This forms a repeating disaccharide of [→ 4)-α-L-idopyranosyl uronic acid-(1 → 3)-β-D-N-acetyl-2-amino-2-deoxy-4-sulfato-β-D-galactopyranosyl-(1 →]. The epimerization reaction is usually incomplete so that there are also some D-glucuronic acid residues (see Fig. 6.14C) [131].

Some of the L-iduronic acid residues are sulfated at C-2. These residues adopt the 1C conformation. The nonsulfated L-iduronic acid residues can have either the C1 or the 1C conformation. Dermatan sulfate represents a repeating two-monosaccharide heteropolysaccharide with α-1 → 3 and β-1 → 4 heterolinkages. Dermatan sulfate is found primarily in skin.

iv. Keratan Sulfate

Keratan sulfate consists of the repeating disaccharide N-acetyl-lactosamine linked β-1 → 3 to the D-galactopyranose unit of the next disaccharide (see Fig. 6.14D). Keratan sulfate is the most heterogeneous of the glycosaminoglycans in that the sulfate content is variable, and it contains small amounts of L-fucose, D-mannose, and N-acetyl-D-neuraminic acid residues [130]. Keratan sulfate proteoglycan is found in the cornea, on the surface of erythrocytes, in cartilage, and in bone.

v. Heparan Sulfate

Heparan sulfate is a glycosaminoglycan that consists of the repeating disaccharide of [→ 4)-β-D-glucopyranosyl uronic acid-(1 → 4)-N-sulfato-2-amino-2-deoxy-α-D-glucopyranosyl-(1 →] (see Fig. 6.14E) [131]. The polysaccharide is linked to a core protein to give a proteoglycan that is found as a matrix component of arterial wall, lung, heart, liver, and skin. Heparan sulfate proteoglycan is most probably the precursor for the anti-blood clotting polysaccharide, heparin sulfate [131].

vi. Heparin Sulfate

Heparin sulfate arises by the action of a C-5 epimerase on heparan sulfate that causes the inversion of the carboxylate group of the D-glucuronic acid residue [131]. This gives rise to α-D-iduronic acid units. Thus, by this inversion, heparin sulfate is a glycosaminoglycan containing all α-1 → 4-linked monosaccharide residues, resulting in a repeating disaccharide of [→ 4)-α-L-idopyranosyl uronic acid-(1 → 4)-N-sulfato-α-D-glucosamine-(1 →]. The polysaccharide contains a variable number of sulfate groups at C-2 of the L-iduronic acid residue and at the C-6 hydroxyl and 2-amino groups of the D-glucosamine residue (see Fig. 6.14F) [131].

Heparin sulfate is released from the proteoglycans of mast cells into blood when there is an injury to blood vessels, particularly of heart, liver, lungs, and skin. The release of heparin near the site of the injury acts as a controlling agent to prevent massive clotting of blood and, hence, run-away clot formation.

c. Chitin

Chitin is a polysaccharide that has a structure very similar to that of cellulose and the bacterial cell wall polysaccharide murein. The structure of chitin is essentially the structure of cellulose, with the hydroxyl group at C-2 of the D-glucopyranose residue substituted with an N-acetylamino group [132] (see Fig. 6.15A). Chitin is the structural polysaccharide that replaces cellulose in the cell wall of many species of lower plants. It is found in fungi, yeast, green algae, and brown and red seaweed cell walls. Chitin is also the major component of the exoskeleton of insects. It is found in the cuticles of annelids, molluscs, and in the shells of crustaceans such as shrimp, crab, and lobster [133].

Chitin forms intermolecular hydrogen bonds, similar to cellulose, to create fibers [134a]. This is a highly ordered, crystalline structure, as evidenced by X-ray diffraction. Chitin exists in three types of intermolecular hydrogen-bonded structures called α-, β-, and γ-chitin [134b]. The chains in α-chitin are bonded in an antiparallel structure. The chains in β-chitin are parallel, and in γ-chitin the chains are bonded with two parallel chains interspersed by an antiparallel chain [134b]. The most abundant form is α-chitin, which also is the most stable. The β-form is converted into the α-form in formic acid solution or by treatment with cold 6 M

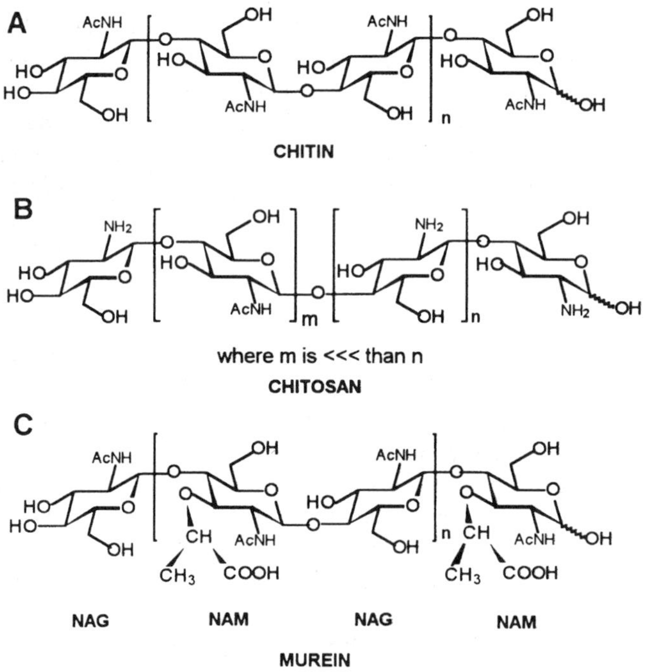

Figure 6.15. Structures of 2-acetamido- and 2-amino-glucans: **A,** chitin; **B,** chitosan; **C,** murein.

HCl [134b,c]. The γ-form is converted into the α-form by treatment with a saturated aqueous solution of lithium thiocyanate [134c]. These transformations appear to be irreversible. β- and γ-Chitin occur where the properties of the polysaccharide require flexibility and toughness [134d].

Because of the intermolecular hydrogen bonding, chitin is very insoluble in water and most other solvents. When chitin is treated with strong alkali, the N-acetyl groups are removed to give a readily water-soluble polysaccharide of β(1 → 4)-poly-2-amino-2-deoxy-D-glucopyranose, which is called *chitosan* (see Fig. 6.15B for the structure). Chitosan is a polycation at pH 6–7. It has been proposed as a chelating agent for heavy metal ions. Chitosan has also been suggested for a number of medical uses such as wound dressings, drug delivery agent, hypocholesterolemic agent, and for use in contact lenses [135]. It is also used in the purification of drinking water and in cosmetics and personal care products [135]. Similar to alginate, it has been used as an encapsulation and absorption agent for the immobilization of enzymes and cells [136]. Chitosan, being a β-glucan, is not digested by humans and can serve as a dietary fiber.

6.4 Microbial Polysaccharides

a. Murein

The major component of all known bacterial cell walls is a polysaccharide composed of N-acetyl-2-amino-2-deoxy-D-glucopyranose units linked β-1 → 4 [137–139]. One of the N-acetyl-D-glucosamine units (NAG) is substituted at C-3 with an O-lactic acid group through an ether linkage to give N-acetyl-D-muramic acid (NAM) [140–143]. The two monosaccharide units are joined together by β-1 → 4 glycosidic bonds to form the repeating sequence of NAG-NAM (see Fig. 6.15C). Chitin and murein are the major naturally occurring compounds that contain amino sugars.

The actual structure of the bacterial cell wall is a highly cross-linked peptidoglycan in which a tetrapeptide is attached to the carboxyl group of the L-lactic acid by a peptide (amide) linkage [144]. In *Staphylococcus,* this peptide is further attached to a pentaglycine chain that cross-links with another peptide chain attached to an adjacent murein chain [144]. Other types of cross-linking occur, using different peptides. See Chapter 10 for a complete discussion of the structure and biosynthesis of the bacterial cell wall peptidoglycans.

The peptidoglycan (murein plus polypeptide) is the component of the bacterial cell wall that is responsible for the shape and rigidity of bacterial cells. It is also the site of attack by the enzyme lysozyme that is involved in the lysis of bacterial cells [145]. There is a possible relationship between the chain length of the murein and the shape of the bacterial cell. This is suggested by observations of the cell wall and the shape of *Arthrobacter crystallopoietes.* This particular bacterium undergoes a reversible coccus-rod transformation that is dependent on the conditions of growth. The murein obtained from the spherical (coccus) form had an average of 40 monosaccharide units, with a significant amount of polydispersity, whereas the murein isolated from the rod form had much less polydispersity and was about three times larger, with an average of 135 monosaccharide units [146]. Nevertheless, because of the cross-linking of the peptidoglycan, the bacterial cell wall is considered to be one giant, bag-shaped macromolecule [147].

b. Dextrans and Related Polysaccharides: Mutan and Alternan

Besides the murein sacculus, many bacteria also produce other polysaccharides that surround and are exterior to the murein cell wall. These polysaccharides serve various purposes in protecting the cell from lysis, virus infection, and changes in the environment including pH, temperature, and concentrations of oxygen. These materials are compact and can be microscopically observed surrounding the cell. They are called *capsules*. Other bacteria produce more diffuse polysaccharides that also are extracellular but are not so intimately associated with the cell. These less-defined polysaccharides are often called *slimes*. The dextrans make up such materials.

In the nineteenth century, there were reports of a mysterious thickening and sometimes gelling of cane and beet sugar solutions. Pasteur [148] reported in 1861, that these "viscous fermentations" of sucrose resulted from microbial action. Sucrose solutions were observed to be converted into viscous solutions, gels, and/or flocculent precipitates [149]. The material that produced such changes in the sucrose solutions was isolated and found to be a polysaccharide that was called *dextran* [150]. Van Tieghem isolated and named the bacterium that produced the polysaccharide, *Leuconostoc mesenteroides* [151].

The synthesis of dextran from sucrose by a cell-free bacterial culture filtrate was first reported by Hehre in 1941 [152]. The genera of bacteria that are recognized to produce enzymes capable of synthesizing polysaccharides from sucrose are principally *Leuconostoc* and *Streptococcus*. These genera are gram positive, facultatively anaerobic cocci that are very closely related to each other. One notable difference between them is that the *L. mesenteroides* strains required sucrose in the growth medium to induce the formation of the enzyme(s), whereas the *Streptococcus* species did not require sucrose in the medium to form the enzymes [153].

In 1954, Jeanes et al. [154] reported the formation of glucans from sucrose by 96 strains of *L. mesenteroides* and *Streptococcus* species. The polysaccharides were characterized by optical rotation, viscosity, periodate oxidation, and physical appearance after alcohol precipitation. The latter were described by Jeanes et al. in somewhat fanciful, qualitative terms, such as pasty, fluid, stringy, tough, long, short, flocculent, gellike, opaque, translucent, and so forth. These descriptions of the appearance of the alcohol precipitates provided an early suggestion that the different polysaccharides had different structures [153]. Both water-soluble and water-insoluble polysaccharides were obtained, and some strains appeared to form more than one kind of polysaccharide, as judged by their water solubility and by differences in the amount of alcohol needed to precipitate them. It was shown that different kinds of polysaccharides could be separated by using differential alcohol precipitation [155].

Methylation and NMR analyses showed that different kinds of polysaccharides, with different structures, were being synthesized by enzymes elaborated by the different strains and species of organisms [153]. It has turned out that not all of the polysaccharides are dextrans. A dextran is now defined as a glucan that is enzymatically synthesized from sucrose and has contiguous α-1 $\rightarrow$ 6-linked D-glucopyranose units in the main chains. All known dextrans are branched. This is one of the important characteristics that distinguish the different kinds of dextrans. Branch linkages have been found to be α-1 $\rightarrow$ 2, α-1 $\rightarrow$ 3, and α-1 $\rightarrow$ 4 and the branches can be single D-glucose units and/or chains of α-1 $\rightarrow$ 6-linked D-glucose units [153]. Further, the percent and the manner in which the branches are arranged give rise to differences in the structures.

The classic dextran that to date is the only one commercially produced is synthesized by *L. mesenteroides* B-512F[1]. It contains 95% α-1 $\rightarrow$ 6 linkages and 5% α-1 $\rightarrow$ 3 branch linkages [156]. The branch chains consist of both single α-1 $\rightarrow$ 3-linked D-glucopyranose units and long chains of α-1 $\rightarrow$ 6-linked D-glucopyranose units attached to the main chains by an α-1 $\rightarrow$ 3 branch linkage (see Fig. 6.16A).).

L. mesenteroides B-742 produces two types of dextrans: fraction L, which is precipitated by 38% ethanol and has 14% α-1 $\rightarrow$ 4 branch linkages [153]; and fraction S, which is precipitated by 44% ethanol and has the highest degree of branching possible, with 50% α-1 $\rightarrow$ 3 branch linkages that primarily consist of single D-glucopyranose units [157]. The latter structure has contiguous α-1 $\rightarrow$ 6-linked D-glucopyranose residues with some α-1 $\rightarrow$ 6-linked chains attached by an α-1 $\rightarrow$ 3 branch linkage, and single D-glucopyranose residues attached by α-1 $\rightarrow$ 3 branch linkages to every glucose residue in the α-1 $\rightarrow$ 6-linked chains (see Fig. 6.16B). The structure is characterized as a bifurcated regular comb dextran in which the single α-1 $\rightarrow$ 3-linked glucose units are like teeth of a comb attached to the α-1 $\rightarrow$ 6-linked glucose backbone of the comb [153].

Streptococcus mutans 6715 elaborates two enzymes that synthesize two polysaccharides from sucrose. The water-soluble polysaccharide is a dextran with a structure similar to that of the *L. mesenteroides* B-742 regular comb dextran. The *S. mutans* dextran has 64% α-1 $\rightarrow$ 6-linked D-glucopyranose residues and 36% α-1 $\rightarrow$ 3 branch linkages that are primarily single D-glucose residues [158]. This gives a structure of α-1 $\rightarrow$ 6-linked D-glucopyranose residues in the main chains, with some α-1 $\rightarrow$ 6-linked branch chains linked by an α-1 $\rightarrow$ 3 branch linkage, and the majority of the branch units as single D-glucopyranose residues attached to every other D-glucopyranose residue in the main chains. This is similar to the B-742 regular comb dextran, except that the single glucose units are attached to

---

[1]The B-numbers refer to the specific bacterium species in the culture collection of the USDA Northern Regional Research Laboratory, Peoria, IL, USA.

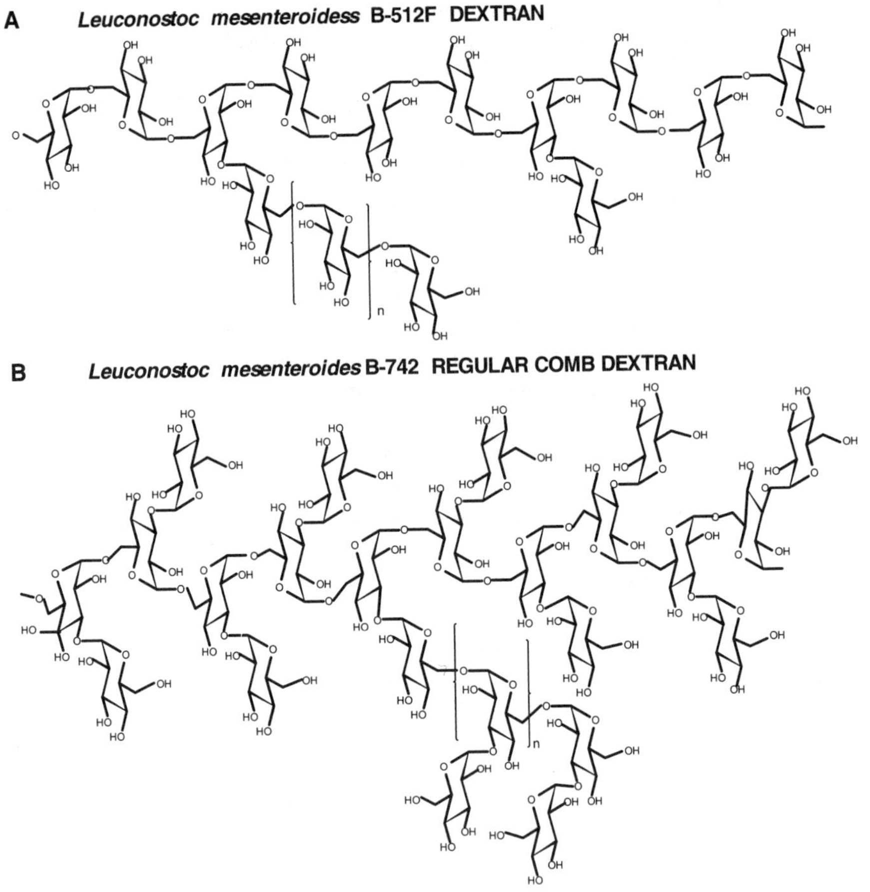

A   *Leuconostoc mesenteroidess* B-512F **DEXTRAN**

B   *Leuconostoc mesenteroides* B-742 **REGULAR COMB DEXTRAN**

Figure 6.16. Structures of dextrans and related glucans: **A,** *L. mesenteroides* B-512F dextran; **B,** *L. mesenteroides* B-742 regular comb dextran.

                                                                    (continued)

every other glucose unit [153]. This dextran is described as an alternating, bifurcated comb dextran (see Fig. 6.16C).

The other glucan synthesized by an enzyme elaborated by *S. mutans* 6715 is an α-1 → 3-linked glucan that is very water insoluble. It does not have contiguous α-1 → 6-linked glucose residues and also is not a dextran, and it is called *mutan* (see Fig. 6.16E). Some strains of *L. mesenteroides* (e.g., B-523 and B-119) also elaborate enzymes that synthesize mutan [153].

   L. mesenteroides B-1355 elaborates an enzyme that synthesizes a dextran identical in structure to B-512F dextran. It also elaborates a second enzyme that synthesizes a glucan from sucrose with an alternating sequence of α-1 → 6- and α-1 → 3-linked D-glucopyranose residues in the main chains [159]. This glucan has

C   *Streptococcus mutans* **ALTERNATING COMB DEXTRAN**

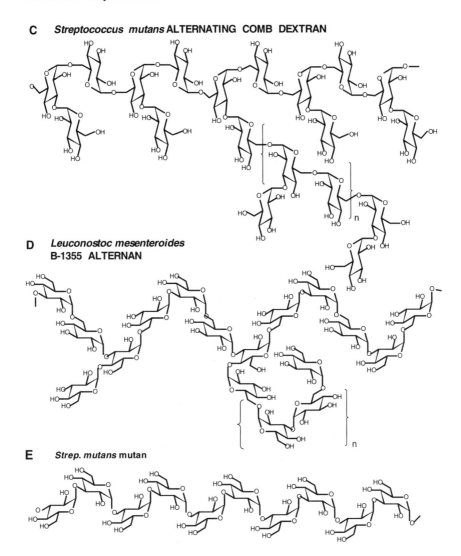

D   *Leuconostoc mesenteroides*
B-1355 ALTERNAN

E   *Strep. mutans* **mutan**

Figure 6.16. *(continued)* **C,** *S. mutans* alternating comb dextran; **D,** *L. mesenteroides* B-1355 alternan; **E,** *S. mutans* mutan.

11% α-1 → 3 branch linkages [160] (see Fig. 6.16D). Because of the absence of contiguous α-1 → 6 linkages, this glucan is not a dextran. It has different properties than the dextrans and, hence, is called *alternan* because of its alternating α-1 → 6- and α-1 → 3-linked structure [161].

B-512F dextran has found a number of commercial applications. Low molecular weight fractions of 50,000–100,000 Da have been used as blood plasma substitutes [162]. B-512F dextran also serves as the starting material for epichlorohydrin cross-linked molecular sieves known as Sephadexes [163] (see Chapter 7).

c. Pullulan

Pullulan is a homopolysaccharide that is elaborated by many species of the fungus *Aureobasidium,* specifically *A. pullulans.* This fungus has also been called *Pullularia pullulans,* and sometimes it is called "black yeast" due to its formation of a black pigment [164]. *A. pullulans* will form pullulan when cultured on a number of sugars such as sucrose, maltose, D-xylose, and starch. D-glucose and sucrose have been reported to be the preferred carbon sources [165]. Pullulan is produced extracellularly in a fermentation process.

Pullulan is a highly water-soluble linear polysaccharide of D-glucopyranose residues, containing α-1 $\rightarrow$ 4 and α-1 $\rightarrow$ 6 glycosidic linkages in the ratio of 2:1 [165,166]. The structure is that of a polymer of maltotriose joined end to end by an α-1 $\rightarrow$ 6 linkage (see Fig. 6.17). It is a homopolysaccharide with heterolinkages of two α-1 $\rightarrow$ 4 linkages followed by an α-1 $\rightarrow$ 6 linkage. In addition to maltotriose units, pullulan also has 5–7% maltotetraose units located in the interior part of the polysaccharide [167]. The molecular weight ranges between 2×10^6 and 1.5×10^5 [168].

Although pullulan is a linear polysaccharide devoid of any branch linkages, the presence of an α-1 $\rightarrow$ 6 glycosidic linkage on every third D-glucose residue combined with the two α-1 $\rightarrow$ 4 glycosidic linkages imparts an enhanced flexibility to the pullulan chain and produces a highly water-soluble, linear polysaccharide. Most linear polysaccharides have low water solubility, for example, amylose, an α-1 $\rightarrow$ 4-linked glucan, and mutan, an α-1 $\rightarrow$ 3-linked glucan, the latter extremely so.

Pullulan readily dissolves in water to give stable, viscous solutions that do not gel. Pullulan will form fibers and films. It is nontoxic and has low digestibility because human salivary and pancreatic α-amylases will hydrolyze only the α-1 $\rightarrow$ 4 glycosidic linkage of the maltotetraose unit; the α-1 $\rightarrow$ 4 linkages of maltotriose is too short to be bound in the active sites of the α-amylases. Pullulan can be used as a noncaloric food ingredient. Films formed from pullulan are water soluble and impervious to oxygen. It can be used as a suitable packaging and coating material for foods and pharmaceuticals [169]. These films have particular potential as edible packaging agents of noncaloric value and have particular use in protection against oxidation. Its incorporation into foods is reported to improve taste, flavor, and texture. Its addition to foods prevents drying and the retrogradation of starch.

Pullulan fibers have a high gloss that resembles rayon, and after stretching they have a tensile strength similar to nylon [169]. Use of pullulan in the coating of paper gives a smooth coating of a very fine texture with desirable gloss and ink absorption properties [170a]. It has been proposed that pullulan has uses in ethical pharmaceuticals such as hand and facial lotions, shampoos, and cosmetics. Pullulan is being produced by the Hayashibara Biochemical Laboratories in Okayama, Japan for the development of commercial applications.

d. Bacterial Fructans

Two gram-positive bacterial genera, *Bacillus* and *Streptococcus,* have species that elaborate fructansucrases that synthesize fructans from sucrose. *Bacillus levani-*

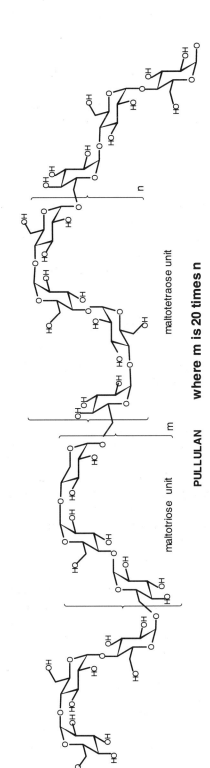

Figure 6.17. Structure of pullulan.

cans produces a levansucrase that synthesizes levan from sucrose [170b]. *Streptococcus salivarius* also produces a levansucrase that synthesizes levan [170c]. Several *S. mutans* strains elaborate inulinsucrase that synthesizes inulin from sucrose [170d]. The bacterial fructans have the same structures as the plant fructans (see Fig. 6.11).

e. Xanthan

In the 1950s, the U.S. Department of Agriculture Northern Regional Research Laboratory in Peoria, Illinois conducted a survey of their bacterial culture collection to find extracellular, water-soluble gums. They discovered that several *Xanthanomonas* species produced polysaccharide gums. In particular, they found that *Xanthanomonas campestris* B-1459 produced a polysaccharide gum that had properties that complimented other naturally occurring water-soluble gums. They called the polysaccharide *xanthan.*

Xanthan is elaborated in a submerged aerobic fermentation process. The primary carbon source is 2–3% (w/v) D-glucose, although sucrose and starch can also be used. During the fermentation, the viscosity of the culture liquor progressively increases. After the fermentation is complete, the liquor is diluted with water to lower the viscosity, and the cells are removed. The xanthan is then precipitated by the addition of methanol or 2-propanol to 50% (v/v) to the cell-free liquor in the presence of 2% (w/v) potassium chloride.

Xanthan consists of a repeating pentasaccharide unit containing two D-glucose residues, two D-mannose residues, and one D-glucuronic acid residue. The main chain is essentially cellulose with the D-glucopyranosyl units linked β-1 $\rightarrow$ 4. A trisaccharide is attached α-1 $\rightarrow$ 3 to every other D-glucose unit in the main chain, giving the pentasaccharide repeating unit. The trisaccharide consists of D-mannopyranosyl linked β-1 $\rightarrow$ 4 to D-glucopyranosyl uronic acid that is linked β-1 $\rightarrow$ 2 to a D-mannopyranosyl unit [171]. The nonreducing terminal D-mannopyranosyl unit of the trisaccharide has a pyruvic acid linked to positions 4 and 6, forming a cyclic, six-membered pyruvic acid ketal. C-6 of the first D-mannopyranosyl unit in the trisaccharide is *O*-acetylated. See Fig. 6.18 for the structure of xanthan.

The pyruvic acid content can vary significantly for different strains of *X. campestris,* providing xanthan solutions with different viscosities. Those with high pyruvate content have high viscosity on the addition of KCl and an enhanced thermal stability. One strain, ATCC[2] 31313, produced a xanthan without pyruvate that had a relatively low viscosity [172]. The molecular weight of xanthan has been reported between 3×10^5 and 8×10^6 Da [173]. X-ray diffraction studies indicate that xanthan has a helical structure [174]. Molecular modeling suggests that the branched trisaccharide units are roughly parallel to the axis of the helix [174].

Xanthan has found a number of commercial applications due to its emulsion-stabilizing and particle-suspending properties, the small variation in its viscosity with changes in temperature, and its tolerance for high salt concentrations. It is the

---

[2]ATCC refers to the American Type Culture Collection, Rockville, MD, USA.

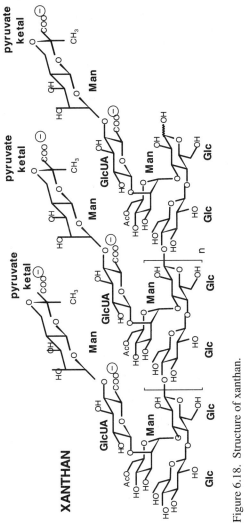

Figure 6.18. Structure of xanthan.

second bacterial polysaccharide to be produced commercially, B-512F dextran being the first. Xanthan has been used in oil well drilling and pipeline cleaning. It reduces the mobility of water and decreases its permeability so that xanthan-water solutions can be used as a lubricant fluid in oil well drilling. Xanthan also is used as a thickener in textile printing, where it controls the rheological properties of the dye solution and gives sharp, clean patterns by preventing migration of the dye. Xanthan is compatible with components of ceramic glazes and prevents their aggregation, giving uniform suspensions during glazing. It can be used with either acid or alkaline materials because its viscosity is not affected by changes in pH.

Xanthan is used in foods, where it has good solubility in either hot or cold solutions. It can provide high viscosities at low concentrations, thus acting as a thickener and bulking agent. It has excellent suspending properties, excellent thermal stability, and provides stability during freezing and thawing. Thus, it is useful in frozen-food preparations. It is used as a thickener and suspending agent in salad dressings, syrups, toppings, relishes, sauces, dairy products, and baked goods. Xanthan can be used to suspend pharmaceuticals such as antibiotics, steroids, and so forth and is used in ethical pharmaceuticals such as tooth pastes, creams, lotions, and makeup [172,175].

f. Bacterial Gels

As already mentioned in the section on plant polysaccharides, two genera of bacteria produce alginic acid that readily forms gels on contact with divalent metal ions. They are *Pseudomonas aeruginosa* [92] and *Azotobacter vinelandii* [93]. One of the major differences in the alginic acids produced by the brown seaweeds and these bacteria is the presence of *O*-acetylation in the bacterial alginates. This acetylation inhibits gelling and produces inferior gels. Perhaps this undesirable structural feature in the formation of gels might be eliminated by mutation and selection, and the production of bacterial alginates by fermentation, under controlled conditions, might become a viable alternative to the obtaining of alginates from seaweeds.

Pseudomonas elodea elaborates an extracellular polysaccharide that forms thermal-reversible gels with characteristics similar to those made from seaweed agaran [176]. The polysaccharide is called *gellan.* It is a linear heteropolysaccharide containing D-glucopyranose, D-glucopyranose uronic acid, and L-rhamnopyranose in the ratio of 2:1:1. The structure of gellan is a repeating tetrasaccharide that is partially acetylated (50%) at C-6 and esterified by L-glyceric acid at C-2 of the first D-glucose residue in the repeating sequence of the tetrasaccharide [177,178a]. Gellan has the following structure:

L-glyceryl
$\downarrow$
2
$[\rightarrow 3)\text{-}\beta\text{-}D\text{-Glc}p\text{-}(1 \rightarrow 4)\text{-}\beta\text{-}D\text{-Glc}pUA\text{-}(1 \rightarrow 4)\text{-}$
6 $\beta\text{-}D\text{-Glc}p\text{-}(1 \rightarrow 4)\text{-}\alpha\text{-}L\text{-Rha}p\text{-}(1 \rightarrow]_n$
$\uparrow$
$Ac_{0.5}$

Solutions of gellan undergo thermal-reversible viscosity changes. The viscosity decreases exponentially with an increase in temperature from 20–121°C. The viscosity returns to the original value on cooling. Gellan requires the presence of either monovalent or divalent cations to gel. Gels with different properties can be obtained by changing the concentration, the type of cation, and the degree of acylation [172]. The potassium salt of gellan exists as a double helix [178b].

Gellan has been used primarily as a substitute for seaweed agaran in the culturing of microorganisms and in plant tissue culturing. Gellan gels provide a higher degree of clarity than do agaran gels. It has no toxicity to particularly sensitive microorganisms and provides gel strength equivalent to agaran at approximately one-half the concentration.

Another bacterial heteropolysaccharide, related in structure to gellan, but with different properties, is elaborated by *Alcaligenes* sp. ATCC 31555. The polysaccharide is composed of a repeating pentasaccharide that has an identical structure as gellan, but with 66% single α-L-rhamnopyranose or 34% single α-L-mannopyranose branch side chain attached to C-3 of the β-D-glucopyranose that is third in the repeating tetrasaccharide [179], giving the following structure:

$$[\rightarrow 3)\text{-}\beta\text{-}\mathrm{D}\text{-}\mathrm{Glc}p\text{-}(1 \rightarrow 4)\text{-}\beta\text{-}\mathrm{D}\text{-}\mathrm{Glc}p\mathrm{UA}\text{-}(1 \rightarrow 4)\text{-}$$

$$\beta\text{-}\mathrm{D}\text{-}\mathrm{Glc}p\text{-}(1 \rightarrow 4)\text{-}\alpha\text{-}\mathrm{L}\text{-}\mathrm{Rha}p\text{-}(1 \rightarrow]_n$$
$$3$$
$$\uparrow$$
$$1$$
$$(\alpha\text{-}\mathrm{L}\text{-}\mathrm{Rha}p)_{66\%} \text{ or } (\alpha\text{-}\mathrm{L}\text{-}\mathrm{Man}p)_{34\%}$$

The polysaccharide is called *welan*. Welan gives high viscosity solutions at very low polysaccharide concentrations (1 mg/mL gives 100 cP). In contrast with gellan, temperature has very little effect on the viscosity of welan solutions between 37 and 140°C. The viscosity is maintained at 140°C over at least 15 hr. The viscosity is also not affected by changes in pH between 2 and 12. Welan is very stable in solutions containing high salt concentrations and maintains its solubility and viscosity in seawater and salt brines [172].

Yet another polysaccharide that is elaborated by a different *Alcaligenes* species, ATCC 31961, is related to the structures of gellan and welan, but with some different properties. It has the same repeating tetrasaccharide structure in the main chain as gellan, but with a disaccharide branch side chain of β-D-glucopyranosyl-(1 → 6)-α-D-glucopyranosyl linked 1 → 6 onto the second β-D-glucopyranosyl residue in the repeating tetrasaccharide unit of the main chain [180]. The polysaccharide is called *rhamsan* and has the following structure:

$$[\rightarrow 3)\text{-}\beta\text{-}\mathrm{D}\text{-}\mathrm{Glc}p\text{-}(1 \rightarrow 4)\text{-}\beta\text{-}\mathrm{D}\text{-}\mathrm{Glc}p\mathrm{UA}\text{-}(1 \rightarrow 4)\text{-}$$

$$\beta\text{-}\mathrm{D}\text{-}\mathrm{Glc}p\text{-}(1 \rightarrow 4)\text{-}\alpha\text{-}\mathrm{L}\text{-}\mathrm{Rha}p\text{-}(1 \rightarrow]_n$$
$$6$$
$$\uparrow$$
$$1$$
$$\beta\text{-}\mathrm{D}\text{-}\mathrm{Glc}p\text{-}(1 \rightarrow 6)\text{-}\alpha\text{-}\mathrm{D}\text{-}\mathrm{Glc}p$$

Rhamsan produces solutions of extremely high viscosity at very low concentrations (2.5 mg/mL gives 2,000 cP) [181]. The viscosities of rhamsan solutions increase only slightly between 20 and 100°C and then begin to drop. Rhamsan solutions are also very stable in high salt concentrations, and it is a very effective suspending agent [172]. Rhamsan has found uses in water-based paints, where it provides excellent covering and brushing or spraying properties. It has also found use as a carrier for fertilizers.

g. Pneumococcal Capsule Polysaccharides

The pneumococci are pathogenic bacteria that produce infections of the lungs and respiratory tract, causing bacterial pneumonia. The bacteria were originally called *Diplococcus pneumoniae* but they were reclassified in the 1970s as *Streptococcus pneumoniae*. They are ovoid, gram-positive cocci that occur in pairs enveloped by a capsule. The capsule is a well-defined morphological feature of the organism that can be easily observed under the light microscope. All virulent strains elaborate a capsule and it is most voluminous when the organism is at its most virulent stage. When the organisms are grown under unfavorable culture conditions they loose their capsule and become nonvirulent. The primary function of the capsule apparently is to delay or prevent phagocytosis, thereby allowing the organism to survive and multiply in the host tissues. Although the pneumococcal capsules can be considered a classic example of the bacterial capsule, not all encapsulated bacterial species are pathogenic.

Most bacterial capsules are polysaccharides with a specific structure. In many respects, the capsular polysaccharide can be considered an identifying feature of the organism. The pneumococci are divided into types (or groups) that are defined by a particular immunological precipitin reaction of the capsular polysaccharide with an antibody preparation obtained by inoculation of the polysaccharide into an animal host such as a rabbit. The reaction is usually highly specific and depends on the structure of the polysaccharide. *S. pneumoniae* has been found to have over 80 different immunological types and, therefore, 80 different polysaccharide structures.

The capsules are all heteropolysaccharides of relatively complex and diverse structure. Their monosaccharide composition contains an array of D-glucopyranose, D-glucopyranose uronic acid, α-L-rhamnopyranose, and N-acetyl-D-glucosamine. In addition, however, the pneumococcal polysaccharides have some unusual sugars in some of their repeating structures, such as the sugar alcohols, glycerol, erythritol, D-threitol, and ribitol; N-acetyl-L-fucosamine, N-acetyl-mannosamine, N-acetyl-galactosamine, and an unusual aminosugar, 2-acetamido-2,6-dideoxy-L-talose (N-acetyl-L-pneumosamine); D-galactose in the furanose form; and phosphodiesters. Many of the structures are repeating tetra-, penta-, or hexasaccharides. One polysaccharide was reported to have a repeating unit with a sequence of 14 monosaccharide residues [182]. The polysaccharides have glycosidic linkages at different positions, with α- and β-configurations. Some are linear and others are branched, with a single residue or with a disaccharide unit.

Not all of the pneumococcal capsular polysaccharides have had their structures completely determined. Because of their antigenicity and use in producing vaccines, there has been great interest in this field of carbohydrate chemistry. We will consider only the structures of six types of *S. pneumoniae* capsular polysaccharides that have had their structures definitively determined.

Type 2 polysaccharide has an unusual structure with a repeating tetrasaccharide unit containing a sequence of three α- and β-linked L-rhamnopyranose residues linked to position 3, and a single α-D-glucopyranose linked at position 4. The polysaccharide has a branched disaccharide unit of α-D-glucopyranose uronic acid-$(1 \rightarrow 6)$-α-D-glucopyranose linked $1 \rightarrow 2$ to the second L-rhamnose residue to form a repeating hexasaccharide of the following structure [183]:

$$[\rightarrow 3)\text{-}\alpha\text{-L-Rha}p\text{-}(1 \rightarrow 3)\text{-}\alpha\text{-L-Rha}p\text{-}(1 \rightarrow 3)\text{-}$$
$$\qquad\qquad\qquad\qquad\qquad 2 \qquad\qquad\quad \beta\text{-L-Rha}p\text{-}(1 \rightarrow 4)\text{-}\alpha\text{-D-Glc}p\text{-}(1 \rightarrow]_n$$
$$\qquad\qquad\qquad\qquad\qquad \uparrow$$
$$\qquad\qquad\qquad\qquad\qquad 1$$
$$\alpha\text{-D-Glc}p\text{UA-}(1 \rightarrow 6)\text{-}\alpha\text{-D-Glc}p$$

Type 3 polysaccharide is one of the first pneumococcal capsule polysaccharides to have its structure determined [184]. It has a simpler structure than type 2 polysaccharide, with only two monosaccharide residues, D-glucopyranose and D-glucopyranose uronic acid, with alternating β-$1 \rightarrow 3$ and β-$1 \rightarrow 4$ glycosidic linkages in a repeating disaccharide of the following structure [184]:

$$[\rightarrow 4)\text{-}\beta\text{-D-Glc}p\text{-}(1 \rightarrow 3)\text{-}\beta\text{-D-Glc}p\text{UA-}(1 \rightarrow]_n$$

Type 8 polysaccharide consists of a repeating tetrasaccharide with β-D-glucopyranosyl uronic acid linked $1 \rightarrow 4$ to β-D-glucopyranose and with α-D-glucopyranosyl linked $1 \rightarrow 4$ to α-D-galactopyranose. The polysaccharide contains three kinds of monosaccharide residues, with two β-$1 \rightarrow 4$ glycosidic linkages followed by two α-$1 \rightarrow 4$ glycosidic linkages to give the following structure [185]:

$$[\rightarrow 4)\text{-}\beta\text{-D-Glc}p\text{UA-}(1 \rightarrow 4)\text{-}\beta\text{-D-Glc}p\text{-}(1 \rightarrow 4)\text{-}$$
$$\qquad\qquad\qquad\qquad\qquad \alpha\text{-D-Glc}p\text{-}(1 \rightarrow 4)\text{-}\alpha\text{-D-Gal}p\text{-}(1 \rightarrow]_n$$

Type 6 polysaccharide contains four different sugar residues in a repeating tetrasaccharide structure. This polysaccharide has the unusual structural feature of having a sugar alcohol, ribitol, linked to a phosphate that then joins the repeating tetrasaccharide units together by phosphodiester linkages, and it has $1 \rightarrow 2$ and $1 \rightarrow 3$ glycosidic linkages [182,186], forming the following structure:

$$[\rightarrow 2)\text{-}\alpha\text{-D-Gal}p\text{-}(1 \rightarrow 3)\text{-}\alpha\text{-D-Glc}p\text{-}(1 \rightarrow 3)\text{-}$$
$$\qquad\qquad\qquad\qquad\qquad \alpha\text{-L-Rha}p\text{-}(1 \rightarrow 3)\text{-ribitol-}(5\text{-}O\text{-PO}_3 \rightarrow]_n$$

The presence of the ribitol phosphodiester is similar to the structures found in the teichoic acids (see section 6.5).

Type 13 polysaccharide is similar to type 6 in that it also has a ribitol phosphodiester linkage joining a repeating pentasaccharide unit. This pneumococcal polysaccharide has the additional feature of having *N*-acetyl-β-D-glucosamine and β-D-galacofuranose residues, forming the following structure [187]:

$$[\rightarrow 4)\text{-}\alpha\text{-}D\text{-}Galp\text{-}(1 \rightarrow 4)\text{-}\alpha\text{-}D\text{-}Glcp\text{-}(1 \rightarrow 3)\text{-}$$
$$2 \text{ or } 3 \quad \alpha\text{-}D\text{-}Galf\text{-}(1 \rightarrow 3)\text{-}D\text{-}GlcNAcp\text{-}(1 \rightarrow 4)\text{-ribitol-}(1\text{-}O\text{-}PO_3 \rightarrow]_n$$
$$\uparrow$$
$$OAc$$

Type 34 polysaccharide is a repeating pentasaccharide that contains two sugar residues with D-ribitol-5-phosphate, two β-D-galactofuranose units, an α-D-galactopyranose unit, and two α-D-glucopyranose units, along with ribitol-5-phosphate. The polysaccharide contains an unusual alternating sequence of β-D-galactofuranose linked $1 \rightarrow 3$ to α-D-galactopyranose linked $1 \rightarrow 2$ to give the following structure [188]:

$$[\rightarrow 3)\text{-}\beta\text{-}D\text{-}Galf\text{-}(1 \rightarrow 3)\text{-}\alpha\text{-}D\text{-}Galp\text{-}(1 \rightarrow 2)\text{-}$$
$$\beta\text{-}D\text{-}Galf\text{-}(1 \rightarrow 3)\text{-}\alpha\text{-}D\text{-}Galp\text{-}(1 \rightarrow 2)\text{-ribitol-}(5\text{-}O\text{-}PO_3 \rightarrow]_n$$

h. *Salmonella* O-Antigen Polysaccharides

Gram-negative bacteria have a cellular, inner membrane that surrounds the cytoplasm, the peptidoglycan cell wall, and also has the so-called outer membrane that consists of complex lipopolysaccharides, proteins, and phospholipids. The outermost surface of gram-negative bacteria is often coated with capsular polysaccharides that uniquely characterize the bacterial species or strain and are known as *O-antigens.* These gram-negative bacteria are very often pathogenic. The observation that mutant strains of pathogenic bacteria that lack the O-antigen are nonpathogenic suggests that the O-antigen polysaccharides are also the mechanism by which the bacteria invade the host. This mechanism by which the invasion occurs is thought to involve the binding of the repeating unit of the polysaccharide with receptors on the surface of the host cells. To prevent infection, the host produces antibodies to the O-antigen. In the continuing duel between pathogens and their hosts, the structures of the O-antigens are constantly changing by rapid mutation so as to generate new bacterial strains that the host does not initially recognize and destroy.

The *Salmonella* are examples of such pathogens. There are hundreds of *Salmonella* species and strains that produce immunologically distinct O-antigen polysaccharides. This apparent complexity, results from relatively small variations of a single basic structure.

The O-antigen polysaccharides of *Salmonella* serogroups A, B, and D are closely related chemically and immunochemically and consist of a branched tetrasaccharide of the following structure [189]:

$$
\begin{array}{c}
\mathrm{X} \\
1 \\
\downarrow \\
3
\end{array}
$$

$$[\rightarrow)\alpha\text{-}D\text{-}Manp\text{-}(1 \rightarrow 4)\text{-}\alpha\text{-}L\text{-}Rhap\text{-}(1 \rightarrow 3)\text{-}\alpha\text{-}D\text{-}Galp\text{-}(1 \rightarrow]_n$$

where X is a 3,6-dideoxyhexopyranose that is immunologically specific for one of the serogroups. D-Paratose (3,6-dideoxy-D-glucopyranose) is specific for group A; D-abequeose (3,6-dideoxy-D-galactopyranose) is specific for group B; D-tyvelose (3,6-dideoxy-D-mannopyranose) is specific for group D [190] (see Fig. 6.19 for the structures of the 3,6-dideoxyhexoses). This general structure for *Salmonella* O-antigens represents a typical heteropolysaccharide that contains four monosaccharide residues in the repeating unit, with variation in the branch residue. Branched tetra- and pentasaccharides appear to be common, although a simple linear trisaccharide occurs for some species. In some of the structures, a pentasaccharide occurs with a fifth monosaccharide residue attached to one of the residues in the main chain. Great diversity can be obtained by just changing the configuration of one of the glycosidic linkages from α- to β- or by changing the position of the linkage for one of the residues, for example, a change from $(1 \rightarrow 4)$ to $(1 \rightarrow 3)$, and so on. The following are some of the examples of the variations that have been observed in *Salmonella* sp. O-antigens:

$$[\rightarrow 6)\text{-}\alpha\text{-}D\text{-}Manp\text{-}(1 \rightarrow 4)\text{-}\alpha\text{-}L\text{-}Rhap\text{-}(1 \rightarrow 3)\text{-}\alpha\text{-}D\text{-}Galp\text{-}(1 \rightarrow]_n \quad \textit{S. anatum}$$
$$[\rightarrow 6)\text{-}\alpha\text{-}D\text{-}Manp\text{-}(1 \rightarrow 4)\text{-}\alpha\text{-}L\text{-}Rhap\text{-}(1 \rightarrow 3)\text{-}\beta\text{-}D\text{-}Galp\text{-}(1 \rightarrow]_n \quad \textit{S. newington}$$

These two repeating trisaccharides differ by having an α-linked D-galactose and a β-linked D-galactose [191].

Group A *Salmonella* species have D-paratose attached to D-mannopyranose by a $1 \rightarrow 3$ linkage. Variations can occur as to the configuration (α- or β-) and the position of attachment to the D-mannose. *Salmonella typhimurium* belongs to group B and has D-abequose attached to the D-mannose. At least three different repeating units are known. The first two differ in the position of attachment of D-galactose to the D-mannose, $1 \rightarrow 4$ and $1 \rightarrow 6$ [192].

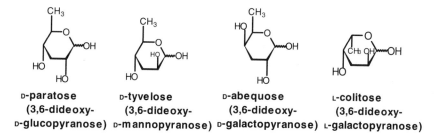

Figure 6.19. Structures of 3,6-dideoxy sugars found in O-antigen polysaccharides.

α-D-Abep
1
↓

3
[→ 4)-β-D-Manp-(1 → 4)-α-L-Rhap-(1 → 3)-α-D-Galp(1 →]$_n$

α-D-Abep
1
↓

3
[→ 6)-β-D-Manp-(1 → 4)-α-L-Rhap-(1 → 3)-α-D-Galp-(1 →]$_n$

The third structure has, in addition, an α-D-glucopyranose unit attached 1 → 6 to the D-galactopyranose unit [192]:

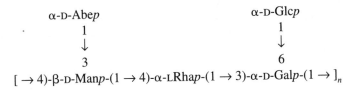

Escherichia coli, a gram-negative rod closely related to *Salmonella,* has been divided into more than 100 serogroups. Among other variations found in some of the *E. coli* types is an unusual 3,6-dideoxyhexose, L-colitose (3,6-dideoxy-L-galactopyranose).

i. Capsular Polysaccharides of Other Gram-Negative Bacteria

Neisseria meningitidis is a pathogen that causes meningitis in humans. It has an unusual capsular homopolysaccharide that is composed of N-acetyl-D-neuraminic acid. The monosaccharide residue has nine carbons (see Fig. 6.20A). It makes up a family of sugar acids called *sialic acids* that are N- and O-substituted derivatives (see Chapter 9). The sialic acids are widely distributed and are primarily found at the ends of the oligosaccharide chains attached to glycoproteins that are especially prevalent in bovine and ovine submaxillary mucins, erythrocytes, and nerve tissues. N-acetyl-D-neuraminic acid is biosynthesized by an aldol condensation of pyruvic acid with N-acetyl-D-mannosamine (see Chapter 10).

N. meningitidis (sero group B) and *E. coli* K1 both have capsular polysaccharides of poly(N-acetyl-D-neuraminic acid) linked 2 → 8 [193], and both cause meningitis in humans. There is evidence that the polysaccharide is responsible for the infection of tissues [194], possibly by specific cell recognition. The polysaccharide is known as *colominic acid* (see Fig. 6.20B for the structure).

A variation in the colominic acids for *N. meningitidis* (serogroup C) and *E. coli* K92 is a 2 → 9 linkage for colominic acid instead of 2 → 8 [193] (see Fig. 6.20C). Most of the 2 → 9 colominic acids are O-acetylated at C-7 or C-8. Al-

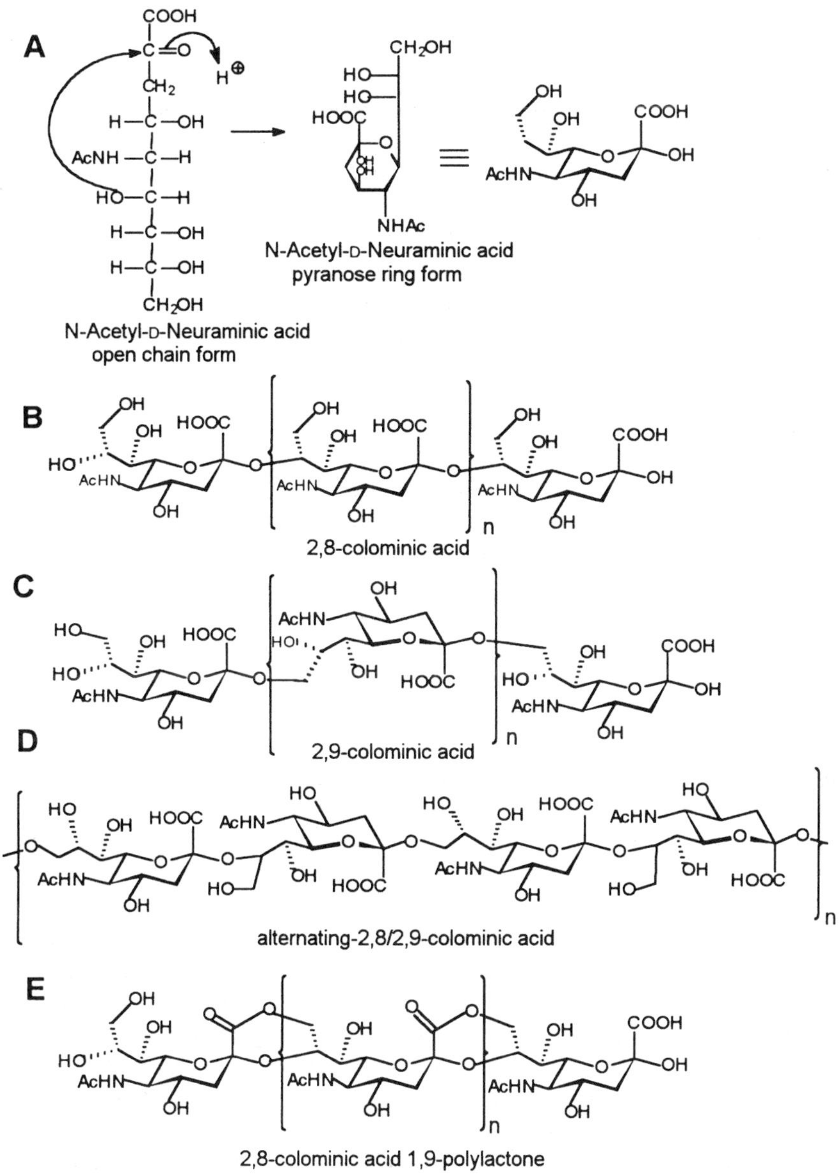

Figure 6.20. Structures of **A,** *N*-acetyl-D-neuraminic acid; **B,** 2,8-colominic acid; **C,** 2,9-colominic acid; **D,** 2,8/2,9-alternating colominic acid; **E,** 2,8-colominic acid 1,9-polylactone.

though chemically similar, the B and C serogroup polysaccharides give distinct immunological reactions.

The colominic acid from *E. coli* Bos-12 has an interesting variation in that it has an alternating α-2 → 8/α-2 → 9-heterolinked homopolysaccharide of *N*-acetyl-D-neuraminic acid [195] (see Fig. 6.20D). The α-2 → 8 colominic acid further differs from the α-2 → 9 colominic acid by readily undergoing intramolecular esterification between the C-1 carboxyl group of one residue and the C-9 hydroxyl group of an adjacent residue, forming interresidue lactones [196] (see Fig. 6.20E for the structure of 2,8-colominic acid-1,9-lactone). The carboxyl group of the 2,9-linked colominic acid does not undergo the formation of an interresidue lactone by reaction with the C-8 hydroxyl group.

6.5 Teichoic Acids

The teichoic acids (from the Greek word *teichos,* meaning "walls") are polymers of sugar alcohols and phosphate that are found particularly in the cell walls of gram-positive bacteria [197]. They consist primarily of glycerol or ribitol units joined together by phosphodiester linkages between phosphoric acid and the primary alcohol groups of the sugar alcohols [198]. Glycerol units are connected through a phosphodiester linkage between C-1-OH and C-3-OH of two glycerol units. The C-2 position is often acylated with D-alanine or it can be glycosylated by other carbohydrates [198] (see Fig. 6.21A). Phosphodiester linkages have also been observed between C-1-OH and C-2-OH in a teichoic acid from *Bacillus subtilis* [199a] and *Streptomyces antibioticus* [199b]. Ribitol units are joined together by phosphodiester linkages between C-1-OH and C-5-OH. The C-3 or C-4 OH groups can be acylated by D-alanine, and the C-2-OH can be substituted by either *N*-acetyl-2-amino-2-deoxy-β-D-glucopyranosyl or α-D-glucopyranosyl units, depending on the bacterium (see Fig. 6.21B). A significant amount of variation occurs for the different species in the amount and nature of the carbohydrate residues that are attached to the sugar alcohols. There is a range from those that have little or no substituted carbohydrate residues to those that have di- and trisaccharide units attached to each sugar alcohol.

More complex teichoic acids occur that have an alternating structure of glycerol joined to the C-4-OH of *N*-acetyl-2-amino-2-deoxy-α-D-glucopyranosyl-1-phosphate by a phosphodiester linkage, with the α-1-phosphate joined to C-1-OH of the next glycerol unit [200] (see Fig. 6.21C).

The number of repeating units in the teichoic acids is usually relatively small, ranging between 10 and 100. The teichoic acid chains are linked covalently to the peptidoglycan through a terminal *N*-acetyl-D-glucosaminyl-1-phosphodiester to C-6-OH of the *N*-acetyl-D-muramic acid residues in the murein chain [201] (see Fig. 6.21D). Although gram-negative bacteria have much less teichoic acid than gram-positive bacteria, the teichoic acids of gram-negative bacteria are primarily linked to lipids found in membranes.

A polysaccharide that is very closely related in structure to the teichoic acids is an *O*-phospho-D-mannan, elaborated by yeasts of the *Hansenula* genus. The polysaccharide differs from the usual teichoic acids in that the carbohydrate unit is an

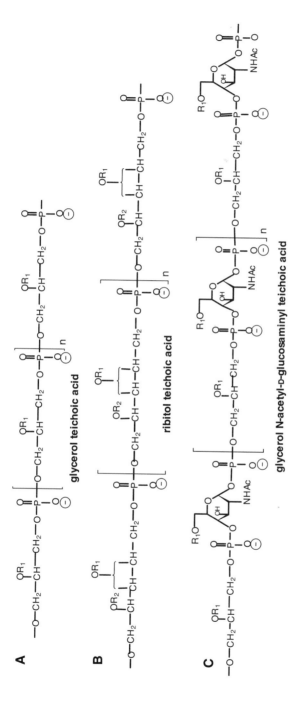

Figure 6.21. Structures of teichoic acids: **A**, glycerol teichoic acid; **B**, ribitol teichoic acid; **C**, glycerol *N*-acetyl-D-glucosaminyl teichoic acid.

(*continued*)

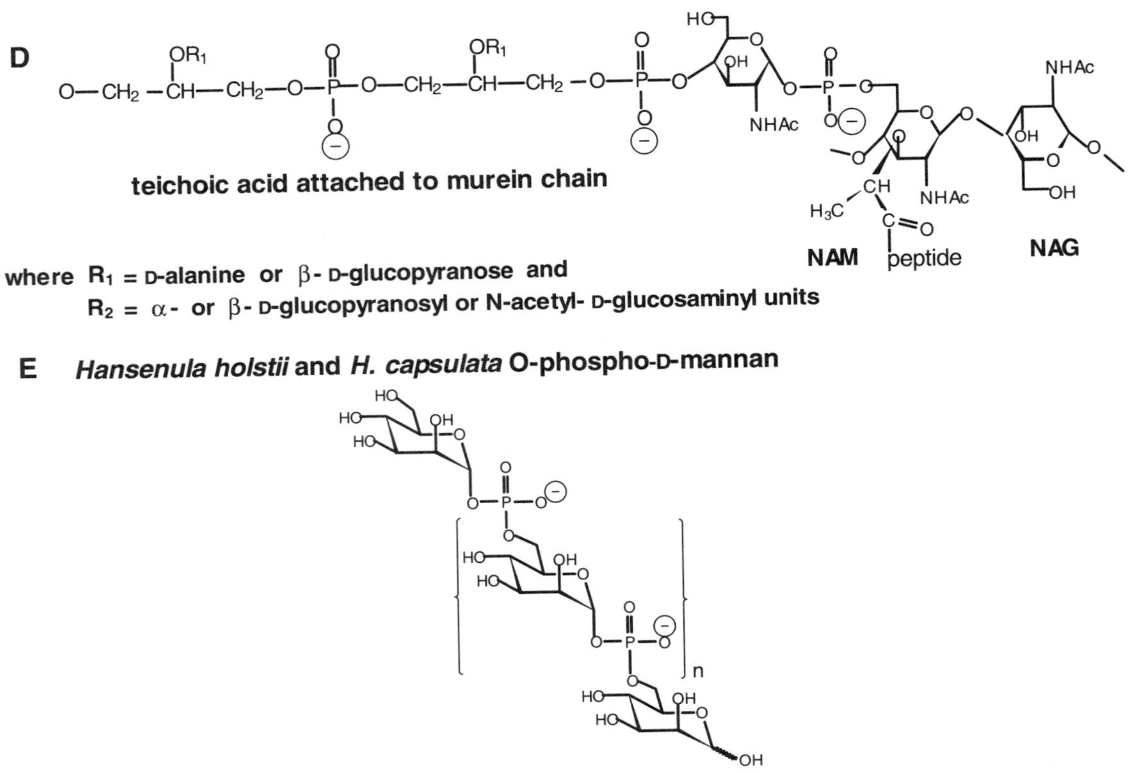

Figure 6.21. *(continued)* **D**, glycerol teichoic acid attached to murein chain at C-6 of *N*-acetyl-D-muramic acid; **E**, *O*-phospho-D-mannan.

aldose rather than an alditol. The polysaccharides that are elaborated extracellularly by *Hansenula holstii* NRRL[3] Y-2448 and *Hansenula capsulata* NRRL Y-1842 contain phosphate and D-mannose in a ratio of 1:1. The phosphate joins α-D-mannopyranose by a phosphodiester linkage between the α-anomeric, hemiacetal hydroxyl group of one D-mannose to the C-6 primary hydroxyl group of an adjoining D-mannose [202] (see Fig. 6.21E).

The teichoic acids are water soluble and obtained from the cell walls of gram-positive bacteria by hydrolysis of the acid-labile glucosaminyl-1-phosphate linkage. The bacterial cells are washed with saline and then treated with cold (4°C) 10% trichloroacetic acid in a blender for 1 min. The hydrolysis is then allowed to proceed for 15 hr at 0°C. The solution of solubilized teichoic acid solution is neutralized, and the teichoic acids are precipitated by the addition of one or two volumes of ethanol. The ethanol solution is allowed to stand 12 hr at 0°C, and the precipitated teichoic acid is removed by centrifugation [197].

The function of the teichoic acids in the bacterial cell wall has not been definitely determined. Shortly after their discovery in the late 1950s, it was suggested that they function in the control of the concentration of cations in the cell wall. There was some evidence that they played a role in the balance of divalent cations in membranes and thereby provided membrane integrity and stability to the enzymes in the membrane and cell wall [203].

Teichoic acids have also been implicated in the extracellular secretion of enzymes by bacteria. Treatment of *Bacillus amyloliquefaciens* with tunicamycin, an antibiotic known to inhibit cell wall synthesis, produced cells that were deficient in teichoic acid and did not secrete α-amylase. The α-amylase was found to be cell bound. The accumulation of α-amylase in the cell was also observed when teichoic acid synthesis was inhibited by culturing the organism in a medium that was limited in phosphate. Further evidence of the involvement of teichoic acids in protein secretion was obtained by using mutants that did not synthesize teichoic acids. These mutants also had a marked decrease in the amount of α-amylase that was secreted [204].

It has been suggested that there is a lipoteichoic acid complex formed with glucan that is involved in the pathogenicity of sucrose-produced dental plaque formation by *S. mutans* [205]. This is possibly the result of the involvement of teichoic acids in the extracellular secretion of glucansucrases.

The teichoic acids, similar to capsular polysaccharides of gram-negative bacteria, produce strong and distinctive immunological reactions. They may, therefore, simply be an important structural component for the cell walls of gram-positive bacteria and provide some degree of protection against lysis and unfavorable changes in the environment.

The teichoic acids have structures that have a common feature with the nucleic acids. The nucleic acids are polymers composed of D-ribofuranose or 2-deoxy-D-ribofuranose residues joined together by phosphodiester linkages between C-3-

---

[3]NRRL refers to the microbial culture collection of the Northern Regional Research Laboratory, Peoria, IL, USA; the Y-numbers refers to the specific strains of yeast.

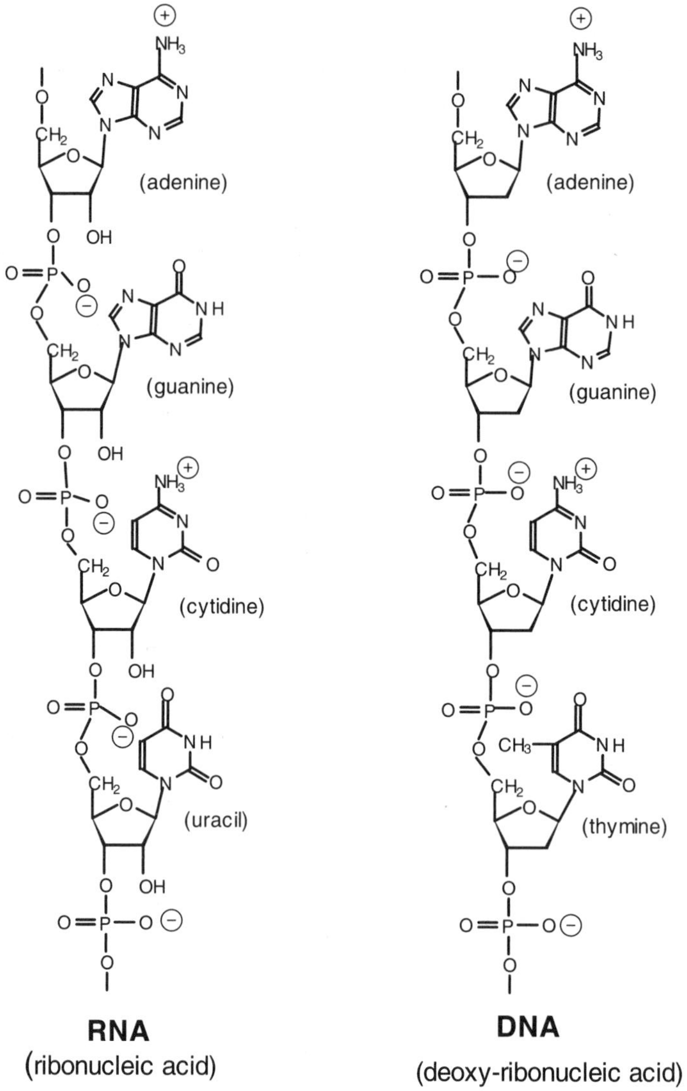

Figure 6.22. Structures of the nucleic acids.

OH and C-5-OH, similar to the kinds of phosphodiester linkages found in the teichoic acids. The nucleic acids have their ribose-reducing carbon substituted by purines and pyrimidines in *N*-glycosidic linkages (see Fig. 6.22). The two polymers, RNA and DNA, are very similar in structure; the former is composed of D-ribofuranose, and the latter is composed of 2-deoxy-D-ribofuranose. The latter also has the pyrimidine, thymine, substituted for uracil, the pyrimidine found in RNA. Even though it is widely recognized that the purine and pyrimidine bases play an important role in preserving and transmitting the genetic information of the cell, it is not so well recognized that the two polymers are essentially carbohydrate polymers that have secondary and tertiary structures and properties that are similar to those displayed by polysaccharides, such as the formation of double helices, and that they undergo chemical reactions that are essentially those of a carbohydrate polymer.

DNA is well known for its function of carrying the genetic information (genes) for organisms and for its formation of a specific antiparallel double helix in which a purine of one chain specifically hydrogen bonds to a pyrimidine of the second chain-adenine with thymine and guanine with cytosine, giving complimentary binding. RNA has multiple roles in the transfer of the genetic information from the DNA to give the biosynthesis of proteins. There are three types of RNA: (1) transfer RNA (tRNA), which forms a high-energy covalent linkage with a specific amino acid; (2) messenger RNA (mRNA), which carries a sequence of codes, copied from the DNA gene, to the ribosome; (3) ribosomal RNA (rRNA), which is located in the ribosome along with proteins and enzymes necessary for carrying out the biosynthesis of the specific proteins coded for by the genes. The specific tRNA, carrying a specific amino acid, pairs up its code (anticodon) to a complimentary code (codon) on the mRNA. This specific aligning of the different tRNAs, carrying specific amino acids, with the mRNA creates a specific sequence of amino acids. These amino acids are joined together by enzyme-catalyzed formation of amide (peptide) linkages to eventually form a protein with a specific sequence of amino acids that has been determined by the three-base (purine/pyrimidine) sequence (codons) of the DNA gene.

6.6 Simplified Representation of Oligosaccharide and Polysaccharide Structures

The structures that have been used primarily in this chapter are the conformational formulas (C1 and 1C) that illustrate the configuration of each carbon as well as the position of each substituent in space relative to each other and to the pyranose ring. All of this information may not always be necessary when representing the structures of oligo- and polysaccharides. Often the important information is the nature of the monosaccharide residue(s) and how they are joined together. The method of representing the structure need only show the sequence of the monosaccharide residues, the types of glycosidic bonds, and the type of branching. These facets of oligo- and polysaccharide structure can be very simply illustrated by representing the pyranose ring as a circle with the five hydroxyl groups represented as lines extending from the circle. Those hydroxyl groups that are *above* the plane of the pyranose ring in the Haworth or conformational formulas are represented by placing the line *inside* the circle, and those hydroxyl groups that are *below* the plane of the ring are represented by placing the line *outside* the circle [206]. The orientation of the circle is the same as the Haworth or conformational formulas, with the hemiacetal hydroxyl to the right and the other hydroxyl groups at posi-

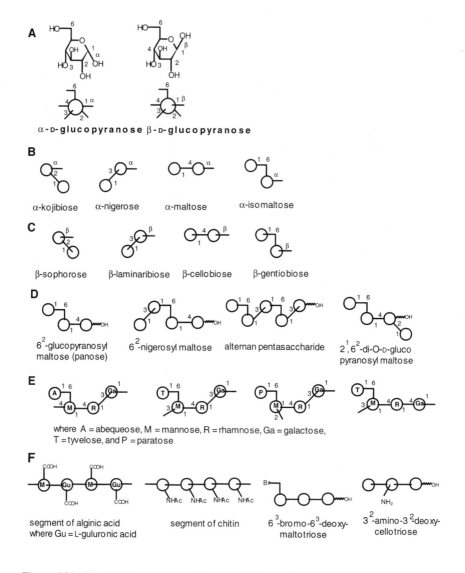

Figure 6.23. Simplified structures of oligosaccharides and polysaccharides.

tions 2, 3, 4, and 6, following clockwise around the circle. Compare the Haworth structure and the simplified circle structure in Fig. 6.23A for α- and β-D-glucopyranose.

When carbohydrate residues are combined together, only the glycosidic bond need be shown. This is illustrated for the eight α- and β-glucodisaccharides that have the $1 \to 2$, $1 \to 3$, $1 \to 4$, and $1 \to 6$ glycosidic linkages (see Fig. 6.23B and C).

More complex tri, tetra-, and pentasaccharides, each with more than one type of glycosidic linkage, can also be represented, as shown in Fig. 6.23D. When

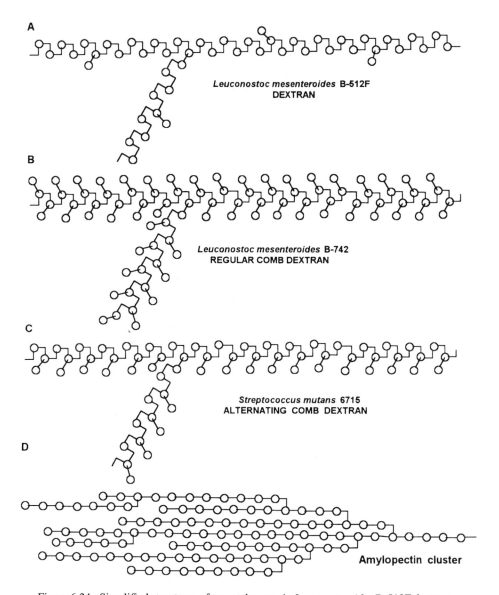

Figure 6.24. Simplified structures of some glucans. **A,** *L. mesenteroides* B-512F dextran; **B,** *L. mesenteroides* B-742 regular comb dextran; **C,** *S. mutans* alternating comb dextran; **D,** amylopectin cluster.

more than one type of monosaccharide residue are linked together to form heterosaccharide structures, they may most easily be distinguished by using a single upper case letter, or possibly two letters, to represent each different monosaccharide residue. Fig. 6.23E illustrates this type of structure for some *Salmonella* O-antigen repeating units, and Fig. 6.23F illustrates the repeating structure of alginic acid. For some monosaccharide residues that have distinguishing substituents such as acetamido, carboxyl, halogen, and so on, the groups can simply be added to the particular line where they are substituted onto the pyranose ring. This is illustrated in Fig. 6.23F for alginic acid, chitin, and two substituted trisaccharides. This simplified method for representing polysaccharide structures is particularly useful in illustrating various branching patterns. This is shown for a number of branched glucans in Fig. 6.24.

6.7 Literature Cited

1. N. P. Badenhuizen, *The Biogenesis of Starch Granules in Higher Plants,* pp. 5–25, Appleton-Century-Crofts, New York (1969).
2. J.-L. Jane, T. Kasemsuwan, S. Leas, H. Zobel, and J. F. Robyt, *Starch/Stärke,* **46** (1994) 121–129.
3. Personal communication, National Starch & Chem. Co., Bridgewater, NJ, USA (1995).
4. K. H. Meyer, W. Brentano, and P. Bernfeld, *Helv. Chim. Acta,* **23** (1940) 845–853.
5. K. H. Meyer and P. Bernfeld, *Helv. Chim. Acta,* **24** (1941) 359E.
6. K. H. Meyer, P. Bernfeld, R. A. Boissonnas, P. Guertler, and G. Noelting, *J. Phys. Chem.,* **53** (1949) 319–325.
7. S. A. Watson, A. H. A. de Willigen, and N. P. Badenhuizen, *Methods Carbohydr. Chem.,* **4** (1964) 3–32.
8. T. J. Schoch, *Cereal Chem.,* **18** (1941) 121–125.
9. T. J. Schoch, *J. Am. Chem. Soc.,* **64** (1942) 2957–2960; *Adv. Carbohydr. Chem.,* **1** (1942) 247–278.
10. R. L. Whistler and G. E. Hilbert, *J. Am. Chem. Soc.,* **67** (1945) 1161–1162.
11. W. N. Haworth, S. Peat, and P. E. Sargott, *Nature,* **157** (1946) 19–20.
12. B. Zaslow, *Biopolymers,* **1** (1963) 165–169.
13. T. D. Simpson, F. R. Dintzis, and N. W. Taylor, *Biopolymers,* **11** (1972) 2591–2600.
14. J. Szejtli and S. Augustat, *Stärke,* **18** (1966) 38–44; J. Szejtli, S. Augustat, and M. Richter, *Biopolymers,* **5** (1967) 5–10.
15. J. Holló and J. Szejtli, *Stärke,* **10** (1958) 49–54.
16. W. Banks and C. T. Greenwood, *Stärke,* **23** (1971) 300–305.
17. R. E. Rundle and R. R. Baldwin, *J. Am. Chem. Soc.,* **65** (1943) 554–557; R. E. Rundle and D. French, *J. Am. Chem. Soc.,* **65** (1943) 558–560; **65** (1943) 1707–1709.
18. D. L. Mould and R. L. M. Synge, *Biochem. J.,* **58** (1954) 585–591; J. M. Bailey and W. J. Whelan, *J. Biol. Chem.,* **236** (1961) 969–973.
19. R. E. Rundle and F. C. Edwards, *J. Am. Chem. Soc.,* **65** (1943) 2200–2203; R. S. Bear, *J. Am. Chem. Soc.,* **66** (1994) 2122–2123; F. F. Mikus, R. M. Hixon, and R. E. Rundle, *J. Am. Chem. Soc.,* **68** (1946) 1115–1123.
20. R. St. J. Manley, *J. Polym. Sci. Part A,* **2** (1964) 4503–4512.
21. Y. Yamashita, *J. Polym. Sci., Part A,* **3** (1965) 3251–3260.
22. Y. Yamashita and N. Hirai, *J. Polym. Sci., Part A-2,* **4** (1966) 161–171.
23. Y. Yamashita and K. Monobe, *J. Polym. Sci., Part A-2,* **9** (1971) 1471–1481.
24. J.-L. Jane and J. F. Robyt, *Carbohydr. Res.,* **132** (1984) 105–118.
25. K. H. Meyer and P. Bernfeld, *Helv. Chim. Acta,* **23** (1940) 875–884.
26. D. French, *J. Jpn. Soc. Starch Sci.,* **19** (1972) 8–25.

27. Z. Nikuni, *J. Jpn. Soc. Starch Sci.,* **22** (1975) 78–92.

28. S. Hizukuri, *Carbohydr. Res.,* **147** (1986) 342–347.

29. T. J. Schoch and E. C. Maywald, in *Starch Chemistry and Technology,* Vol. 2, pp. 637–685 (R. L. Whistler, E. G. Paschall, J. N. BeMiller, and H. J. Roberts, eds.) Academic, New York (1967); L. E. Fitt and E. M. Snyder, in *Starch: Chemistry and Technology,* 2nd edn., pp. 675–689 (R. L. Whistler J. N. BeMiller, and E. G. Paschall, eds.) Academic, New York (1984).

30a. T. J. Schoch and E. Maywald, *Stärke,* **20** (1968) 362–365.

30b. R. Gracza, in *Starch: Chemistry and Technology,* 1st edn., Vol. 1, pp. 110–115 (R. L. Whistler, E. F. Paschall, J. N. BeMiller, and H. J. Roberts, eds.) Academic, New York (1965).

30c. D. French, in *Starch Chemistry and Technology,* 2nd edn., pp. 139–167 (R. L. Whistler, J. N. BeMiller, and E. G. Paschall, eds.) Academic, New York (1984).

30d. H. F. Zobel, *Stärke,* **40** (1988) 1–7.

31. C. Sterling, in *Starch and Its Derivatives,* 4th edn., pp. 139–167 (J. A. Radley, ed.) Chapman and Hall, London (1968).

32. D. J. Manners, *Carbohydr. Polym.,* **11** (1989) 87–96; P. A. M. Steeneken and E. Smith, *Carbohydr. Res.,* **209** (1991) 239–248.

33. M. Yamaguchi, K. Kainuma, and D. French, *J. Ultrastruc. Res.,* **69** (1979) 249–261.

34. D. J. Gallant and A. Guilbot, *Stärke,* **21** (1969) 156–163.

35. D. J. Gallant, C. Mercier, and A. Guilbot, *Cereal Chem.,* **49** (1972) 354–365.

36. D. J. Gallant, C. Mercier, and A. Guilbot, *Stärke,* **25** (1973) 335–342.

37. G. Hollinger and R. H. Marchessault, *Biopolymers,* **14** (1975) 265–276.

38. K. Kainuma and D. French, *Biopolymers,* **11** (1972) 2241–2250.

39. J. P. Robbin, C. Mercier, R. Charbonniere, and A. Guilbot, *Cereal Chem.,* **51** (1974) 389–398.

40. M. J. Gidley and S. M. Bociek, *J. Am. Chem. Soc.,* **107** (1985) 7040–7041; **110** (1988) 3820–3822.

41. P. Nordin, H. Moser, G. Rao, N. Giri, and T. Liang, *Stärke,* **22** (1970) 256–261.

42. H. W. Leach and T. J. Schoch, *Cereal Chem.,* **38** (1961) 34–46.

43. J.-C. Valentudie, P. Colonna, B. Bouchet, and D. J. Gallant, *Starch/Stärke,* **45** (1993) 270–276.

44. D. M. Hall and J. G. Sayre, *Textile Res. J.,* **39** (1969) 1044–1052; **40** (1970) 256–266.

45. G. H. Lathe and C. R. T. Ruthven, *Biochem. J.,* **62** (1956) 665–674.

46. S. A. Brown and D. French, *Carbohydr. Res.,* **59** (1977) 203–212.

47. A. Kimura and J. F. Robyt, *Carbohydr. Res.,* **277** (1995) 87–107; **287** (1996) 255–261; **288** (1996) 233–240.

48a. P. Ross, R. Mayer, and M. Benziman, *Microbiol. Rev.,* **55** (1991) 35–58.

48b. L. Glaser, *J. Biol. Chem.,* **232** (1958) 627–636; S. Klungsöyr, *Nature,* **185** (1960) 104; J. R. Colvin, *Nature,* **183** (1959) 1135–1136; A. M. Brown, *Nature,* **187** (1960) 1010–1011; M. Swissa, Y. Aloni, H. Weinhouse, and M. Benziman, *J. Bacteriol.,* **143** (1980) 1142–1150; Y. Aloni, R. Cohen, M. Benziman, and D. Delmer, *J. Biol. Chem.,* **258** (1983) 4419–4423; D. P. Delmer, *Adv. Carbohydr. Chem. Biochem.,* **41** (1983) 105–154; M. E. Embuscado, J. S. Marks, and J. N. BeMiller, *Food Hydrocoll.,* **8** (1994) 407–418; **8** (1994) 419–430.

48c. K. Ward, Jr. and P. A. Seib, in *The Carbohydrates,* 2nd edn., Vol. 2A pp. 413–415, (W. Pigman, D. Horton, and A. Herp, eds.) Academic, New York (1983).

49. R. H. Marchessault and P. R. Sundararajan, in *The Polysaccharides*, Vol. 2, pp. 25–65 (G. O. Aspinall, ed.) Academic, New York (1983).

50. K. Hess, G. Mahl, and E. Gütter, *Kolloid. Z.*, **155** (1957) 1–9.

51. W. M. Corbett, *Methods Carbohydr. Chem.*, **3** (1963) 3–4.

52. J. W. Green, *Methods Carbohydr. Chem.* **3** (1963) 9–20.

53. G. S. Kasbekar, *Curr. Sci.*, **9** (1940) 411–413.

54. G. Jayme and K. Neuschaffer, *Makromol. Chem.*, **28** (1957) 71–77.

55. M. R. Ladisch, C. M. Ladisch, and G. T. Tsao, *Science*, **201** (1978) 743–745.

56a. C. M. Ladisch, *Methods Enzymol.*, **160** (1988) 16–19.

56b. S. Hakamori, *J. Biochem. (Tokyo)*, **55** (1964) 205–208.

57. G. O. Aspinall, J. A. Molloy, and J. W. T. Craig, *Can. J. Biochem.*, **47** (1969) 1063–1070.

58. W. D. Bauer, K. W. Talmadge, K. Keestra, and P. Albersheim, *Plant Physiol.*, **51** (1973) 174–187.

59. P. Kooiman, *Recl. Trav. Chim. Pays-Bas*, **80** (1961) 849–865.

60. S. E. B. Gould, D. A. Rees, and N. J. Wright, *Biochem. J.*, **124** (1971) 47–53.

61. D. S. Hus and R. E. Reeves, *Carbohydr. Res.*, **5** (1967) 202–209.

62. G. O. Aspinall, T. N. Krishnamurthy, and K.-G. Rosell, *Carbohydr. Res.*, **55** (1977) 11–19.

63. T. E. Timell, *Adv. Carbohydr. Chem.*, **19** (1964) 247–302; **20** (1965) 409–483.

64. G. O. Aspinall, *Adv. Carbohydr. Chem.* **14** (1959) 429–468.

65. G. A. Towle and R. L. Whistler, in *Phytochemistry*, Vol. 1, 198–248 (L. P. Miller, ed.) Van Nostrand Reinhold, New York (1973).

66. Y. S. Ovodov, *Pure Appl. Chem.*, **42** (1975) 351–369.

67. A. M. Stephen, in *The Polysaccharides*, Vol. 2, pp. 166–169 (G. O. Aspinall, ed.) Academic, San Diego (1983).

68. D. A. Rees, *MTP Int. Rev. Sci.: Org. Chem., Ser. One*, **7** (1973) 251–283.

69. B. V. McCleary, N. K. Matheson, and D. M. Small, *Phytochemistry*, **15** (1976) 1111–1117.

70. P. M. Dey, *Adv. Carbohydr. Chem. Biochem.*, **35** (1978) 341–392.

71. H. Maier, M. Anderson, C. Karl, K. Magnuson, and R. L. Whistler, in *Industrial Gums: Polysaccharides and Their Derivatives*, pp. 181–226 (R. L. Whistler and J. N. BeMiller, eds.) Academic, San Diego (1993).

72. B. V. McCleary, *Carbohydr. Res.*, **71** (1979) 205–212; H. Grasdalen and T. J. Painter, *Carbohydr. Res.*, **81** (1980) 59–66; S. B. Bociek, M. J. Izzard, A. Morrison, and D. Welti, *Carbohydr. Res.*, **93** (1981) 279–288.

73. B. R. Vijayendran and T. Bone, *Carbohydr. Polym.*, **4** (1984) 299–304.

74. R. L. Whistler, in *Industrial Gums: Polysaccharides and Their Derivatives*, pp. 302–305 (R. L. Whistler and J. N. BeMiller, eds.) Academic, San Diego (1993).

75. C. Rolin, in *Industrial Gums: Polysaccharides and Their Derivatives*, pp. 257–282 (R. L. Whistler and J. N. BeMiller, eds.) Academic, San Diego (1993).

76. E. L. Pippen, R. M. McCready, and H. S. Owens, *J. Am. Chem. Soc.*, **72** (1950) 813–815.

77. W. J. Kim, V. M. N. Rao and C. J. B. Smit, *J. Food Sci.*, **43** (1978) 572–578.

78. D. Thom, G. T. Grant, E. R. Morris, and D. A. Rees, *Carbohydr. Res.*, **100** (1982) 29–36.

79. D. A. Rees and A. W. Wright, *J. Chem. Soc., B* (1971) 1366–1369.

80. K. J. Palmer and M. B. Hartzog, *J. Am. Chem. Soc.*, **67** (1945) 2122–2124.

81. M. D. Walkinshaw and S. Arnott, *J. Mol. Biol.*, **153** (1981) 1055–1062; **153** (1981) 1075–1083.

82. D. Oakenfull and A. Scott, *J. Food Sci.*, **49** (1984) 1093–1100.

83. D. M. W. Anderson, M. C. L. Gill, A. M. Jeffrey, and F. J. McDougall, *Phytochemistry*, **24** (1985) 71–77.

84. G. O. Aspinall, E. L. Hirst, and A. Wickstrom, *J. Chem. Soc.* (1955) 1160–1164; G. O. Aspinall, B. J. Auret, and E. L. Hirst, *J. Chem. Soc.* (1958) 4408–4412.

85. G. O. Aspinall and J. Baille, *J. Chem. Soc.* (1963) 1702–1706.

86. A. D. French and A. L. Waterhouse, in *Science and Technology of Fructans*, pp. 41–82 (M. Suzuki and N. J. Chatterton, eds.) CRC, Boca Raton (1993).

87. G. A. F. Hendry and R. K. Wallace, in *Science and Technology of Fructans*, pp. 119–140 (M. Suzuki and N. J. Chatterton, eds.) CRC, Boca Raton (1993).

88. T. J. Painter, in *The Polysaccharides*, Vol. 2, pp. 263–264 (G. O. Aspinall, ed.) Academic, San Diego (1983).

89. O. Smidsrød and A. Haug, *Acta Chem. Scand.*, **22** (1968) 1989–1997.

90. K. Clare, in *Industrial Gums, Polysaccharides and Their Derivatives*, 3rd edn., p. 108, (R. L. Whistler and J. N. BeMiller, eds.) Academic, San Diego (1993).

91. I. W. Cottrell and P. Kovacs, in *Handbook of Water-Soluble Gums and Resins*, pp. 2.2–2.4 (R. L. Davidson, ed.) McGraw-Hill, New York (1980).

92. A. Linker and R. S. Jones, *J. Biol. Chem.*, **241** (1966) 3845–3849; L. R. Evans and A. Linker, *J. Bacteriol.*, **116** (1973) 915–924.

93. P. A. J. Gorin and J. F. T. Spencer, *Can. J. Chem.*, **44** (1966) 993–1000; D. F. Pindar and C. Bucke, *Biochem. J.*, **152** (1975) 617–622.

94. W. L. Nelson and L. H. Cretcher, *J. Am. Chem. Soc.*, **51** (1929) 1914–1920.

95. A. Haug, *Methods Carbohydr. Chem.*, **5** (1965) 69–71.

96. E. L. Hirst, E. Percival, and J. K. Wold, *J. Chem. Soc.* (1964) 1493–1498; E. L. Hirst and D. A. Rees, *J. Chem. Soc.* (1965) 1182–1188.

97. A. Haug and B. Larsen, *Acta Chem. Scand.*, **16** (1962) 1908–1918.

98. B. Larsen and A. Haug, *Carbohydr. Res.*, **17** (1971) 287–297.

99. A. Haug, B. Larsen, and O. Smidsrød, *Acta Chem. Scand.*, **20** (1966) 183–190; **21** (1967) 691–700.

100. A. Haug, B. Larsen, O. Smidsrød, and T. J. Painter, *Acta Chem. Scand.*, **23** (1969) 2955–2964.

101. E. D. T. Atkins, W. Mackie, and E. E. Smolko, *Nature*, **225** (1970) 626–627.

102. W. H. McNeely and D. J. Pettitt, in *Industrial Gums*, 2nd edn., pp. 74–75 (R. L. Whistler and J. N. BeMiller, eds.) Academic, San Diego (1973).

103. G. Skjåk-Bræk, O. Smidsrød, and B. Larsen, *Int. J. Biol. Macromol.*, **8** (1986) 330–337.

104. O. Smidsrød, A. Haug, and S. Whittington, *Acta Chem. Scand.*, **26** (1972) 2563–2564.

105. R. G. Schwiger, *J. Org. Chem.*, **27** (1962) 1789–1791.

106. C. A. Steginsky, J. M. Beale, H. G. Floss, and R. M. Mayer, *Carbohydr. Res.*, **225** (1992) 11–26.

107. G. A. King, A. J. Daugulis, P. Faulkner, and M. F. A. Goosen, *Biotech. Prog.*, **3** (1987) 231–240; Y. Fukushima, K. Okamura, K. Imai, and H. Motai, *Biotech. Bioeng.*, **32** (1988) 584–594; O. Smidsrød and G. Skjåk-Bræk, *TIBTECH*, **8** (1990) 71–78; K. D. Vorlop and H. J. Steinert, *Ann. N.Y. Acad. Sci.*, **501** (1987) 339–342.

108a. H. Tanaka, H. Kurosawa, E. Kokufuta, and I. A. Veliky, *Biotech. Bioeng.*, **26** (1984) 1393–1394; T. Kato and K. Horikoshi, *Biotech. Bioeng.*, **26** (1984) 595–598; E.

Kokufuta, N. Shimizu, H. Tanaka, and I. Nakamura, *Biotech. Bioeng.,* **32** (1988) 756–759; P. Bajpai and A. Margaritis, *Enzyme Microb. Technol.,* **7** (1985) 34–36.

108b. K. D. Vorlop, H. J. Steinert, and J. Klein, *Ann. N.Y. Acad. Sci.,* **501** (1987) 339–342.

108c. H. Tanaka, H. Kurosawa, E. Kokufuta, and I. A. Veliky, *Biotech. Bioeng.,* **26** (1984) 1393–1394.

108d. M. M. Daly and D. Knorr, *Biotech. Prog.,* **4** (1988) 76–81.

109. C. J. Gray and J. Dowsett, *Biotech. Bioeng.,* **31** (1988) 607–612.

110. C. D. Scott, *Enzyme Microb. Technol.,* **9** (1987) 66–73.

111. G. H. Therkelsen, in *Industrial Gums, Polysaccharides and Their Derivatives,* 3rd edn., pp. 146–148 (R. L. Whistler and J. N. BeMiller, eds.) Academic, San Diego (1993).

112. T. J. Painter, *Methods Carbohydr. Chem.,* **5** (1965) 98–100.

113. E. R. Morris, D. A. Rees, and E. J. Welsh, *J. Chem. Soc. Perkin Trans.,* **2** (1978) 793–800.

114. A. A. McKinnon, D. A. Rees, and F. B. Williamson, *J. Chem. Soc., Chem. Commun.,* (1969) 701–707; D. A. Rees, *J. Chem. Soc. B,* (1969) 217–222; I. C. M. Dea, A. A. McKinnon, and D. A. Rees, *J. Mol. Biol.,* **68** (1972) 153–159; S. Arnott and W. E. Scott, *J. Mol. Biol.,* **90** (1974) 253–260.

115. E. R. Morris, D. A. Rees, and G. Robinson, *J. Mol. Biol.,* **138** (1980) 349–358; O. Simdsrød and H. Grasdalen, *Carbohydr. Polym.,* **2** (1982) 270–278.

116. L.-G. Ekstroem and J. Kuivinen, *Carbohydr. Res.,* **116** (1983) 89–94; H. J. Vreeman, T. H. M. Snoeren, and T. A. J. Payens, *Biopolymers,* **19** (1980) 1357–1364; H.-K. Wong, K.-H. Lee, and H.-A. Wong, *Carbohydr. Res.,* **81** (1980) 1–7.

117. S. Hjerten, *J. Chromatog.,* **61** (1971) 73–80.

118. N. R. Krieg and P. Gerhardt, in *Manual of Methods for General Bacteriology,* pp. 143–150 (P. Gerhardt, R. G. E. Murray, R. N. Costilow, E. W. Nester, W. A. Wood, N. R. Krieg, and G. Briggs Phillips, eds.) American Society of Microbiology, Washington, D.C. (1981).

119a. E. F. W. Pfluger, *Arch. Gen. Physiol.,* **131** (1910) 201–202.

119b. M. Somogyi, *Methods Enzymol.,* **3** (1957) 3–4.

120. W. A. J. Bryce, C. T. Greenwood, and I. G. Jones, *J. Chem. Soc.* (1958) 3845–3850.

121. M. R. Stetten, H. M. Katzen, and D. Stetten, Jr., *J. Biol. Chem.,* **222** (1956) 587–599.

122. R. Laskov and E. Margoliash, *Bull. Res. Counc. Israel, Sect. A,* **11** (1963) 351–362.

123. W. N. Haworth, E. L. Hirst, and F. Smith, *J. Chem. Soc.* (1939) 1914–1922.

124. J. J. Marshall, *Adv. Carbohydr. Chem.,* **30** (1974) 257–370.

125. R. Geddes, J. D. Harvey, and P. R. Wills, *Biochem. J.,* **163** (1977) 201–209.

126. R. R. Cardell, *Int. Rev. Cytol.,* **48** (1977) 221–279; R. N. Margolis, R. R. Cardell, and R. T. Curnow, *J. Cell Biol.,* **83** (1979) 348–356.

127. J-C. Wanson and P. Drochmans, *J. Cell Biol.,* **38** (1968) 130–150; **54** (1972) 206–224.

128. S. A. Orrell and E. Bueding, *J. Biol. Chem.,* **239** (1964) 4021–4026.

129. R. Geddes, in *The Polysaccharides,* Vol. 3, p. 316 (G. O. Aspinall, ed.) Academic, San Diego (1985).

130. L.-Å. Fransson, in *The Polysaccharides,* Vol. 3, pp. 338–386 (G. O. Aspinall, ed.) Academic, New York (1985).

131. B. Casu, *Adv. Carbohydr. Chem. Biochem.,* **43** (1985) 51–70.

132. P. Karrer and G. Francois, *Helv. Chim. Acta,* **12** (1929) 986–996; R. H. Hackman, *Aust. J. Biol. Sci.,* **7** (1954) 168–173.

133. K. Ward, Jr. and P. A. Seib, in *The Carbohydrates,* Vol. 2A, pp. 435–437 (W. Pigman, D. Horton, and A. Herp, eds.) Academic, New York (1970).

134a. K. H. Meyer and G. W. Pankow, *Helv. Chim. Acta,* **18** (1935) 589–595; W. Lotman and L. E. R. Picken, *Experienta,* **6** (1950) 58–67.

134b. K. M. Rudall, *Adv. Insect Physiol.,* **1** (1963) 257–313.

134c. K. M. Rudall and W. Kenchington, *Biol. Rev.,* **48** (1973) 597–636.

134d. G. A. F. Roberts, *Chitin Chemistry,* p. 22, Macmillian, New York (1992).

135. P. A. Sanford, in *Chitin and Chitosan,* pp. 51–69 (G. Skjåk-Braek, T. Anthonsen, and P. A. Sanford, eds.) Elsevier Applied Science, London (1989).

136. "Chitosan for Cell Immobilization," PLI-004, Protan Laboratories, Redmond, WA (1987); "Chitosan for Immobilization of Enzymes," PLI-003, Protan Laboratories (1987).

137. R. W. Jeanloz, N. Sharon, and H. M. Flowers, *Biochem. Biophys. Res. Commun.,* **13** (1963) 20–25.

138. D. J. Tipper, J.-M. Ghuysen, and J. L. Strominger, *Biochemistry,* **4** (1965) 468–473.

139. N. Sharon, T. Osawa, H. M. Flowers, and R. W. Jeanloz, *J. Biol. Chem.,* **242** (1966) 223–230.

140. D. J. Tipper and J. L. Strominger, *Biochem. Biophys. Res. Commun.,* **22** (1966) 48–56.

141. M. Leyh-Bouille, J.-M. Ghuysen, D. J. Tipper, and J. L. Strominger, *Biochemistry,* **5** (1966) 3079–3090.

142. H. P. Browder, W. A. Zygmunt, J. R. Young, and P. A. Tavormina, *Biochem. Biophys. Res. Commun.,* **19** (1965) 383–389.

143. D. J. Tipper, J. L. Strominger, and J. C. Ensign, *Biochemistry,* **6** (1967) 906–920.

144. J.-M. Ghuysen, *Bacteriol. Rev.,* **32** (1968) 425–464.

145. E. Munoz, J.-M. Ghuysen, M. Leyh-Bouille, J. F. Petit, and R. Tinelli, *Biochemistry,* **5** (1966) 3091–3098.

146. T. A. Krulwich, J. C. Ensign, D. J. Tipper, and J. L. Strominger, *J. Bacteriol.,* **94** (1967) 734–740.

147. W. Weidel and H. Pelzer, *Adv. Enzymol.,* **26** (1964) 193–232.

148. L. Pasteur, *Bull. Soc. Chim. Paris,* (1861) 30–36.

149. M. E. Durin, *Compt. Rend.,* **83** (1876) 128–137.

150. C. Scheubler, *Z. Ver. Dtsch. Zuker-Ind.,* **24** (1874) 309–335.

151. P. van Tieghem, *Ann. Sci. Nat. Bot. Biol. Veg.,* **7** (1878) 180–203.

152. E. J. Hehre, *Science,* **93** (1941) 237–238.

153. J. F. Robyt, *Adv. Carbohydr. Chem. Biochem.,* **51** (1995) 133–168.

154. A. Jeanes, W. C. Haynes, C. A. Wilham, J. C. Rankin, E. H. Melvin, M. J. Austin, J. E. Cluskey, B. E. Fisher, H. M. Tsuchiya, and E. E. Rist, *J. Am. Chem. Soc.,* **76** (1954) 5041–5047.

155. C. A. Wilham, B. H. Alexander, and A. Jeanes, *Arch. Biochem. Biophys.,* **59** (1955) 61–69.

156. J. W. Van Cleve, W. C. Schaefer, and E. E. Rist, *J. Am. Chem. Soc.,* **78** (1956) 4435–4440; A. Jeanes and F. R. Seymour, *Carbohydr. Res.,* **74** (1979) 31–40.

157. F. R. Seymour, E. C. M. Chen, and S. H. Bishop, *Carbohydr. Res.,* **68** (1978) 245–255.

158. A. Shimamura, H. Tsumori, and H. Mukasa, *Biochim. Biophys. Acta,* **702** (1982) 72–80.

159. F. R. Seymour, R. L. Julian, A. Jeanes, and B. L. Lamberts, *Carbohydr. Res.,* **86** (1980) 227–246.

160. F. R. Seymour, R. D. Knapp, E. C. M. Chen, and S. H. Bishop, *Carbohydr. Res.,* **74** (1979) 41–47.

161. G. L. Côté and J. F. Robyt, *Carbohydr. Res.,* **101** (1982) 57–74.

162. A. J. T. Grönwall and B. G. A. Ingleman, *Acta Physiol. Scand.,* **7** (1944) 97–100; **9** (1945) 1–8.

163. P. Flodin and B. Ingleman, Swed. Pat. 169,293 (1959); P. Flodin and J. Porath, U.S. Pat. 3,002,823 (1961); J. C. Jansson, *Chromatographia,* **23** (1987) 361–366; P. Flodin, Swed. Pat., 358,894 (1961).

164. Y. Tsiyisaka and M. Mitsuhashi, in *Industrial Gums: Polysaccharides and Their Derivatives,* 3rd edn., pp. 447–460 (R. L. Whistler and J. N. BeMiller, eds.) Academic, San Diego (1993).

165. H. Bender, J. Lehmann, and K. Wallenfels, *Biochim. Biophys. Acta,* **36** (1959) 309–318.

166. K. Wallenfels, H. Bender, G. Keilich, and G. Bechtler, *Angew. Chem.,* **73** (1961) 245–251; H. O. Bouveng, H. Kiessling, B. Lindberg, and J. McKay, *Acta Chem. Scand.,* **16** (1962) 615–620; **17** (1963) 797–802; K. Wallenfels, G. Keilich, G. Bechtler, and D. Freudenberger, *Biochem. Z.,* **341** (1965) 433–437.

167. B. J. Catley, J. F. Robyt, and W. J. Whelan, *Biochem. J.,* **100** (1966) 5p; B. J. Catley and W. J. Whelan, *Arch. Biochem. Biophys.,* **143** (1971) 138–142.

168. B. J. Catley, *FEBS Lett.,* **20** (1972) 174–176.

169. S. Yuen, *Process Biochem.,* **9** (11) (1974) 7–9.

170a. S. Nakashio, N. Sekine, N. Toyota, F. Fujita, and M. Dohmoto, *Chem. Abstr.,* **83** (1975) 30247s.

170b. G. Avigad, in *Encyclopedia of Polymer Science and Engineering,* Vol. 8, pp. 71–78 (H. F. Mark, N. G. Gaylord, and N. M. Bikales, eds.) Wiley Interscience, New York (1968).

170c. S. M. Garzozynski and J. R. Edwards, *Arch. Oral Biol.,* **18** (1973) 239–251; J. Eshrlich, S. S. Stivala, W. S. Bahary, S. K. Garg, L. W. Long, and E. Newbrun, *J. Dent. Res.,* **54** (1975) 290–297; R. A. Hancock, K. Marshall, and H. Weigel, *Carbohydr. Res.,* **49** (1976) 351–360.

170d. J. K. Baird, V. M. C. Longyear, and D. C. Ellwood, *Microbios,* **8** (1973) 143–150; K.-G. Rossel and D. Birkhead, *Acta Chem. Scand., Ser. B,* **28** (1974) 589–592; S. Ebisu, K. Kato, S. Kotani, and A. Misaki, *J. Biochem. (Tokyo),* **78** (1975) 879–887; A. J. Corrigan and J. F. Robyt, *Infect. Immun.,* **26** (1979) 387–389.

171. B. Lindberg, J. Lorngren, and J. L. Thompson, *Carbohydr. Res.,* **28** (1973) 351–359; H. Bjorndal, C. G. Hellerquist, B. Lindberg, and S. Svensson, *Angew. Chem. Int. Ed. Engl.,* **9** (1970) 610–618; L. D. Melton, L. Mindt, D. A. Rees, and G. R. Sanderson, *Carbohydr. Res.,* **46** (1976) 245–257.

172. K. S. Kang and D. J. Pettitt, in *Industrial Gums: Polysaccharides and Their Derivatives,* 3rd edn. pp. 341–397 (R. L. Whistler and J. N. BeMiller, eds.) Academic, San Diego (1993).

173. I. T. Norton D. M. Goodall, S. A. Frangon, E. R. Morris, and D. A. Rees, *J. Mol. Biol.,* **175** (1984) 371–380; T. Sato, S. Kojima, T. Norisuye, and H. Fujita, *Polym. J.,* **16** (1984) 423–429.

174. R. Moorhouse, S. Arnott, and M. D. Walkinshaw, in *Extracellular Microbial Polysaccharides,* pp. 90–102 (P. A. Sanford and A. Laskin, eds.) ACS Symposium Series 45, Washington, D.C. (1977).

175. T. R. Andrew, in *Extracellular Microbial Polysaccharides,* pp. 231–241 (P. A. Sanford and A. Laskin, eds.) ACS Symposium Series 45, Washington, D.C. (1977).

176. K. S. Kang, G. T. Veeder, P. J. Mirrasoul, T. Kaneko, and I. W. Cottrell, *Appl. Environ. Microbiol.,* **43** (1982) 1086–1092.

177. P.-E. Jansson, B. Lindberg, and P. A. Sanford, *Carbohydr. Res.,* **124** (1984) 135–139.

178a. M.-S. Kuo, A. J. Mort, and A. Dell, *Carbohydr. Res.,* **156** (1986) 173–187.

178b. R. Chandrasekaran, L. C. Puigjaner, K. L. Joyce, and S. Arnott, *Carbohydr. Res.,* **181** (1988) 23–32.

179. P. E. Jansson, B. Lindberg, G. Widmalm, and P. A. Sanford, *Carbohydr. Res.,* **139** (1985) 217–223.

180. P.-E. Jansson, B. Lindberg, J. Lindberg, E. Maekawa, and P. A. Sanford, *Carbohydr. Res.,* **156** (1986) 157–163.

181. J. A. Peik, S. M. Steenbergen, and H. R. Hayden, U.S. Pat. 4,401,760 (1983).

182. O. Larm and B. Lindberg, *Adv. Carbohydr. Chem. Biochem.,* **33** (1976) 295–322.

183. L. Kenne, B. Lindberg, and S. Svensson, *Carbohydr. Res.,* **40** (1975) 69–75.

184. R. E. Reeves and W. F. Goebel, *J. Biol. Chem.,* **40** (1941) 511–519.

185. J. N. K. Jones and M. B. Perry, *J. Am. Chem. Soc.,* **79** (1957) 2787–2793.

186. P. A. Rebers and M. Heidelberger, *J. Am. Chem. Soc.,* **83** (1961) 3056–3059.

187. M. J. Watson, J. M. Tyler, J. G. Buchanan, and J. Baddiley, *Biochem. J.,* **130** (1972) 45–54.

188. G. J. F. Chittenden, W. K. Roberts, J. G. Buchanan, and J. Baddiley, *Biochem. J.,* **109** (1968) 597–602; E. V. Rao, M. J. Watson, J. G. Buchanan, and J. Baddiley, *Biochem. J.,* **111** (1969) 547–556.

189. O. Lüderitz, O. Westphal, A. M. Staub, and H. Nikaido, in *Microbial Toxins,* Vol. 4, pp. 158–162 (G. Weinbaum, S. Kadis, and S. J. Ajl, eds.) Academic, New York (1971); L. LeMinor and A. M. Staub, *Ann. Inst. Pasteur,* **110** (1966) 834–848.

190. O. Lüderitz, A. M. Staub, and O. Westphal, *Bacteriol. Rev.,* **30** (1966) 193–256.

191. P. W. Robbins and T. Uchida, *Biochemistry,* **1** (1962) 323–325.

192. A. M. Staub and G. Bagdian,*Ann. Inst. Pasteur,* **110** (1966) 849–860; G. Bagdian, O. Lüderitz, and A. M. Staub, *Ann. N.Y. Acad. Sci.,* **133** (1966) 849–860.

193. E. J. McGuire and S. B. Binkley, *Biochemistry,* **3** (1964) 247–251; A. K. Bhattacharjee, H. J. Jennings, C. P. Kenny, A. Martin, and I. C. P. Smith, *J. Biol. Chem.,* **250** (1975) 1926–1932.

194. J. B. Robbins, G. H. McCracken, Jr., E. C. Gotschlich, F. Orskov, I. Orskov, and L. A. Hanson, *N. Engl. J. Med.,* **290** (1974) 1216–1220.

195. W. Egan, T.-Y. Liu, D. Dorow, J. S. Cohen, J. D. Robbins, E. C. Gotschlich, and J. B. Robbins, *Biochemistry,* **16** (1977) 3687–3692.

196. M. R. Lifely, A. S. Gilbert, and C. Moreno, *Carbohydr. Res.,* **94** (1981) 193–203.

197. J. J. Armstrong, J. Baddiley, J. G. Buchanan, B. Carss, and G. R. Greenberg, *J. Chem. Soc.* (1958) 4344–4350.

198. A. R. Archibald and J. Baddiley, *Adv. Carbohydr. Chem.,* **21** (1966) 323–375.

199a. W. R. DeBoer, F. J. Kruyssen, and J. M. T. Wouters, *Eur. J. Biochem.,* **62** (1976) 1–6.

199b. A. S. Shashkov, M. S. Zaretskaya, S. V. Yarotsky, I. B. Naumova, O. S. Chizhov, and Z. A. Sabarova, *Eur. J. Biochem.,* **102** (1979) 477–481.

200. A. R. Archibald, J. Baddiley, and D. Button, *Biochem. J.,* **110** (1968) 543–557; A. R. Archibald, J. Baddiley, J. E. Heckels, and S. Heptinstall, *Biochem. J.,* **125** (1971) 353–359.

201. J. Coley, A. R. Archibald, and J. Baddiley, *FEBS Lett.,* **80** (1977) 405–407; J. Heptinstall, J. Coley, P. J. Ward, A. R. Archibald, J. Baddiley, *Biochem. J.,* **169** (1978) 329–336.

202. M. E. Slodki, *Biochim. Biophys. Acta,* **57** (1962) 525–533.
203. A. H. Hughes, I. C. Hancock, and J. Baddiley, *Biochem. J.,* **132** (1971) 83–93.
204. B. H. Song and N. S. Lee, *Chem. Abstr.,* **103** (1985) 157090x.
205. R. Gunnar, in *Bacterial Surface Amphiphiles,* pp. 365–379 (G. D. Shockman and A. J. Wicken, eds.) Academic, New York (1981).
206. J. F. Robyt, *J. Chem. Educ.,* **63** (1986) 560–561.

6.8 References for Further Study

Starch and Its Components, W. Banks and C. T. Greenwood, Edinburgh University (1975).

"Organization of starch granules," D. French in *Starch Chemistry and Technology,* 2nd edn., pp. 184–248 (R. L. Whistler, J. N. BeMiller, and E. F. Paschall, eds.) Academic, New York (1984).

The Polysaccharides, Vol. 1 (G. O. Aspinall, ed.) Academic, New York (1982). Chap. 1, General introduction; Chap. 2, Isolation and fractionation of polysaccharides; Chap. 5, Shapes and interaction of carbohydrate chains; Chap. 6, Immunology of polysaccharides.

The Polysaccharides, Vol. 2 (G. O. Aspinall, ed.) Academic, New York (1983). Chap. 1, Classification of polysaccharides; Chap. 2, Cellulose; Chap. 3, Other plant polysaccharides; Chap. 4, Algal polysaccharides; Chap. 5, Bacterial polysaccharides; Chap. 6, The chemistry of polysaccharides of fungi and lichens.

The Polysaccharides, Vol. 3 (G. O. Aspinall, ed.) Academic, New York (1985). Chap. 3, Starch; Chap. 4, Glycogen: a structural viewpoint; Chap. 5, Mammalian glycosaminoglycans; Chap. 6, Chitin.

Industrial Gums: Polysaccharides and Their Derivatives, 3rd edn. (R. L. Whistler and J. N. BeMiller, eds.) Academic, San Diego (1993).

"Reserve carbohydrates other than starch in higher plants," H. Meier and J. S. G. Reid, in *Plant Carbohydrates, I. Intracellular Carbohydrates, Encyclopedia of Plant Physiology,* Vol. 13A pp. 418–450 (F. A. Loewus and W. Tanner, eds.) Springer-Verlag, Berlin (1982).

"Fructans," H. G. Pontis and E. del Campillo, in *Biochemistry of Storage Carbohydrates in Green Plants,* pp. 205–255 (P. M. Dey and R. A. Dixon, eds.) Academic, New York (1985).

Science and Technology of Fructans, M. Suzuki and N. Jerry Chatterton, eds., CRC, Boca Ration, FL (1993).

Proteoglycans-Biological and Chemical Aspects in Human Life, J. F. Kennedy, Elsevier Scientific, New York (1979).

Proteoglycans, P. Jollès, ed., Birkhäuser, Basel (1994).

Functions of the Proteoglycans, (Ciba Foundation Symposium 124) (D. Evered and J. Whelan, eds.) Wiley Interscience, New York (1986).

Heparin (and Related Polysaccharides) (Polymer Monographs, Vol. 7) W. D. Comper, Gordon and Breach Science, New York (1981).

Chitin and Chitosan: Sources, Chemistry, Biochemistry, Physical Properties, and Applications, G. Skjåk-Bræk, T. Anthonsen, and P. Sanford, eds., Elsevier Applied Science, New York (1989).

"Dextran," J. F. Robyt, in *Encyclopedia of Polymer Science and Engineering,* 2nd edn., Vol. 4, pp. 752–767 (H. F. Mark, N. N. Bikales, C. G. Overberger, and G. Menges, eds.) J. Wiley & Sons, New York (1986).

"Structure and some reactions of cellulose," D. M. Jones, *Adv. Carbohydr. Chem.*, **19** (1964) 219–246.

"Wood hemicelluloses. Part I," T. E. Timell, *Adv. Carbohydr. Chem.*, **19** (1964) 247–299; "Part II," **20** (1965) 409–483.

"The pneumococcal polysaccharides," M. J. How, J. S. Brimacombe, and M. Stacey, *Adv. Carbohydr. Chem.*, **19** (1976) 303–358.

"The pneumococcal polysaccharides: a reexamination," O. Larm and B. Lindberg, *Adv. Carbohydr. Chem. Biochem.*, **33** (1976) 295–322.

"Extracellular Microbial Polysaccharides," P. A. Sanford, *Adv. Carbohydr. Chem. Biochem,* (1979) 266–314.

"Structural Chemistry of Polysaccharides from Fungi and Lichens," E. Barreto-Bergter and P. A. J. Gorin, *Adv. Carbohydr. Chem. Biochem.,* **41** (1983) 68–104.

"Capsular polysaccharides as human vaccines," H. J. Jennings, *Adv. Carbohydr. Chem. Biochem.,* **41** (1983) 155–208.

"Plant cell walls," P. M. Dey and K. Brimson, *Adv. Carbohydr. Chem. Biochem.,* **42** (1984) 266–382.

"Plant cell walls," P. M. Dey and K. Brinson, *Adv. Carbohydr. Chem. Biochem.,* **42** (1984) 266–382.

"Structure and biological activity of heparin," B. Casu, *Adv. Carbohydr. Chem. Biochem.,* **43** (1985) 51–134.

"The macrostructure of mucus glycoproteins in solution," S. E. Harding, *Adv. Carbohydr. Chem. Biochem.,* **47** (1989) 345–381.

"Components of bacterial polysaccharides," B. Lindberg, *Adv. Carbohydr. Chem. Biochem.,* **48** (1990) 279–318.

Developments in Carbohydrate Chemistry (R. J. Alexander and H. F. Zobel, eds.) American Association of Cereal Chemists, St. Paul, MN (1988).

Food Polysaccharides and Their Applications (A. M. Stephen, ed.) Marcel Dekker, New York (1995).

Another important cellulose ester is cellulose acetate. It is produced by treating purified cotton linters or wood pulp with acetic acid at 50°C to bring about swelling and increased reactivity. The acetic acid acts as the solvent and reduces the intermolecular hydrogen bonding between the cellulose chains. Acetylation is obtained by the addition of acetic anhydride and sulfuric acid as a catalyst [2] (reaction 7.3). Water (5–10%) is added at the conclusion of the reaction to hydrolyze the excess acetic anhydride. The mixture is usually held at 40°C to reduce the d.s. to 2.1–2.6 by acid hydrolysis [3]. Partial deacetylation is desirable because cellulose triacetate is soluble in only a few organic solvents such as glacial acetic acid, chloroform, and dichloromethane, whereas cellulose acetate, with a lower d.s., is soluble in low-cost commercial organic solvents such as acetone. After holding at 40°C, the mixture is poured into water to achieve precipitation of the cellulose acetate, which is washed with water to remove acetic acid.

$$\text{cellulose} \xrightarrow[\substack{\text{AcOH} \\ \text{H}_2\text{SO}_4}]{\text{Ac}_2\text{O}} \text{cellulose acetate, d.s.} = (7 \div 9) \times 3 = 2.33 \tag{7.3}$$

Cellulose acetate is much less flammable than cellulose nitrate. One of the major applications of cellulose acetate is the formation of textiles. Cellulose acetate, dissolved in acetone, can be forced through fine holes into a stream of warm air, resulting in the evaporation of the acetone solvent, and the formation of solid filaments of cellulose acetate. These filaments are used in making cellulose acetate cloth, commonly known as *acetates, acetate rayon, or acetate silk.* Cellulose acetate can be mixed with various plasticizers such as tricresyl phosphate, triphenyl phosphate, or dibutyl phthalate to make photographic films and various plastic items by injection molding. Artificial leather can be made by coating cotton fabrics with cellulose-ester lacquers. Films can be made into sheets as thin as 0.025 mm or as thick as 6 mm. Solutions of cellulose acetate in acetone or dioxane produce liquid crystals that have a large number of applications such as digital numbers, light- and heat-sensitive detectors, and so on.

Mixed esters of cellulose acetate and cellulose propionate, and cellulose acetate and cellulose butyrate can be formed by using mixtures of the anhydride of one acid and the acid of the other [4]. These mixed esters (60–65% acetate and 40–35% propionate or butyrate) give improved properties of toughness and shock resistance necessary for injection molding and special-purpose lacquers.

Polysaccharides react with carbon disulfide and sodium hydroxide to give dithiocarbonic acid esters called *xanthates.* The polysaccharide, such as cellulose, is first treated with 18% sodium hydroxide to give "alkaline cellulose" (reaction 7.4). The alkaline cellulose is usually aged several hours to achieve lower d.p. values. It is then reacted with carbon disulfide for 1–3 hr at 20–30°C (reaction 7.4).

The excess carbon disulfide is removed by vacuum, leaving the sodium salt of cellulose xanthate, which is known as *viscose* [5–7].

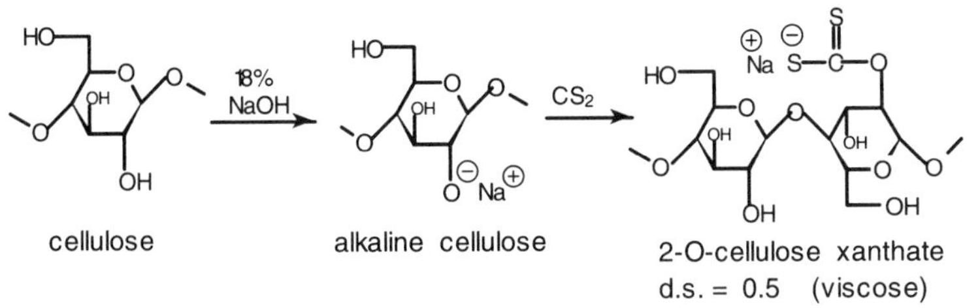

cellulose   alkaline cellulose   2-O-cellulose xanthate
                                               d.s. = 0.5   (viscose)

$$(7.4)$$

Viscose can be forced through fine holes in a spinerette into an acid bath, giving the removal of the xanthate group and the regeneration of cellulose in the form of a fine filament or thread known as *rayon*. If instead, the viscose is forced through a narrow slit into an acid bath, cellulose is regenerated as a thin, transparent sheet that can be softened with glycerol to make *cellophane.*

7.3 Formation of Polysaccharide Ethers

Ethers can be formed with the alcohol groups of polysaccharides by reacting an alkaline solution or suspension of the polysaccharide with alkyl halides or epoxides [8]. The ethers produce polysaccharide derivatives that are usually soluble in water. This is in contrast to polysaccharide esters, which are usually insoluble in water and soluble in organic solvents.

Methyl or ethyl ethers can be formed by the reaction of the "alkaline polysaccharide" with methyl chloride (reaction 7.5), or ethyl chloride at 70–120°C to give d.s. values of 1.2–2.3 [9].

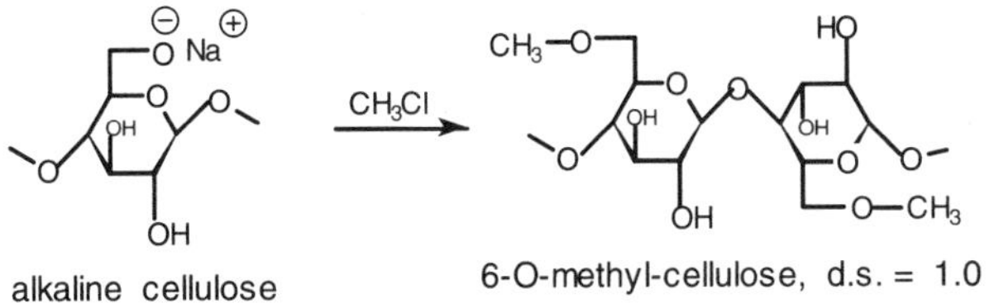

alkaline cellulose   6-O-methyl-cellulose, d.s. = 1.0

$$(7.5)$$

Carboxymethyl ethers can be prepared by reaction of alkaline polysaccharide with chloroacetate at 30–80°C to achieve d.s. values of 0.8 to 1.1 [8,10] (reaction 7.6). Carboxymethyl cellulose (CM-cellulose) was first developed in Germany in 1916, and then in 1935, it was found that it improved the performance of synthetic detergents [11]. CM-cellulose has since found wide application in textiles, detergents, foods, pharmaceuticals, cosmetics, oil and gas drilling "muds," paper, adhesives, coatings, ceramics, suspending agents, and emulsions [11].

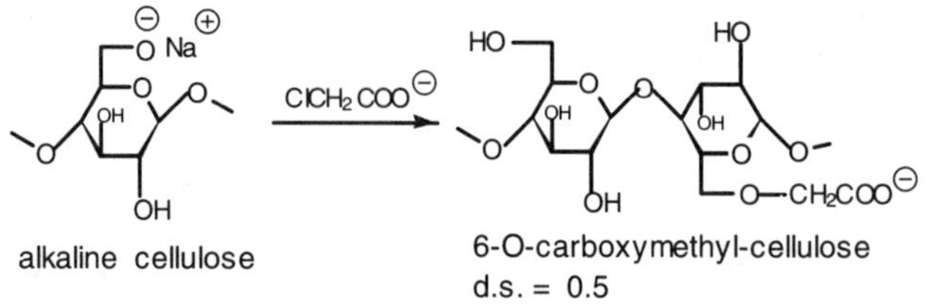

alkaline cellulose

6-O-carboxymethyl-cellulose
d.s. = 0.5

(7.6)

In the 1960s, CM-cellulose was prepared with low d.s. values to form a water-insoluble product that could be used as an anion exchanger for the separation of proteins according to differences in charge [12]. In a similar manner, reaction of alkaline cellulose with *N,N*-diethyl aminoethyl chloride gave a cation exchanger, diethylaminoethyl cellulose (DEAE-cellulose, reaction 7.7).

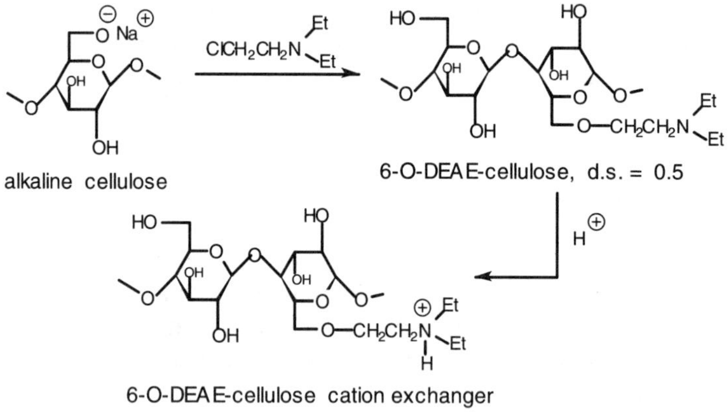

alkaline cellulose

6-O-DEAE-cellulose, d.s. = 0.5

(7.7)

6-O-DEAE-cellulose cation exchanger

The commercial availability of these two ion exchangers provided the biochemist with powerful tools for the separation of proteins and nucleic acids under very mild conditions. DEAE-derivatives of agaran have been prepared as a matrix for protein electrophoresis and immunoelectrophoresis. DEAE-derivatives of cross-linked dextran (see section 7.4) have been prepared with the idea of combining ion exchange and molecular sieving of charged macromolecules.

Cyanoethyl ethers of polysaccharides can be prepared by the reaction of alkaline polysaccharide with acrylonitrile [8,13] (reaction 7.8), and the formation of sulfoethyl ethers can be prepared by the reaction of alkaline polysaccharide with sodium vinylsulfonate [8] (reaction 7.9).

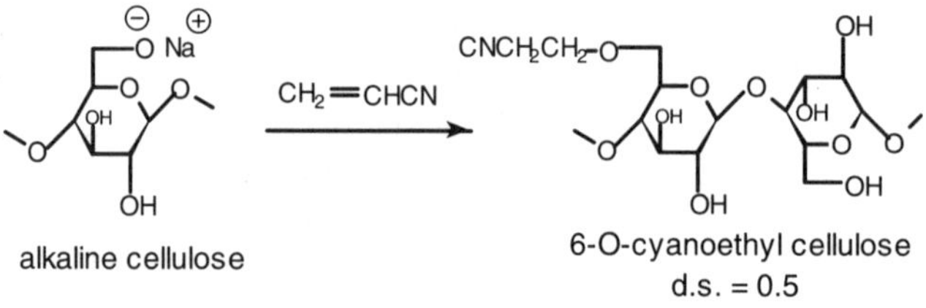

alkaline cellulose

6-O-cyanoethyl cellulose
d.s. = 0.5

(7.8)

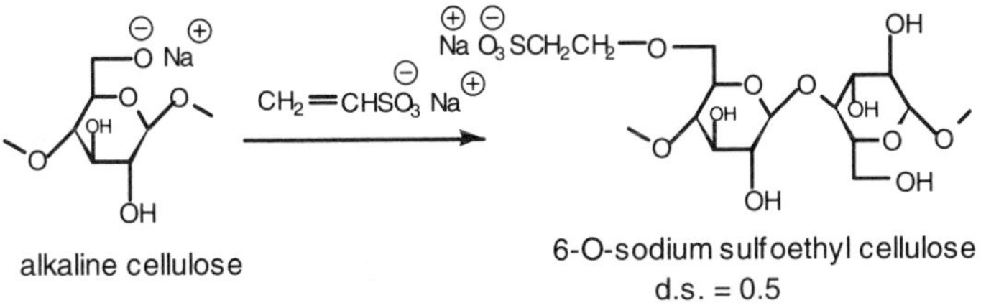

alkaline cellulose

6-O-sodium sulfoethyl cellulose
d.s. = 0.5

(7.9)

Hydroxyethers of polysaccharides are prepared by the reaction of alkaline polysaccharide with epoxides, for example, ethylene epoxide and propylene epoxide [8,14] (reactions 7.10 and 7.11). The reaction is usually run on a 40–45% (w/v) suspension of starch granules in water under strongly alkaline conditions at 40–50°C under nitrogen. The hydroxyethyl starches are used primarily as binders for pigmented coatings and as surface sizing agents in paper manufacturing [15]. Hydroxypropyl starches are of importance in food applications, where they are used as an edible, water-soluble film coating [15]. The formation of hydroxyalkyl derivatives of polysaccharides increases their water solubility. Hydroxypropylation of starch decreases or prevents retrogradation of the starch chains.

The reaction of these epoxides with hydroxyl groups of polysaccharides provides derivatives with new hydroxyl groups that themselves can react with the epoxide to give ether chains attached to the polysaccharides [15] (reactions 7.10 and 7.11). This chaining reaction can be reduced by conducting the reaction in the presence of methyl or ethyl chloride [8] (reaction 7.12).

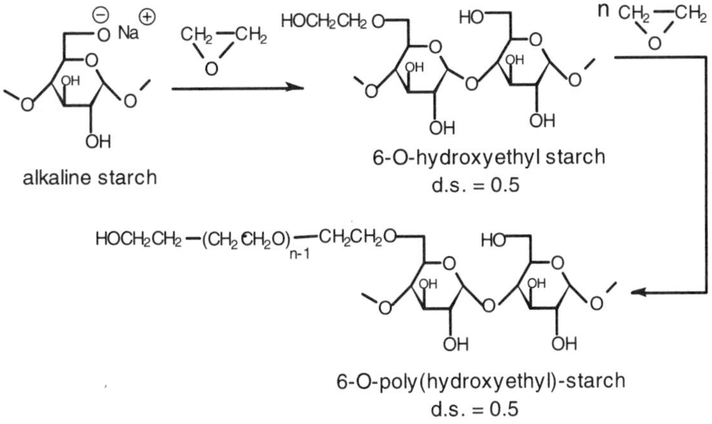

alkaline starch

6-O-hydroxyethyl starch
d.s. = 0.5

(7.10)

6-O-poly(hydroxyethyl)-starch
d.s. = 0.5

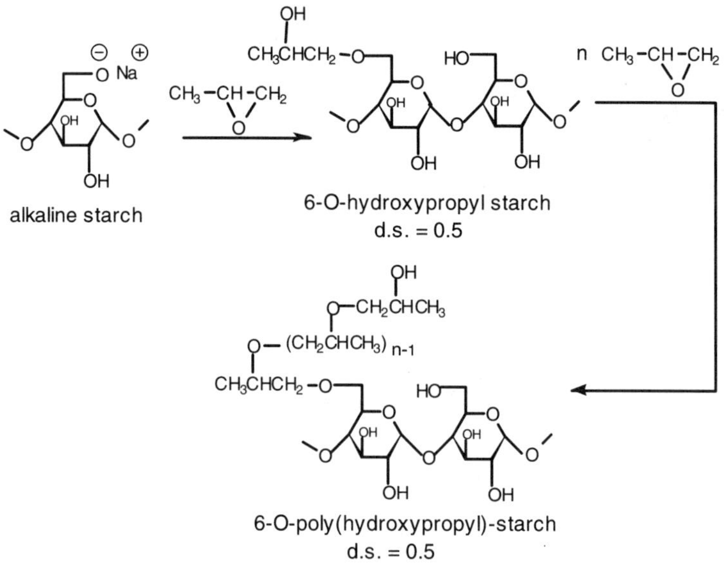

6-O-hydroxypropyl starch
d.s. = 0.5

alkaline starch

(7.11)

6-O-poly(hydroxypropyl)-starch
d.s. = 0.5

alkaline starch

6-O-(methyl-O-hydroxyethyl) starch
d.s. = 0.5

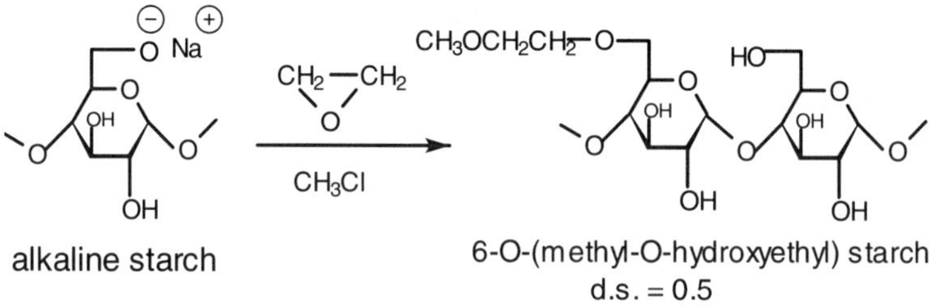

(7.12)

The addition of two ether-forming reagents will often give products with completely new properties, compared with the individual polysaccharide ethers, or compared with a physical mixture of the two polysaccharide ether derivatives [8].

7.4 Formation of Cross-linked Polysaccharide Ethers

The reaction of alkaline polysaccharides with epichlorohydrin results in the formation of ethers and the cross-linking of polysaccharide chains. This reaction has been most successfully applied to B-512F dextran to give a molecular sieve or gel-filtration product [16,17] (reaction 7.13).

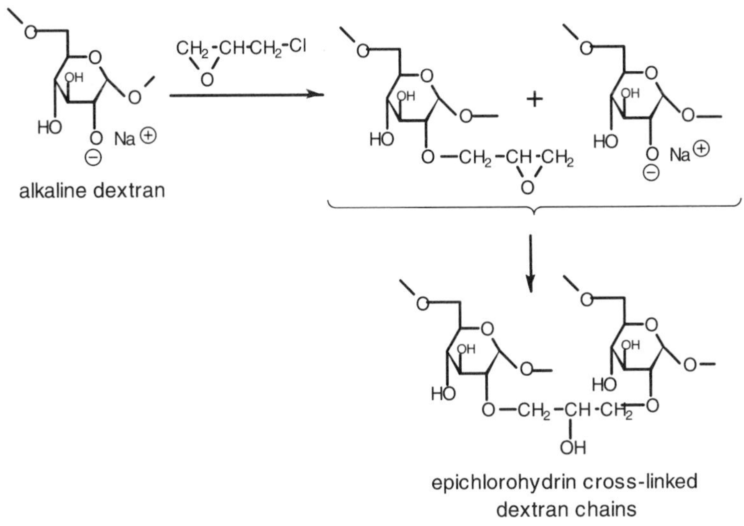

$$(7.13)$$

epichlorohydrin cross-linked
dextran chains

The B-512F dextran cross-linked gels were commercialized by Pharmacia AB (Uppsala, Sweden) in 1959 as a gel matrix called Sephadex for gel-permeation chromatography. This material was greatly improved by conducting the cross-linking reaction in a water-immiscible organic solvent such as poly(vinyl acetate) in toluene to give the formation of gels in a beaded form [18]. By varying the degree of cross-linking, gel beads with different pore sizes are obtained and have different molecular weight cutoffs. The cross-linked dextran gels provide a chromatographic material that is capable of separating proteins and other macromolecules according to their molecular size. Fig. 7.1 shows the cross-linked B-512F dextran chains after reaction with epichlorohydrin.

7.5 Polysaccharide Phosphates

The hydroxyl groups of polysaccharides can be phosphorylated by reaction with tripolyphosphate at pH 5–6.5; the reaction is conducted at 110–130°C in the dry state and creates products containing 0.07–0.1% phosphorus [15] (reaction 7.14). Reaction of starch granules with tripolyphosphate at 150°C resulted in 0.3% covalently linked phosphorus that was distributed 9% at O-3, 28% at O-2, and 63% at O-6 [19].

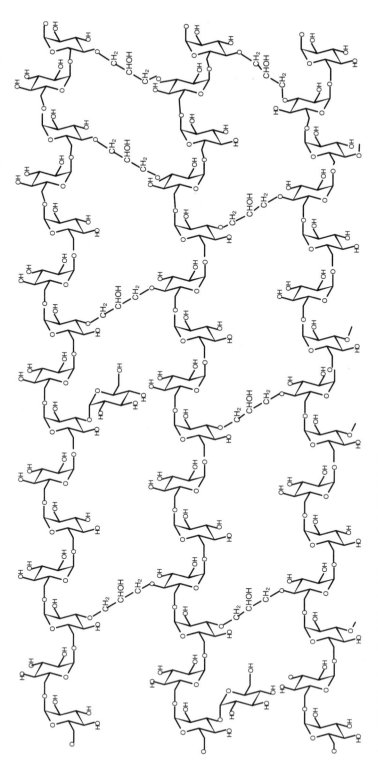

Figure 7.1. Structure of epichlorohydrin cross-linked B-512F dextran.

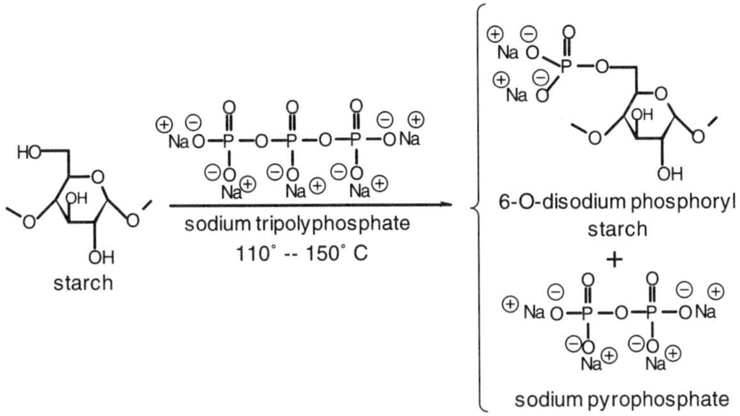

6-O-disodium phosphoryl (7.14)
starch
+

sodium pyrophosphate

Granular starch has been cross-linked with phosphate by reaction of an aqueous alkaline suspension at pH 8–12 with 0.005–0.25% phosphorus oxychloride [15] (reaction 7.15). If starch granules are treated with 1% or higher concentrations of phosphorus oxychloride, the granules become very resistant to gelatinization. Trimetaphosphate has also been used to produce phosphate cross-linkages [15] (reaction 7.16). Phosphorylation of O-6 is predominant, and that of O-3 is minimal [19].

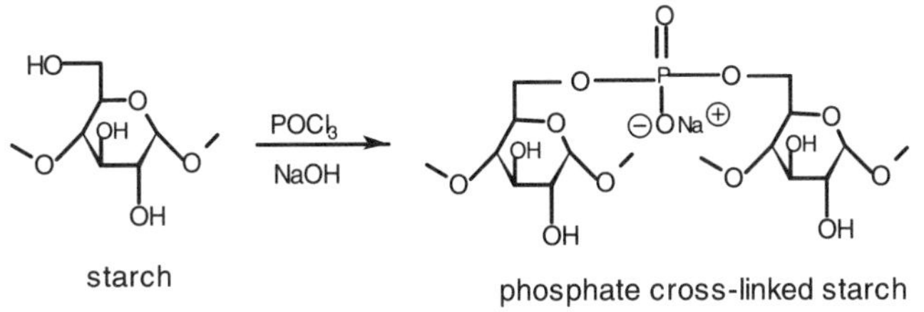

starch   phosphate cross-linked starch   (7.15)

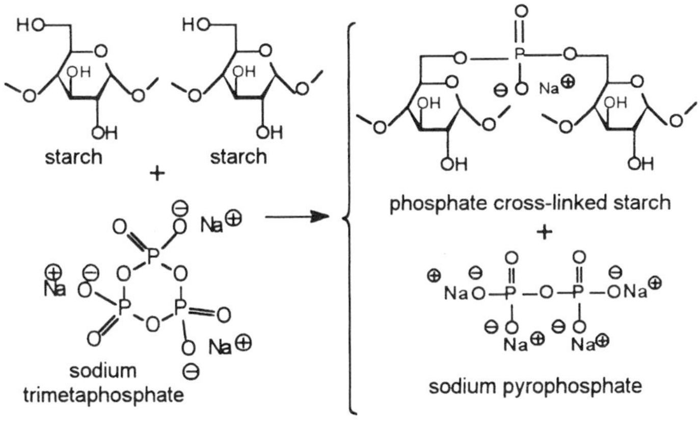

phosphate cross-linked starch   (7.16)
+

sodium pyrophosphate

7.6 Polysaccharide Sulfates

Several reagents can be used to form polysaccharide sulfates. A common reagent that has been used is triethylamine sulfur trioxide in *N,N*-dimethylformamide [20,21]. Sulfate esters of amylose, amylopectin, cellulose, and guaran have been prepared in which O-6 is the predominant position sulfated. Dimethylsulfoxide/sulfur trioxide is a mild sulfating reagent that causes minimal depolymerization and has become the preferred sulfating reagent for polysaccharides [22,23]. Another reagent that has been used to give higher degrees of substitution, with no degradation, is *N,N*-dimethylforamide/sulfur trioxide [24]. It has been used to sulfate acidic polysaccharides such as alginic acid and pectic acid [25]. The secondary alcohol groups of dextran have been sulfated with 1-piperidine sulfonic acid [26].

Dextran sulfate is of particular interest because its low molecular weight preparations have anticoagulant properties similar to those of heparin, although it is not as potent as heparin. A certain minimum number of sulfate groups per glucose residue is required for anticoagulant activity. The activity increases sharply between d.s. of 1.0 and 1.3. Good anticoagulant activity was found for dextran sulfates of molecular weights of 7,300–10,000 Da and a sulfate d.s. of 1.9 [27].

Dextran sulfate seems to have specific affinity for binding nucleic acids and is a potent inhibitor for ribonuclease [28]. Dextran sulfate also inhibits the replication of some viruses [29], including HIV [30], and it has been considered as a possible drug in the treatment of AIDS.

7.7 Polysaccharide Dye Derivatives

Procion dyes are chemically reactive dyes that have been formed by reaction of a dye with 2,4,6-trichlorotriazine or 2,4-dichlorotriazine [31]. A chloro group on the triazine ring of the dye is replaced by a polysaccharide hydroxyl group to form a stable, covalent linkage between the dye and the polysaccharide (reaction 7.17). Cibachron Blue, a monochlorotriazinyl dye, has been used to produce dyed complexes (reaction 7.18) with amylose, amylopectin, starch [32,33], cellulose [34], chitin [35], and dextran [32,36].

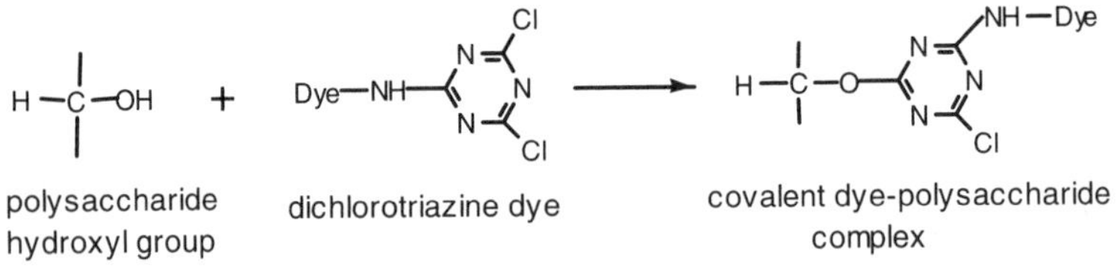

$$(7.17)$$

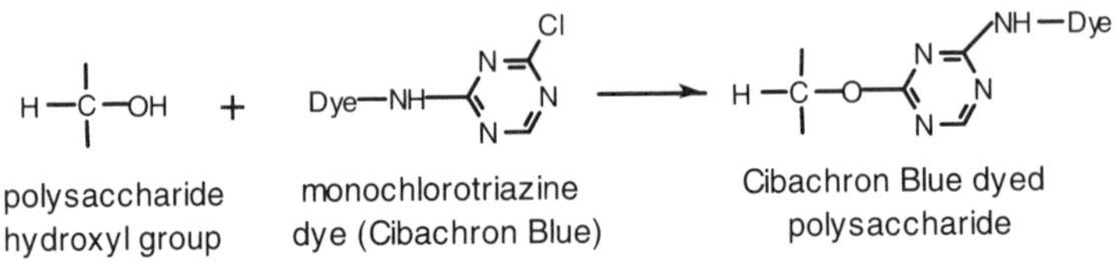

polysaccharide hydroxyl group   dichlorotriazine dye   covalent dye-polysaccharide complex

$$(7.18)$$

polysaccharide hydroxyl group   monochlorotriazine dye (Cibachron Blue)   Cibachron Blue dyed polysaccharide

Various colors of chlorotriazine dyes are available, including Cibacron Blue, Procion Brilliant Blue, Procion Brilliant Red, Procion Brown, Procion Green, and Procion Yellow.

The dyeing of polysaccharides has been used for three purposes: (1) the formation of dyefast fabrics, (2) the visible identification of polysaccharides for study in electrophoresis and gel chromatography [32], and (c) to provide substrates for identification of enzymes in enzyme screening tests and in assaying of specific enzymes such as α-amylase [33], cellulase [34], chitinase [35], dextranases [36], and so forth.

7.8 Activation of Polysaccharides for Covalently Attaching Ligands and Proteins

Polysaccharides and cross-linked polysaccharides can be activated by reaction with alkaline cyanogen bromide (reaction 7.19). Polysaccharides such as cross-linked dextran, agaran, and cellulose have been activated by this reaction [37,38]. The activated polysaccharides can then react with amino groups of various ligands that can be used in affinity chromatography [39], or they can react with amino groups on the surface of enzymes to give immobilized enzymes [40] (reaction 7.20).

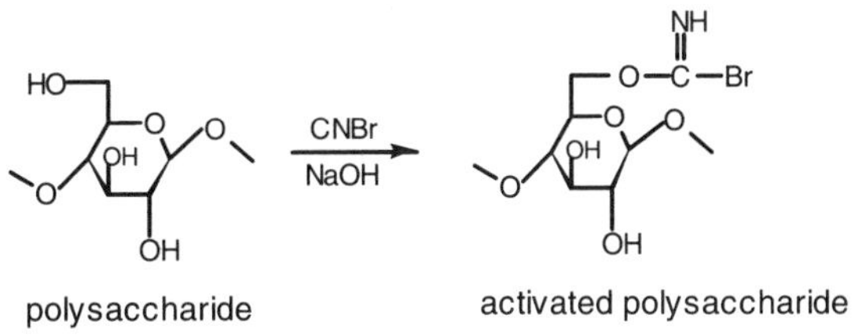

$$(7.19)$$

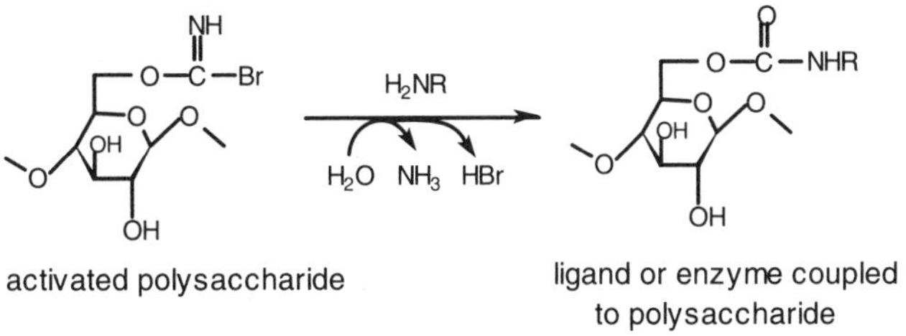

polysaccharide   activated polysaccharide

$$(7.20)$$

activated polysaccharide   ligand or enzyme coupled
                                  to polysaccharide

A spacer arm between the polysaccharide and the ligand or enzyme can be introduced by the sequential reactions of the activated polysaccharide with diaminoalkane and glutaraldehyde. An amino group on the ligand or protein can then react with the free aldehyde group (reaction 7.21). The resulting imines can be stabilized by specific reduction with sodium cyanoborohydride [41] (reaction 7.21). If the ligand has a reactive aldehyde group, it can be coupled directly with the amino group of the spacer arm.

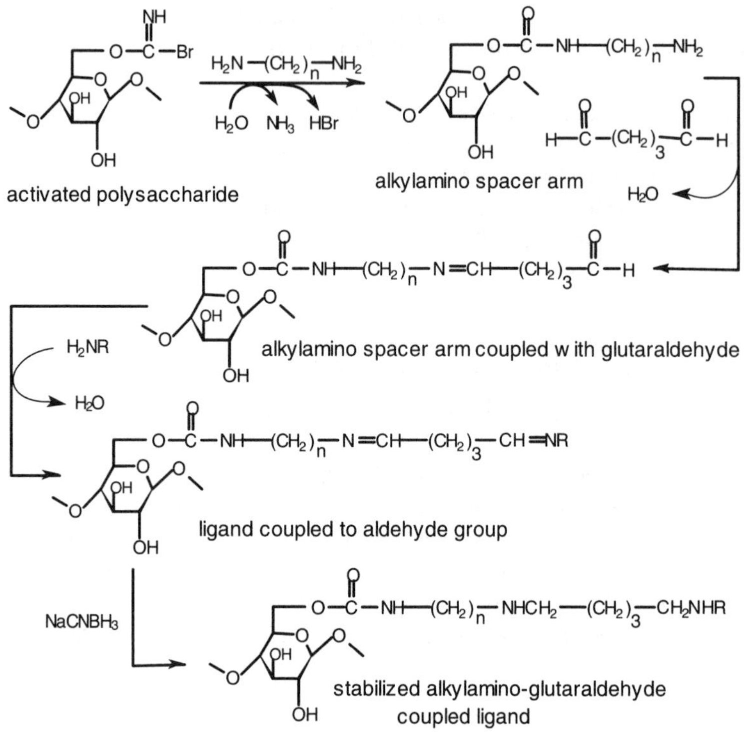

$$(7.21)$$

7.9 Oxidation of Primary Alcohols of Polysaccharides to Carboxyl Groups

Primary alcohol groups of polysaccharides can be specifically oxidized to carboxyl groups by reaction with 2,2,6,6-tetramethyl-1-piperidine (TEMPO) [42] (see reaction 4.140 in Chapter 4). Several polysaccharides (amylose, amylopectin, cellulose, chitin, chitosan, pullulan, alternan, regular comb dextran, carboxymethyl cellulose, and starch granules) have been oxidized under very mild conditions, resulting in high yields and high selectivity for the primary alcohol groups. The oxidation produced new polyuronic acids that had greatly increased water solubility. The water-insoluble polysaccharides (amylose, cellulose, chitin, and starch granules) became soluble during the course of the reaction [42].

7.10 Periodate Oxidation of Polysaccharides

Vicinal alcohol groupings in polysaccharides can be oxidized by periodate to cleave the carbon-carbon bond and produce two aldehyde groups. Starch is the principal polysaccharide that has been oxidized to produce "dialdehyde starch" (reaction 7.22) [15]. The degree of oxidation can be controlled by the amount of periodate added. The resulting aldehydes can react with other compounds such as alcohols, amines, hydrazines, and hydrazides to give other polysaccharide derivatives. Chlorous acid will oxidize the aldehyde groups to give polycarboxylic acid polymers [43] (reaction 7.22). Other polysaccharides with vicinal hydroxyl groups, such as dextran, pullulan, and cellulose, can also be oxidized by periodate in a similar manner.

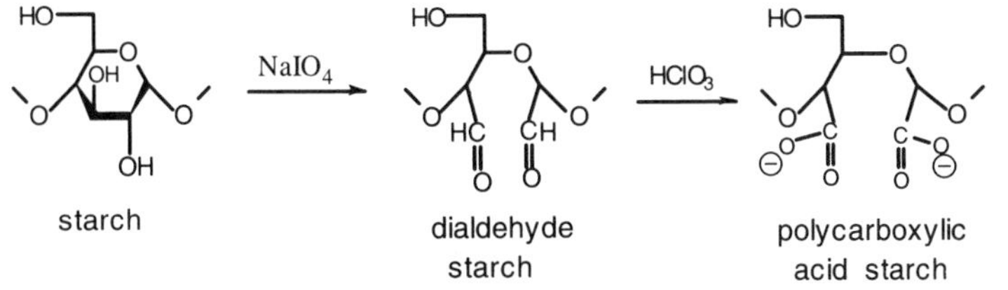

$$(7.22)$$

starch   dialdehyde   polycarboxylic
                           starch   acid starch

7.11 Miscellaneous Modifications of Alcohol Groups of Polysaccharides

The primary alcohol groups of polysaccharides can be modified by many of the reactions proposed in section 4.10 of Chapter 4 for the modification of the C-6 primary alcohol group. The primary alcohol could be iodinated using triphenyl phosphine, iodine, and imidazole. The 6-iodo group could then be catalytically reduced to give 6-deoxypolysaccharide. The 6-iodo group could be displaced by thio acetate to give the 6-acetylthio derivative that can have the acetyl group removed by reaction with $NaOCH_3/CH_3OH$ to give 6-deoxy-6-thiopolysaccharide. The 6-O-tosyl derivative could be made and the tosyl group displaced by various nucleophiles such as azide to give the 6-azido derivative that can be reduced to give 6-amino-6-deoxypolysaccharide. The 6-O-tosyl derivative can be treated with base to give a 3,6-anhydropolysaccharide [44].

The D-glucose residues of amylose have been specifically oxidized at C-3 with phosphorus pentoxide in dimethylsulfoxide, followed by reduction with sodium borohydride, to produce D-allose residues in the amylose chain [45].

7.12 Modification of Starch Granules by Acid Hydrolysis

Amylose and amylopectin components in starch granules are of high molecular weight. The molecular weights can be lowered by acid hydrolysis. In 1874, Nägeli reported [46] that treatment of native starch granules with 15% (w/v) sulfuric acid for one month produced an acid-resistant fraction that was readily soluble in hot water. This material is called *Nägeli amylodextrin* and has a low molecular weight with an average d.p. of 25–30 [47].

A hot-water-soluble starch was prepared by treatment of starch granules with 7.5% (w/v) hydrochloric acid in a water suspension for 1 week at 22–24°C, or for

3 days at 40°C, followed by washing of the granules until they were free of acid [48]. The method is called the *Lintner procedure* and is the method currently used to produce commercial soluble starch. The product is highly heterogeneous and has a relatively high reducing value.

Another method for producing soluble starches is the method reported by Small in which starch granules are refluxed in 95% ethanol containing 0.2–1.6% (w/v) hydrochloric acid for 6–15 min [49]. This method gives the maximum conversion of raw starch into soluble starch, with minimal production of low molecular weight dextrins.

In 1987, Ma and Robyt [50] reported the preparation and characterization of soluble starches that had different average molecular weights and different proportions of amylose and amylopectin. Starch granules were treated with 0.36% (w/v) hydrochloric acid in anhydrous methanol, ethanol, 2-propanol, and 1-butanol at 65°C for 1 hr. The starch granule was maintained, but the molecular sizes of the starch molecules progressively decreased in the different alcohols in the order: methanol > ethanol > 2-propanol > 1-butanol. The amylose component was completely absent from the starch granules treated in 2-propanol and 1-butanol. It was found that the hydrolysis was occurring with the water (10–12%, w/w) that was inside the granule. It was also found that the amount of acid inside the granules increased as the size of the alcohol increased.

In further study of the treatment of starch granules in these alcohols with 0.36 and 6.0% (w/v) hydrochloric acid at 25°C, Fox and Robyt [51] reported the kinetics of the reaction (d.p. vs. time of reaction). It was found that different limiting d.p. values, different chain lengths, and different proportions of amylose and amylopectin were obtained by reaction in the four different alcohols and the two acid concentrations. The procedures create new kinds of soluble limit dextrins.

Robyt, et al. [52] further studied the hydrolysis in terms of the parameters of temperature (5 to 65°C), acid concentration (0.36–5.0%, w/v), and starch concentration. The d.p. values of the limit dextrins were dependent on all three parameters. By varying the acid concentration and the temperature, combinations can be obtained to give a wide range of limit dextrins with average d.p. values of 12–1,800. As the concentration of the starch granules increases, lower degrees of hydrolysis and a higher limiting d.p. were obtained. This latter result confirmed that the mechanism of hydrolysis of the starch granules suspended in alcohol involves the hydrolysis of glycosidic bonds by the water contained inside the granules.

The acid hydrolysis of starch granules from potato, waxy maize, and amylomaize-7 was studied in different ratios of mixtures of two of the four alcohols, from 100/0 to 0/100% in 10% (v/v) intervals [53]. Again, a series of soluble limit dextrins was obtained in 48–72 hr of reaction. The average d.p. of the limit dextrins decreased as the content of the higher alcohol was increased. Plots of the d.p. of the limit dextrins vs. the ratio of each alcohol mixture produced curves that resembled titration curves, with different plateau regions in which the d.p. of the limit dextrins changed very slowly or not at all. These plateau regions were characteristic and dependent on the types of alcohols, the volume ratios of the two alcohols in the mixture, and the type of starch. It was proposed that the formation of limit dextrins and the formation of plateaus in the plots reflected the presence of physically different "kinds" of α-1 $\rightarrow$ 4 and α-1 $\rightarrow$ 6 glycosidic linkages in the starch granules. It was further proposed that the mechanism by which the alcohols or combination of alcohols produce the different kinds of linkages that become more susceptible to acid hydrolysis, involves the conversion of crystalline regions in the granule into amorphous regions.

7.13 Literature Cited

1. J. W. Green, *Methods Carbohydr. Chem.*, **3** (1963) 213–217.
2. L. J. Tanghe, L. B. Genung, and J. W. Mench, *Methods Carbohydr. Chem.*, **3** (1963) 193–197.
3. L. J. Tanghe, L. B. Genung, and J. W. Mench, *Methods Carbohydr. Chem.*, **3** (1963) 198–200.
4. D. Gray, *J. Appl. Polym. Sci.*, **37** (1983) 179–192.
5. C. Y. Chen, R. E. Montonna, and C. S. Grove, *Tappi*, **34** (1951) 420–428.
6. A. K. Sanyal, E. L. Falconer, D. L. Vincent, and C. B. Purves, *Can. J. Chem.*, **35** (1957) 1164–1173.
7. T. E. Muller and C. B. Purves, *Methods Carbohydr. Chem.*, **3** (1963) 238–251.
8. U.-H. Felcht, in *Cellulose and Its Derivatives*, pp. 273–284 (J. F. Kennedy, G. O. Phillips, D. J. Wedlock, and P. A. Williams, eds.) J. Wiley & Sons, New York (1985).
9. I. Croon and R. St. J. Manley, *Methods Carbohydr. Chem.*, **3** (1963) 271–273.
10. J. W. Green, *Methods Carbohydr. Chem.*, **3** (1963) 322–326.
11. R. L. Feddersen and S. N. Thorp, in *Industrial Gums*, pp. 537–575 (R. L. Whistler and J. N. BeMiller, eds.) Academic, San Diego (1993).
12. E. A. Peterson and H. A. Sober, *A Laboratory Manual of Analytical Methods in Protein Chemistry*, pp. 88–102, Pergamon, New York (1960).
13. J. Compton, *Methods Carbohydr. Chem.*, **3** (1963) 317–321.
14. E. D. Klug, *Methods Carbohydr. Chem.*, **3** (1963) 315–316.
15. M. W. Rutenberg and D. Solarek, in *Starch Chemistry and Technology*, 2nd edn., pp. 344–349 (R. L. Whistler, J. N. BeMiller, and E. F. Paschall, eds.) Academic, New York (1984).
16. P. Flodin and B. Ingelman, Swed. Pat. 169,293 (1959).
17. P. Flodin and J. Porath, U.S. Pat. 3,002,823 (1961).
18. P. Flodin, Swed. Pat. 358,894 (1961).
19. R. E. Gramera, J. Heerema, and F. W. Parrish, *Cereal Chem.*, **43** (1966) 104–113.
20. R. L. Whistler and W. W. Spenser, *Arch. Biochem. Biophys.*, **95** (1961) 36–41.
21. W. W. Doane and R. L. Whistler, *Arch. Biochem. Biophys.*, **101** (1963) 436–438.
22. R. L. Whistler, D. G. Unrau, and G. Ruffini, *Arch. Biochem. Biophys.*, **126** (1968) 647–652.
23. R. L. Whistler, *Methods Carbohydr. Chem.*, **3** (1963) 426–428.
24. R. G. Schweiger, *Carbohydr. Res.*, **21** (1972) 219–228.
25. R. G. Schweiger and T. R. Andrew, *Carbohydr. Res.*, **21** (1972) 275–281.
26. K. Nagasawa, H. Harada, S. Hayashi, and T. Misawa, *Carbohydr. Res.*, **21** (1972) 420–426.
27. C. R. Ricketts, *Biochem. J.*, **51** (1959) 129–133.
28. L. Philipson and M. Kaufman, *Biochim. Biophys. Acta*, **80** (1964) 151–154.

29. K. K. Takemoto and H. Liebhaber, *Virology,* **17** (1962) 499–501.

30. M. Ito, M. Baba, A. Sator, A. Pauwels, R. DeClercq, and S. Shigeta, *Antiviral Res.,* **7** (1987) 361–366.

31. W. F. Dudman and C. T. Bishop, *Can. J. Chem.,* **46** (1968) 3079–3084.

32. A. F. Pavlenko and Y. S. Ovodov, *J. Chromatog.,* **52** (1970) 165–168.

33. J. J. Marshall, *Anal. Biochem.,* **37** (1970) 466–470.

34. H. N. Fernley, *Biochem. J.,* **87** (1963) 90–95.

35. R. H. Hackman and M. Goldberg, *Anal. Biochem.,* **8** (1964) 397–401.

36. K. K. Mäkinen and I. K. Paunio, *Anal. Biochem.,* **39** (1971) 202–207.

37. R. Axen, J. Porath, and S. Ernback, *Nature,* **214** (1967) 1302–1304.

38. J. Porath, R. Axen, and S. Ernback, *Nature,* **215** (1967) 1491–1492.

39. P. Cuatrecasas and C. Anfinsen, *Methods Enzymol.,* **22** (1971) 355–378.

40. W. B. Jakoby and M. Wilchek, in *Methods Enzymol.,* **34** (1974) 13–305.

41. C. F. Lane, *Synthesis,* (1975) 135–146.

42. P. S. Chang and J. F. Robyt, *J. Carbohydr. Chem.,* **15** (1996) 819–830.

43. C. L. Mehltretter, *Stärke,* **15** (1963) 313–319.

44. R. L. Whistler and S. Hirase, *J. Org. Chem.,* **26** (1961) 4600–4605.

45. P. J. Braun, D. French, and J. F. Robyt, *Carbohydr. Res.,* **141** (1985) 265–271.

46. W. Nägeli, *Justus Liebigs Ann. Chem.,* **173** (1874) 218–227.

47. T. Watanabe and D. French, *Carbohydr. Res.,* **84** (1980) 115–123.

48. C. J. Lintner, *J. Prakt. Chem.,* **34** (1886) 378–386.

49. J. C. Small, *J. Am. Chem. Soc.,* **41** (1919) 113–120.

50. W.-P. Ma and J. F. Robyt, *Carbohydr. Res.,* **166** (1987) 283–297.

51. J. D. Fox and J. F. Robyt, *Carbohydr. Res.,* **227** (1992) 163–170.

52. J. F. Robyt, J.-Y. Choe, R. S. Hahn, and E. B. Fuchs, *Carbohydr. Res.,* **281** (1996) 203–218.

53. J. F. Robyt, J.-Y. Choe, J. D. Fox, R. S. Hahn, and E. B. Fuchs, *Carbohydr. Res.,* **283** (1996) 141–150.

Chapter 8

Cyclodextrins

8.1 Cyclomaltodextrins

Nonreducing dextrins that form crystals from alcohol solutions have been known for over 100 years. They were first reported by Villers [1] in 1891. In 1903, during studies on food spoilage, Schardinger isolated heat-resistant bacteria and found that some of them fermented starch and formed two kinds of nonreducing, crystalline dextrins [2]. The organism was named *Bacillus macerans* because of its ability to macerate starchy vegetables [3]. Schardinger found that his dextrins gave very characteristic triiodide crystalline products that appeared as blue-purple hexagons and brown, fan-shaped needles. He called his two dextrins α- and β-dextrins, respectively [4]. Later, these dextrins were called *Schardinger dextrins* in his honor.

Controversy developed over their structure, apparently because of impure preparations. An important contribution was made by Freudenberg and Jacobi [5] when they developed a relatively straightforward procedure to obtain the cyclodextrins in pure form. In considering the properties of the dextrins, such as nonreducing character, resistance to α- and β-amylase hydrolysis, formation of a single methylation product (2,3,6-tri-*O*-methyl-D-glucose), and their high positive optical rotations, Freudenberg, et al. [6] concluded that the dextrins were large, cyclic D-glucose molecules linked α-1 → 4. Freudenberg's laboratory also isolated a third dextrin (γ-dextrin). Several procedures have been developed over the years to isolate the three Schardinger dextrins in pure form. For summaries of these procedures, see French [7] and Thoma and Stewart [8].

The definitive structures of the α- and β-dextrins were shown by French and Rundle [9] to be composed of six and seven D-glucose residues, each linked α-1 → 4 in a cyclic structure, respectively. Freudenberg and Cramer [10] and French, et al. [11] showed that γ-dextrin had eight D-glucose residues. Figures 8.1 and 8.2 show the structural formulas and molecular models of these cyclic dex-

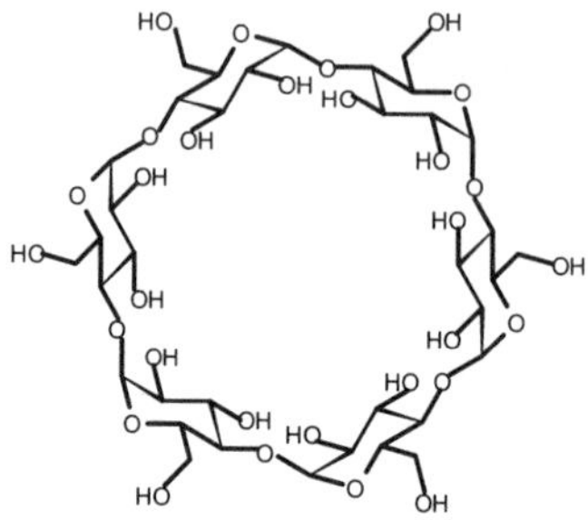

cyclomaltohexaose
(α-cyclodextrin)

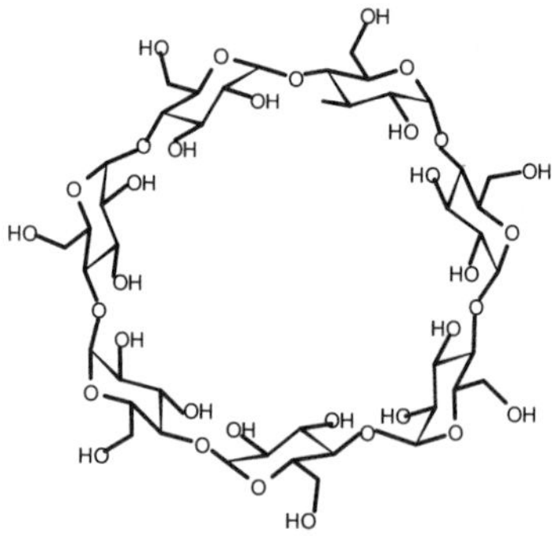

cyclomaltoheptaose
(β-cyclodextrin)

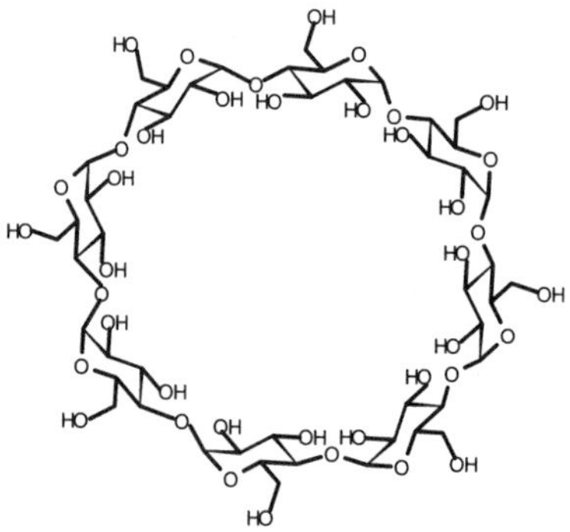

cyclomaltooctaose
(γ-cyclodextrin)

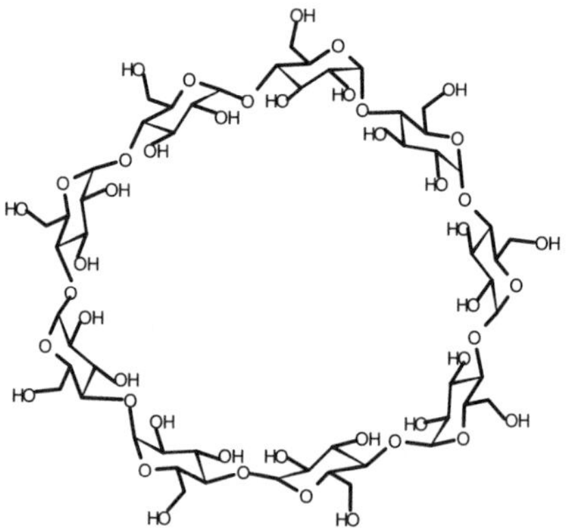

cyclomaltononaose
(δ-cyclodextrin)

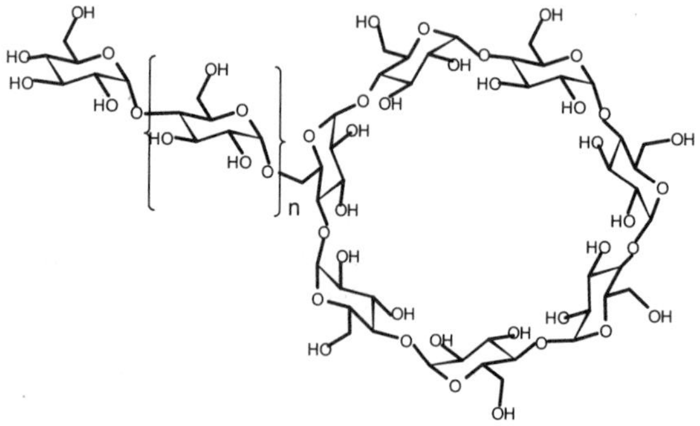

maltodextrinyl α–1,6-branched cyclomaltoheptaose
n = 0 to 5

Figure 8.1. Structural formulas for the cyclomaltodextrins.

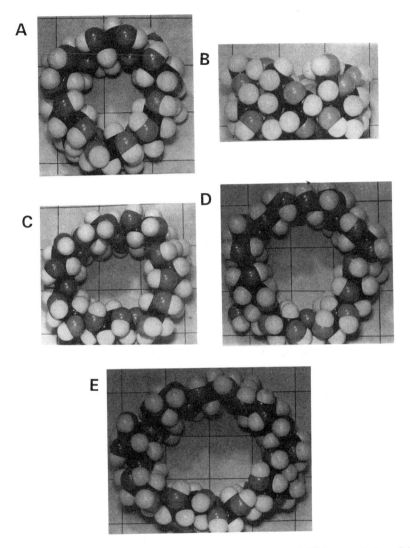

Figure 8.2. Molecular models of the cyclomaltodextrins from ref. [7] by permission of the publisher. **A,** model of cyclomaltohexaose from the top; **B,** model of cyclomaltohexaose from the side; **C,** model of cyclomaltoheptaose from the top; **D,** model of cyclomaltooctaose from the top; **E,** model of cyclomaltononaose from the top.

trins. Modern nomenclature for these dextrins is cyclomaltohexaose (α-cyclodextrin), cyclomaltoheptaose (β-cyclodextrin), and cyclomaltooctaose (γ-cyclodextrin). The enzyme elaborated by *B. macerans* and other related bacilli that convert starch into these cyclomaltodextrins was originally called *macerans amylase.* The enzyme is not, however, an amylase, since it does not catalyze hydrolysis, but, rather it catalyzes an intramolecular transglycosylation reaction, and is correctly called *cyclomaltodextrin glucanosyltransferase.*

There were early indications [10] that even higher cyclomaltodextrins are produced by the action of this enzyme. The actual demonstration of these higher cyclodextrins was reported in 1961 by Pulley and French [12], who showed the formation of cyclomaltodextrins with 9, 10, 11, and 12 D-glucose residues. They performed a *B. macerans* enzyme digest of starch, followed by a solvent fractionation and β-amylase treatment. This preparation was chromatographed on a cellulose column at high temperature to give separation of the higher cyclic dextrins. The size and structure of these higher cyclomatodextrins were confirmed by French, et al. [13]. Cyclomaltodextrins having 9–13 D-glucose residues were subsequently isolated and their properties determined by Kobayashi et al. [14,15].

The cyclomaltodextrins have a rigid structure with a somewhat conical shape. They all have a depth of 7.8 Å, with different outside and cavity diameters that are dependent on the number of D-glucose residues in the molecule. Also, as the number of glucose residues increases, there is a dissymmetric shape to the molecule, with a long and a short diameter (see the structure of cyclomaltooctaose and cyclomaltononaose in Figs. 8.1 and 8.2). The top and the bottom of the cyclomaltodextrins are also asymmetric. The primary alcohol groups project from the narrower bottom of the molecules, and the C-2 and C-3 secondary alcohols project from the broader top of the molecule. The interiors of the cavities contain the glycosidic oxygen atoms and a high percentage of the carbon and hydrogen atoms of the glucose rings, resulting in a high electron density and hydrophobic environment. This is similar to the environment found in the interior of the single helix of amylose.

The cyclomaltodextrins have a curious water solubility in which the molecules with an even number of D-glucose residues are much more water soluble than are the molecules with an odd number of D-glucose residues [7,14]. The properties and characteristics of the cyclomaltodextrins are summarized in Table 8.1.

One of the most impressive properties of the cyclomaltodextrins is their ability to form inclusion complexes with a wide variety of organic and inorganic compounds. Many of these complexes form crystals, and they have been the basis for the development of separation and purification schemes for the

Table 8.1. Characteristics and Properties of Cyclomaltodextrins

	α-CD	β-CD	γ-CD	δ-CD	ε-CD
No. of glc.	6	7	8	9	10
Mol. wt.	972	1134	1296	1458	1620
Depth	7.8Å	7.8Å	7.8Å	7.8Å	7.8Å
Diam. top	13.7Å	15.3Å	16.9Å	18.5Å	19.6Å
Diam. cavity	5.7Å	7.8Å	9.5Å	11.0Å	12.1Å
Sp. opt. rot. $[\alpha]_D$	+150.5°	+162.5°	+177.4°	+191°	+197°
H_2O solubility[a]	10.1	1.60	23.2	2.7	–
I_2/I^- color	Blue	Brown	Yellow	–[b]	–[b]

[a]g/100 mL at 20°C
[b]Colorless

Table 8.2. Solubility[a] of Cyclomaltodextrins in Aqueous Solutions with Various Organic Compounds

Organic compound	α-CD	β-CD	γ-CD
Benzene	1.24	0.075	1.45
Fluorobenzene	1.14	0.000	1.60
Chlorobenzene	1.32	0.000	0.36
Bromobenzene	1.62	0.01	1.64
p-Cymene	2.92	0.04	0.15
Cyclohexanol	1.85	0.99	4.77
1,1,2,2-Tetrachloroethane	0.57	0.12	ND[b]
Tetrachloroethylene	0.7	0.004	ND
Anthracene	—[c]	—[c]	2.76

[a]g/100 mL
[b]Not determined
[c]Does not form a complex

cyclomaltodextrins (see refs. [7,8]). With some organic molecules, the water solubility is reduced to less than 1%. Table 8.2 gives the water solubility of α- and β-, and γ-cyclomaltodextrins in the presence of various organic molecules.

The internal hydrophobic cavity is the key structural feature of the cyclodextrins; it provides their ability to complex and hold a wide variety of inclusion molecules. Differences between the cyclomaltodextrins arise from differences in the diameters of their cavities. The organic inclusion molecules have alkyl, aromatic, alcohol, and ester groups. To bind with the cyclomaltodextrin, the inclusion molecule must have a size that fits, at least partially, into the cavity, creating the complex. The inclusion compound, however, does not have to be completely contained in the cavity. Complexes can be formed by the insertion of some specific functional group or part of the molecule to bind in the hydrophobic cavity. This partial insertion of the inclusion compound gives a "cap" or "lid" on the cavity. Sometimes more than one cyclodextrin will bind with a single inclusion molecule.

The formation of inclusion complexes can stabilize labile compounds and provide protection for light- or oxygen-sensitive compounds. Inclusion complexes can also be used to control the release, stability, solubility, and utilization of biologically active compounds such as drugs, vitamins, flavors, odors, insecticides, herbicides, and so forth. The cyclomaltodextrins can also be used to mask, alter, and/or eliminate undesirable flavors or odors, and they can be used to increase the water solubility of compounds that otherwise have low solubility [16].

The cyclomaltodextrins also serve a special role as models for enzyme catalysis [17]. This arose from their property of forming complexes with various molecules in their central cavity. The binding of an inclusion molecule in the central cavity is analogous with the formation of an enzyme-substrate complex in which the substrate is specifically bound at the active site of the enzyme. Cyclomaltodextrins have been observed to accelerate hydrolysis of carboxylic acid esters and pyrophosphate linkages [17,18]. The reaction with diphenylpyrophosphate produced phenol, phenyl phosphate, and 2-*O*-phosphorylated cyclomaltodextrin [18]. The formation of the cyclomaltodextrin phosphate suggested that the C-2 hydroxyl group on the cyclodextrin was acting as a catalytic nucleophile, similar to a catalytic group at an enzyme active site. The hydroxyl group makes a nucleophilic attack on a phosphorus atom in the pyrophosphate group, cleaving the linkage between the two phosphates. At 61°C and pH 10, cyclomaltoheptaose had a rate enhancement of approximately fivefold.

To further increase the rate of catalysis by cyclomaltodextrins, more reactive nucleophiles have been introduced onto the cyclodextrin ring. Mono- or di-6-*O*-tosyl groups can be formed, and imidazole groups introduced, by displacement of the 6-*O*-tosyl groups with aminomethyl imidazole [19a]. An effective catalyst for the hydrolysis of *p*-nitrophenyl acetate was cyclomaltoheptaose containing two imidazole groups. The enhancement of the rate probably resulted from one imidazole acting as a nucleophile and the other acting as a proton donor to give concerted acid-base catalysis.

A large number and variety of chemically modified cyclodextrins have been prepared. Derivatives at both the primary and/or secondary alcohol groups have been reported. Some of the derivatives can be specifically substituted onto or for either the primary or secondary alcohol groups [19b,20]. Derivatives with aryl, alkyl, tosyl, ester, hydroxypropyl, azido, amino, halo, and carboxyl groups, and phosphorus-, imidazole-, pyridine-, and sulfur-containing groups are known [20]. Rigidly "capped" cyclodextrins have been prepared on the primary alcohol side of the cyclodextrin rings by reaction with bifunctional reactants such as aryl disulfonyl or dicarboxyl chlorides. The number of aromatic rings can vary from two to three, with two predominating. The rings are often joined together by azo, methylene ether, carbonyl, or unsaturated alkyl groups.

Branched cyclomaltodextrins with single D-glucose residues linked α-1 $\rightarrow$ 6 to the cyclodextrin have been prepared [21]. Cyclomaltodextrins with branch chains of maltodextrins having 2 to 6 D-glucose residues have also been prepared by using the reverse condensation reaction catalyzed by *Pseudomonas amyloderamosa* isoamylase between cyclomaltodextrin and the individual linear maltodextrins [22]. Branched cyclomaltodextrins have also been prepared by a similar reaction of cyclomaltodextrin with maltose and panose, catalyzed by pullulanase [23]. The addition of branch chains onto the cyclomaltodextrins greatly increases their water solubility, especially for the cyclomaltodextrins with an odd number of glucose residues, and it may also alter the specificity for binding inclusion compounds. See, Fig. 8.1 for the structures of the branched cyclomaltodextrins.

Cyclomaltohexaose forms high yields of a crystalline complex with poly(ethylene glycol) [24]. It was shown that the complex consists of several cyclodextrin molecules threaded onto a single poly(ethylene glycol) chain [25]. Cyclomaltohexaose did not form a complex with poly(propylene glycol). Cyclomaltoheptaose and cyclomaltooctaose formed complexes with poly(propylene glycol), but not with poly(ethylene glycol) [26]. The cyclodextrins can be trapped on the polymer by capping both ends of the polymer chain with bulky groups. As many as 37 cyclodextrin rings were permanently threaded on poly(iminotrimethylene imin-odecamethylene) [27].

8.2 Cyclic β-1 → 2 Glucans (Cyclosophorans)

After the discovery of the cyclomaltodextrins, the next report of cyclic dextrins was not for nearly fifty years, in 1942. Cyclic β-glucans were reported in the culture supernatants of *Agrobacterium tumefaciens* [28]. These glucans were originally called *crown gall polysaccharides* and were described as low molecular weight glucans with d.p.'s of around 22. They were subsequently observed in the culture supernatants of all species of *Agrobacterium, Rhizobium,* and *Bradyrhizobium.* The glucans gave a single, methylated D-glucose product (3,4,6-tri-*O*-methyl-D-glucose) [29–33]. This indicated that the glucose residues were all linked by a 1 → 2 glycosidic bond, and that the molecule was cyclic. The configuration at C-1 was suggested to be β, on the basis of the negative optical rotation, which was confirmed by ^{1}H and ^{13}C-NMR [34–36]. The molecules were cyclic β-1 → 2-linked dextrins (cyclosophorans) and were mixtures of molecules with d.p. values ranging from 17 to 40. These cyclic dextrins are considerably larger than the cyclomaltodextrins and are actually macrocyclic molecules. A comparison of the distribution of ring sizes that are elaborated by a variety of *Agrobacterium* and *Rhizobium* species indicated that the distribution of ring size is dependent on the species of bacteria [36]. The cyclic sophorans produced by *Rhizobium leguminosarum, Rhizobium japonicum,* and *Agrobacterium tumefaciens* strains consist of a mixture of molecules with d.p.'s ranging from 17 to 25, with a majority in the range of 19–22. The cyclic sophorans of *Rhizobium meliloti* have a broad range from 17 to 40, with a majority having d.p.'s of 21–22. *Rhizobium phaseoli, Rhizobium trifolii,* and *Rhizobium lupini* elaborated cyclic sophorans with d.p. values in the range of 17–21, with d.p. 17 predominating. Rings with d.p. values less than 17 have not been found. It has been speculated that there may be too much strain in the glycosidic linkages for β-1 → 2-linked rings with less than 17 D-glucose residues [37,38]. From modeling studies it was postulated that ring strain energy would decrease exponentially as the d.p. increases from 17 to 24. The cavity diameters for the cyclic β-1 → 2 dextrins have been estimated to range from 8.8Å for d.p. 18 to 13Å for d.p. 24 [37]. The structures for the cyclic β-1 → 2 dextrins are shown in Fig. 8.3.

The cyclosophorans are secreted into the culture medium but are also found associated with the cell [31]. Cellular concentrations range between 5 and 20% of the total dry cell weight. They are located primarily in the periplasmic space. Their secretion from the cell varies greatly for the different species and is influenced by the stage of growth and conditions of growth such as the composition of the medium and the temperature [39,40]. Large amounts of extracellular cyclosophorans are found after the stationary phase of the culture has been reached [39–41].

The cyclosophorans are sometimes substituted with anionic substituents such as 1-phosphoglycerol, succinic acid, and methyl malonic acid [42–46]; 1-phos-

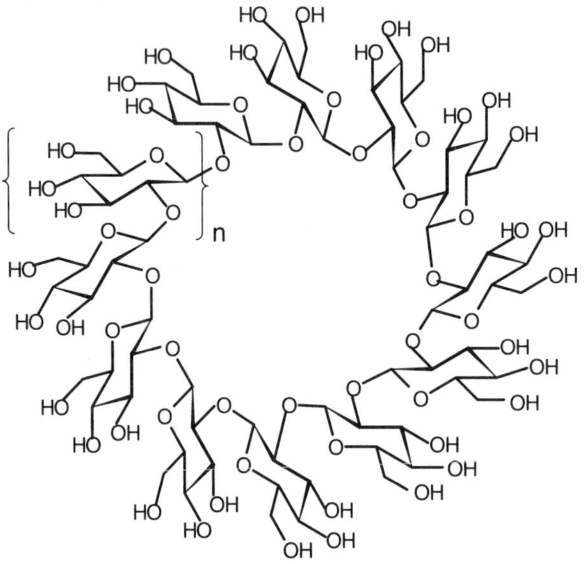

cyclosophoran
(cyclo-β-1,2-D-glucan)
n = 6 --13

Figure 8.3. Structural formula for the cyclosophorans.

phoglycerol is derived from phosphatidyl glycerol [44,45]. The degree of substitution of the cyclosophorans varies greatly for the different species of bacteria and the stage of growth. *R. leguminosarium* strains synthesize only neutral, unsubstituted cyclosophorans [46].

It has been proposed that the cyclosophorans play a role during osmotic adaptation of the cell [47]. The highest levels of synthesis occur during growth in low-osmotic media [48,49]. The cyclosophorans may also have a role in plant infection by *A. tumefaciens* and in the formation of nitrogen-fixing nodules by *Rhizobium* species [49,50]. The majority of the evidence for these roles comes from studies of mutants that do not produce the cyclosophorans. Other evidence comes from the observation that nodulation is stimulated when exogenous cyclosophorans are added to *Rhizobium* sp.–legume systems [49,51]. There has also been speculation that the different size distributions of the cyclosophorans that are produced by different *Rhizobium* species imparts a specificity for their formation of nitrogen-fixing nodules with roots of specific kinds of legumes.

8.3 Cyclic β-1 → 6 and β-1 → 3 D-Glucans

Several species of *Bradyrhizobium* synthesize cyclic dextrins that have both β-1 → 6 and β-1 → 3-linked D-glucose residues [52–54]. These cyclodextrins are similar to the cyclic β-1 → 2 D-glucose dextrins elaborated by *A. tumefaciens* and

Rhizobia species. They are found in the periplasmic space of the bacterial cell and are also secreted into the culture supernatant. They are branched, with d.p. values between 10 and 13 [53,54]. *Bradyrhizobium japonicum* USDA 110 elaborates a single cyclic dextrin that has 13 D-glucose residues. The cyclic structure contains three contiguous β-1 → 6-linked D-glucose residues followed by three contiguous β-1 → 3-linked D-glucose residues. This sequence repeats itself to give 12 D-glucose residues in the cyclic structure. One of the β-1 → 3-linked sequences has a single branched D-glucose residue substituted β-1 → 6 onto the center D-glucose residue (see Fig. 8.4 for the structural formula). This β-1 → 6/β-1 → 3 cyclo D-glucan apparently has similar roles for the bacteria, as do the cyclosophorans.

A cyclodextrin containing only β-1 → 3-linked D-glucose residues (a cyclo-laminarinose) has been found to be elaborated by a recombinant strain of *Rhizobium meliloti* TY7, a cyclic β-1 → 2 D-glucan-deficient mutant that carries a genetic locus for the cyclo β-1 → 6/β-1 → 3 D-glucan elaborated by *B. japonicum* USDA 110. This cyclo β-1 → 3 D-glucan contains 10 D-glucose residues in the ring, with a single laminaribiose disaccharide substituted β-1 → 6 onto the ring [55]. See Fig. 8.5 for the structure.

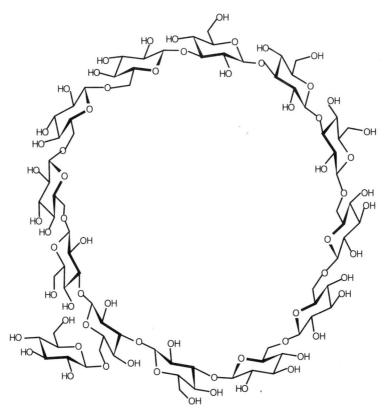

Figure 8.4. Structural formula for β-D-glucopyranosyl-1 → 6-cyclo-β-1 → 6/β-1 → 3-D-dodecaose.

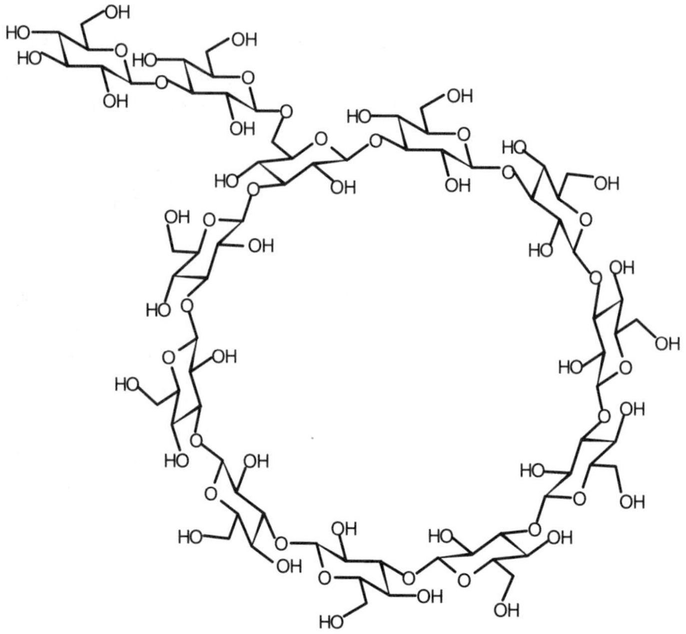

Figure 8.5. Structural formula for β-laminaribiosyl-1 → 6-cyclolaminaridecaose (cyclo-β-1 → 3-D-glucodecaose), reprinted from ref. [55] with kind permission from Elsevier Science, Ltd, The Boulevard, Langford Lane, Kidlington OX5 1GB, UK and the authors.

The β-1 → 3 cyclodextrin is very different in structure, cavity size, and type of substitution from the cyclo-β-D-glucan elaborated by either one of the wild type precursors, *B. japonicum* USDA 110 and *R. meliloti* 102F34. Its elaboration by a negative cyclo-β-D-glucan mutant restored the wild type properties of hypoosmotic adaptation and symbiotic competence with alfalfa. For another negative cyclo-β-D-glucan mutant of *B. japonicum* whose wild type was specific for soybean nodulation, synthesis of the cyclolaminarinose restored the hypoosmotic competence but did not restore the symbiosis with soybeans [55]. This latter result supports the hypothesis that specific cyclo-β-D-glucans, specific in either size and/or in structure, are required in the symbiotic nitrogen-fixing process between the specific bacterium and plant.

8.4 Cycloisomaltodextrins

The newest of the cyclodextrins is produced by an enzyme elaborated by *Bacillus* sp. T-3040 in a reaction with B-512F dextran to give cyclic isomaltodextrins [56]. Three kinds of cyclic isomaltodextrins have been obtained: cycloisomaltoheptaose with seven glucose* residues, cycloisomaltooctaose with eight glucose residues, and cycloisomaltononaose with nine glucose residues, all with α-1 → 6 linkages [57]. The formation of these cycloisomaltodextrins is by the cyclic trans-

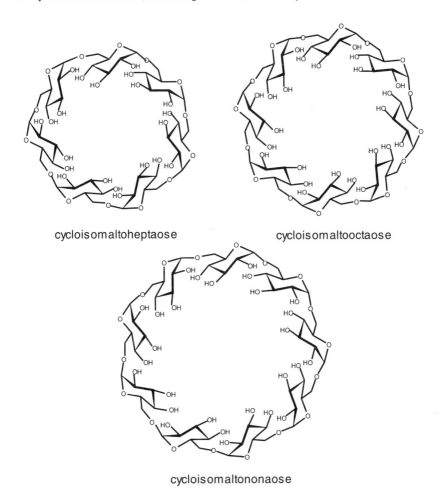

cycloisomaltoheptaose   cycloisomaltooctaose

cycloisomaltononaose

Figure 8.6. Structural formulas for cycloisomaltodextrins.

glycosylation reaction from the nonreducing ends of the dextran chains. The enzyme would thus be called *cycloisomaltodextrin glucanosyltransferase.* See Fig. 8.6 for the structural formulas of these new cyclic dextrins.

8.5 Cycloalternanotetraose (Alternating α-1 → 6/α-1 → 3 Cyclotetraose)

The smallest naturally occurring cyclic dextrin contains four D-glucose residues, with alternating α-1 → 6 and α-1 → 3 glycosidic linkages. The cyclic tetrasaccharide is produced by the action of a bacterial enzyme on alternan [58]. Other saccharide products that are produced include isomaltose and 3^2-α-D-glucopyra-

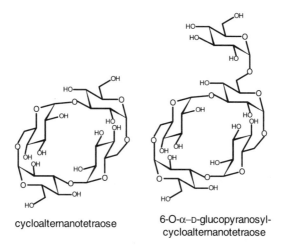

cycloalternanotetraose   6-O-α–D-glucopyranosyl-
cycloalternanotetraose

Figure 8.7. Structural formulas for cycloalternanotetraoses, from ref. [58] by permission of the authors, and publisher.

nosyl isomaltose. The cyclic tetrasaccharide product was produced in the greatest proportion. Oligosaccharides of higher degree of polymerization were also formed and included α-D-glucopyranosyl linked α-1 → 6 onto the cyclic tetrasaccharide. See Fig. 8.7 for the structures of cycloalternanotetraose and α-D-glucopyranosyl-(1 → 6)-cycloalternanotetraose.

8.6 Cycloinulodextrin

Cycloinulohexaose was obtained by the action of an enzyme from *Bacillus circulans* OKUMZ 31B on inulin [59]. Small amounts of cycloinuloheptaose and cycloinulooctaose were also obtained. The linkages in these cyclic dextrins are β-2 → 1, as is found in inulin. The formation of these cyclic dextrins by this enzyme is most probably similar to that of cyclomaltodextrin glucanosyltransferase and cycloisomaltodextrin glucanosyltransferase. The structure for cycloinulohexaose is given in Fig. 8.8.

8.7 Chemical Synthesis of Cyclodextrins

Several chemical syntheses of cyclodextrins have recently been developed. Although the syntheses are usually long, with many steps, and have low yields, a number of unusual and heretofore unknown cyclodextrins have been obtained.

The total synthesis of cylomaltohexaose (α-cyclodextrin) has been achieved starting with maltose. Maltose derivatives were condensed two times to give the key intermediate 2,3,6-octa-o-benzyl-β-maltohexaosyl fluoride, which was cy-

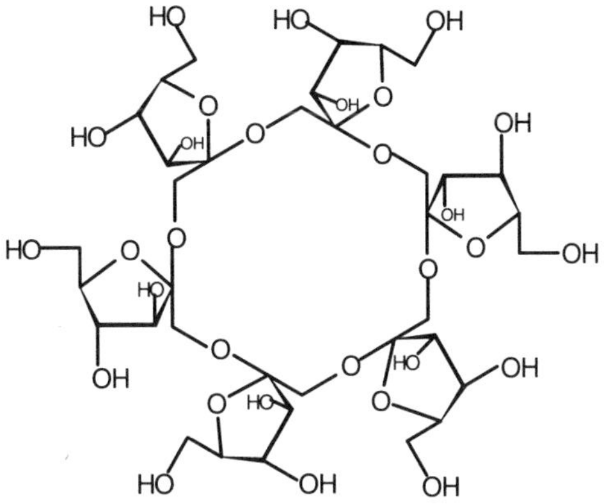

cycloinulohexaose

Figure 8.8. Structural formulas for cycloinulohexaose, reprinted from ref. [59] with kind permission from Elsevier Science Ltd, The Boulevard, Langford Lance, Kidlington OX5 1GB, UK and the authors.

clized using silver triflate as a catalyst [60,61]. The benzyl protective groups were removed to give cyclomaltohexaose. The synthesis required 21 steps and had an overall yield of 0.3%.

The total synthesis of cyclomaltooctaose (γ-cyclodextrin) has been achieved in a similar manner in 21 steps from maltose with an overall yield of 0.02% [62]. A similar synthesis was also conducted to form a hexasaccharide that was cyclized to give an α-1 → 6 linkage and the formation of an isomer of cyclomaltohexaose with one α-1 → 6 linkage and five α-1 → 4 linkages in the ring [62]. Specific introduction of an azide group for one of the primary hydroxyl groups of the hexasaccharide has also been achieved to give cyclomaltohexaose substituted at one C-6 by an azido group [62].

Another cyclomaltohexaose isomer was synthesized in which the D-glucose residues were replaced by D-mannose residues [63,64]. The key intermediate was the formation of a mannohexaosyl methylthioglycoside that was cyclized by an intramolecular glycosylation reaction using phenylselenyl triflate as a catalyst.

The synthesis of β-1 → 3-linked cyclo-D-glucohexaose was achieved by using 1,2-o-benzilidene-4,6-ethylidene-α-D-glucopyranose and the formation of 2-o-benzoyl-4,6-ethylidene-D-glucopyranosyl bromide by photobromination. This bromide was sequentially condensed to give a β-1 → 3-linked-α-hexaosyl bromide, which was cyclized by silver triflate catalyst to produce the protected β-1 → 3 cyclolaminarihexaose [65].

Another unique cyclodextrin has been synthesized to give the first cyclodextrin with L-monosaccharide residues [66]. L-Rhamnose was converted into an α-1 → 4-linked L-rhamnohexasaccharide derivative (2,3-dodeca-o-benzyl-1-methyl-α-thio-L-rhamnohexaoside) that was cyclized using phenylselenyl triflate to form α-1 → 4-linked cyclo-L-rhamnohexaose (see Fig. 8.9 for the structure).

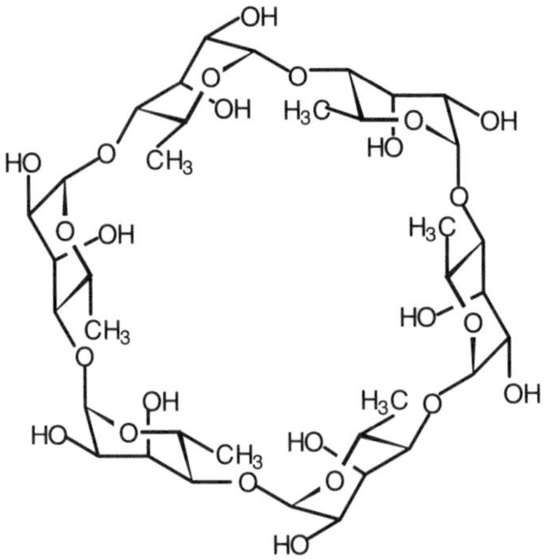

cyclo-L-rhamnohexaose

Figure 8.9. Structural formulas for cyclo-α-1 $\rightarrow$ 4-L-rhamnohexaose.

8.8 Macrocyclic Maltodextrins

Reaction of amylose (d.p. 2000) with potato disproportionating enzyme (D-enzyme) resulted in a decrease in the formation of the blue amylose triiodide complex [67]. The product was devoid of reducing and nonreducing ends and was resistant to the action of both exo- and endo-acting hydrolyzing enzymes. D-enzyme catalyzed an intramolecular transglycosylation reaction to produce a cyclic α-1 $\rightarrow$ 4-D-glucan (cycloamylose). The d.p. was found to range from 17 to several hundred, with an average of 90. The cycloamylose was highly soluble in cold water and formed inclusion complexes with several inorganic and organic compounds [67].

The action of a bacterial starch-branching enzyme elaborated by *Bacillus stearothermophilus* with amylose also produced a nonreducing cyclic α-1 $\rightarrow$ 4-D-glucan (cycloamylose) [68]. The product in this reaction had one α-1 $\rightarrow$ 6 glycosidic linkage in the ring at the site of the transglycosylation.

The *B. stearothermophilus* branching enzyme also formed large cyclic glucans when reacting with amylopectin. The average d.p. was 900. The macrocyclic product was joined together by an α-1 $\rightarrow$ 6 linkage at the site of the transglycosylation and had several noncyclic maltodextrinyl side chains that were linked α-1 $\rightarrow$ 6 to the macro ring [69].

8.9 Literature Cited

1. A. Villers, *Compt. rend.*, **112** (1891) 536–541
2. F. Schardinger, *Wein. Klin. Wochschr.*, **16** (1903) 468–479; **17** (1904) 207–214.
3. F. Schardinger, *Zentr. Bakteriol. Parasitenk, Abt. II*, **14** (1905) 772–778.

4. F. Schardinger, *Zentr. Bakteriol. Parasitenk, Abt. II,* **29** (1911) 188–193.

5. K. Freudenberg and R. Jacobi, *Ann. Chem.* **518** (1935) 102–108.

6. K. Freudenberg, G. Blomquist, L. Ewald, and K. Soff, *Chem. Ber.,* **69** (1936) 1258–1266.

7. D. French, *Adv. Carbohydr. Chem.,* **12** (1957) 189–260.

8. J. A. Thoma and L. Stewart, in *Starch: Chemistry and Technology,* pp. 209–249 (R. L. Whistler, E. F. Paschall, J. N. BeMiller, and H. J. Roberts, eds.) Academic, New York (1965).

9. D. French and R. E. Rundle, *J. Am. Chem. Soc.,* **64** (1942) 1651–1653.

10. K. Freudenberg and F. Cramer, *Z. Naturforsch.,* **3b** (1948) 464–470.

11. D. French, D. W. Knapp, and J. H. Pazur, *J. Am. Chem. Soc.,* **72** (1950) 5150–5152.

12. A. O. Pulley and D. French, *Biochem. Biophys. Res. Commun.,* **5** (1961) 11–15.

13. D. French, A. O. Pulley, J. A. Effenberger, M. A. Rougvie, and M. Abdullah, *Arch. Biochem. Biophys.,* **111** (1965) 153–160.

14. T. Endo, H. Ueda, S. Kobayashi, and T. Nagai, *Carbohydr. Res.,* **269** (1995) 369–373.

15. T. Endo, H. Nagase, H. Ueda, S. Kobayashi, and T. Nagai, *Chem. Pharm. Bull.* **45** (1997) 532–536.

16. J. Szejtli, *Stärke,* **29** (1976) 26–33; **30** (1978) 427–431; **33** (1981) 387–390.

17. D. W. Griffiths and M. L. Bender, *Adv. Catal.,* **23** (1973) 209–261.

18. N. Hennrich and F. Cramer, *J. Am. Chem. Soc.,* **78** (1965) 1121–1123.

19a. F. Cramer and G. Mackensen, *Angew. Chem. Int. Ed. Engl.,* **5** (1966) 601–606; *Chem. Ber.,* **103** (1970) 2138–2142.

19b. S. P. Tian, P. Forgo, and V. T. D'Souza, *Tetrahedron Lett.,* **37** (1996) 8309–8312.

20. A. P. Croft and R. A. Bartsch, *Tetrahedron,* **39** (1983) 1417–1474.

21. Y. Okada, Y. Kubota, K. Koizumi, S. Hizukuri, T. Ohfuji, and K. Ogata, *Chem. Pharm. Bull.,* **36** (1988) 2176–2185.

22. S. Hizukuri, J.-I. Abe, K. Koizumi, Y. Okada, Y. Kubota, S. Sakai, and T. Mandai, *Carbohydr. Res.,* **185** (1989) 191–198.

23. S. Kobayashi, K. Nakashima, and M. Arahira, *Carbohydr. Res.,* **192** (1989) 223–231.

24. A. Harada and M. Kamachi, *Macromolecules,* **23** (1990) 2821–2923.

25. A. Harada, J. Li, and M. Kamachi, *Nature,* **356** (1992) 325–327.

26. A. Harada and M. Kamachi, *J. Chem. Soc. Chem. Commun.,* (1990) 1322–1323.

27. G. Wenz and B. Keller, *Angew. Chem. Int. Ed. Engl.,* **31** (1992) 197–199.

28. F. C. McIntire, W. H. Peterson, and A. J. Riker, *J. Biol. Chem.,* **143** (1942) 491–496.

29. E. W. Putman, A. L. Potter, R. Hodgson, W. Z. Hassid, *J. Am. Chem. Soc.,* **72** (1950) 5024–5026.

30. P. A. J. Gorin, J. F. T. Spencer, and D. W. S. Westlake, *Can. J. Chem.,* **39** (1961) 1067–1073.

31. L. P. T. M. Zevenhuizen and H. J. Scholten-Koerselman, *Antonie Leewenhoek,* **45** (1979) 165–175.

32. W. S. York, M. McNeil, A. G. Darvill, and P. Albersheim, *J. Bacteriol.,* **142** (1980) 243–248.

33. J. M. DaCastro, M. Bruneteau, S. Mutaftshiev, G. Truchet, and G. Michel, *FEMS Microbiol. Lett.,* **18** (1983) 269–275.

34. A. Amemura, M. Hisamatsu, H. Mitani, and T. Harada, *Carbohydr. Res.,* **114** (1983) 277–285.

35. A. Dell, W. S. York, M. McNeil, A. G. Darvill, and P. Albersheim, *Carbohydr. Res.,* **117** (1983) 185–200.

36. M. Hisamatsu, A. Amemura, K. Koizumi, T. Utamura, and Y. Okada, *Carbohydr. Res.,* **121** (1983) 31–40.

37. A. Palleschi and V. Crescenzi, *Gazz. Chim. Ital.,* **115** (1985) 243–245.

38. W. S. York, J. U. Thomsen, and B. Meyer, *Carbohydr. Res.,* **248** (1993) 55–80.

39. M. W. Breedveld, L. P. T. M. Zevenhuizen, and A. J. B. Zehnder, *Appl. Environ. Microbiol.,* **56** (1990) 2080–2086.

40. M. W. Breedveld, L. P. T. M. Zevenhuizen, and A. J. B. Zehnder, *J. Bacteriol.,* **174** (1992) 6336–6342.

41. M. Hisamatsu, *Carbohydr. Res.,* **231** (1992) 137–146.

42. M. Batley, J. W. Redmond, S. P. Djordjevic, and B. G. Rolfe, *Biochim. Biophys. Acta,* **901** (1987) 119–126.

43. M. Hisamatsu, T. Yamada, T. Higashiura, and M. Ikeda, *Carbohydr. Res.,* **163** (1987) 115–122.

44. K. J. Miller, R. S. Gore, and A. J. Benesi, *J. Bacteriol.,* **170** (1988) 4569–4575.

45. K. J. Miller, V. N. Reinhold, A. C. Weissborn, and E. P. Kennedy, *Biochim. Biophys. Acta,* **901** (1987) 112–118.

46. L. P. T. M. Zevenhuizen, A. Van Veldhuizen, and R. H. Fokkens, *Antonie Leewenhoek,* **57** (1990) 173–178.

47. K. J. Miller, E. P. Kennedy, and V. N. Reinhold, *Science,* **231** (1986) 48–51.

48. A. Zorreguieta, S. Cavaignac, R. A. Geremia, and R. A. Ugalde, *J. Bacteriol.,* **172** (1990) 4701–4704.

49. T. Dylan, D. R. Helinski, and G. S. Ditta, *J. Bacteriol.,* **172** (1990) 1400–1408.

50. G. A. Cangelosi, L. Hung, V. Puvanesarajah, G. Stacey, D. A. Ozga, J. A. Leigh, and E. W. Nester, *J. Bacteriol.,* **169** (1987) 2086–2091.

51. M. Abe, A. Amemura, and S. Higashi, *Plant Soil,* **64** (1982) 315–324.

52. W. F. Dudman and A. J. Jones, *Carbohydr. Res.,* **84** (1980) 358–364.

53. K. J. Miller, R. S. Gore, R. Johnson, A. J. Benesi, and V. N. Reinhold, *J. Bacteriol.,* **172** (1990) 136–142.

54. D. B. Rolin, P. E. Pfeffer, S. F. Osman, B. S. Szwergold, F. Kappler, and A. J. Benesi, *Biochim. Biophys. Acta,* **1116** (1992) 215–225.

55. P. E. Pfeffer, S. F. Osman, A. Hotchkiss, A. A. Bhagwat, D. L. Keister, and K. M. Valentine, *Carbohydr. Res.,* **296** (1996) 23–37.

56. T. Oguma, K. Tobe, and M. Kobayashi, *FEBS Lett.,* **345** (1994) 135–138.

57. T. Oguma, T. Horiuchi, and M. Kobayashi, *Biosci. Biotech. Biochem.,* **57** (1993) 1225–1227.

58. G. L. Côté and P. Biely, *Eur. J. Biochem.,* **226** (1994) 641–648.

59. M. Kawamura, T. Uchiyama, T. Kuramoto, Y. Tamura, and K. Mizutani, *Carbohydr. Res.,* **192** (1980) 83–90.

60. T. Ogawa and Y. Takahashi, *Carbohydr. Res.,* **138** (1985) C5–C9.

61. Y. Takahashi and T. Ogawa, *Carbohydr. Res.,* **164** (1987) 277–296.

62. Y. Takahashi and T. Ogawa, *Carbohydr. Res.,* **169** (1987) 127–149.

63. M. Mori, Y. Ito, and T. Ogawa, *Tetrahedron Lett.,* **30** (1989) 1273–1276.

64. M. Mori, Y. Ito, and T. Ogawa, *Carbohydr. Res.,* **192** (1989) 131–146.

65. P. M. Collins and H. A. Mezher, *Tetrahedron Lett.,* **31** (1990) 4517–4520.

66. M. Nishizawa, H. Imagawa, Y. Kan, and H. Yamada, *Tetrahedron Lett.,* **32** (1991) 5551–5554.

67. T. Takaha, M. Yanase, H. Takata, S. Okada, and S. M. Smith, *J. Biol Chem.,* **271** (1996) 2902–2908.

68. H. Takata, T. Takaha, S. Okada, M. Takagi, and T. Imanaka, *J. Bacteriol.,* **178** (1996) 1600–1606.
69. H. Takata, T. Takaha, S. Okada, S. Hizukuri, M. Takagi, and T. Imanaka, *Carbohydr. Res.,* **295** (1996) 91–101.

8.10 References for Further Study

"The Schardinger dextrins," D. French, *Adv. Carbohydr. Chem.,* **12** (1957) 189–259.
"Cycloamyloses as catalysts," D. W. Griffiths and M. L. Bender, *Adv. Catal.,* **23** (1973) 209–261.
"Synthesis of chemically modified cyclodextrins," A. P. Croft and R. A. Bartsch, *Tetrahedron,* **39** (1983) 1414–1474.
"Production, characterization, and applications of cyclodextrins," H. Bender, *Adv. Biotech. Processes,* **6** (1986) 31–71.
"Inclusion complexes of the cyclomalto-oligosaccharides (cyclodextrins)," R. J. Clarke, J. H. Coates, and S. F. Lincoln, *Adv. Carbohydr. Chem. Biochem.,* **46** (1988) 205–249.
"A collection of invited papers dealing with cyclodextrins," *Carbohydr. Res.,* **192** (1989).
"Cyclic β-glucans of members of the family Rhizobiaceae," M. W. Breedveld and K. J. Miller, *Microbiol. Rev.,* **58** (1994) 145–161.

Chapter 9

Glycoconjugates

9.1 Introduction

The discovery of a covalent linkage between a carbohydrate and a protein was made in 1938 by Albert Neuberger. He reacted crystalline hen ovalbumin with a proteinase and isolated a low molecular weight glycopeptide and showed that it contained aspartic acid and a saccharide composed of D-mannose and D-glucosamine. Because of the lack of methods at the time for the determination of structures of biological materials, the nature of the linkage between the carbohydrate and the peptide was not determined. Twenty years later, Neuberger returned to the problem and showed that the linkage was an N-glycoside formed between the hemiacetal hydroxyl of N-acetyl-D-glucosamine and the nitrogen of the amide group of L-asparagine.

In the first half of the twentieth century, proteins were purified until all nonproteinaceous materials had been removed, otherwise they were regarded as impure mixtures. Likewise, polysaccharides were purified until all proteinaceous material had been removed. Despite Neuberger's discovery in 1938, proteins were still considered in the 1950s to be just polypeptides composed of a specific sequence of amino acids, and the presence of carbohydrate was considered to be due to contamination. As analytical techniques improved, it appeared that carbohydrate and proteinaceous materials could not be completely removed from each other.

Since 1960, it has been shown by many laboratories that carbohydrates are covalently linked through their hemiacetal hydroxyl to proteins primarily in two ways: (1) linkage to the amide nitrogen of L-asparagine to give N-glycosides, and (2) linkage to the hydroxyl groups of L-serine and L-threonine to give O-glycosides. It also was recognized in the 1960s that proteins with covalently bound carbohydrates are widely distributed, being found in all types of living organisms. Most proteins from eukaryotic organisms have covalently linked carbohydrates and are called *glycoproteins*.

In Chapter 6, we briefly mentioned the polysaccharide-protein covalent complexes of the *proteoglycans*. In proteoglycans, there are many relatively long polysaccharide chains of the glycosaminoglycan class, such as hyaluronic acid, chondroitin sulfate, dermatan sulfate, keratan sulfate, and heparan sulfate that are attached to a single protein chain. Although the types of linkages between the protein and the carbohydrate in proteoglycans and glycoproteins are the same, glycoproteins differ from proteoglycans by having much less carbohydrate. Thus, the chemical properties of proteoglycans are primarily those of polysaccharides, whereas the chemical properties of glycoproteins are primarily those of proteins.

Different proteins are glycosylated to different degrees. They can have single or several L-asparagine, L-serine, or L-threonine sites of glycosylation. The carbohydrate can be a single monosaccharide residue or an oligosaccharide unit of several monosaccharide residues. As analytical methods improved and new ones were developed, it was found that only a select group of monosaccharide residues were present in glycoproteins. These included N-acetyl-D-glucosamine (GlcNAc), N-acetyl-D-galactosamine (GalNAc), D-mannose (Man), L-fucose (6-deoxy-L-galactose [Fuc]), D-xylose (Xyl), and the sialic acids (substituted D-neuraminic acids [Neu]) [1]. Figure 9.1 gives the structural formulas for these monosaccharide residues.

The sialic acids make up a family of carbohydrates that is composed of D-neuraminic acid substituted at different positions by a limited number of substituents.

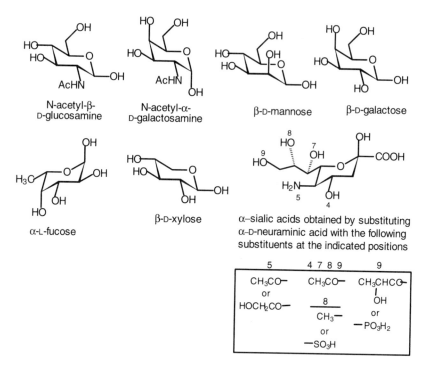

Figure 9.1. Structural formulas for the most common monosaccharide residues found in glycoproteins.

The amino group is substituted by an acetyl or glycoloyl group, and the hydroxyl groups at C-4, -7, -8, and -9 may be methylated or esterified by acetyl, lactyl, sulfate, or phosphate groups [2]. See Fig. 9.1 for the position specificity for neuraminic acid substitution. Sometimes several of the substituents are present on the same molecule. Unsubstituted D-neuraminic acid has not been found in nature. Twenty different sialic acids have been observed, showing a wide degree of structural diversity [2].

In the middle of the twentieth century, it was believed by many biological scientists that carbohydrates were essentially dull and uninteresting molecules, serving as structural and protective materials in plants and as a storage form of chemical energy. With the discovery that most proteins were glycosylated, it was assumed that there was a reason for the glycosylation. Indeed, it has developed that the dull carbohydrate molecules attached to proteins play very important and sometimes subtle roles in living systems. From studies of human blood group carbohydrates, it was shown that the carbohydrates attached to glycoproteins specify four types of human blood, and that the four types have different mutual compatabilities, an important consideration in administrating blood transfusions. Many other diverse functions have been discovered. To date, they include the establishment and stabilization of protein conformation, the secretion of proteins through membranes, control of protein turnover, protection of proteins against proteinase hydrolysis, and increasing protein solubility. It has also been found that carbohydrate residues attached to both proteins and lipids located on the surface of cells specify the biological recognition that is involved in cell-cell interactions that control and influence growth, differentiation, formation of organs, fertilization, processes of bacterial and viral infection, formation of tumors (absence of differentiation), tumor metastasis, and the prevention of autoimmune reactions.

Among some biological scientists, the carbohydrates have acquired as much importance and interest as proteins and nucleic acids, and many structures of the oligosaccharide chains of glycoproteins and glycolipids have been determined.

With six different monosaccharide residues found in glycoproteins, there is the possibility of over one billion different structural combinations of linear hexasaccharides, and this does not include any branching or monosaccharide duplication. Nature, however, is quite conservative and does not construct a totally different structure for every need. Instead, basic structures are produced and are conservatively modified by the addition of specific monosaccharide residue(s) to the ends of the chains or by the formation of a branch chain. The sequence, the types of linkages, and branching have a definite affect on specifying the biological properties.

9.2 Structures of *N*-Linked Glycosides in Glycoproteins

Three families of glycosides that are attached to the nitrogen of the amide group of L-asparagine have been observed. All of the *N*-glycosides have a relatively fixed pentasaccharide structure (there are a few minor deviations) called the "inner core."

It consists of chitobiosyl that is attached to the amide nitrogen by a β-linkage. β-D-Mannose is attached 1 → 4 to the chitobiose, and two α-D-mannose residues are attached to this mannose residue by 1 → 6 and 1 → 3 glycosidic linkages. Figures 9.2A and 9.3A give the structure of this inner core pentasaccharide [3]. There are three families of *N*-linked glycosides that result from the attachment of monosaccharide residues to the pentasaccharide core. These monosaccharide units make up the *variable region*. In the first family, the pentasaccharide is substituted by other α-D-mannose residues linked 1 → 6 and 1 → 3 to give further bifurcation of the terminal mannose residues of the core. The chains are terminated by the addition of one or two α-1 → 2-linked D-mannose residues. This structural family makes up the "high mannose" *N*-linked oligosaccharides. See Fig. 9.3B–F for their structures.

In the second family of *N*-linked oligosaccharides, the two terminal D-mannose residues of the core pentasaccharide are substituted by a variable number of *N*-acetyl-lactosamine (β-Gal-1 → 4-GlcNAc) units. The position of substitution of the *N*-acetyl-lactosamine can be (1 → 2), (1 → 4), and/or (1 → 6). This structural family makes up the "lactosamine" oligosaccharides. The structures for three such *N*-acetyl-lactosamine *N*-glycosides are given in Fig. 9.4A–C [4,5]. The family of *N*-glycosides is frequently substituted at the terminal monosaccharide residues by sialic acids (see Fig. 9.4D,E). In addition, α-L-fucose is also frequently found substituted onto the first α-D-GlycNAc residues of the core pentasaccharide [5,6] (see Figs. 9.4E and 9.5A). The oligosaccharide of human immunoglobulin IgG has two *N*-acetyl-lactosamine units attached α-1 → 2 to the two terminal D-mannose residues of the core pentasaccharide, with an additional β-GlcNAc attached 1 → 4 to the first D-mannose residue of the core pentasaccharide (Fig. 9.5A) [7,8]. α-L-Fucose has also been found substituted 1 → 2 onto the *N*-acetyl-lactosamine units of oligosaccharides that are excreted into the urine of patients with fucosidosis [9]. One such oligosaccharide structure is shown in Fig. 9.5C. The *N*-linked oligosaccharides of the neural cell adhesion molecule (NCAM) have a poly(*N*-acetyl-neuraminic acid) chain attached to the *N*-acetyl-lactosamine unit (Fig. 9.5D) [10]. The polysialic acid chain is apparently involved in development and differentiation, as shown by its presence on embryonic NCAM and its absence in adult NCAM. It is reexpressed on the surface of cells of the highly metastatic Wilm's tumor [10].

The third family of *N*-linked glycosides has a mixed structure of the high mannose family and the *N*-acetyl-lactosamine family. These structures have been observed for *N*-linked glycosides of ovalbumin glycoproteins [11,12]. Many of the structures have the β-D-galactose residue of the *N*-acetyl lactosamine missing, leaving the *N*-acetyl-D-glucosamine residue. Representative structures for this third family of *N*-linked oligosaccharides are shown in Fig. 9.6.

Variants of the core pentasaccharide have been observed in which there is only a single *N*-acetyl-D-glucosamine residue instead of a chitobiosyl unit, resulting in a tetrasaccharide attached to L-asparagine, as found in human myeloma IgM immunoglobulin [13] (Fig. 9.5A). Another variant has the insertion of a β-D-mannose residue, with a branched mannobiosyl unit between the two *N*-acetyl-D-glucosamine residues of the chitobiosyl unit, as found in human myeloma IgE immunoglobulin [7] (Fig. 9.5B). Variants of the *N*-acetyl lactosamine units also

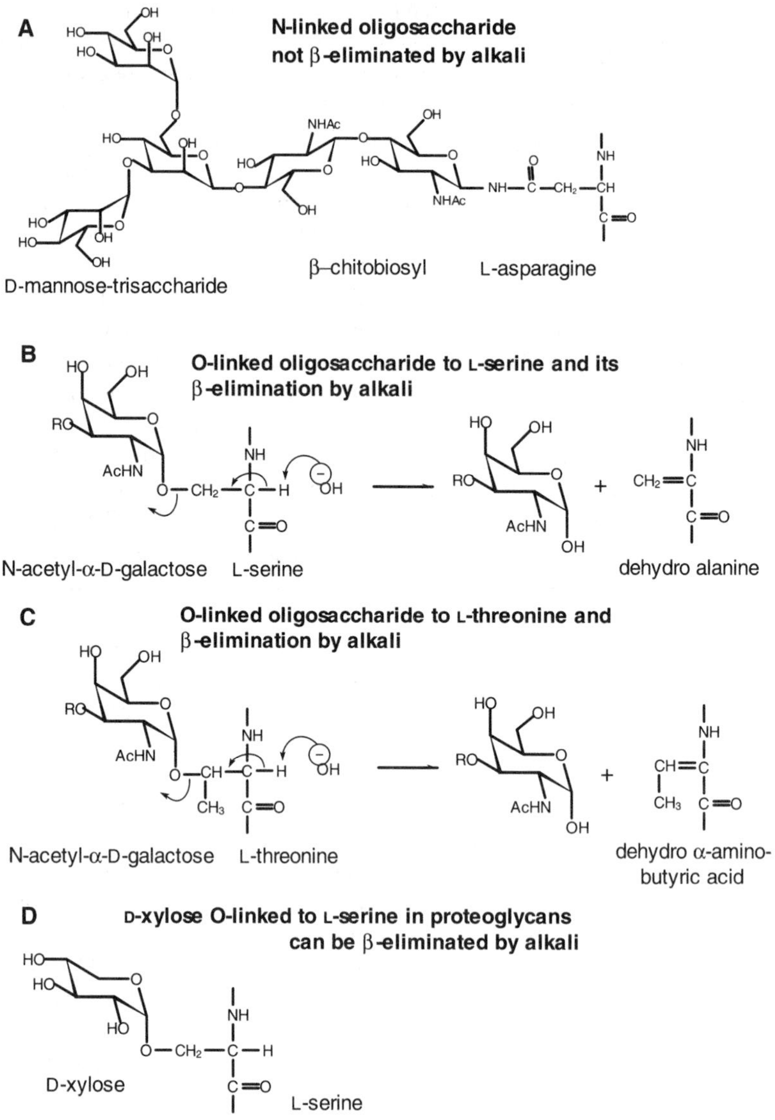

Figure 9.2. Structures for the attachment of glycosides to glycoproteins.

(continued)

E β-D-galactose O-linked to 5-hydroxy-L-lysine in collagen
not eliminated by alkali

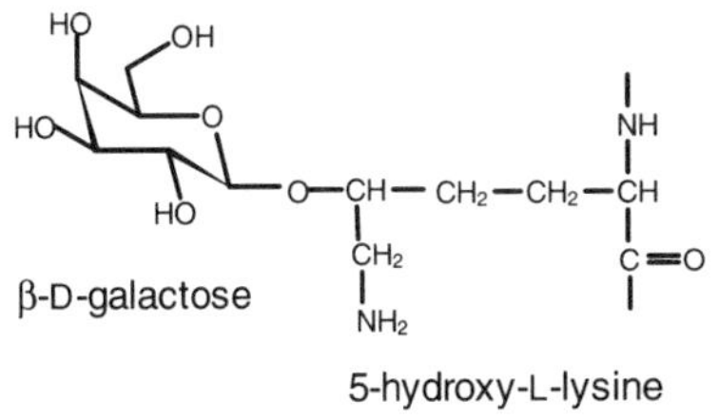

β-D-galactose

5-hydroxy-L-lysine

F β-D-glucose oligosaccharide S-linked to L-cysteine, β-eliminated by alkali

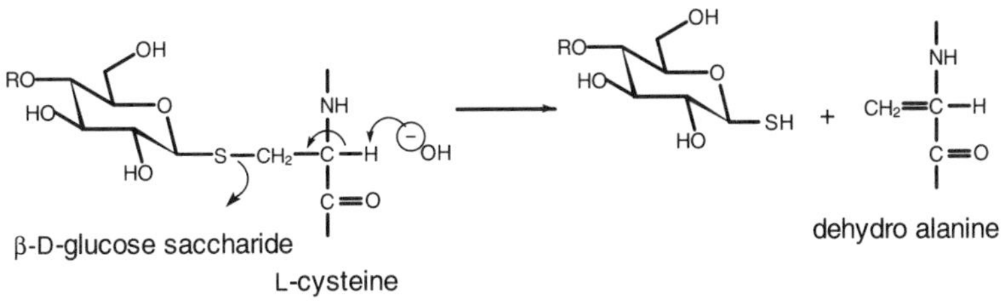

β-D-glucose saccharide

L-cysteine

dehydro alanine

G N-acetyl-neuraminic acid O-linked to L-tyrosine in chicken ovomucoid

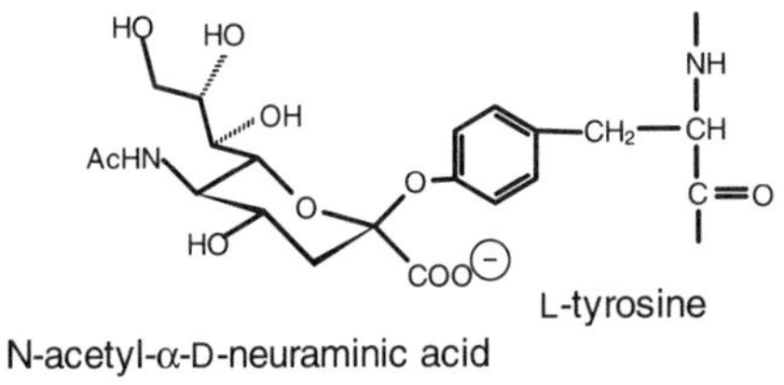

N-acetyl-α-D-neuraminic acid

L-tyrosine

Figure 9.2. *(continued)*

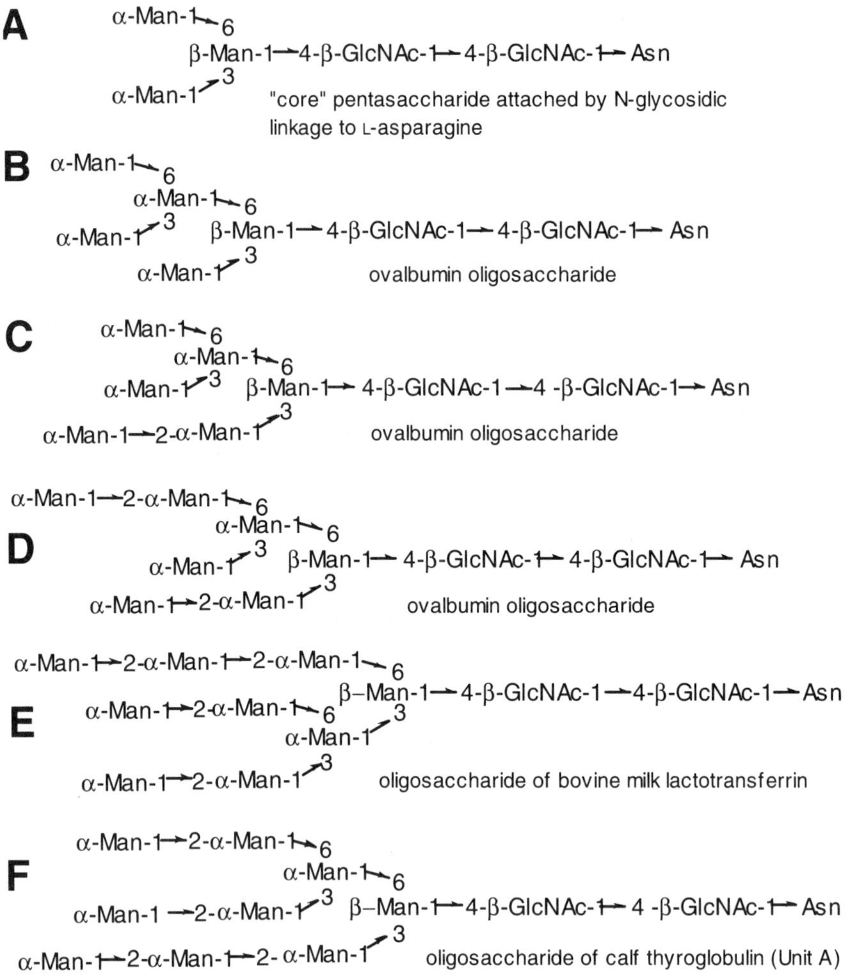

A
α-Man-1↖6
    β-Man-1→4-β-GlcNAc-1→4-β-GlcNAc-1→Asn
α-Man-1↗3
        "core" pentasaccharide attached by N-glycosidic
        linkage to L-asparagine

B
α-Man-1↖6
    α-Man-1↖6
α-Man-1↗3 β-Man-1→4-β-GlcNAc-1→4-β-GlcNAc-1→Asn
    α-Man-1↗3
            ovalbumin oligosaccharide

C
α-Man-1↖6
    α-Man-1↖6
α-Man-1↗3 β-Man-1→4-β-GlcNAc-1→4-β-GlcNAc-1→Asn
α-Man-1→2-α-Man-1↗3
            ovalbumin oligosaccharide

D
α-Man-1→2-α-Man-1↖6
    α-Man-1↖6
    α-Man-1↗3 β-Man-1→4-β-GlcNAc-1→4-β-GlcNAc-1→Asn
α-Man-1→2-α-Man-1↗3
            ovalbumin oligosaccharide

E
α-Man-1→2-α-Man-1→2-α-Man-1↖6
    α-Man-1→2-α-Man-1↖6 β-Man-1→4-β-GlcNAc-1→4-β-GlcNAc-1→Asn
        α-Man-1↗3
α-Man-1→2-α-Man-1↗3
            oligosaccharide of bovine milk lactotransferrin

F
α-Man-1→2-α-Man-1↖6
    α-Man-1↖6
α-Man-1→2-α-Man-1↗3 β-Man-1→4-β-GlcNAc-1→4-β-GlcNAc-1→Asn
α-Man-1→2-α-Man-1→2-α-Man-1↗3
            oligosaccharide of calf thyroglobulin (Unit A)

Figure 9.3. Structures for the high-mannose, bifurcated *N*-linked oligosaccharides.

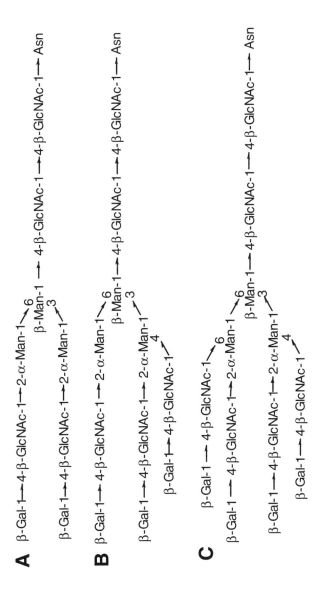

Figure 9.4. *N*-acetyl-lactosamine family of *N*-linked oligosaccharides.

(*continued*)

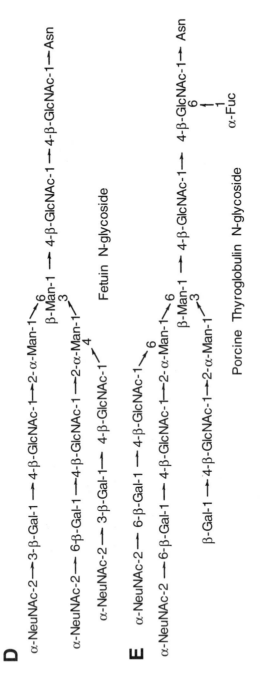

D

α-NeuNAc-2→3-β-Gal-1 → 4-β-GlcNAc-1→2-α-Man-1⟍6

α-NeuNAc-2→ 6β-Gal-1 →4-β-GlcNAc-1→2-α-Man-1⟍4
                                                          ⟍4
α-NeuNAc-2→ 3-β-Gal-1→ 4-β-GlcNAc-1⟋

β-Man-1 → 4-β-GlcNAc-1 → 4-β-GlcNAc-1→ Asn
          ⟋3

Fetuin N-glycoside

E   α-NeuNAc-2→ 6-β-Gal-1 → 4-β-GlcNAc-1⟍6

α-NeuNAc-2→ 6-β-Gal-1 → 4-β-GlcNAc-1→2-α-Man-1⟍6

β-Gal-1 → 4-β-GlcNAc-1→2-α-Man-1⟋3

β-Man-1 → 4-β-GlcNAc-1→ 4-β-GlcNAc-1→ Asn
          ⟋3   6
                                      ↑
                                      1
                                    α-Fuc

Porcine Thyroglobulin N-glycoside

Figure 9.4. (*continued*)

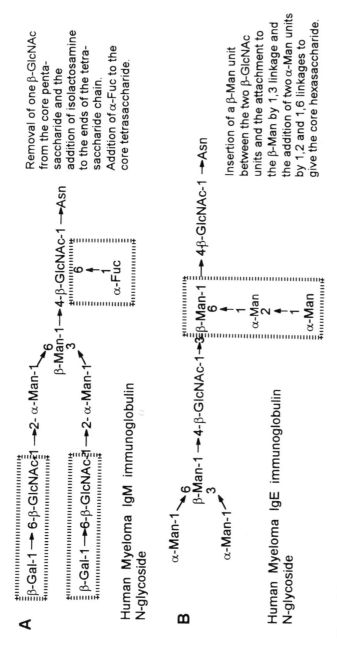

Figure 9.5. Variants in the structures of the inner core pentasaccharide and terminal N-acetyl lactosamine units of N-linked oligosaccharides: **A** and **B**, variants in the inner core pentasaccharides.

(continued)

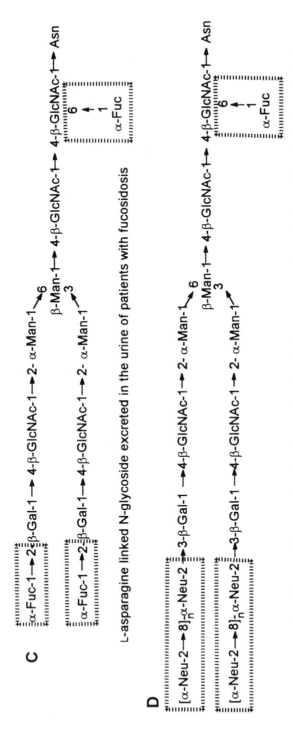

C

α-Fuc-1 → 2-β-Gal-1 → 4-β-GlcNAc-1 → 2- α-Man-1
                                                              ↘6
                                          β-Man-1 → 4-β-GlcNAc-1 → 4-β-GlcNAc-1 → Asn
                                                              ↗3   ↑6
α-Fuc-1 → 2-β-Gal-1 → 4-β-GlcNAc-1 → 2- α-Man-1   1
                                                                              α-Fuc

L-asparagine linked N-glycoside excreted in the urine of patients with fucosidosis

D

[α-Neu-2 → 8]ₙ-α-Neu-2 → 3-β-Gal-1 → 4-β-GlcNAc-1 → 2- α-Man-1
                                                                              ↘6
                                                β-Man-1 → 4-β-GlcNAc-1 → 4-β-GlcNAc-1 → Asn
                                                                              ↗3   ↑6
[α-Neu-2 → 8]ₙ-α-Neu-2 → 3-β-Gal-1 → 4-β-GlcNAc-1 → 2- α-Man-1   1
                                                                                          α-Fuc

L-asparagine N-linked oligosaccharide of neural cell adhesion molecule terminated with poly sialic acid chains, where n = 20-100.

Figure 9.5. (*continued*) **C** and **D**, variants in the terminal units of *N*-acetyl-lactosamine units.

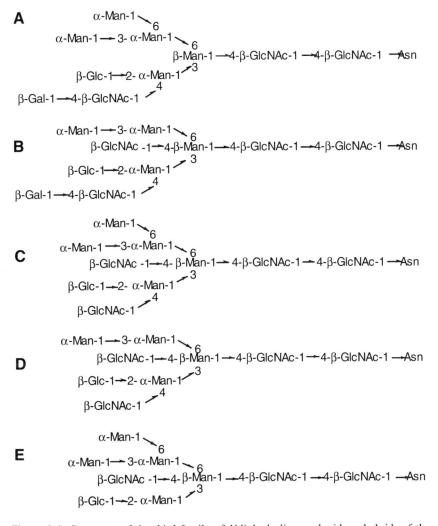

Figure 9.6. Structures of the third family of *N*-linked oligosaccharides—hybrids of the high mannose and *N*-acetyl-lactosamine families, observed in ovalbumin oligosaccharides.

occur in which *N*-acetyl-isolactosamine [β-Gal-(1 → 6)-GlcNAc] is substituted for *N*-acetyl-lactosamine, as observed for human myeloma IgM immunoglobulin [13] (Fig. 9.5A).

9.3 Structures of *O*-Linked Glycosides in Glycoproteins

The common structural feature of *O*-linked glycoside chains is the attachment of *N*-acetyl-α-D-galactosamine to the hydroxyl groups of either L-serine or L-threonine (see Fig. 9.2B and C). There are reports of D-xylose being *O*-linked in fungal α-amylase [14], and the linkage of proteoglycans to L-serine in human thyroglobulin and cartilage [15]. Further classification of *O*-linked oligosaccharides is more difficult [16]. Many of them have *N*-acetyl-lactosamine or an α-1 → 3-linked *N*-acetyl-lactosamine isomer. In addition, the ends of the chains are often high in *N*-acetyl-D-neuraminic acid residues. See Fig. 9.7 for representative *O*-linked oligosaccharides.

A common *O*-linked glycoprotein is the antifreeze protein found in blood sera of almost all fish living in the antarctic and the arctic. This highly glycosylated

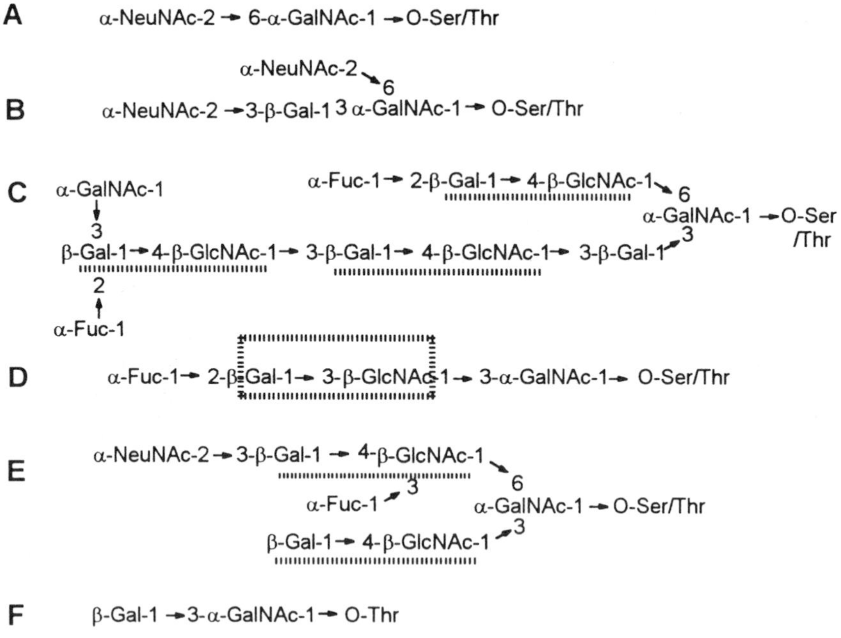

A   α-NeuNAc-2 → 6-α-GalNAc-1 → O-Ser/Thr

α-NeuNAc-2
                                          6
B   α-NeuNAc-2 → 3-β-Gal-1 3 α-GalNAc-1 → O-Ser/Thr

C α-GalNAc-1   α-Fuc-1 → 2-β-Gal-1 → 4-β-GlcNAc-1 6
             ↓   α-GalNAc-1 → O-Ser
             3   3   /Thr
β-Gal-1 → 4-β-GlcNAc-1 → 3-β-Gal-1 → 4-β-GlcNAc-1 → 3-β-Gal-1
             2
             ↑
        α-Fuc-1

D   α-Fuc-1 → 2-β-Gal-1 → 3-β-GlcNAc-1 → 3-α-GalNAc-1 → O-Ser/Thr

E   α-NeuNAc-2 → 3-β-Gal-1 → 4-β-GlcNAc-1
                                              6
                 α-Fuc-1   3   α-GalNAc-1 → O-Ser/Thr
            β-Gal-1 → 4-β-GlcNAc-1   3

F   β-Gal-1 → 3-α-GalNAc-1 → O-Thr

Figure 9.7. Structural features of *O*-linked oligosaccharides. All structures are attached to serine or threonine by an *N*-acetyl-α-D-galactosamine residue. The lactosamine structures are underlined and 1 → 3 isolactosamine structure is boxed.

protein lowers the freezing temperature of the blood of these fish. The protein has a simple repeating structure of L-alanyl-L-alanyl-L-threonine with β-D-galactopyranosyl-1 → 3-*N*-acetyl-α-D-galactosamine attached to the hydroxyl group of the L-threonine [17] (see Fig. 9.7F).

Another group of well-characterized *O*-linked glycosides is the human blood group substances, known as the *ABO blood group*. They are composed of oligosaccharides that are attached to glycolipids and glycoproteins and are present in the plasma membranes of many human tissues. In approximately 80% of the human population, the ABO blood group substances are secreted as glycoproteins into various body fluids such as saliva, milk, semen, gastric juice, and urine. They are also abundant on the surface of erythrocytes as glycolipids. The ABO system is characterized by three types of glycoside chains, the A-, B-, and H-glycosides. The three glycosides are closely related in structure and give rise to four blood types [18,19]. The H-glycoside composes the basic or core structure that is common to all of the ABO blood group oligosaccharides and gives what is known as "O-type blood." The H-glycoside is a bifurcated saccharide with nine monosaccharide residues *O*-linked to L-serine or L-threonine in glycoproteins, and to sphingosine in glycolipids (see section 9.4b). There are two types of branch chains, type 1 and type 2, depending on whether the β-D-galactose residues are attached 1 → 3 or 1 → 4 to the *N*-acetyl-β-D-glucosamine residue (see Fig. 9.8A) [20]. The H-glycoside becomes the A-glycoside by the addition of *N*-acetyl-D-galactosamine to C-3 of the ends of the two branched chains (see Fig. 9.8B) and gives A-type blood. The H-glycoside becomes the B-glycoside by the addition of α-D-galactose to C-3 of the ends of the two branched chains (see Fig. 9.8C) and gives B-type blood. The AB blood group is obtained when both *N*-acetyl-α-D-galactosamine and α-D-galactose are added to C-3 of the ends of the two branch chains of the H-glycoside (see Fig. 9.8D) and gives AB-type blood [20].

The terminal monosaccharide residues on the oligosaccharides impart specific immunological reactions. The immunodominant sugar for the H-glycoside is α-L-fucose; for the A-glycoside, it is *N*-acetyl-α-D-galactosamine; for the B-glycoside, it is α-D-galactose. Individuals with O-type blood (i.e., having the H-glycoside) produce antibodies that will agglutinate erythrocytes from A-, B-, and AB-type blood and can therefore accept blood only from individuals with O-type. Individuals with A-type blood produce antibodies that will agglutinate erythrocytes from B-type and AB-type individuals and can accept blood of A-type and O-type. Individuals with B-type blood produce antibodies that will agglutinate erythrocytes from A-type and AB-type individuals and can accept blood of B-type and O-type. Individuals with AB-type blood do not produce antibodies against A-, B-, or O-type blood and can accept blood from all three blood types and are known as *universal blood acceptors*. They, however, can donate blood only to AB-type individuals. Individuals with O-type blood can donate blood to all three of the other blood types and are known as *universal blood donors*.

Two other blood type determinants are formed from the H-glycoside. These depend on the position of attachment of the α-L-fucose. When α-L-fucose is attached 1 → 4 to the *N*-acetyl-D-glucosamine residue of the type 1 branch chain in-

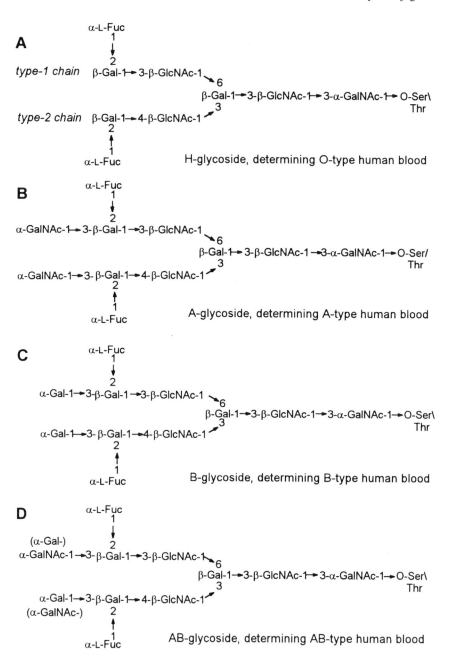

Figure 9.8. Structures of human blood group *O*-linked oligosaccharides that specify the ABO human blood groups.

stead of 1 → 2 to the D-galactose residue, the Lewis-a, or Le^a blood type is formed (see Fig. 9.9A). When there are two α-L-fucose residues attached 1 → 2 to the two D-galactose residues and one attached 1 → 4 to the N-acetyl-D-glucosamine residue of the type 1 chain, the Lewis-b, or Le^b, blood type is formed (see Fig. 9.9B).

Another well-characterized group of oligosaccharides comprises the O- and N-linked saccharides of the transmembrane glycoprotein, glycophorin A, found in the membrane of human erythrocytes. The part of the protein that protrudes from the membrane has 15 identical O-linked oligosaccharides. It consists of a disac-charide, β-D-galactopyranosyl-1 → 4-N-acetyl-α-D-galactosamine, to which there are attached two N-acetyl-D-neuraminic acid residues linked 2 → 3 and 2 → 6 to the galactose and the galactosamine residues, respectively (see Fig. 9.10A) [21]. There is one N-linked saccharide, with the structure given in Fig. 9.10B [21]. The oligosaccharides located on the surface of human erythrocytes are the sites of in-fluenza virus and malarial parasite receptors. Both the virus and the parasite have a neuraminidase enzyme that hydrolyzes the N-acetyl-D-neuraminic acid residues from the saccharides and allows them to obtain entrance into the cell. It has been speculated that glycophorin A's numerous negatively charged neuraminic acid residues prevent the adhering of the closely packed erythrocytes to each other in the blood stream [22].

Besides the N-linkage of carbohydrates to L-asparagine and the O-linkage to L-serine and L-threonine, a few linkages to other kinds of amino acids residues

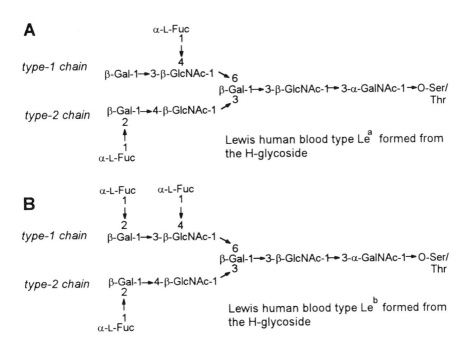

Figure 9.9. Lewis human blood groups.

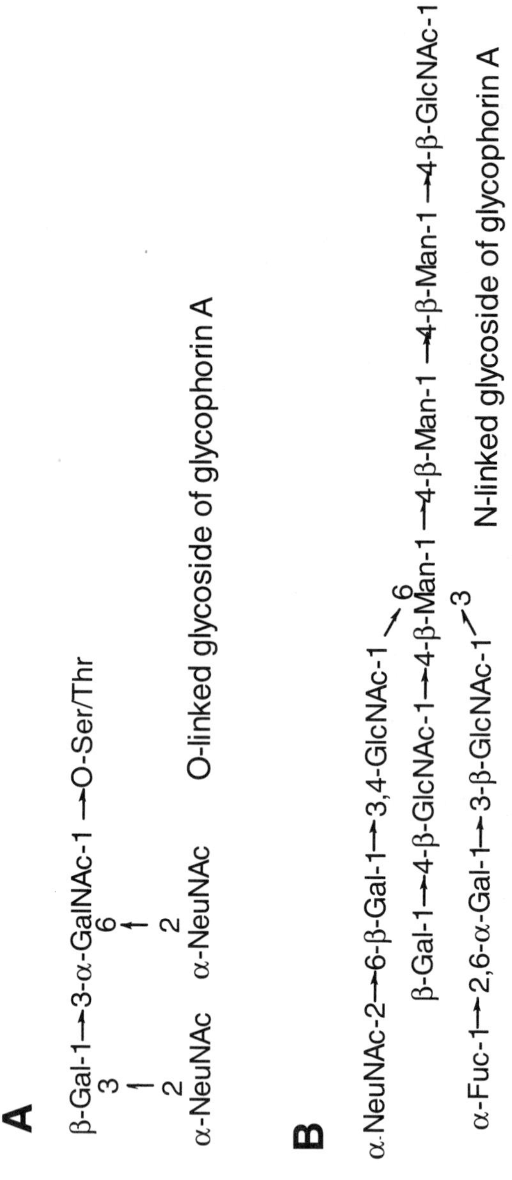

A

β-Gal-1→3-α-GalNAc-1 →O-Ser/Thr   O-linked glycoside of glycophorin A
         3   6
         ↑   ↑
         2   2
    α-NeuNAc   α-NeuNAc

B

α-NeuNAc-2→6-β-Gal-1→3,4-GlcNAc-1
                                    ↘6
   β-Gal-1→4-β-GlcNAc-1→4-β-Man-1 →4-β-Man-1 →4-β-GlcNAc-1
                                    ↗3
   α-Fuc-1→2,6-α-Gal-1→3-β-GlcNAc-1   N-linked glycoside of glycophorin A

Figure 9.10. Structures of the *O*-linked and *N*-linked oligosaccharides of glycophorin *A*.

have been observed. β-D-Galactose has been reported to be linked to the hydroxyl group of 5-hydroxy-L-lysine (see Fig. 9.2E) found in topocollagen of guinea pig skin [23,24], and the disaccharide, β-D-glucopyranosyl-1 → 2-β-D-galactopyranose attached to hydroxy-L-lysine of cuttlefish and leech collagen [25,26]. L-Arabinofuranose has been reported to be attached to hydroxy-L-proline in plant cell walls [27,28]. A galactose disaccharide and a glucose trisaccharide have been reported to be attached to the thiol group of L-cysteine in human urine [29,30] (Fig. 9.2F). Sialic acid has been reported to be attached to the phenolic hydroxyl group of L-tyrosine in chicken ovomucoid [31] (see Fig. 9.2G).

The O-linked oligosaccharides of L-serine and L-threonine are alkali labile. When treated with alkali, these O-linked oligosaccharides are released by a β-elimination reaction to form unsaturated amino acids and the free sugar chain (see Fig. 9.2B,C and section 12.11 in Chapter 12). S-Linked oligosaccharides to L-cysteine are also released by an identical reaction (see Fig. 9.2F). O-linked oligosaccharides to hydroxy-L-lysine, hydroxy-L-proline, or phenolic hydroxyl of L-tyrosine are not readily released by alkali.

9.4 Structures of Glycolipids

There are four major categories of glycolipids: (1) glycoglycerol lipids, (2) glycosphingolipids, (3) glycosyl polyisoprenol pyrophosphates, and (4) carbohydrates esterified by fatty acids found in gram-negative bacteria. The carbohydrates are oligosaccharides containing one, two, or higher number of monosaccharide units, similar to the oligosaccharides found in glycoproteins. In addition to the seven kinds of monosaccharide residues found in glycoproteins, glycolipids also contain D-glucose, D-quinovose, D-glucuronic acid, and N-acetyl-L-fucosamine. In some instances the monosaccharide residues are sulfated or phosphorylated. A number of different kinds of linkages are found and, in contrast to glycoproteins, some of the oligosaccharides have only one kind of monosaccharide residue.

a. Glycoglycerolipids

The simplest glycolipid is a glycosylated diacylglycerol in which the carbohydrate moiety is attached to the free hydroxyl group of a diglyceride by a glycosidic linkage. Examples of carbohydrate oligosaccharide containing only D-glucose are α-kojitriosyl diacyl glycerol, obtained from *Streptococcus hemolyticus* [32], and β-kojitriosyl diacyl glycerol, from *Bifidobacterium bifidum* [33]. Mono-D-galactosyl-, digalactosyl-, and trigalactosyldiacyl glycerols have been found in a number of plant tissues, such as chloroplasts, wheat flour, runner bean leaves, potato tubers, rice bran, chlorella, marine diatoms, and spinach leaves [34–38]

where R^1 and R^2 are hydrocarbon chains, giving fatty acids with 16 to 24 carbons

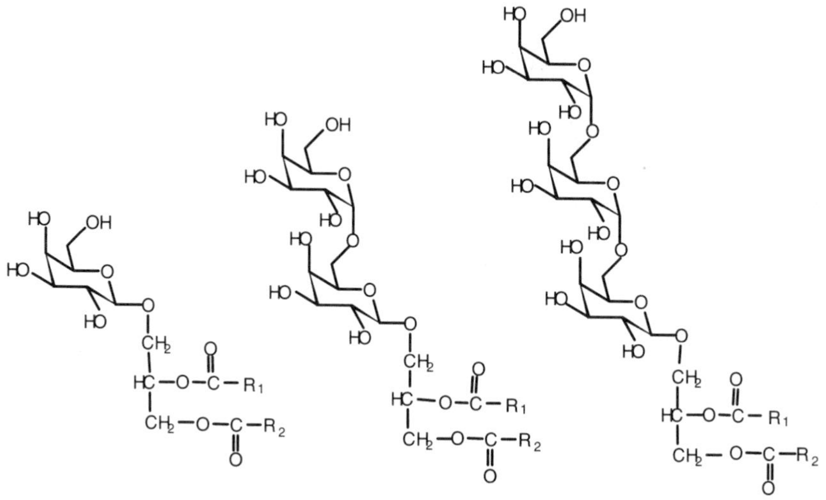

Figure 9.11. Mono-, di-, and trigalactosyl diacyl glycerols found widely distributed in plants, animals, and microbes.

and in a number of animal tissues such as rat brain, human brain, human testis, and human sperm [39–41]. See Fig. 9.11 for representative structures of the glycodiacylglycerol lipids.

Sulfated sugars attached to diacylglycerol are common in the membranes of chloroplasts, for example, 6-sulfo-α-D-glucopyranosyl diglyceride. Diacyl glycerol with higher saccharides such as α-isomaltopentaose, α-isomaltohexaose, and α-isomaltooctaose have been found in pulmonary lavage, human gastric juice, and human saliva [42,43].

Phosphodiacylglycerol (phosphatidic acid) has also been found to have carbohydrates attached to the phosphate group. A common sugar is the cyclic sugar alcohol myoinositol (see Fig. 9.12A). Complex glycophosphatidyl glycerol structures have been found in bacterial culture supernatants [44] (see Fig. 9.12B). Uronic acids have been found in glycoglycerol lipids of *Pseudomonas* species [45,46], and sialic acids have been found in glycoglycerol lipids of *Meningococcus* species [47] (see Fig. 9.13).

b. Glycosphingolipids

Glycosphingolipids are composed of sphingosine, a hydrocarbon alcohol amine (see Fig. 9.14A) that has a fatty acid attached to the amino group and a carbohydrate attached to a primary alcohol group at the end of the hydrocarbon chain (see Fig. 9.14B for the structure). Glycosphingolipids, called *cerebrosides,* are very widely distributed among living organisms. They occur primarily as components

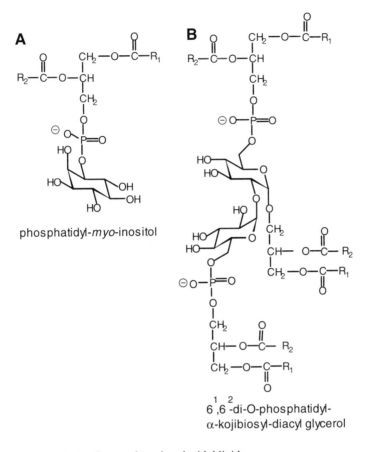

Figure 9.12. Structures of some glycophosphatidyl lipids.

of cell membranes. The hydrophobic part, the acyl sphingosine (called a ce-
ramide), is embedded in the lipid bilayer of the membrane, leaving the carbohydrate protruding from the surface of the membrane. These carbohydrate oligosaccharides serve a number of important biological functions as specific recognition
sites. We have already discussed the carbohydrate structures of the human blood
group oligosaccharides (see Figs. 9.8 and 9.9) that are attached to ceramides and
cover the surface of human erythrocytes, giving a specific chemical character to
the blood, as discussed under the *O*-linked blood group oligosaccharides of glycoproteins.

There are over 50 known glycosphingolipids based on the composition and sequence of the carbohydrate oligosaccharide. D-Galactocerebrosides are prevalent
in the neuronal cell membranes of the brain (see Fig. 9.14B). Glucocerebrosides
occur in the membranes of other nerve tissues. In many instances the carbohydrate moieties are oligosaccharides. A representative structure, known as lacto-
cerebroside is shown in Fig. 9.14C.

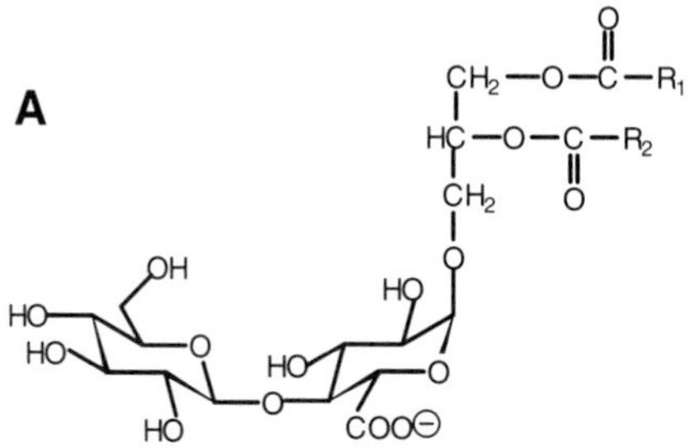

β-D-glucopyranosyl-1→4-α-
D-glucuronpyranosyl-diacyl glycerol
elaborated by *Pseudomonas species* [45,46]

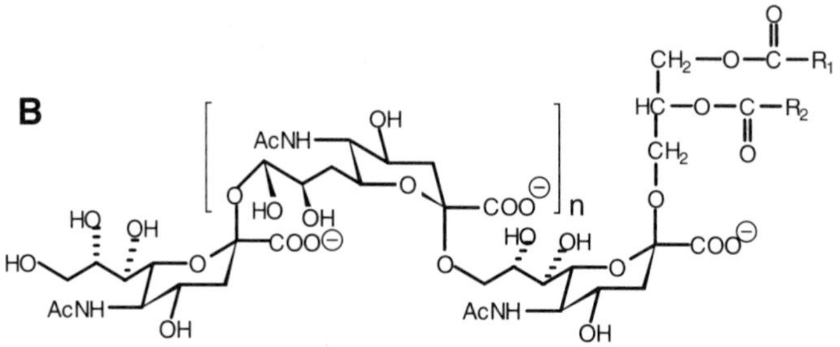

poly-α-D-neuraminic acid 2→9-α-neuraminyl-2→
diacyl glycerol from *Meningococcus sp.* [47]

Figure 9.13. Uronic and sialic acid diacylglycerol glycolipids.

When the oligosaccharide contains a sialic acid residue, the cerebroside is given a special name, *ganglioside.* There are over 60 known gangliosides. A representative pentasaccharide ganglioside is shown in Fig. 9.14D. Gangliosides play significant physiological roles that are determined by the structure of the carbohydrate group. They have roles in cell-cell recognition and are involved in growth and differentiation of cells. In some instances, when the carbohydrates are altered, the cells become undifferentiated and a tumor results. Specific enzymatic breakdown of the oligosaccharide chains is required during different stages of development. Absence of specific enzymes required for ganglioside modification are responsible for a number of hereditary diseases. One of these is *Tay-Sachs disease* which is characterized by neurological deterioration. Another hereditary neurological disorder, *multiple sclerosis,* is produced by an autoimmune reaction in which antibodies are produced against the carbohydrates of brain and nerve tissue gangliosides. Many toxins effect their action by having binding sites for the carbohydrate moieties of gangliosides. Also, many infective agents, both bacterial and viral, gain entrance into their host cells by the removal of the sialic acid units on the ends of the carbohydrate chains of cell surface gangliosides.

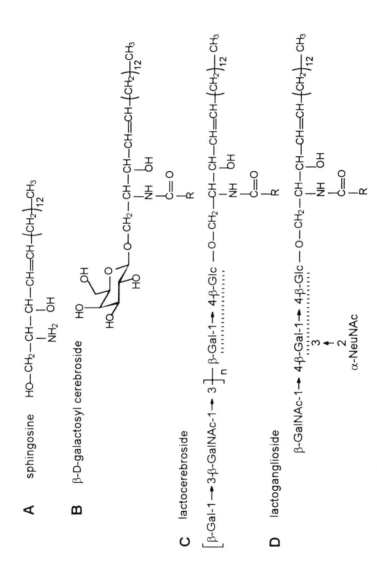

Figure 9.14. Glycosphingosine glycolipids.

c. Glycosyl Pyrophospho Polyprenols

The polyprenols are polyisoprenoid alcohols that form phosphate esters with car-bohydrates. These phosphoesters act as coenzymes in the transfer of carbohydrate units in the biosynthesis of polysaccharides and in the biosynthesis of N-linked oligosaccharides for glycoproteins. The linkage between the polyisoprenol and the carbohydrate is a high-energy pyrophosphate group. The polyisoprenols have the general structure given in Fig. 9.14 in which n can vary from 5 to 24. For any one source, the variance in n is relatively small. Most of the isoprene units have the *cis*-configuration for the double bond. Only two or three of the units at the end of the polyisoprenoid chain, opposite from the alcohol group, have the *trans*-con-figuration. These chains are sometimes called poly-*cis*-isoprenoid alcohols be-cause the majority of the configurations are *cis*. A more common name is just *polyprenols*. One of the best-characterized polyprenols is *bactoprenol,* produced by many bacteria. It has 55 carbons, 11 isoprenoid units and one alcohol group. Bactoprenol phosphate (Fig. 9.15B) is the coenzyme involved in the biosynthesis of many bacterial polysaccharides, including the murein chain of the bacterial cell wall, the *Salmonella* O-antigen, and colominic acid capsule of *E. coli* (Chapter 6, Fig. 6.20, and Chapter 10). In the biosynthesis of these polysaccharides, the re-peating units are added to bactoprenol phosphate to produce repeating unit bacto-prenol pyrophosphate, for example N-acetyl-β-D-glucosaminyl-1 $\rightarrow$ 4-N-acetyl-α-D-muramyl bactoprenol pyrophosphate for the synthesis of murein, and the tetrasaccharide repeating unit bactoprenol pyrophosphate for the synthesis of *Sal-monella* O-antigen polysaccharide (see Fig. 9.15C,D). These repeating units are then transferred to the growing polysaccharide chain. See Chapter 10 for the de-tails of the biosynthesis of these polysaccharides.

Another well-characterized polyprenol phosphate is *dolichol.* It is found in an-imal tissues and is involved as a coenzyme in the biosynthesis and transfer of the oligosaccharides found attached to glycoproteins. Dolichol differs from bacto-prenol by having a larger number of isoprenoid units (20–24), and the unit carry-ing the primary alcohol group is saturated (see Fig. 9.15E). The oligosaccharides are assembled and attached to dolichol pyrophosphate and then transferred to the protein. Figure 9.15E shows dolichol pyrophosphate "core pentasaccharide" that is transferred to L-asparagine to form N-linked oligosaccharides.

d. Lipopolysaccharide of Gram-Negative Bacteria

Gram-negative bacteria have the same peptidoglycan (peptidomurein) for their cell walls as do gram-positive bacteria. See Chapters 6 and 10 for the structures of murein and peptidomurein. But, in addition, gram-negative bacteria have a second outer membrane between the peptidomurein cell wall and the outermost O-antigen polysaccharide. This outer membrane is a lipopolysaccharide that is composed of two parts: the so-called Lipid A part and a core polysaccharide. Lipid A is a highly

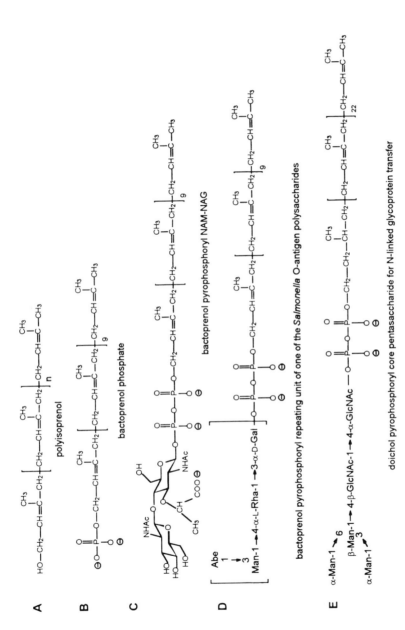

Figure 9.15. Polyisoprenol glycosyl glycolipids.

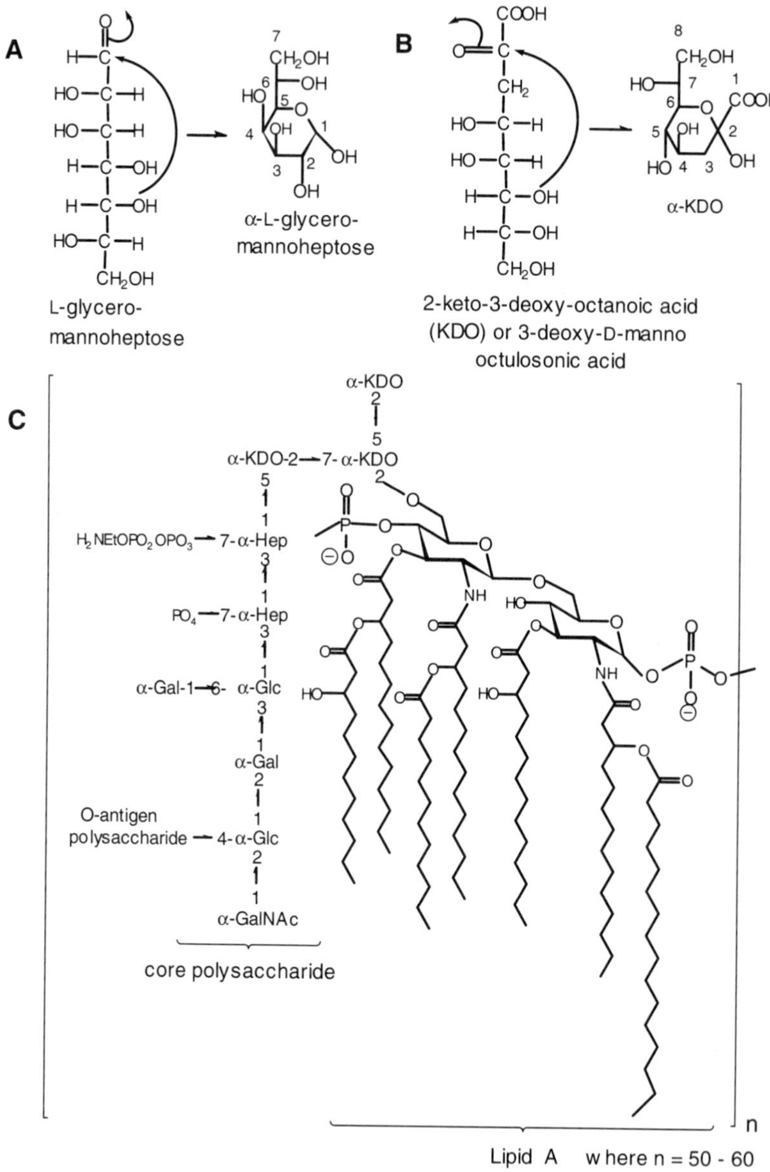

Figure 9.16. Structural features of bacterial lipopolysaccharides: **A,** structure of the hep-tose and its conversion into a pyranose ring; **B,** structure of KDO and its conversion into a pyranose ring; **C,** structure of lipid A and the core polysaccharide.

acylated disaccharide repeating unit of β-D-glucosaminyl-1 → 6-α-D-glucosamine joined together 1 → 4 by a pyrophosphate group [48]. Attached to the amino groups and to 3^1 and 3^2 is 3-hydroxy-D-myristic acid [49]. In addition, lauric acid and palmitic acid are esterified with the hydroxyl groups of the two amide-linked 3-hydroxy-D-myristic acids for at least one bacterial species, *Salmonella minnesota* [49]. The composition and distribution of the fatty acids do vary for different bacterial species. There are usually between 5 and 7 fatty acids in the Lipid A repeating unit. Attached to position 6^2 is a complex decasaccharide that makes up the so-called core polysaccharide. It contains two unusual sugars that are found only in gram-negative bacterial cells. The sugars are a seven-carbon heptose (L-glycerol-mannoheptose) and an eight-carbon α-keto-sugar acid (2-keto-3-deoxyoctanoic acid or 3-deoxy-D-mannooctulosonic acid), commonly known as *KDO* [50–52]. See Fig. 9.16A,B for the structures and formation of pyranose rings of the heptose and KDO.

KDO has chemical properties similar to those of neuraminic acid as it has a carboxylate group, a 3-deoxy group, and a similar biosynthesis (see Chapter 10). The monosaccharide composition, sequence, and linkages, along with the position of attachment of phosphate and ethanolamine pyrophosphate substituted onto the heptose residues of the core polysaccharide have been determined [53–58] and are shown in Fig. 9.16C. The fatty acids in the lipopolysaccharide participate in the formation of a lipid bilayer for the outer membrane. The carbohydrate is hydrophilic and is located on the outer faces of the lipid bilayer membrane. The O-antigen polysaccharides and other capsular polysaccharides are attached to position-4 of the next to last monosaccharide residue, α-D-glucopyranose of the core polysaccharide.

9.5 Literature Cited

1. J. Montreuil, *Adv. Carbohydr. Chem. Biochem.*, **37** (1980) 158–223.
2. R. Schauer, *Adv. Carbohydr. Chem. Biochem.*, **40** (1982) 132–234.
3. J. Montreuil, *Pure Appl. Chem.*, **42** (1975) 431–437.
4. B. Fournet, G. Strecker, J. Montreuil, L. Dorland, J. Haverkamp, J. F. G. Vliegenthart, K. Schmid, and J. P. Binette, *Biochemistry*, **17** (1978) 5206–5214.
5. B. Nilsson, N. E. Nordén, and S. Svensson, *J. Biol. Chem.*, **254** (1979) 4545–4553.
6. T. Kondo, M. Fukuda, and T. Osawa, *Carbohydr. Res.*, **58** (1977) 405–414.
7. J. Baenziger, S. Kornfeld, and S. Kochwa, *J. Biol. Chem.*, **249** (1974) 1889–1896; 1897–1903.
8. J. Baenziger and S. Kornfeld, *J. Biol. Chem.*, **249** (1974) 7260–7269; 7270–7281.
9. T. Krusius, J. Finne, and H. Rauvala, *Eur. J. Biochem.*, **92** (1978) 289–297.
10. U. Rutishauser, A. Acheson, A. K. Hall, D. M. Mann, and J. Sunshine, *Science*, **240** (1988) 53–57.
11. T. Tai, K. Yamashita, I. Setsuko, and A. Kobata, *J. Biol. Chem.*, **252** (1977) 6687–6694.
12. K. Yamashita, Y. Tachibana, and A. Kobata, *J. Biol. Chem.*, **253** (1978) 3862–3869.
13. F. Miller, *Immunochemistry*, **9** (1972) 217–228.
14. A. Tsugita and S. Akabori, *J. Biochem.*, **46** (1959) 695–704.
15. N. Katsura and E. A. Davidson, *Biochim. Biophys. Acta*, **121** (1966) 120–127.

16. J. C. Paulson, *Trends Biol. Sci.,* **14** (1989) 272–276.
17. R. E. Feeney and Y. Yeh, *Adv. Protein Chem.,* **32** (1978) 191–282.
18. W. M. Watkins, *Science,* **152** (1966) 172–181.
19. V. Ginsberg, *Adv. Enzymol.,* **36** (1972) 131–149.
20. K. O. Lloyd and E. A. Kabat, *Proc. Natl. Acad. Sci. U.S.,* **61** (1968) 1470–1479.
21. R. C. Hughes, *Prog. Biophys. Mol. Biol.,* **26** (1973) 189–268.
22. R. Schauer, *Trends Biol. Sci.,* **10** (1985) 357–360.
23. W. T. Butler and L. W. Cunningham, *J. Biol. Chem.,* **241** (1966) 3882–3888.
24. R. Spiro, *J. Biol. Chem.,* **242** (1967) 4813–4823.
25. M. Isemura, T. Ikenaka, and Y. Matsushima, *J. Biochem.,* **74** (1973) 11–21.
26. T. Biswas and A. K. Mukherjee, *Carbohydr. Res.,* **63** (1978) 173–181.
27. D. T. A. Lamport, *Nature,* **216** (1967) 1322–1324.
28. D. T. A. Lamport and D. H. Miller, *Plant Physiol.,* **48** (1971) 454–456.
29. C-J. Lote and J. B. Weiss, *FEBS Lett.,* **16** (1971) 81–85.
30. J. B. Weiss C.-J. Lote, and H. Bobinski, *Nature,* **234** (1971) 25–26.
31. M. A. Krysteva, I. N. Mancheva, and I. D. Dobrew, *Eur. J. Biochem.,* **40** (1973) 155–161.
32. W. Fischer, *Lipids,* **1** (1976) 255–266; I. Ishizuka and T. Yamakawa, *J. Biochem.,* **64** (1968) 13–23.
33. J. H. Veerkamp, *Biochim. Biophys. Acta,* **273** (1972) 359–367.
34. R. Pincelot, *Arch. Biochem. Biophys.,* **159** (1973) 134–142.
35. Y. Fujino and T. Miyazawa, *Biochim. Biophys. Acta,* **572** (1979) 442–451.
36. H. P. Siebertz, E. Hein, M. Linscheid, J. Joyard, and R. Douce, *Eur. J. Biochem.,* **101** (1979) 429–438.
37. T. Galliard, *Phytochemistry,* **7** (1968) 1907–1914.
38. T. Galliard, *Biochem. J.,* **115** (1969) 335–339.
39. G. Rouser, G. Kritchevshy, G. Simon, and G. J. Nelson, *Lipids,* **2** (1967) 37–40.
40. T. Inoue, S. S. Desmukh, and R. A. Pieringer, *J. Biol. Chem.,* **246** (1971) 5688–5694.
41. M. Levine, J. Bain, R. Narasimhan, B. Palmer, A. J. Yates, and R. K. Murray, *Biochim. Biophys. Acta,* **441** (1976) 134–145.
42. B. L. Slomiany, A. Slomiany, and G. B. J. Glass, *Eur. J. Biochem.,* **84** (1978) 53–59.
43. B. L. Slomiany, F. B. Smith, and A. Slomiany, *Biochim. Biophys. Acta,* **574** (1979) 480–486.
44. W. Fischer, *Biochim. Biophys. Acta,* **487** (1977) 89–104.
45. S. G. Wilkinson, *Biochim. Biophys. Acta,* **164** (1968) 148–156.
46. S. G. Wilkinson, *Biochim. Biophys. Acta,* **187** (1969) 492–500.
47. E. C. Gotschlich, B. A. Fraser, O. Nishimura, J. B. Robbins, and T.-Y. Liu, *J. Biol. Chem.,* **256** (1981) 8915–8921.
48. J. Gmeiner, O. Lüderitz, and O. Westphal, *Eur. J. Biochem.,* **7** (1969) 370–379.
49. U. Zähringer, B. Lindner, and E. T. Rietschel, *Adv. Carbohydr. Chem. Biochem.,* **50** (1994) 211–276.
50. W. Weidel, Z. *Physiol. Chem.,* **299** (1955) 253–258.
51. E. C. Heath and M. A. Ghalambor, *Biochem. Biophys. Res. Commun.,* **10** (1963) 340–343.
52. P. D. Rick and M. J. Osborn, *Proc. Natl. Acad. Sci. USA,* **69** (1972) 3756–3760.
53. W. Dröge, O. Lüderitz, and O. Westphal, *Eur. J. Biochem.,* **4** (1968) 126–133.
54. M. Berst, C. G. Hellerquist, B. Lindber, O. Lüderitz, S. Svensson, and O. Westphal, *Eur. J. Biochem.,* **11** (1969) 353–359.

55. W. Dröge, V. Lehmann, O. Lüderitz, and O. Westphal, *Eur. J. Biochem.*, **14** (1970) 175–184.
56. G. Hammerling, O. Lüderitz, and O. Westphal, *Eur. J. Biochem.*, **15** (1970) 48–56.
57. G. Hammerling, O. Lüderitz, O. Westphal, and P. H. Mäkelä, *Eur. J. Biochem.*, **22** (1971) 331–344.
58. V. Lehmann, O. Lüderitz, and O. Westphal, *Eur. J. Biochem.*, **21** (1971) 339–347.

9.6 References for Further Study

"Primary structure of glycoprotein glycans: basis for the molecular biology of glycoproteins," J. Montreuil, *Adv. Carbohydr. Chem. Biochem.*, **37** (1980) 158–223.

"The carbohydrates of glycoproteins," A. Kobata, in *Biology of Carbohydrates,* Vol. 2, Chap. 2, pp. 87–161, (V. Ginsburg and P. W. Robbins, eds.) J. Wiley, New York (1984).

"Biosynthesis of glycoproteins: formation of N-linked Oligosaccharides," M. D. Snider, in *Biology of Carbohydrates,* Vol. 2, Chap. 3, pp. 163–197, (V. Ginsburg and P. W. Robbins, eds.) J. Wiley, New York (1984).

"Biosynthesis of glycoproteins: formation of O-linked oligosaccharides," J. E. Sadler, in *Biology of Carbohydrates,* Vol. 2, Chap. 4, pp. 199–287, J. Wiley, New York (V. Ginsburg and P. W. Robbins, eds.) (1984).

"Enzymatic basis for blood groups in man," V. Ginsburg, *Adv. Enzymol.*, **36** (1972) 131–149.

"Glycophorin A: a model membrane glycoprotein," H. Furthmayr, in *Biology of Carbohydrates,* Vol. 1, Chap. 4, pp. 123–197, (V. Ginsburg and P. W. Robbins, eds.) J. Wiley, New York (1981).

Glycolipids, New Comprehensive Biochemistry, Vol. 10 (H. Wiegandt, ed.) Elsevier, London (1985).

"Glycoproteins: what are the sugar chains for?" J. C. Paulson, *Trends Biol. Sci.*, **14** (1989) 272–276.

"Sialic acids and their role as biological masks," R. Schauer, *Trends Biol. Sci.*, **10** (1985) 357–360.

"The chemistry and biological significance of 3-deoxy-D-manno-2-octulosonic acid (KDO)," F. M. Unger, *Adv. Carbohydr. Chem. Biochem.*, **38** (1981) 324–388.

Chapter 10

Biosynthesis

10.1 Photosynthesis and the Formation of Carbohydrates

In Chapter 1, it was stated that carbohydrates are produced on the Earth during the process of photosynthesis. A very simple reaction was written (reaction 1.1) in which 6 CO_2 combined with 6 H_2O to give carbohydrate, $C_6H_{12}O_6$, and six molecules of molecular oxygen, O_2. It was further indicated that the energy necessary to form the carbon-carbon bonds in the synthesis of carbohydrate came from the sun. While all of this is true, the process is far more complex than the simple reaction given in Chapter 1.

Photosynthesis takes place in organisms (usually eukaryotic plants) in chloroplasts. Photosynthesis also takes place in some prokaryotes, principally the blue-green algae, green bacteria, and purple bacteria. The chloroplasts in eukaryotes are membraneous organelles with specialized pigments (mainly chlorophylls) that absorb the photons of light. For discussion purposes, the photosynthetic process is divided into two parts, the *light-dependent reactions* and the *light-independent reactions*, sometimes called the *light reactions* and the *dark reactions*, respectively. Actually the light-independent reactions depend on the light reactions [1].

In the light reactions, the photons from the sun hit the pigments in what is called *Photosystem I* and are absorbed. In doing so, the energy of the light is transferred to electrons in the chlorophyll molecule and they become "excited." Two excited electrons are passed to two ferric ions contained in an enzyme, *ferrodoxin NADP+ reductase*, leaving the chlorophyll with two positive charges in a high-energy, excited state (refer to Fig. 10.1). The ferric ion of *ferrodoxin reductase* is reduced to ferrous ion. Two ferrous ions reduce nicotinamide adenine dinucleotide phosphate ($NADP^+$) by transferring two electrons with a hydrogen ion to give NADPH [2] (see Figs. 10.1 and 10.2A). The ferrous ions thereby become ox-

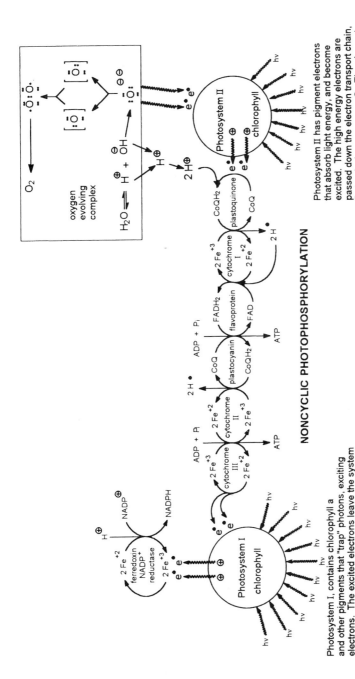

Figure 10.1. Noncyclic photophosphorylation electron transport involved in the light reactions of photosynthesis—formation of NADPH, ATP, and O_2; hv represents photons of light.

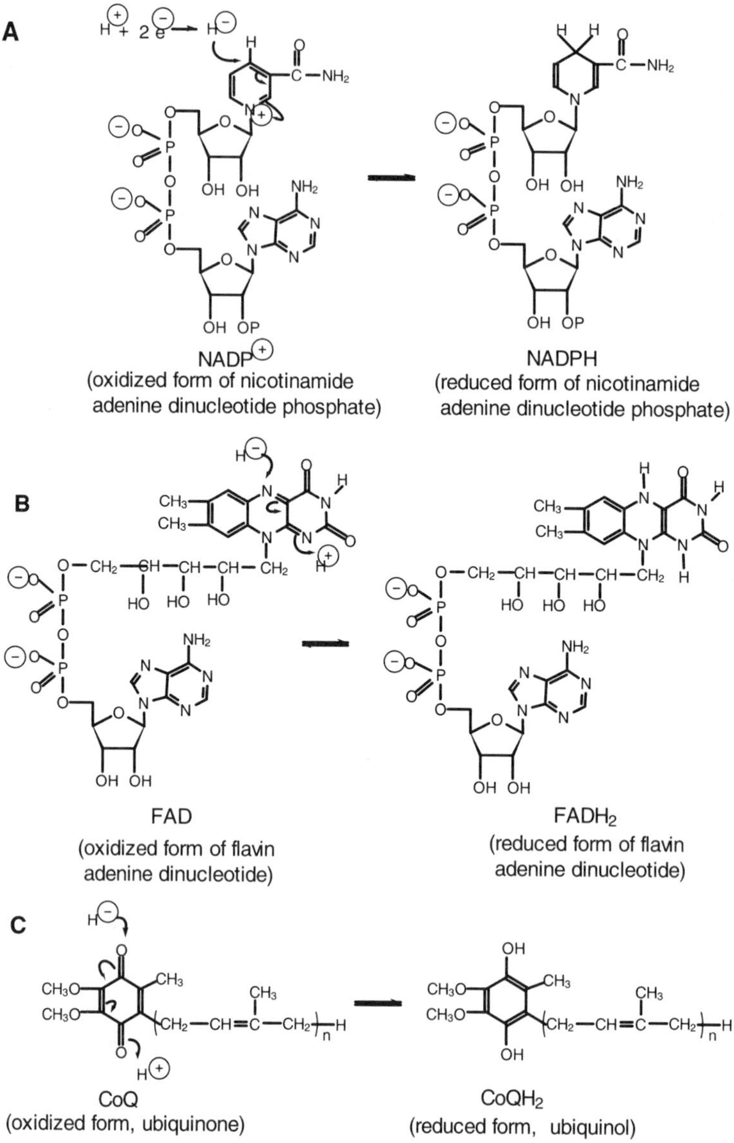

Figure 10.2. Oxidized and reduced structures of the coenzymes, $NADP^+$, FAD, and CoQ involved in photosynthetic electron transport.

idized to ferric ions, ready for the next pair of electrons from Photosystem I chlorophyll. The NADPH is a *required* compound in the formation of carbohydrate in the dark reactions.

The positively charged, excited chlorophyll in Photosystem I is brought back to its ground state by picking up two electrons from an electron transport system. The electron transport system is a series of redox enzymes that pass high-energy electrons from one enzyme to the next in a series of reductions and oxidations. The electrons are generated by light hitting a second photosystem, *Photosystem II* (see Fig. 10.1) [3]. As the electrons pass down the transport chain, ultimately bringing the chlorophyll of Photosystem I back to its ground state, they lose energy by reducing the various cytochromes, flavoprotein, and plastocyanins. Some of this energy is used to give the synthesis of 2 ATP from 2 ADP and 2 inorganic phosphate (Pi). This high-energy compound (ATP) is used to provide the energy for the formation of the carbon-carbon bonds in converting CO_2 into carbohydrate in the dark reactions. The process, however, leaves Photosystem II in a high-energy, positively charged state. Nature has evolved a very unusual reaction to bring Photosystem II back to the ground state. Water is used as a reducing agent to provide electrons to neutralize the positive charges on the pigments. Most chemists do not think of water as a reducing agent. When they think of reducing agents, they think of molecular hydrogen (H_2), lithium aluminum hydride ($LiAlH_4$), or sulfhydryl compounds (RSH). In some bacterial photosynthetic systems, hydrogen and hydrogen sulfide (H_2S) are used as the reducing agent, as shown in the following reactions:

$$6\ CO_2 + 12\ H_2 \xrightarrow{light} C_6H_{12}O_6 + 6\ H_2O \tag{10.1}$$

$$6\ CO_2 + 12\ H_2S \xrightarrow{light} C_6H_{12}O_6 + 12\ S + 6\ H_2O \tag{10.2}$$

But nature has chosen the unlikely candidate, water, as the primary reducing agent for most photosynthesizing systems, probably because of its abundance when the process of photosynthesis was evolving.

Water ionizes into a hydrogen ion and a hydroxide ion. If a proton (hydrogen ion) is removed from the hydroxide ion, an oxygen atom with two negative charges (an oxide ion) results (see *oxygen-evolving complex* in Fig. 10.1). This oxide ion gives up two electrons to Photosystem II, bringing Photosystem II back to the ground state. The oxygen atom becomes electron deficient in its outer electron orbit and joins with another electron deficient oxygen atom to form electronically stable molecular oxygen. This process takes place in the chloroplast, catalyzed by a protein-enzyme complex called the *oxygen-evolving complex*. The whole process is called *noncyclic photophosphorylation* and generates NADPH, 2 ATP, and O_2 [4]. The first two compounds are used in the fixing of CO_2 and the forming of carbohydrate. Molecular oxygen (O_2) is passed into the atmosphere and is available to nonphotosynthesizing organisms for the oxidization of carbohydrates in the process of *respiration* (see Chapter 11) to obtain the energy of the sun that was stored in the carbohydrates as chemical energy during photosynthesis.

The reactions involved in the fixing of CO_2 and its incorporation into carbohydrates by photosynthesizing organisms are complicated. These reactions were studied by Melvin Calvin and his associates James Bassham and Andrew Benson over a ten-year period from 1946 to 1956. They used ^{14}C-labeled CO_2 and traced the metabolic fate of the CO_2 by exposing it to growing cultures of algae (*Chlorella*) for various periods of time. They analyzed what radioactive compounds were formed and where the labeled carbon was located in the compounds. From this information, they deduced how CO_2 was being fixed and converted into carbohydrate [5,6a].

The first radioactively labeled compound to be formed was 3-phospho-D-glyceric acid, and it was determined that only half of the 3-phospho-D-glyceric acid molecules were labeled. It was further determined that, initially, only the carboxyl group was labeled. This suggested that the CO_2 was being added to a two-carbon compound. They searched unsuccessfully for the two-carbon compound. Using pulse and chase techniques, they eventually found that CO_2 was being added to a five-carbon compound, D-ribulose-1,5-bisphosphate (see reaction 2 in Fig. 10.3). This should have yielded a six-carbon compound, but again it was not observed. Eventually it was found that a six-carbon compound, 2-carboxy-3-keto-D-ribitol-1,5-bisphosphate, was formed, but it was very unstable and as soon as it was formed it was cleaved into two molecules of 3-phospho-D-glyceric acid [5,6a,6b]. Because only one of the carbons in the six-carbon compound was labeled, only one of the two 3-phospho-D-glyceric acid molecules was labeled (see Fig. 10.4).

Many other photosynthetic carbohydrates were identified and, by means of labeling and chemical degradation studies of the labeled compounds, their labeling patterns were determined, and the various reactions involved in the formation of several different carbohydrates were obtained. For example, the formation of D-fructose-1,6-bisphosphate showed that the label from CO_2 was initially located only at C-3 and C-4, indicating that it was formed by the condensation of 3-phospho-D-glyceraldehyde and dihydroxyacetone phosphate in a reverse aldolase-catalyzed reaction (reaction 7 in Fig. 10.3).

3-Phospho-D-glyceraldehyde, dihydroxyacetone phosphate, D-erythrose-4-phosphate, D-ribose-5-phosphate, D-ribulose-5-phosphate, D-xylulose-5-phosphate, D-sedoheptulose-7-phosphate, and D-sedoheptulose-1,7-bisphosphate were identified and isolated and the initial locations of the label determined. From these studies a series of reactions was worked out that has become known as the *Calvin Cycle* (see Fig. 10.3) [5,6b,7a]. Most of the Calvin Cycle reactions also occur in other metabolic pathways (for example, the *pentose phosphate pathway* [7b]), except for the carboxylation reaction, which is unique to the Calvin Cycle. The fixing of CO_2 starts with the phosphorylation of D-ribulose-5-phosphate by ATP, catalyzed by *D-ribulose-5-phosphate kinase*[1] (reaction 1 of Fig. 10.3), to form D-ribulose-1,5-bisphosphate. This is the first use of the ATP that was generated in the light reactions of photosynthesis. D-Ribulose-1,5-bisphosphate is isomerized to 2,3-enol-D-ribose-1,5-bisphosphate, to which the CO_2 is added by the nucleophilic attack of the enediol onto CO_2, catalyzed by *D-ribulose-1,5-bisphosphate carboxylase*. See Fig. 10.4 for the mechanism of addition of CO_2 by this enzyme. The resulting six-carbon β-keto acid is unstable and is rapidly hydrolyzed (reaction 3 of Fig. 10.3) to give cleavage of the carbon-carbon bond and the formation of two molecules of 3-phospho-D-glyceric acid.

The carboxyl group of the 3-phospho-D-glyceric acid is phosphorylated by ATP (reaction 4 of Fig. 10.3), catalyzed by *3-phospho-D-glyceric acid kinase*, pro-

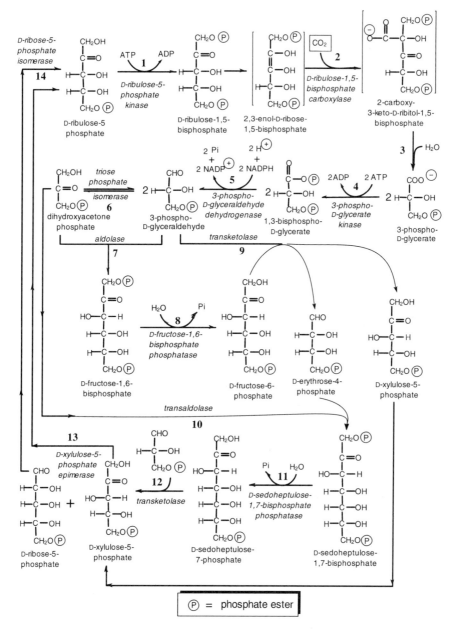

Figure 10.3. The fixation of CO_2 and the carbohydrates formed in the light-independent reactions of the photosynthetic Calvin cycle.

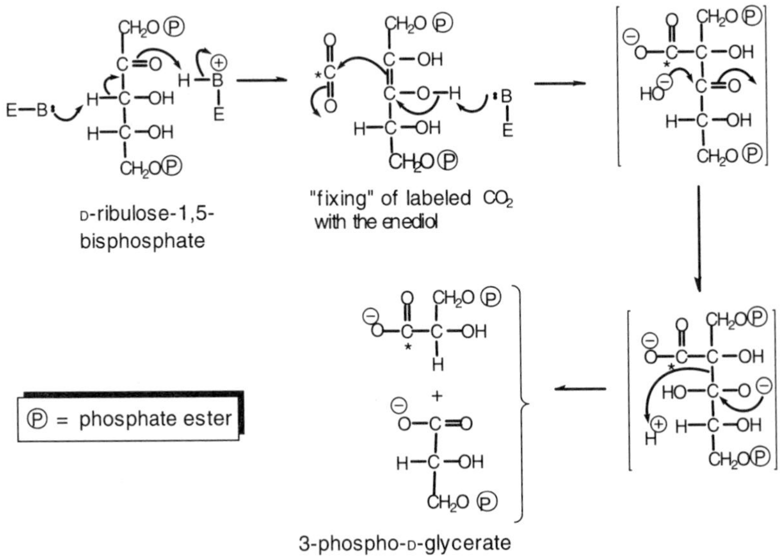

Figure 10.4. Mechanism for the fixing of CO_2 by addition to D-ribulose-1,5-bisphosphate catalyzed by *D-ribulose-1,5-bisphosphate carboxylase.*

ducing 1,3-bisphospho-D-glycerate. This is the second use of the ATP that is generated in the light reactions of photosynthesis. The phosphorylation of the carboxyl group increases the energy of this group by forming a carboxyl phosphoryl anhydride. The high-energy, mixed acid anhydride is then able to be reduced by NADPH (generated in the light reactions) to give the release of the phosphate group and the reduction of the carboxyl group to an aldehyde, producing 3-phospho-D-glyceraldehyde [6b].

3-Phospho-D-glyceraldehyde is isomerized by *triose phosphate isomerase* to dihydroxyacetone phosphate (reaction 6). The two trioses, 3-phospho-D-glyceraldehyde and dihydroxyacetone phosphate, are then condensed together in a aldol-type condensation, catalyzed by *aldolase,* to give D-fructose-1,6-bisphosphate (reaction 7). This is a key intermediate in the photosynthetic process for the formation of many other carbohydrates, including D-glucose (see Fig. 10.5).

Thus, the overall stoichiometry for the fixing of one CO_2 requires three ATP (one for D-ribulose-5-phosphate and two for the two 3-phospho-D-glycerate) and two NADPH (for reducing the two 1,3-bisphospho-D-glycerate), generated in photophosphorylation (light reactions). The biosynthesis of a complete six-carbon sugar, $C_6H_{12}O_6$, does not occur by the fixing of 6 CO_2 molecules into a single molecule, but by the sequential addition of 6 CO_2 to 6 D-ribulose-1,5-bisphosphates, requiring 18 ATP and 12 NADPH molecules. The overall reactions of photosynthesis then are the following:

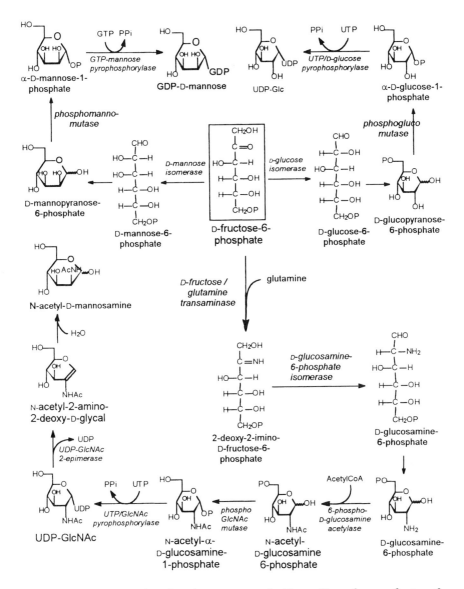

Figure 10.5. Biosynthesis of various monosaccharide residues from D-fructose-6-phosphate.

6 CO_2 + 6 D-ribulose-5-phosphate + 18 ATP
    + 12 NADPH + 12 H^+ → 6 D-fructose-1-6-bisphosphate   (10.3)
    + 12 $NADP^+$ + 18 ADP + 12 Pi + 6 O_2

6 D-fructose-1,6-bisphosphate + 6 H_2O
$$→ 6 \text{ D-fructose-6-phosphate} + 6 \text{ Pi}\quad(10.4)$$

Sum of Reactions 10.3 and 10.4

---

6 CO_2 + 6 D-ribulose-5-phosphate + 18 ATP + 12 NADPH
    + 12 H^+ + 6 H_2O → 6 D-fructose-6-phosphate + 12 $NADP^+$
    + 18 ADP + 18 Pi + 6 O_2 (10.5)

10.2 Biosynthesis of Sugar Nucleotides

Uridine diphospho-D-glucose (UDP-Glc) was discovered by Leloir et al. in 1950 [8]. After Leloir's discovery, nucleotide diphospho sugars were then rapidly observed by many investigators. The nucleotide diphospho sugars have been found to be essential in the conversion of one monosaccharide into another and in the activation of monosaccharides for their incorporation into oligosaccharides and polysaccharides [9–11]. The nucleotide diphospho sugars are formed by an enzyme-catalyzed reaction of sugar-1-phosphate with a nucleotide triphosphate, as shown in the following reactions:

$$\text{Glc-1-P} + \text{UTP} \rightarrow \text{UDP-Glc} + \text{PPi}\qquad(10.6)$$

$$\text{PPi} + H_2O \rightarrow 2 \text{ Pi}\qquad(10.7)$$

The reaction is catalyzed by a group of enzymes called *nucleotide pyrophosphorylases.* Different purine and pyrimidine nucleotide triphosphates such as adenosine triphosphate (ATP), guanidine triphosphate (GTP), cytosine triphosphate (CTP), uridine triphosphate (UTP), and deoxythymidine triphosphate (dTTP), provide many different kinds of nucleotide diphospho sugars such as UDP-Glc, ADP-Glc, GDP-Man, GDP-Fuc, UDP-Gal, UDP-GlcNAc, UDP-GalNAc, CDP-Abe, CDP-Tyv, dTDP-L-Rha, and dTDP-Glc. The higher-sugar acids, *N*-acetyl-D-neuraminic acid and KDO, occur as the sugar nucleotide monophosphates, CMP-NeuNAc and CMP-KDO. The different purine and pyrimidine bases, along with the specific monosaccharide residues, confer specificity for recognition and binding by the different enzymes involved in their biosynthesis and in their use as high-energy monosaccharide donors in forming other monosaccharides, oligosaccharides, and polysaccharides. The energy for their formation ultimately comes from ATP, which forms the other nucleotide triphosphates, as shown in reaction 10.8.

$$\text{ATP} + \text{UDP} \rightarrow \text{ADP} + \text{UTP}\qquad(10.8)$$

The equilibrium for the formation of the nucleotide diphospho sugar, reaction (10.6), is shifted in the direction of nucleotide diphospho sugar formation by the hydrolysis of the pyrophosphate product, catalyzed by the enzyme, *pyrophosphatase* (reaction 10.7). The configuration of the C-1 of the monosaccharide residue attached to the nucleotide diphosphate is the configuration of the monosaccharide-1-phosphate, which is usually α-because the sugar retains its phosphate group in the formation of the nucleotide diphospho sugar.

10.3 Biosynthesis of Different Monosaccharides by Epimerization, Oxidation, and Decarboxylation

In section 10.1, we saw that the first hexose to be formed in the fixation of CO_2 was D-fructose-1,6-bisphosphate. This sugar phosphate is converted to D-fructose-6-phosphate by a specific phosphatase (reaction 10.3). The keto group can be epimerized by two different enzymes, *D-fructose-6-phosphate/D-glucose epimerase* or *D-fructose-6-phosphate/D-mannose epimerase,* to form D-glucose-6-phosphate and D-mannose-6-phosphate, respectively (see Fig. 10.5). Both of these sugar phosphates can be converted into their 1-phosphates by reaction with specific *mutase* enzymes. α-D-Glucose-1-phosphate and α-D-mannose-1-phosphate can then react with UTP and GTP to form UDP-Glc and GDP-Man (see Fig. 10.5).

D-Fructose-6-phosphate can also react with L-glutamine in a *transaminase*-catalyzed reaction in which the α-amino group of the amino acid is transferred to the keto group of D-fructose-6-phosphate to provide an imine that is epimerized to give D-glucosamine-6-phosphate (see Fig. 10.5). The amino group is then acetylated by an enzyme involving acetyl coenzyme A (see Chapter 11 for the formation of acetyl CoA), N-acetyl-D-glucosamine-6-phosphate (GlcNAc-6-P). This can be converted into N-acetyl-D-glucosamine-1-phosphate by a *mutase* and then into UDP-GlcNAc by reaction with UTP. UDP-GlcNAc can be converted into N-acetyl-D-mannosamine by a 2-epimerase that has a unique reaction mechanism in which UDP is removed by a β-elimination reaction giving N-acetyl-2-deoxy-D-glycal. The glycal is stereospecifically hydrated to give N-acetyl-D-mannosamine [12] (see Fig. 10.5).

N-acetyl-mannosamine is an important precursor for the biosynthesis of the nine carbon sugar acid, N-acetyl-D-neuraminic acid (see section 10.5 and Fig. 10.8A) that is frequently found as a monosaccharide in glycoproteins and glycolipids (see Chapter 9).

UDP-Glc is converted into UDP-Gal by a reaction catalyzed by *UDP-Gal-4-epimerase.* The enzyme is NAD^+ dependent and the first reaction that it catalyzes is the oxidation of C-4 of UDP-Glc to give the 4-keto intermediate and NADH. The complex undergoes a conformational change and the NADH-product stereospecifically reduces the keto group to give the D-galactose structure [13–15] (see Fig. 10.6).

UDP-Glc is also converted into UDP-glucuronic acid (UDP-GlcUA) by a reaction catalyzed by an enzyme requiring two moles of NAD^+ per mole of UDP-Glc [16] (see Fig. 10.6). In plants, there is an enzyme, *UDP-GlcUA decarboxy-*

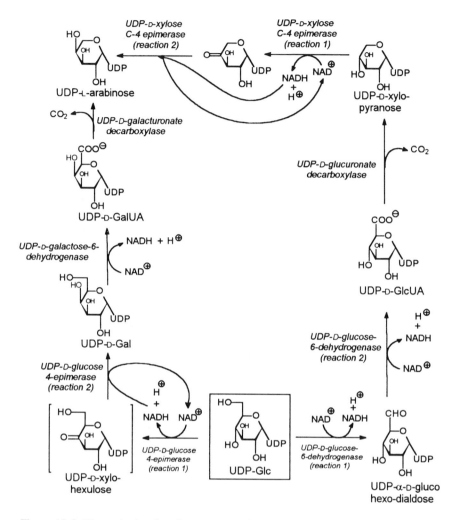

Figure 10.6. Biosynthesis of various monosaccharide units (UDP-D-galactose, UDP-D-glucuronic acid, UDP-D-galacturonic, UDP-D-xylopyranose, and UDP-L-arabinose) from UDP-D-glucose.

lase, that catalyzes the decarboxylation and removal of C-6 of the uronic acid to give UDP-D-xylopyranose (Fig. 10.6) [17–19]. UDP-D-xylopyranose serves as the high-energy D-xylose donor for the biosynthesis of the xylose-containing hemicelluloses (see Fig. 6.9 in Chapter 6 for these structures). A similar decarboxylation reaction of UDP-GalUA forms UDP-L-arabinose. UDP-L-arabinose is also formed by a C-4 epimerase reaction with UDP-D-xylopyranose. L-Arabinose is also a common constituent in hemicelluloses. See Fig. 10.6 for a summary of these reactions and conversions.

10.4 Biosynthesis of D- and L-6-Deoxyhexoses

The biosynthesis of 6-deoxyhexoses takes place in three distinct reactions, starting with nucleotide diphospho sugars. The first reaction is the oxidation of C-4 by a dehydrogenase, using the coenzyme NAD$^+$. This reaction labilizes the hydrogen atom on C-5 due to the electron-withdrawing effect of the C-4 carbonyl group. Water is eliminated between C-5 and C-6 by the removal of the C-5 hydrogen and the C-6 hydroxyl group in a β-elimination reaction that forms the 5,6-ene. C-6 is then stereospecifically reduced by NADH-*dehydrogenase* to give either the D- or the L-6-deoxy-sugar, depending on the enzyme. An enediol is then formed between C-3 and C-4 and the carbon-carbon double bond is stereospecifically reduced by NADH-dehydrogenase to give a specific 6-deoxyhexose.

These reactions are shown in Fig. 10.7 for the conversion of dTDP-D-glucose to dTDP-L-rhamnose [17] and for the conversion of GDP-D-mannose into GDP-D-rhamnose and GDP-L-fucose [20–22]. The initial reaction to form the 4-keto sugar has a close similarity to the reaction for the conversion of UDP-Glc into UDP-Gal.

10.5 Biosynthesis of Eight- and Nine-Carbon Sugars: *N*-Acetyl-D-Neuraminic Acid, *N*-Acetyl-D-Muramic Acid, and 2-Keto-3-Deoxy-D-Mannooctulosonic Acid (KDO)

N-acetyl-D-neuraminic acid is biosynthesized from *N*-acetyl-D-mannosamine and phosphoenol pyruvate, catalyzed by *N-acetyl-D-neuraminic acid synthase*. The first step involves the addition of an electron pair from the double bond of the phosphoenol pyruvate to the aldehyde group to give an aldol-type condensation (see Fig. 10.8A). The product is the nine-carbon sugar acid, *N*-acetyl-D-neuraminic acid [23]. In some instances the enzyme requires *N*-acetyl-D-mannosamine-6-phosphate as the substrate and forms *N*-acetyl-D-neuraminic acid-9-phosphate. Various hydroxyl groups on C-4, -7, -8, and -9 can be acetylated by specific acetyl transferases using acetyl CoA as the donor. KDO (2-keto-3-deoxy-D-mannooctulosonic acid) is biosynthesized by a very similar condensation between D-arabinose-5-phosphate and pyruvic acid, catalyzed by *KDO synthase* (see Fig. 10.8B) [24].

The nine-carbon sugar, *N*-acetyl-D-muramic acid, is biosynthesized from UDP-GlcNAc by reaction with phosphoenol pyruvate, catalyzed by *N-acetyl-D-muramic acid synthase*. The enzyme catalyzes the attack of the C-3 hydroxyl group of UDP-GlcNAc onto C-2 of phosphoenol pyruvate, displacing the enol phosphate ester and forming an *O*-ether with the D-glucosamine residue. The carbon-carbon double bond is then reduced by NADPH to give a C-3-*O*-lactyl ether of UDP-*N*-acetyl-D-glucosamine, which is UDP-*N*-acetyl-D-muramic acid (UDP-NAM) [25,26] (see Fig. 10.8C).

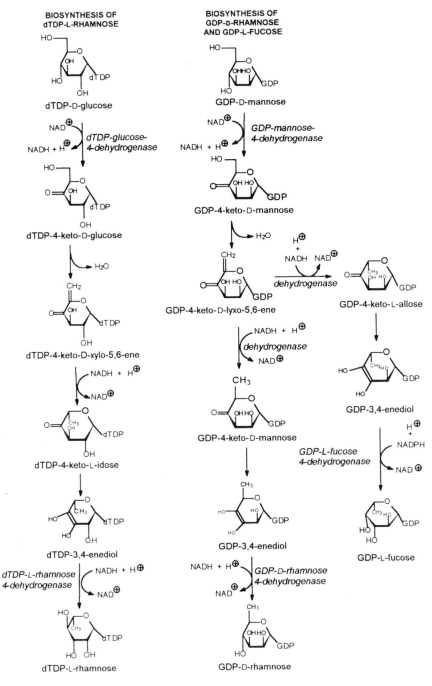

Figure 10.7. Biosynthesis of 6-deoxyhexoses: dTDP-L-rhamnose, GDP-D-rhamnose, and GDP-L-fucose.

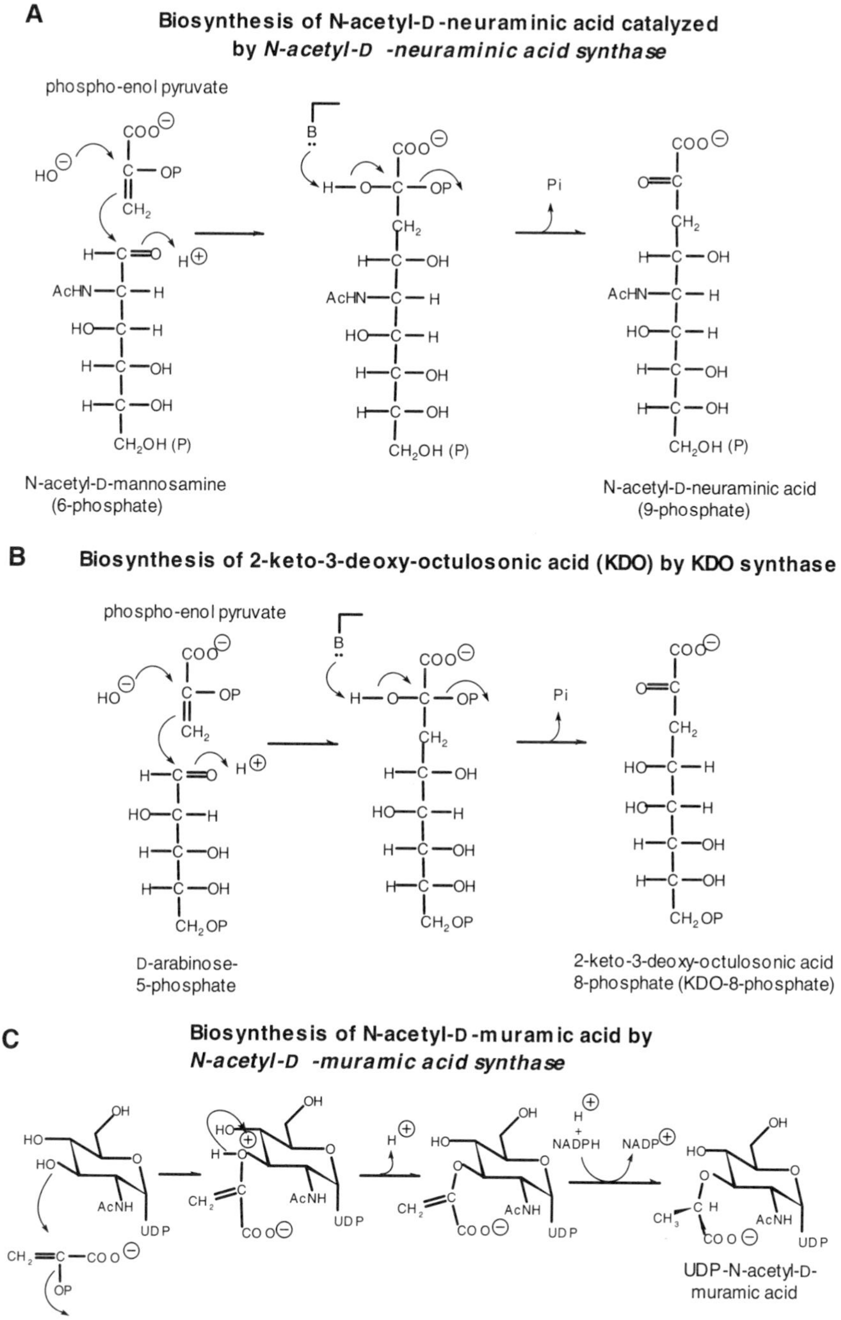

A

Biosynthesis of N-acetyl-D-neuraminic acid catalyzed
by *N-acetyl-D-neuraminic acid synthase*

B

Biosynthesis of 2-keto-3-deoxy-octulosonic acid (KDO) by KDO synthase

C

Biosynthesis of N-acetyl-D-muramic acid by
N-acetyl-D-muramic acid synthase

Figure 10.8. Biosynthesis of sugar acids with eight and nine carbons; **A,** *N*-acetyl-D-neuraminic acid; **B,** 2-keto-3-deoxyoctulosonic acid (KDO); **C,** *N*-acetyl-D-muranic acid.

10.6 Biosynthesis of the Naturally Occurring Disaccharides: α,α-Trehalose, Sucrose, and Lactose

α,α-Trehalose was the first of the three naturally occurring disaccharides whose biosynthesis was described. Leloir and Cabib [27,28] showed that enzymes (α,α-*trehalose-6-phosphate synthase*) from yeasts and insects catalyzed the biosynthesis of α,α-trehalose-6-phosphate from UDP-Glc and D-glucose-6-phosphate. The phosphate group is removed by the action of a *phosphatase*, as shown in the following reactions:

$$\text{UDP-Glc} + \text{Glc-6-P} \xrightarrow{\text{synthase}} \text{α,α-trehalose-6-P} + \text{UDP} \qquad (10.9)$$

$$\text{α,α-trehalose-6-P} + \text{H}_2\text{O} \xrightarrow{\text{phosphatase}} \text{α,α-trehalose} + \text{Pi} \qquad (10.10)$$

Leloir, et al. [29,30] and Bean and Hassid [31] showed that sucrose was biosynthesized in a number of plants by a similar reaction scheme involving UDP-Glc and D-fructose-6-phosphate according to the following reactions:

$$\text{UDP-Glc} + \text{Fru-6-P} \xrightarrow{\text{sucrose synthase}} \text{sucrose-6'-P} + \text{UDP} \qquad (10.11)$$

$$\text{sucrose-6'-P} + \text{H}_2\text{O} \xrightarrow{\text{sucrose phosphatase}} \text{sucrose} + \text{Pi} \qquad (10.12)$$

Lactose was also shown to be synthesized in an identical manner, starting with UDP-Gal and Glc-6-P [32]:

$$\text{UDP-Gal} + \text{Glc-6-P} \xrightarrow{\text{lactose synthase}} \text{lactose-6'-P} + \text{UDP} \qquad (10.13)$$

$$\text{lactose-6'-P} + \text{H}_2\text{O} \xrightarrow{\text{lactose phosphatase}} \text{lactose} + \text{Pi} \qquad (10.14)$$

10.7 Biosynthesis of Sucrose and Starch in Plants

After the fixing of carbon dioxide in the dark reactions of photosynthesis and the formation of D-fructose-6-phosphate, the D-fructose is isomerized to D-glucose-6-phosphate by the action of D-*fructose-6-phosphate isomerase*. The D-glucose-6-phosphate is then further isomerized to α-D-glucopyranosyl-1-phosphate by the action of *phosphoglucomutase*. α-D-Glucose-1-phosphate then reacts with uridine triphosphate (UTP) to give uridine diphospho-D-glucose (UDPG) and inorganic pyrophosphate catalyzed by *UTP/glucose-1-phosphate pyrophosphorylase*. The pyrophosphate that is released is hydrolyzed by *pyrophosphatase* to two inorganic phosphates, driving the reaction toward the formation of the high-energy D-glucose compound (UDPG), UDPG and β-D-fructofuranose-6-phosphate then react with *sucrose synthase* to give sucrose-6'-phosphate and UDP.

Sucrose-6'-phosphate is biosynthesized in the leaves of plants and acts as a transport sugar in plants. In sugarcane it is transported into the stems where *sucrose-6'-phosphate phosphatase* hydrolyzes the phosphate group allowing the deposit of sucrose (cane sugar). In sugar beets the sucrose-6'-phosphate is transported to the tubers where the same kinds of reactions occur to deposit sucrose.

In other plants such as potato or maize, sucrose-6'-phosphate is transported to the tuber or seeds where it reacts with UTP to give UDPG and D-fructose-6-phosphate. The D-fructose-6-phosphate is converted into D-glucose-6-phosphate and then into α-D-glucose-1-phosphate, which reacts with UTP to give UDPG. UDPG is a high-energy glucose donor that is used to synthesize starch. The D-glucose of UDPG is transferred to C-4 of the nonreducing ends of a starch primer chain to give the synthesis of α-1 → 4 glycosidic linkage and the elongation of the chain and the formation of amylose.

The starch primer can be as small as three D-glucose residues. When the amylose chains are sufficiently long (40–50 glucose residues), two chains interact with *starch branching enzyme,* and one of the chains has an α-1 → 4 glycosidic linkage cleaved by the enzyme and the C-6 hydroxyl group of a D-glucose residue on the second chain then makes a nucleophilic attack onto C-1 of the D-glucose unit that has been cleaved from the first chain, thereby making an α-1 → 6 branch linkage. Several such transfers ultimately produce the amylopectin molecule. Branching enzyme is an *α-1 → 4/α-1 → 6 glucanosyl transferase* that catalyzes the transfer of a segment of one amylose chain to the C-6 position of a D-glucose unit on another amylose chain (see Fig. 10.16A, B for the details of the transfer reaction). The reactions for the biosynthesis of sucrose and starch are summarized in Fig. 10.9.

10.8 Biosynthesis of the Bacterial Cell Wall: Peptidomurein

Isolation of fragments of the bacterial cell wall from *Staphylococcus aureus* and *Micrococcus lysodeikiticus* showed that they were composed of *N*-acetyl-D-glucosamine and *N*-acetyl-D-muramic acid residues, along with a few specific amino acids, glycine, D- and L-alanine, D-glutamic acid, and L-lysine. The presence of only a few of the 20 usual amino acids found in proteins, and the presence of two of the unnatural D-amino acids suggested that the amino acids were not components of contaminating proteins. It was shown that the fragment was a linear polymer of alternating *N*-acetyl-D-glucosamine (NAG) and *N*-acetyl-D-muramic acid (NAM), linked β-1 → 4 [33–35]. Analyses of cell walls from a wide variety of bac-

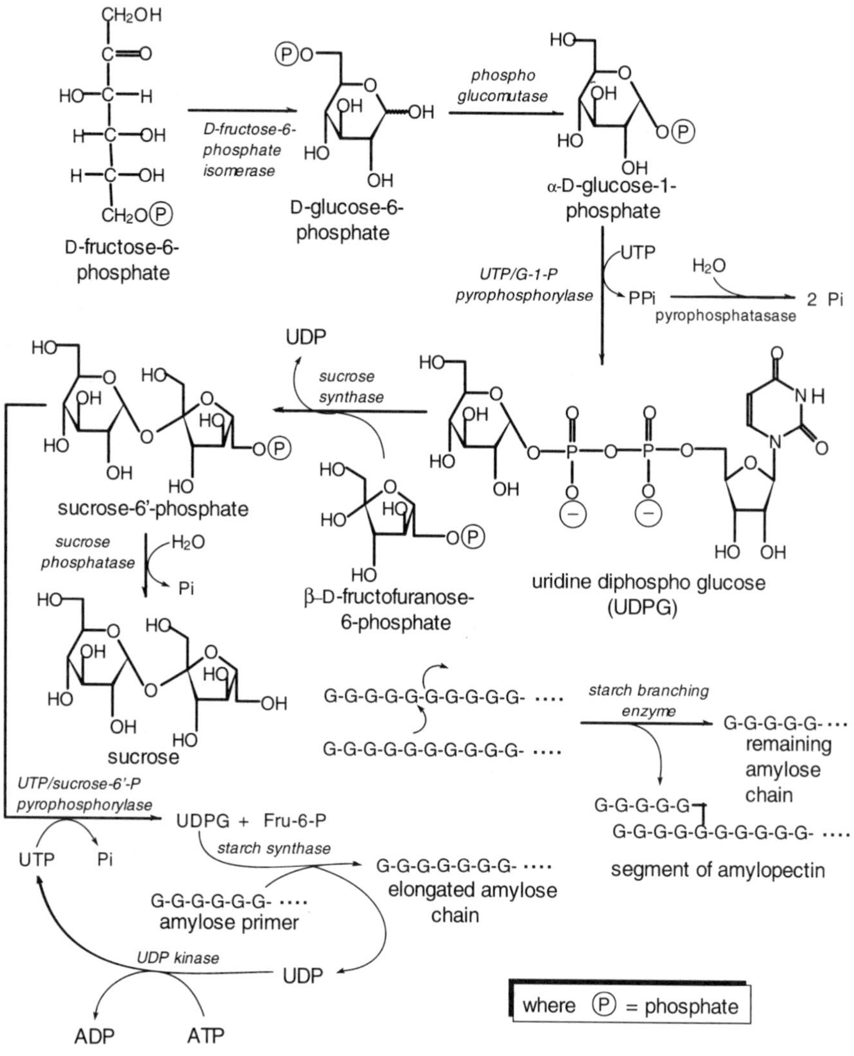

Figure 10.9. Reaction pathways for the biosynthesis of sucrose and starch from D-fructose-6-phosphate in plants.

teria indicated an almost invariable structure, with the exception of substitutions of diaminopimelic acid or L-ornithine for L-lysine in the peptide chain.

In 1949, Park and Johnson [36] observed the accumulation of large amounts of uridine diphospho sugars in the cultures of *Staph aureus* that had been treated with penicillin. The structures of these compounds were unknown and they were called the "Park-Johnson compounds." Much later they were shown to be UDP-

NAG and UDP-NAM-pentapeptide [37,38]. The pentapeptide was attached to the carboxyl group of the O-lactyl group of NAM and found to have the sequence, L-alanyl-D-isoglutamyl-L-lysyl-D-alanyl-D-alanine [39–41]. The peptide had one unusual linkage: the D-glutamic acid was linked through its γ-carboxyl group instead of its α-carboxyl group. The two UDP derivatives (Park-Johnson compounds) were apparently precursors for the biosynthesis of peptidomurein chain.

Strominger and colleagues worked out the reaction sequence for the biosynthesis of the peptidomurein chain [42–45]. The amino acids are added stepwise to UDP-NAM, each addition being catalyzed by a specific enzyme, requiring ATP to form the peptide bonds. It was also discovered that a lipid fraction was necessary for the biosynthesis of the NAG-NAM-pentapeptide polymer. This was shown to be a C_{55}-polyisoprenoid phosphate [46,47]. It was called *bactoprenol phosphate* (see Fig. 9.15 in Chapter 9 for the structure) and has been found to be a coenzyme in the biosynthesis of many bacterial polysaccharides. The biosynthesis of the peptidomurein by *Staph. aureus* and *M. lysodeikiticus* requires five distinct reactions [48–54] (see Fig. 10.10A). In the first reaction, UDP-NAM-pentapeptide (UDP-NAM-p_5) reacts with bactoprenol phosphate to form bactoprenol pyrophosphate-NAM-p_5 and splits out UMP, as shown in Fig. 10.10B. This product then reacts with UDP-GlcNAc to form the repeating disaccharide bactoprenol pyrophosphate (β-NAG-1 $\rightarrow$ 4-NAM-p_5-PP-Bpr) [55]. At this stage, five glycine residues are added to the ϵ-amino group of L-lysine to form pentaglycine chain (in some other species, other kinds of amino acids are involved), resulting in a decapeptide attached to the disaccharide unit. The glycine residues are added enzymatically, one residue at a time, from glycyl-t-RNA. The bactoprenol wraps around the hydrophilic NAG-NAM-p_{10}, forming a hydrophobic coat. The complete unit is transported from inside the cell, through the hydrophobic cellular membrane, to the outside of the cell, where the NAG-NAM-p_{10} is added to the cell wall peptidomurein chain (see Fig. 10.11A). It was shown that the repeating disaccharide decapeptide, NAG-NAM-p_{10}, is added to the reducing end of the growing chain [56], splitting out the bactoprenol pyrophosphate unit that was holding the chain in the membrane [55]. This bactoprenol pyrophosphate is then hydrolyzed by a pyrophosphatase to bactoprenol phosphate and inorganic phosphate. The hydrolysis apparently signals that the bactoprenol phosphate should return through the membrane to the cytoplasm, where it takes up new NAG-NAM-p_{10} units for transport through the membrane and the addition to the reducing end of the cell wall chain [49].

This does not end the process. The last stage in the assembly of the bacterial cell wall is the cross-linking of the chains. The amino group of the terminal glycine residue makes a nucleophilic attack onto the carboxyl group of the second from the end (C-terminus) D-alanyl residue of a pentapeptide on another chain in its vicinity, displacing the terminal D-alanine residue (see Fig. 10.11B for the reactions and the cross-linked structure). This cross-linking reaction is catalyzed by a *transpeptidase* enzyme [57,58]. Both penicillin G and ampicillin are potent inhibitors of this cross-linking reaction [57,58]. Vancomycin and ristocetin inhibit the incorporation of the disaccharide decapeptide lipid intermediate into the poly-

A   **Reactions in the biosynthesis of bacterial cell wall peptidoglycan**

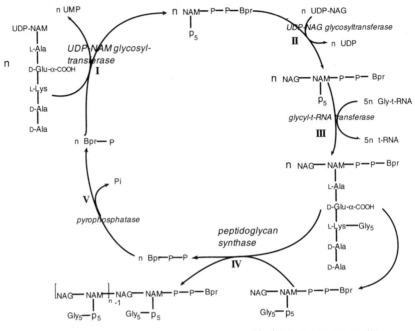

B   **Mechanism for reaction I**

C   **Mechanism for the additon of the donor unit to the reducing-end of the growing peptidoglycan chain (Reaction IV)**

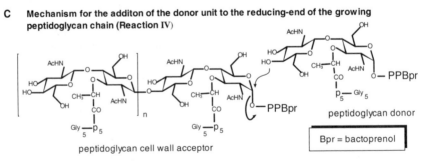

Figure 10.10. Biosynthesis of peptidomurein bacterial cell wall; Bpr, bactoprenol.

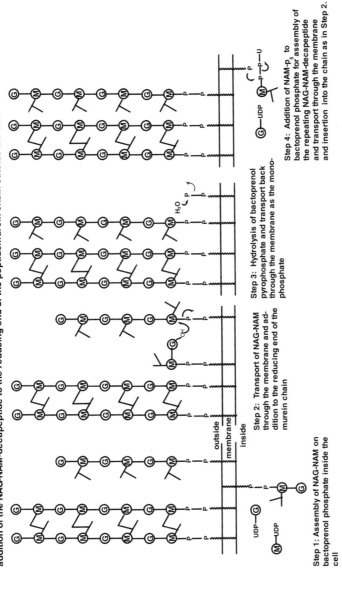

Figure 10.11. Assembly of the bacterial cell wall: **A**, transport of NAG-NAM-decapeptide through the cell membrane and incorporation into peptidomurein chain by "insertion" at the reducing end; **B**, cross-linking of peptidomurein chains by reaction of pentaglycine chain with the second D-alanyl residue of another chain.

(continued)

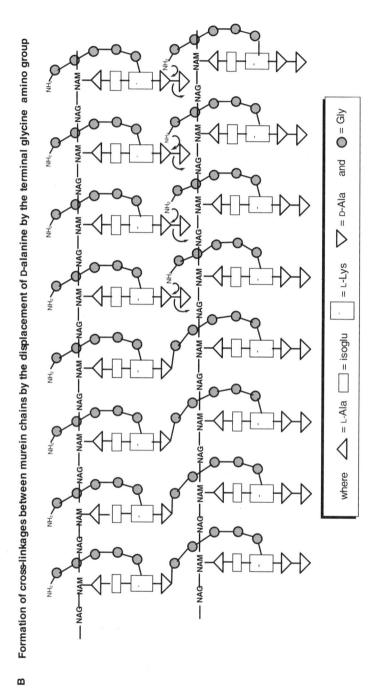

B   Formation of cross-linkages between murein chains by the displacement of D-alanine by the terminal glycine amino group

where △ = L-Ala   ▢ = isoglu   ▢ = L-Lys   ▽ = D-Ala   and   ● = Gly

Figure 10.11. *(continued)*

mer chain [58]. Cross-linking peptides other than pentaglycine occur for some other species of bacteria [59].

10.9 Biosynthesis of the *Salmonella* O-Antigen Outer Capsular Polysaccharide

It has also been found that bactoprenol phosphate was required as a coenzyme carbohydrate carrier in the biosynthesis of *Salmonella* O-antigen polysaccharide. Robbins et al. [60–62] and Osborn et al. [63,64] demonstrated that the biosynthetic reactions of the O-antigen polysaccharide followed a pattern similar to that of the murein chain. The O-antigen polysaccharide repeating unit is assembled on the bactoprenol pyrophosphate coenzyme, which is transported to the outer surface for polymerization by the displacement of the bactoprenol pyrophosphate unit from the chain and the addition of a new repeating unit to the reducing end of the chain [65,66]. This latter aspect of the mechanism of the elongation of the chain was reported before that of the peptidomurein. *Salmonella* O-antigen polysaccharide was the first polysaccharide to be definitively shown to have its elongation from the reducing end of the polysaccharide chain.

This type of polysaccharide elongation mechanism is known as an *insertion mechanism* in which the monomer residue or repeating unit is apparently inserted between the reducing end of the polysaccharide chain and a lipid pyrophosphate coenzyme carrier or an enzyme-protein carrier. Actually, it is not a real insertion, but rather the transfer of the polysaccharide chain from one carrier to the carbohydrate moiety of a monomer or repeating unit attached to another carrier (see Figs. 10.12A section 10.12 on dextran biosynthesis).

An interesting aspect of the biosynthesis of the O-antigen polysaccharide is that different nucleotide triphosphates are the activating agents for each of the monosaccharide residues in the repeating unit. The reaction of UDP-Gal with bactoprenol phosphate is the first reaction in the process, followed by reactions with TDP-Rha, GDP-Man, and CDP-Abe as shown for the biosynthesis of *Salmonella typhimurium* O-antigen polysaccharide summarized in Fig. 10.12A.

10.10 Biosynthesis of *Escherichia coli* and *Neisseria meningitidis* Colominic Acid

The biosynthesis of poly-2 → 8 or -2 → 9-(*N*-acetyl-D-neuraminic acid) capsular polysaccharide (colominic acid) from *E. coli* or *N. meningitidis* (see Chapter 6, Fig. 6.20 for the structures) has also been shown to involve a polyprenol phosphoryl *N*-acetyl-D-neuraminic acid lipid intermediate formed from CMP-NeuNAc [67,68]. This is an example of the biosynthesis of a homopolysaccharide that requires a polyprenol phosphate coenzyme lipid carrier.

A Biosynthesis of *Salmonella typhimurium* O-antigen polysaccharide

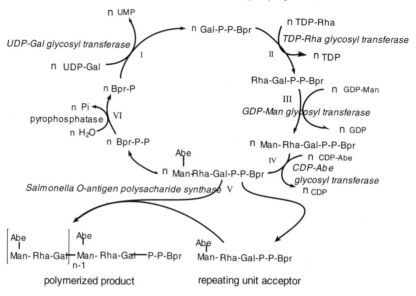

B Mechanism for the biosynthesis of glycerol teichoic acid by glycerol teichoic acid synthase

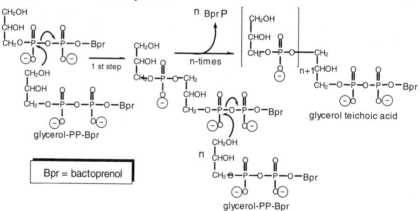

Figure 10.12. Biosynthetic reactions for the formation of **(A)** *Salmonella typhimurium* O-antigen polysaccharide, and **(B)** mechanism for the biosynthesis of glycerol teichoic acid. Bpr, bactoprenol.

10.11 Biosynthesis of Teichoic Acids

The two homoteichoic acids, poly(glycerol phosphate) and poly(ribitol phosphate) from *Bacillicheniformis* and *Bacillus subtilis* were shown to be biosynthesized from CDP-glycerol and CDP-ribitol, respectively [69–71]. The heteroteichoic acid poly(→ 3-α-GlcNAc-1-phosphoryl-1-glycerol-3 → phosphoryl), (see Chapter 6, Fig. 6.21C for the structure) was shown to be biosynthesized from UDP-GlcNAc and CDP-glycerol [72]. These syntheses also require a phospholipid coenzyme carrier. Similarly, it was shown that the homoteichoic acid, poly(→ 3-α-GlcNAc-1-phosphoryl-) is biosynthesized by a transfer of GlcNAc from UDP-GlcNAc to a phospholipid coenzyme [73,74].

The mechanism for the polymerization of the teichoic acids has not been definitively established, but it is most probably very similar to the biosyntheses of the other lipid-coenzyme-dependent polysaccharides, the peptidomurein and *Salmonella* O-antigen polysaccharide, in which the monomer unit or repeating unit attached to the lipid polyprenol pyrophosphate carrier is inserted between the growing chain and the lipid coenzyme carrier. See Fig. 10.12B for the proposed mechanism for teichoic acid biosynthesis.

10.12 Biosynthesis of Dextrans and Related Polysaccharides

A whole family of bacterial polysaccharides is biosynthesized by enzymes utilizing sucrose as a high-energy donor of glucose or fructose instead of nucleotide diphospho sugars. The energy of the gluco-fructo glycosidic linkage is the same as the energy of sugar-phosphate linkage in the nucleotide diphospho sugars (about 5 Kcal/mole). The principal bacteria elaborating the enzymes are various strains of *Leuconostoc mesenteroides* and *Streptococcus* species [75]. The polysaccharides are known primarily as *dextrans* and are produced extracellularly in relatively large quantities in sucrose media. Dextrans are α-1 → 6-linked glucans with varying degrees of branch linkages that are primarily α-1 → 3, although some have α-1 → 2 or α-1 → 4 branch linkages [75] (see Chapter 6, Fig. 6.16 for the structures). There are glucans other than the dextrans, such as mutan and alternan, that are biosynthesized from sucrose. There are also the fructans, levan and inulin, that are biosynthesized using sucrose as the fructofuranosyl donor.

The enzymes responsible for catalyzing the synthesis of dextrans and related polysaccharides form covalent glucose and polysaccharide complexes during synthesis of the glucans [76–79]. Using pulse and chase techniques with [U-^{14}C] sucrose and *L. mesenteroides* B-512FM dextransucrase, and *S. mutans* 6715 dextransucrase and mutansucrase immobilized on Bio-Gel P-2 beads, Robyt et al. [76,77] found that the polysaccharides were synthesized by the addition of glucose to the reducing end of the polysaccharide chains. The results were explained by a two-site insertion mechanism in which the glucose moiety of the glucosyl enzyme intermediate is added to the reducing end of the growing chain. The enzyme mechanisms for the biosynthesis of these glucans are given in Fig. 10.13A, B.

The dextrans are the first polysaccharides considered in this chapter that are branched. Branching enzymes for these polysaccharides have been sought, but without any success. Robyt and Taniguchi [80] showed that branch linkages are formed for the B-512F(M) dextran by the same enzyme that catalyzes the polymerization reaction. They found that exogenous dextran chains could act as acceptors in which their C-3 hydroxyl groups can make nucleophilic attacks onto either the glucosyl enzyme intermediate, or on the glucanyl enzyme intermediate to give single α-1 $\rightarrow$ 3-linked glucose branches or α-1 $\rightarrow$ 3-linked dextran chain branches. See Fig. 10.14 for the mechanism of forming B-512F(M) dextran branches by the dextran chain acceptor reaction.

For the highly branched dextrans formed by dextransucrases from *S. mutans* 6715 and *L. mesenteroides* B-742, another mechanism most probably is involved to give the high degrees (35 and 50%, respectively) of α-1 $\rightarrow$ 3-linked single glucose branches. Robyt [75] proposed that these branches are formed during the course of the polymerization reactions by a two-site mechanism (see Fig. 10.15A, B). In these mechanisms, when a single glucosyl intermediate makes a nucleophilic attack onto C-1 of the opposite glucosyl intermediate, its C-6 hydroxyl group is *always* in stereochemical position to make the attack and form an α-1 $\rightarrow$ 6 linkage. The next reaction, however, is by the C-3 hydroxyl group of the same glucosyl unit that had just been added to the reducing end of the growing dextran chain. In this mechanism, it is postulated that the C-3 hydroxyl group is stereochemically put into place to make the attack onto the opposite glucosyl group in the two sites by virtue of the fact that its glucosyl unit had just made an α-1 $\rightarrow$ 6 linkage and was incorporated into the dextran chain. In this mechanism, two reactions occur in sequence from one site: first, to form an α-1 $\rightarrow$ 6 linkage and provide incorporation into the dextran chain, and, second, to form a single α-1 $\rightarrow$ 3 branched glucose unit before the C-6 hydroxyl of the opposite glucosyl unit makes an attack on C-1 of the growing chain. In the case of synthesizing the 50% branched B-742 dextran (a dextran with single glucose branches linked α-1 $\rightarrow$ 3 to each of the glucose residues in the α-1 $\rightarrow$ 6-linked dextran chain), this double reaction occurs at both of the two enzyme sites, while in the case of *S. mutans* dextran (a dextran with single α-1 $\rightarrow$ 3 linked glucose branches on alternating glucose residues of the α-1 $\rightarrow$ 6-linked dextran chain), the double reaction occurs only at one of the two sites, to form the alternating comb dextran (compare Fig. 10.15A, B).

For the biosynthesis of mutan, an α-1 $\rightarrow$ 3–linked linear glucan, the pulse and chase data indicated that it too was being biosynthesized by a two-site mechanism similar to that proposed for the dextrans [77]. The mechanism for the elongation of this polysaccharide was proposed to have two sites that exclusively orient the glucose units so that the C-3 hydroxyl group of the glucosyl units make the nucleophilic attack onto C-1 of the opposite glucosyl intermediate, to form α-1 $\rightarrow$ 3

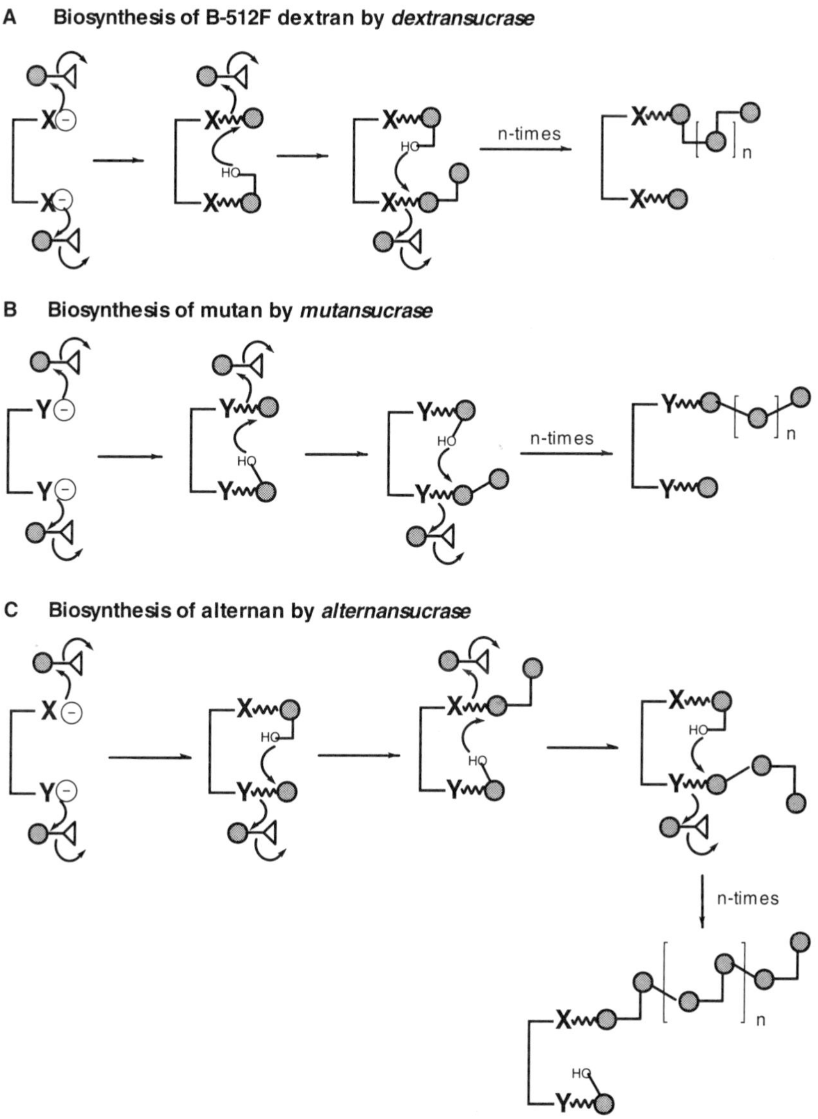

A   Biosynthesis of B-512F dextran by *dextransucrase*

B   Biosynthesis of mutan by *mutansucrase*

C   Biosynthesis of alternan by *alternansucrase*

Figure 10.13. Two-site insertion mechanisms involved in the biosynthesis of (**A**) *L. mesenteroides* B-512F dextran by *dextransucrase*, (**B**) *S. mutans* mutan by *mutansucrase*, and (**C**) *L. mesenteroides* B-1355 alternan by *alternansucrase*. X and Y are nucleophiles at the active site of the enzymes; X orients glucosyl units to make α-1 → 6 linkages, and Y orients glucosyl units to make α-1 → 3 linkages. ○◁ is sucrose; ○ is glucose; and ◁ is fructose. The other symbols are those given in section 6.6 of Chapter 6.

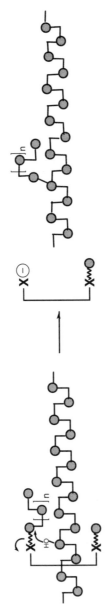

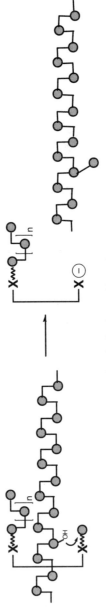

Figure 10.14. Mechanism for the formation of α-1 → 3 branch linkages of *L. mesenteroides* B-512F dextran.

A Two-site insertion mechanism for the biosynthesis of *Leuconostoc mesenteroides* B-742 regular comb dextran by *dextransucrase*

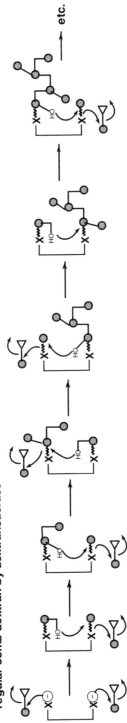

B Two-site insertion mechanism for the biosynthesis of *Streptococcus mutans* alternating comb dextran by *dextransucrase*

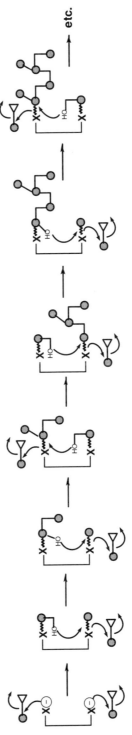

Figure 10.15. Two-site insertion mechanism for the biosynthesis of (**A**) *L. mesenteroides* B-742 regular comb dextran and for the biosynthesis of (**B**) *S. mutans* alternating comb dextran by the respective dextransucrases.

linked glucan (see Fig. 10.13B) [75]. In the biosynthesis of alternan, an alternating α-1 $\rightarrow$ 6 and α-1 $\rightarrow$ 3 glucan, it was proposed that one site orients the glucose unit so that its C-6 hydroxyl group makes the attack to form an α-1 $\rightarrow$ 6 linkage, and the other site orients the glucose unit so that its C-3 hydroxyl group makes the attack to form an α-1 $\rightarrow$ 3 linkage. The elongation then takes place back and forth between the two sites, resulting in an alternating α-1 $\rightarrow$ 6- and α-1 $\rightarrow$ 3-linked structure [75] (see Fig. 10.13C for the mechanism of elongation).

10.13 Biosynthesis of Starch, Glycogen, and Cellulose

Much less is known about the elongation reactions of starch, glycogen, and cellulose biosynthesis than is known about the elongation of the bacterial polysaccharide chains discussed in sections 10.8–10.12. Some of this arises from the fact that starch and glycogen were originally thought to be biosynthesized by *phosphorylase*. In the 1940s, phosphorylase was observed in vitro to catalyze the addition of glucose from α-D-glucose-1-phosphate (α-G-1-P) to the nonreducing ends of a starch or glycogen chain [81,82]. A preformed starch or glycogen chain was required for the addition and was called a *primer*. The synthetic reaction catalyzed by phosphorylase is the following:

$$m\ \alpha\text{-G-1-P} + \underset{\text{primer}}{G_n} \xrightarrow{\text{phosphorylase}} G_{n+m} + m\ \text{Pi} \tag{10.15}$$

It was later shown, however, that the equilibrium ratio of inorganic phosphate to α-G-1-P at pH 6.8 is 3.6 [83]. It was also shown that the concentration of inorganic phosphate (Pi) is many fold higher than α-G-1-P in most plant tissues [84]. In immature maize endosperm, the concentration of Pi is over 100 times higher than that of α-G-1-P [84]. When the reaction is initiated in vitro with α-G-1-P and primer (reaction 10.15), limited elongation of the primer will occur, but as Pi accumulates, the reaction slows down greatly and eventually stops after the addition of only a relatively small amount of glucose to the primer. Under most conditions, especially in vivo, the conditions for synthesis by phosphorylase are very unfavorable, and the favorable reaction is the reverse of reaction 10.15, degradation, rather than synthesis [85].

The concept of a primer for polysaccharide biosynthesis was developed from the early studies of an enzyme that actually was a degradative enzyme whose equilibrium constant was such that synthesis could be induced in vitro by using high concentrations of the glucosyl donor, α-G-1-P, and no Pi. It should then be expected that an enzyme whose primary reaction is the degradation of a polysaccharide (starch or glycogen), should *require* a preformed polysaccharide chain (a primer) for the elongation of the chain by the reverse reaction. Thus, the development of the concept that synthesis of polysaccharides requires a preformed primer chain for synthesis was based on an incorrect premise, derived from the action of an enzyme that was degradative rather than synthetic.

In 1959, a new pathway for the biosynthesis of starch was found in which a nucleotide diphosphoglucose was a glucosyl donor that could add glucose residues to primer chains of starch and glycogen [86–89]. It was found that ADP-Glc was the preferred glucosyl donor for starch chain elongation, and UDP-Glc was the glucosyl donor for elongation of glycogen chains. The reactions were formulated as follows:

$$\underset{\substack{\text{primer}}}{m\ \text{ADP-Glc (for starch)} + (\text{Glc})_n} \overset{\text{starch synthase}}{\longrightarrow} \underset{\substack{\text{chain elongated} \\ \text{by } m \text{ glc units}}}{(\text{Glc})_{n+m}} + m\ \text{ADP} \qquad (10.16)$$

$$m\ \text{UDP-Glc (for glycogen)} + (\text{Glc})_n \overset{\text{glycogen synthase}}{\longrightarrow} (\text{Glc})_{n+m} + m\ \text{UDP} \qquad (10.17)$$

Because these elongation reactions also apparently require a primer, much effort has been spent in trying to determine how the primer is formed. In the 1970s, Krisman and Barengo reported that the synthesis of liver glycogen from UDP-Glc involved the interaction of three proteins [90]. The first was a protein on which glycogen could be synthesized. It was proposed that it was this protein that was the primer, and that an enzyme *glycogen initiator synthase,* adds glucose residues from UDP-Glc to form an oligosaccharide that, when sufficiently long, would act as a classical maltodextrin primer. *Glycogen synthase* along with *glycogen branching enzyme* could then catalyze the synthesis of glycogen from UDP-Glc. Krisman and Barengo proposed that the primer protein had several primer sites, and the product was analogous to a proteoglycan with several glycogen molecules attached to a protein core. They also reported that similar systems were involved in the synthesis of *Escherichia coli* bacterial glycogen [91,92].

Whelan, et al. [93] investigated the problem and found that rabbit liver glycogen was covalently attached to a 37-Da protein, and that the molar ratio of glycogen to protein was 1:1 [94]. They were also able to isolate a glycogen-free protein from heart muscle and skeletal muscle that contained about two glucose residues per mole of protein. These proteins were reported to have autocatalytic properties that could add glucose residues from UDP-Glc to form α-1 $\rightarrow$ 4-linked oligosaccharides with a d.p. of 8, attached to the protein [95]. They called the protein *glycogenin.* It was their hypothesis that glycogenin was the necessary factor for the initiation of glycogen biosynthesis and the formation of a maltodextrin primer chain that could act as the classical primer for glycogen synthase and glycogen branching enzyme [96,97].

Gahan and Conrad [98] reported that *Aerobacter aerogenes* cells and glycogen synthase purified from *A. aerogenes* had an activator protein that could catalyze de novo glycogen synthesis from ADP-Glc. Both the glycogen synthase and the activator were reported to be free of glycogen. Pulse-chase experiments suggested that the synthesis might be occurring by the addition of glucose residues to the reducing ends of the growing glycogen chains.

There also was a report that a lipid carrier might be involved in liver glycogen biosynthesis. Behrens and Leloir [99] showed that a fraction from liver could

A

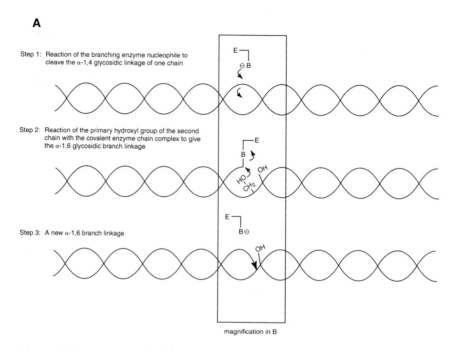

Step 1: Reaction of the branching enzyme nucleophile to cleave the α-1,4 glycosidic linkage of one chain

Step 2: Reaction of the primary hydroxyl group of the second chain with the covalent enzyme chain complex to give the α-1,6 glycosidic branch linkage

Step 3: A new α-1,6 branch linkage

magnification in B

Figure 10.16. Mechanism for forming α-1 → 6 branch linkages by potato starch branching enzyme (Q-enzyme) acting on double helical chains: **A,** mechanism for the cleavage and transfer in a double helix; **B,** magnification of (**A**), showing the reactions on a molecular level.

(continued)

transfer glucose from UDP-Glc to a lipid acceptor, giving dolichol-1-β-D-glucopyranosyl pyrophosphate.

The branching of linear starch and glycogen chains is catalyzed by a *branching enzyme.* Starch branching enzymes have been identified in potatoes, broad beans, spinach leaves, and maize endosperm. The enzyme from potatoes has been called *Q-enzyme,* to distinguish it from glycogen branching enzyme. It is an α-1 → 4/α-1 → 6-transferase that cleaves the α-1 → 4 glycosidic linkages of a linear starch chain and transfers a fragment of the chain to the C-6 primary alcohol of another chain, forming a branch chain [100]. Intrachain transfer does not occur, and two distinct linear chains are required. The minimum chain length that can be transferred is six glucose residues. Potato branching enzyme has been reported to require chain lengths in excess of 35 glucose residues [101]. It was proposed that the transfer of one chain to another could most easily be accomplished if the two chains were part of a double helix, and that this would account for the requirement of relatively long chains, since short chains would not be expected to form double helices [102]. If the substrate is already branched, the branch points would help hold the two chains of the double helix together. A possible mechanism for the transfer reaction involves the formation of a covalent enzyme intermediate that is formed when the α-1 → 4 glycosidic bond is cleaved [103]. It is this intermediate that the primary hydroxyl group of

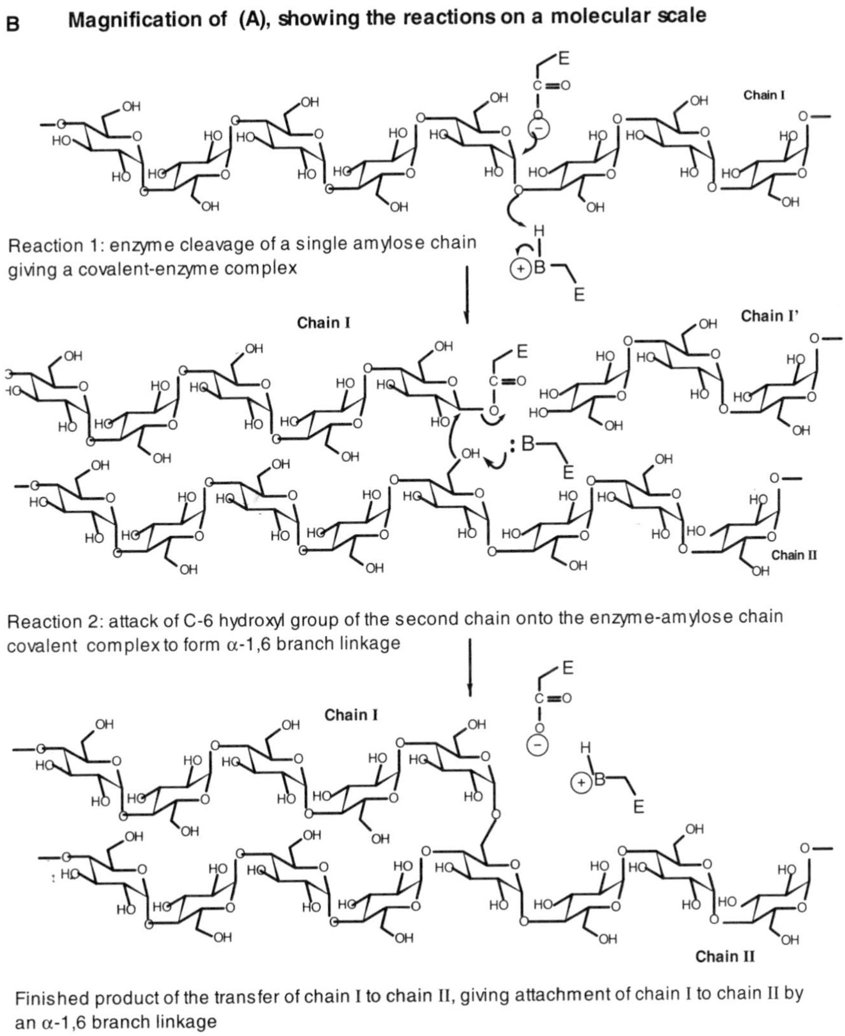
B Magnification of (A), showing the reactions on a molecular scale
Chain I
Reaction 1: enzyme cleavage of a single amylose chain giving a covalent-enzyme complex
Chain I
Chain I'
Chain II
Reaction 2: attack of C-6 hydroxyl group of the second chain onto the enzyme-amylose chain covalent complex to form α-1,6 branch linkage
Chain I
Chain II
Finished product of the transfer of chain I to chain II, giving attachment of chain I to chain II by an α-1,6 branch linkage

Figure 10.16. *(continued)*

the second chain attacks to form the α-1 → 6 branch linkage. See Fig. 10.16A,B for the proposed mechanism for the branching of an amylose double helix.

Two types of branching enzymes have been identified in some plants, one that converts amylose into amylopectin, and the other that converts amylopectin into phytoglycogen [104]. The difference between the two enzymes apparently is in the specificity of the number of glucose residues between the branch points that each enzyme permits.

Particulate enzyme fractions from mung bean seedlings and from cotton bolls were found to incorporate ^{14}C-glucose from UDP-[U-^{14}C]-Glc into an alkali-insoluble polymer that was probably cellulose [105,106]. A cell-free extract obtained from *Acetobacter xylinum* incorporated ^{14}C-glucose from UDP-[U-^{14}C]-Glc into cellulose [107]. A lipid-linked glucose was found to be the precursor of cellulose synthesized by *Acetobacter xylinum* [108], and cellulose synthesis was shown to be a membrane-associated process [109]. The cellulose synthase can be solubilized by treatment of the membranes with digitonin [109], and it was found to have a molecular size of 420 kDa, with catalytically active subunits of 67 and 54 kDa.

^{14}C-Labeled monosaccharide lipid, Glc-PP-dolichol, as well as lipid-linked oliosaccharides (cellodextrins) having 2–10 glucose residues, have been isolated from incubations of a particulate enzyme preparation from the green alga *Prototheca zopfii* with UDP-[U-^{14}C]-Glc [110].

The complete biosynthesis of xanthan, a water-soluble bacterial polysaccharide with a cellulose backbone (see Chapter 6, Fig. 6.18 for the structure), has been worked out by Dankert et al. [111–114]. The pentasaccharide repeating unit is attached to a polyprenol pyrophosphate lipid carrier [111], and the polymerization takes place by the addition of the repeating pentasaccharide from the polyprenol pyrophosphate pentasaccharide to the reducing end of the growing xanthan chain [114]. The mechanism of xanthan chain elongation is an insertion mechanism similar to those of murein, *Salmonella* O-antigen, the dextrans, mutan, and alternan.

10.14 Biosynthesis of Glycoproteins

Dolichol phosphate (Dol-P) is the carrier lipid involved in the assembly of oligosaccharides for transfer to the amide nitrogen of L-asparagine of proteins. The precursor oligosaccharide pyrophosphoryl dolichol is biosynthesized in the endoplasmic reticulum or in the Golgi bodies of eukaryotes from dolichol phosphate and nucleotide diphospho sugars, UDP-GlcNAc, GDP-Man, and UDP-Glc.

$$\text{Dol-P} + \text{UDP-GlcNAc} \rightarrow \text{Dol-PP-GlcNAc} + \text{UMP} \qquad (10.18)$$

Oligosaccharides for *N*-linkage to proteins begins with the transfer of GlcNAc-1-P from UDP-GlcNAc to Dol-P, to give GlcNAc-PP-Dol. The reactions involved in the biosynthesis of the core pentasaccharide are given in Fig. 10.17. Further addition of mannose residues from GDP-Man occurs under the direction of specific enzymes to give the high-mannose oligosaccharide found in many *N*-linked oligosaccharides [115]. The addition of GlcNAc from UDP-GlcNAc and the β-1 → -4 addition of Gal from UDP-Gal formed the lactosamine-type oligosaccharides. The precursor oligosaccharide-PP-Dol is completed with the addition of three glucosyl residues to give Glc$_3$-Man$_9$-[GlcNAc]$_2$ [116]. The addition of the three glucosyl residues apparently signals the termination of the synthesis of the oligosaccharide and its transfer to a protein. The transfer to protein is catalyzed by the enzyme, *oligosaccharide L-asparagine transferase*, located on the luminal

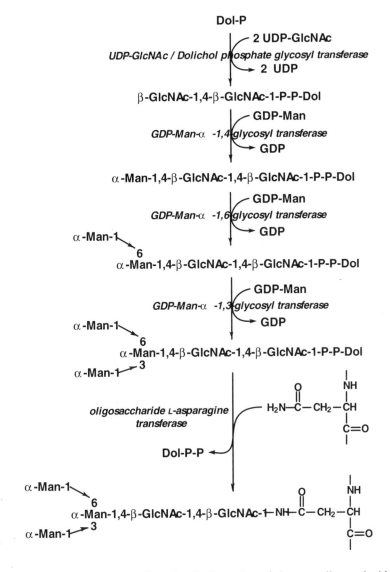

Figure 10.17. Biosynthetic reactions for the formation of the core oligosaccharide involved in attachment to the nitrogen of the amide group of L-asparagine.

face of the endoplasmic reticulum membrane. Mature asparagine-linked oligosaccharides are obtained by glycosidase trimming of the precursor oligosaccharide and by glycosyl additions from nucleotide diphospho sugars. These reactions occur during the transport of the glycoprotein from the site of synthesis and the transfer of the protein to the cytoplasm or to the cell surface. Membrane-bound *glycosidases* and *glycosyltransferases* in both the endoplasmic reticulum and the

Golgi carry out the modifications. The O-linked glycosides are formed directly by the action of specific enzymes with nucleotide diphospho sugars [117].

10.15 Literature Cited

1. R. M. Devlin and A. V. Barker, *Photosynthesis,* Van Nostrand-Reinhold, New York (1971).
2. C. H. Foger, *Photosynthesis,* Vol. 1, Chap. 2, pp. 21–40, J. Wiley Interscience, New York (1984).
3. A. San Pietro, in *Harvesting the Sun, Photosynthesis in Plant Life,* pp. 49–68 (A. San Pietro, F. A. Gerr, T. J. Army, eds.) Academic, New York (1967).
4. G. W. Brudvig, W. F. Beck, and J. C. dePaula, *Annu. Rev. Biophys. Biophys. Chem.,* **18** (1989) 25–46.
5. M. Calvin, *The Photosynthesis of Carbon Compounds,* W. A. Benjamin, New York (1962).
6a. J. A. Bassham and R. B. Jensen, in *Harvesting the Sun, Photosynthesis in Plant Life,* pp. 79–110, (A. San Pietro, F. A. Gerr, T. J. Army, eds.) Academic, New York (1967).
6b. I. Zelitch, *Annu. Rev. Biochem.,* **44** (1975) 123–145.
7a. D. I. Arnon, in *Encyclopedia of Plant Physiol.,* Vol. 5 (New Series) pp. 7–56 (A. Trebst and M. Avron, eds.) Springer-Verlag, Berlin (1977).
7b. C. H. Foger, *Photosynthesis,* Vol. 1, Chap. 4, pp. 79–104, J. Wiley Interscience, New York (1984).
8. R. Caputto, L. F. Lelori, C. E. Cardini, A. C. Paladini, *J. Biol. Chem.,* **184** (1950) 333–341.
9. E. F. Neufeld and W. Z. Hassid, *Adv. Carbohydr. Chem.,* **18** (1963) 309–356.
10. H. Nikaido and W. Z. Hassid, *Adv. Carbohydr. Chem.,* **26** (1971) 352–382.
11. E. Adams, *Adv. Enzymol.,* **44** (1976) 69–138.
12. K. M. Sommar and D. B. Ellis, *Biochim. Biophys. Acta,* **268** (1972) 590–595.
13. A. U. Bertland II, Y. Seyama, and H. M. Kalckar, *Biochemistry,* **10** (1971) 1545–1551.
14. Y. Seyama and H. M. Kalckar, *Biochemistry,* **11** (1972) 40–44.
15. G. L. Nelsestuen and S. Kirkwood, *J. Biol. Chem.,* **246** (1971) 7533–7543.
16. G. L. Nelsestuen and S. Kirkwood, *J. Biol. Chem.,* **246** (1971) 3828–3834.
17. H. Ankel, E. Ankel, D. S. Feingold, and J. S. Schutzback, *Biochim. Biophys. Acta,* **136** (1967) 172–175.
18. A. Bdolah and D. S. Feingold, *Biochem. Biophys. Res. Commun.,* **21** (1965) 543–546.
19. J. E. Silbert and S. DeLuca, *Biochim. Biophys. Acta,* **141** (1967) 193–196.
20. L. Glaser and S. Kornfeld, *J. Biol. Chem.,* **236** (1961) 1795–1799.
21. L. Glaser and H. Zarkowsky, *The Enzymes,* **5** (1971) 465–507.
22. O. Gabriel, *Adv. Chem. Ser.,* **117** (1973) 387–399.
23. R. S. Blacklow and L. Warren, *J. Biol. Chem.,* **237** (1962) 3520–3526.
24. L. Warren and H. Felsenfeld, *Biochem. Biophys. Res. Commun.,* **4** (1961) 232–235.
25. P. D. Rick and M. J. Osborn, *Proc. Natl. Acad. Sci. USA,* **69** (1972) 3756–3760.
26. G. W. Wickus, P. A. Rubenstein, A. D. Wrath, and J. L. Strominger, *J. Bacteriol.,* **113** (1973) 291–294.
27. L. F. Leloir and C. E. Cabib, *J. Am. Chem. Soc.,* **75** (1953) 5445–5448.
28. C. E. Cabib and L. F. Leloir, *J. Biol. Chem.,* **231** (1958) 259–264.
29. C. E. Cardini, L. F. Leloir, and J. Chiriboga, *J. Biol. Chem.,* **214** (1955) 149–155.

30. L. F. Leloir and C. E. Cardini, *J. Biol. Chem.*, **214** (1955) 157–165.
31. R. C. Bean and W. Z. Hassid, *J. Am. Chem. Soc.*, **77** (1955) 5737–5740.
32. W. M. Watkins and W. Z. Hassid, *J. Biol. Chem.*, **237** (1962) 1432–1440.
33. J. M. Ghuysen and J. L. Strominger, *Biochemistry*, **2** (1963) 1119–1125.
34. D. L. Tipper, J. M. Ghuysen, and J. L. Strominger, *Biochemistry,* **4** (1965) 468–473.
35. D. L. Tipper, J. L. Strominger, and J. C. Ensign, *Biochemistry,* **6** (1967) 906–920.
36. J. T. Park and M. Johnson, *J. Biol. Chem.*, **179** (1949) 585–592.
37. R. E. Strange and J. F. Powell, *Biochem. J.,* **58** (1954) 80–86.
38. R. E. Strange and L. H. Kent, *Biochem. J.,* **71** (1959) 333–338.
39. H. P. Browder, W. A. Zygmut, J. R. Young, and P. A. Tavormina, *Biochem. Biophys. Res. Commun.,* **19** (1965) 383–388.
40. C. A. Schindler and V. T. Schuhardt, *Biochim. Biophys. Acta,* **97** (1965) 242–250.
41. D. J. Tipper and J. L. Strominger, *Biochem. Biophys. Res. Commun.,* **22** (1966) 48–53.
42. E. Ito and J. L. Strominger, *J. Biol. Chem.,* **235** (1960) 5–14.
43. E. Ito and J. L. Strominger, *J. Biol. Chem.,* **237** (1962) 2689–2695.
44. E. Ito and J. L. Strominger, *J. Biol. Chem..,* **237** (1962) 2696–2701.
45. S. G. Nathenson, J. L. Strominger, and E. Ito, *J. Biol. Chem.,* **239** (1964) 1773–1780.
46. Y. Higashi, J. L. Strominger, and C. C. Sweeley, *Proc. Natl. Acad. Sci. USA,* **57** (1967) 1878–1884.
47. A. Wright, M. Dankert, P. Fennessey, and P. W. Robbins, *Proc. Natl. Acad. Sci. USA,* **57** (1967) 1798–1803.
48. J. S. Anderson, M. Matsuhashi, M. A. Haskins, and J. L. Strominger, *Proc. Natl, Acad. Sci. USA,* **53** (1965) 881–889.
49. J. S. Anderson and J. L. Strominger, *Biochem. Biophys. Res. Commun.,* **21** (1966) 516–521.
50. J. S. Anderson P. M. Meadow, M. A. Haskins, and J. L. Strominger, *Arch. Biochem. Biophys.,* **116** (1966) 487–515.
51. J. S. Anderson, M. Matsuhashi, M. A. Haskins, and J. L. Strominger, *J. Biol. Chem.* **242** (1967) 3180–3190.
52. M. Matsuhashi, C. P. Dietrich, and J. L. Strominger, *Proc. Natl. Acad. Sci. USA,* **54** (1965) 587–594.
53. R. M. Bumsted, J. L. Dahl, D. Söll, and J. L. Strominger, *J. Biol. Chem.,* **243** (1968) 779–782.
54. G. Siewart and J. L. Strominger, *J. Biol. Chem.,* **243** (1968) 783–790.
55. F. Fiedler and L. Glaser, *Biochim. Biophys. Acta,* **300** (1973) 467–485.
56. J. B. Ward and H. Perkins, *Biochem. J.,* **135** (1973) 721–728.
57. E. M. Wise and J. T. Park, *Proc. Natl. Acad. Sci. USA,* **54** (1965) 75–81.
58. D. J. Tipper and J. L. Strominger, *Proc. Natl. Acad. Sci. USA,* **54** (1965) 1133–1141.
59. J. M. Ghuysen, *Bacteriol. Rev.,* **32** (1968) 425–464.
60. A Wright, M. Dankert, and P. W. Robbins, *Proc. Natl. Acad. Sci. USA,* **54** (1965) 235–241.
61. M. Dankert, A. Wright, W. S. Kelley, and P. W. Robbins, *Arch. Biochem. Biophys.,* **116** (1966) 425–435.
62. P. W. Robbins, A. Wright, and M. Dankert, *J. Gen. Physiol.,* **49** (1966) 331–346.
63. M. J. Osborn and I. M. Weiner, *J. Biol. Chem.,* **243** (1968) 2631–2639.
64. J. L. Kent and M. J. Osborn, *Biochemistry,* **7** (1968) 4419–4422.
65. D. Bray and P. W. Robbins, *Biochem. Biophys. Res. Commun.,* **28** (1967) 334–339.

66. P. W. Robbins, A. Wright, and M. Dankert, *Science,* **158** (1967) 1536–1542.

67. F. A. Troy, I. K. Vijay, and N. Tesche, *J. Biol. Chem.,* **250** (1975) 156–163.

68. L. Masson and B. E. Holbein, *J. Bacteriol.,* **161** (1985) 861–867.

69. M. M. Berger and L. Glaser, *Biochim. Biophys. Acta,* **64** (1962) 575–584.

70. M. M. Berger and L. Glaser, *J. Biol. Chem.,* **239** (1964) 3168–3177.

71. L. Glaser, *J. Biol. Chem.,* **239** (1964) 3178–3188.

72. N. L. Blumson, L. J. Douglas, and J. Baddiley, *Biochem. J.,* **100** (1966) 26C.

73. L. J. Douglas and J. Baddiley, *FEBS Lett.,* **1** (1968) 114–116.

74. D. Brooks and J. Baddiley, *Biochem. J.,* **115** (1969) 307–314.

75. J. F. Robyt, *Adv. Carbohydr. Chem. Biochem.,* **51** (1995) 133–168.

76. J. F. Robyt, B. Kimble, and T. F. Walseth, *Arch. Biochem. Biophys.,* **165** (1974) 634–640.

77. J. F. Robyt and P. J. Martin, *Carbohydr. Res.,* **113** (1983) 301–315.

78. V. K. Parnaik, G. A. Luzio, D. A. Grahame, S. L. Ditson, and R. M. Mayer, *Carbohydr. Res.,* **121** (1983) 257–268.

79. G. Mooser, S. A. Hefta, R. J. Paxton, J. E. Shively, and T. D. Lee, *J. Biol. Chem.,* **266** (1991) 8916–8922.

80. J. F. Robyt and H. Taniguchi, *Arch. Biochem. Biophys.,* **174** (1976) 129–135.

81. C. S. Hanes, *Proc. Royal Soc. B,* **129** (1940) 174–208.

82. M. A. Swanson and C. F. Cori, *J. Biol. Chem.,* **172** (1948) 815–829.

83. W. E. Trevelyan, P. F. E. Mann, and J. S. Harrison, *Arch. Biochem. Biophys.,* **39** (1952) 419–427.

84. T.-T. Liu and J. C. Shannon, *Plant Physiol.,* **67** (1981) 525–533.

85. D. Stetten, Jr. and M. R. Stetten, *Physiol. Rev.,* **40** (1960) 513–519.

86. E. Recondo and L. F. Leloir, *Biochem. Biophys. Res. Commun.,* **6** (1961) 85–88.

87. R. B. Frydman, *Arch. Biochem. Biophys.,* **102** (1963) 242–251.

88. C. E. Cardini and R. B. Frydman, *Methods Enzymol.,* **8** (1966) 387–390.

89. E. Slabnik and R. B. Frydman, *Biochem. Biophys. Res. Commun.,* **38** (1970) 709–712.

90. C. R. Krisman and R. Barengo, *Eur. J. Biochem.,* **52** (1975) 117–123.

91. R. Barengo and C. R. Krisman, *Biochim. Biophys. Acta,* **540** (1978) 190–197.

92. R. Barengo, M. Flawia, and C. R. Krisman, *FEBS Lett.,* **53** (1975) 274–277.

93. N. A. Butler, E. Y. C. Lee, and W. J. Whelan, *Carbohydr. Res.,* **55** (1977) 73–82.

94. L. D. Kennedy, B. R. Kirkman, J. Lomako, I. R. Rodriguez, and W. J. Whelan, in *Membranes and Muscle* (M. C. Berman, W. Gevers, and L. H. Opie, eds.) pp. 65–84 ICSU Press, Oxford (1985).

95. J. Lomako, W. M. Lomako, and W. J. Whelan, *Biochem. Int.,* **21** (1990) 251–260.

96. J. Lomako, W. M. Lomako, and W. J. Whelan, *FASEB J.,* **2** (1988) 3097–3103.

97. J. Lomako, W. M. Lomako, W. J. Whelan, R. S. Dombro, J. T. Neary, and M. D. Norenberg, *FASEB J.,* **7** (1993) 1386–1393.

98. L. C. Gahan and H. E. Conrad, *Biochemistry,* **7** (1968) 3979–3984.

99. N. H. Behrens and L. F. Leloir, *Proc. Natl. Acad. Sci. USA,* **66** (1970) 153–156.

100. D. Borovsky and W. J. Whelan, *Fed. Proc.,* **31** (1972) 477.

101. W. J. Whelan, *Biochem. J.,* **122** (1971) 609–615.

102. D. Borovsky, E. E. Smith, and W. J. Whelan, *Eur. J. Biochem.,* **62** (1976) 307–312.

103. J. F. Robyt, in *Starch, Chemistry and Technology,* 2nd. edn., pp. 87–123 (R. L. Whistler, J. N. BeMiller, and E. F. Paschall, eds.) Academic, San Diego (1984).

104. D. J. Manners, J. J. M. Rowe, and K. L. Rowe, *Carbohydr. Res.,* **8** (1968) 72–78.

105. A. D. Elbein, G. A. Barber, and W. Z. Hassid, *Plant Physiol.,* **43** (1968) 309–310.
106. G. A. Barber, A. D. Elbein, and W. Z. Hassid, *J. Biol. Chem.,* **239** (1964) 4056–4061.
107. L. Glaser, *J. Biol. Chem.,* **232** (1958) 627–636.
108. J. R. Colvin, *Nature,* **183** (1959) 1135–1137.
109. Y. Aloni, D. P. Delmer, and M. Benziman, *Proc. Natl. Acad. Sci. USA,* **77** (1982) 6448–6452.
110. R. Pont Lezica, P. A. Romero, and M. A. Dankert, *Plant Physiol.,* **58** (1976) 675–680.
111. L. Ielpi, R. Courso, and M. A. Dankert, *FEBS Lett.,* **130** (1981) 253–256.
112. L. Ielpi, R. Courso, and M. A. Dankert, *Biochem. Biophys. Res. Commun.,* **102** (1981) 1400–1408.
113. L. Ielpi, R. O. Courso, and M. A. Dankert, *Biochem. Int.,* **6** (1983) 323–333.
114. L. Ielpi, R. O. Courso, and M. A. Dankert, *J. Bacteriol.,* **175** (1993) 2490–2500.
115. H. Schachter and L. Roden, in *Metabolic Conjugation and Metabolic Hydrolysis,* pp. 1–55 (W. H. Fishman, ed.) Academic, New York (1973).
116. M. D. Schnider and P. W. Robbins, in *Basic Mechanisms of Cellular Secretion,* pp. 89–146 (A. R. Hand and C. Oliver, eds.) Academic, New York (1981).
117. E. J. McGuire and S. Roseman, *J. Biol. Chem.,* **242** (1967) 3745–3749.

10.16 References for Further Study

The Photosynthesis of Carbon Compounds, M. Calvin, W. A. Benjamin, New York (1962).
Harvesting the Sun, A. San Pietro, F. A. Geer, and T. J. Army, eds., Academic, New York (1967).
Photosynthesis, R. M. Devlin and A. V. Barker, Van-Nostrand-Reinhold, New York (1971).
"Biosynthesis of saccharides from glycopyranosyl esters of nucleotides ('sugar nucleotides')," E. F. Neufeld and W. Z. Hassid, *Adv. Carbohydr. Chem.,* **18** (1963) 309–356.
"Biosynthesis of saccharides from glycopyranosyl esters of nucleoside pyrophosphates ('sugar nucleotides')," H. Nikaido and W. Z. Hassid, *Adv. Carbohydr. Chem. Biochem.,* **26** (1971) 352–382.
"The lipid pathway of protein glycosylation and its inhibitors: the biological significance of protein-bound carbohydrates," R. T. Schwartz and R. Datema, *Adv. Carbohydr. Chem. Biochem.,* **40** (1982) 287–380.
"Biosynthesis of bacterial polysaccharide chains composed of repeating units," V. N. Shibaev, *Adv. Carbohydr. Chem. Biochem.,* **44** (1986) 277–340.
"Lipid-linked sugars as intermediates in the biosynthesis of complex carbohydrates in plants," R. Pont Lezica, G. R. Dalev, and R. M. Dey, *Adv. Carbohydr. Chem. Biochem.,* **44** (1986) 341–386.

Chapter 11

Biodegradation

11.1 Digestion of Starch

Starch is the storage form of the chemical energy derived from the energy of the sun and is found in many plants. When the plant material, storing the starch, is eaten by nonphotosynthesizing organisms, it is broken down by α-amylases. α-Amylases are very widely distributed and are produced by plants, animals, and microorganisms. In fact, α-amylase is probably a member of an ancient class of enzymes that was produced very early by living organisms. Most α-amylases act best on starch that has had its granules disrupted by heating (cooking).

In humans, as soon as the cooked starch enters the mouth, it encounters an α-amylase that is secreted in saliva. The salivary α-amylase acts on the starch at an optimum pH of 6.7 and 37°C. It catalyzes the hydrolysis of the α-1 → 4 glycosidic linkages of starch, producing primarily maltose, maltotriose, and maltotetraose. Very little, if any, glucose is formed. The food starch and α-amylase are quickly passed into the stomach where the pH is ≈2, and the action of the α-amylase stops. At this stage, only part of the starch is usually hydrolyzed and there are appreciable amounts of incompletely hydrolyzed starch and large-sized amylodextrins. Very little hydrolysis occurs in the stomach. After some time, the food material passes into the small intestine and is neutralized. An α-amylase from the pancreas is secreted into the small intestine and continues the hydrolysis of the starch. The products are essentially those of salivary α-amylase, namely maltose, maltotriose, and maltotetraose. D-Glucose is a very minor product, being formed from a very slow secondary reaction of maltotriose and maltotetraose [1]. There are also a series of α-limit dextrins that have four to eight D-glucose residues. These limit dextrins arise from hydrolysis of the linkages around the branch linkages of amylopectin [2,3]. α-Amylase can hydrolyze linkages only so close to the

328

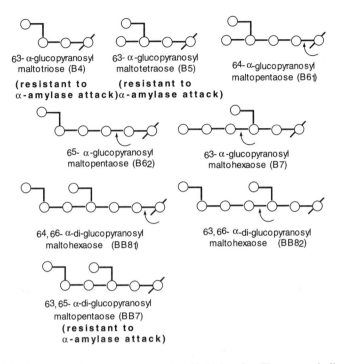

Figure 11.1. Structures of pancreatic α-amylase limit dextrins. The arrows indicate positions of slow hydrolysis.

branch point. The structures for these limit dextrins are given in Fig. 11.1. The smallest limit dextrin is a tetrasaccharide, 6^3-α-glucosyl maltotriose (B4), that is completely resistant to further hydrolysis by salivary- or pancreatic-α-amylase. Some of the higher α-limit dextrins, B6–BB8, are slowly hydrolyzed to give B4. These reactions are indicated by the arrows in Fig. 11.1. The action of the human α-amylases on starch can be formulated by the following reactions:

1. In the mouth

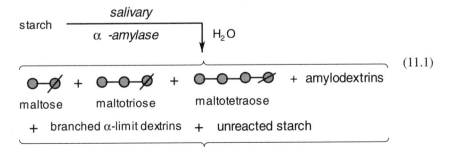

(11.1)

2. Small intestine

$$\left\{ \begin{array}{l} \text{amylodextrins +} \\ \text{unreacted starch} \end{array} \right\} \xrightarrow[\text{α-amylase}]{\text{pancreatic}} \Big\downarrow \text{H}_2\text{O} \tag{11.2}$$

$$\left\{ \begin{array}{l} \text{maltose + maltotriose + maltotetraose} \\ \qquad + \text{α-limit dextrins (B}_4\text{-BB}_8\text{)} \end{array} \right\}$$

The ultimate goal for starch-metabolizing organisms, however, is to convert starch into D-glucose. To convert the α-amylase products into D-glucose in the small intestine, two other enzymes are required, an *α-1,4-glucosidase* and an *α-1,6-glucosidase*. Glucosidases usually act on small dextrins or glycosides, but not on larger starch chains. These enzymes are secreted into the small intestine by the brush border cells that line the small intestinal wall [4].

The α-1,4-glucosidase converts maltose, maltotriose, and maltotetraose into D-glucose by successive action from the nonreducing ends. The α-1,6-glucosidase catalyzes the hydrolysis of the α-1 → 6 linkages of the α-limit dextrins, converting them into linear maltodextrins that are hydrolyzed by the α-1,4-glucosidase. The combined action of these two enzymes completes the task of converting the starch into utilizable D-glucose. The glucose is transported across the small intestine wall, against a concentration gradient, by a specific transporting protein called *glucopermease*. The action of the α-glucosidases in the small intestine is summarized in reactions (11.3) and (11.4).

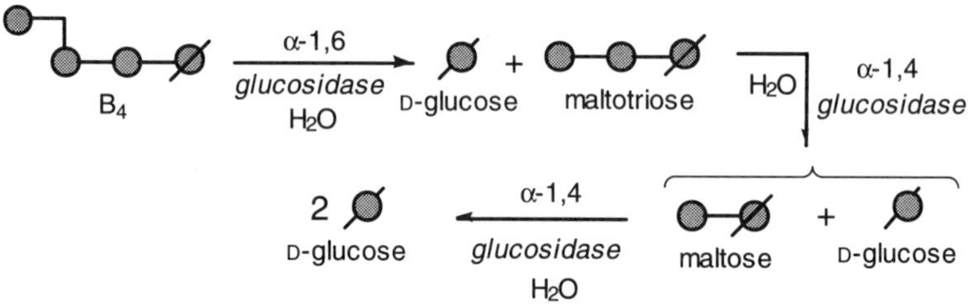

$$(11.3)$$

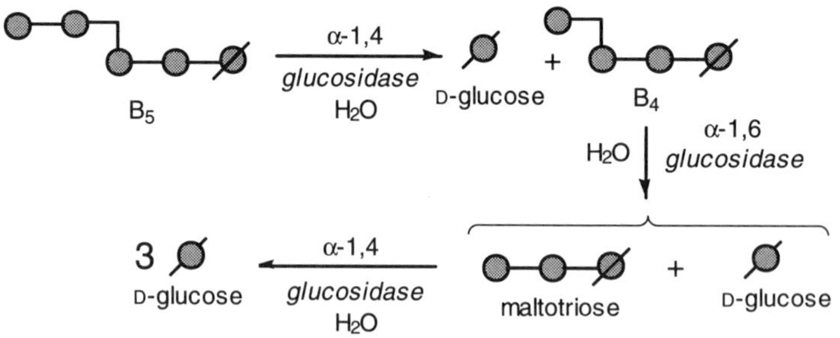

$$(11.4)$$

In humans the D-glucose is actively transported across the small intestine wall into the blood stream. This blood glucose is converted into liver or muscle glycogen for use in maintaining blood glucose levels (liver glycogen) or in obtaining immediate energy for muscle movement (muscle glycogen). Excess D-glucose is converted into fat and stored.

α-Amylases hydrolyze the α-1 → 4 glycosidic linkage, producing products that have the α-configuration at the point of hydrolysis. They are endo-acting enzymes that attack the polymeric starch chains in the inner parts of the polymer, producing amylodextrin fragments. With salivary and pancreatic α-amylases, the low molecular weight products, maltose and maltotriose, however, are produced immediately. It has been shown that these very small products are produced by what is called *multiple attack* [5,6]. A comparison of the rate of decrease of the blue iodine color of an amylose substrate with the rate of increase of the reducing value for α-amylases from different sources showed that different curves were obtained for the different α-amylases, indicating that different α-amylases had different degrees of multiple attack. The α-amylases also had curves that were different from the curve produced by acid-catalyzed hydrolysis [5]. The latter should be close to what would be expected for random hydrolysis. In the multiple attack mechanism, once the enzyme forms a complex with the substrate and produces the first hydrolytic cleavage, one of the fragments remains with the enzyme, most probably by the C-1 of the reducing end glucose residue, forming a covalent linkage with the enzyme [7]. The covalent linkage is hydrolyzed and the amylose fragment then realigns itself at the active site, filling the empty glucose binding sites that were vacated by the other amylose fragment. Hydrolysis occurs again, but this time only a small fragment (maltose or maltotriose) is produced. These events occur several times, forming maltose, maltotriose, or maltotetraose. Porcine pancreatic α-amylase was shown to have an average of seven hydrolytic events per enzyme-amylose encounter [5]. Human pancreatic α-amylase was similar, with seven to eight hydrolytic events per encounter. Human salivary α-amylase was shown to have an average of three hydrolytic events per encounter [5]. Thus, the action of α-amylases is far from a random process, producing specific low molecular weight maltodextrin products by multiple attack. The maltodextrin products are characteristic of the source of the enzyme, but none of the α-amylases form much D-glucose.

Other carbohydrates found in foods, such as sucrose and lactose, are hydrolyzed by specific enzymes, β-*fructosidase (invertase)* and β-*galactosidase (lactase)* that are located in the brush border cells of the human small intestine. There are specific *permeases* in the small intestine wall for transporting D-galactose and D-fructose into the blood.

11.2 Hydrolysis of Starch by Microorganisms

Many kinds of microorganisms produce α-amylases. The enzymes are extracellular and are secreted into the environment of the organism for hydrolysis of starch. One of the early microbial α-amylases to be studied in some detail was *Bacillus amyloliquefaciens* α-amylase (formerly known as *Bacillus subtilis* liquefying α-amylase). The amylase has an unusual action pattern. It produces two categories of products, a low molecular weight product, maltotriose, and higher molecular weight products, maltohexaose and maltoheptaose [8]. The latter two products appeared to be primary end products, although maltoheptaose was slowly hydrolyzed further to give maltohexaose plus glucose, and maltohexaose was hydrolyzed very slowly to give maltopentaose plus glucose. Maltopentaose was not hydrolyzed at all [8]. Its action on branched compounds was much more restrictive than the salivary or pancreatic α-amylases. For example, it produced only maltohexaose and maltoheptaose from the outer chains of amylopectin and glycogen. Maltotriose was produced from the interior chains between the branch linkages of amylopectin and was not formed at all from glycogen [8].

A number of bacterial amylases have been found that produce very specific, single products. *Bacillus polymyxa* produces a β-amylase that has an exo-mechanism specifically producing β-maltose from the nonreducing ends of the starch chains [9]. Before the discovery of *B. polymyxa* β-amylase, β-amylases were thought to be primarily plant enzymes. They have been characterized from sweet potatoes, soybeans, and barley, where they play important roles in modifying starch during cooking or fermentation. For example, sweet potatoes are not sweet until they are cooked. The heating activates sweet potato β-amylase, and the starch is converted into β-maltose, which is much sweeter than starch. The β-amylase in barley and soybean converts the starch in these materials into β-maltose that is then readily hydrolyzed by an α-glucosidase to give D-glucose. The D-glucose is fermented by yeasts and fungi, respectively, to make beer and soy sauce.

There is an exo-acting bacterial α-amylase, elaborated by *Pseudomonas stutzeri*, that exclusively produces maltotetraose from starch [10]. The enzyme hydrolyzes maltotetraose units from the nonreducing ends of the starch chains, and, like β-amylase, it will not bypass the α-1 → 6 branch linkages of amylopectin. From amylopectin, it gives 42% α-maltotetraose and 58% limit dextrin, which is a relatively high molecular weight product with α-1 → 6 branch linkages at the chain ends [10]. Another exo-acting amylase is elaborated by *Aerobacter aerogenes* and exclusively produces maltohexaose from starch [11,12]. A strain of *B. subtilis* elaborates an α-amylase that produces high yields of maltotriose [13], and a strain from *Bacillus circulans* elaborates an α-amylase that produces maltotetraose and maltopentaose [14]. *Bacillus licheniformis* elaborates an α-amylase that is thermophilic and has an optimum temperature of 70–90°C [15]. It gives high yields ($\approx$30%) of maltopentaose.

Pseudomonas amyloderamosa elaborates an *isoamylase* that specifically hydrolyzes the α-1 → 6 branch linkages of starch [16–20]. Another enzyme that also will hydrolyze the α-1 → 6 branch linkages of starch is pullulanase, which is elaborated by a strain of *A. aerogenes* [20]. This enzyme is a hydrolase that specifically hydrolyzes the α-1 → 6 linkages of the linear polysaccharide, pullulan [21,22]. Although it will also hydrolyze the α-1 → 6 branch linkages of amylopectin, it hydrolyzes the α-1 → 6 branch linkage of chains longer than 2–3 glucose residues much more slowly than *P. amylodermosa* isoamylase [20]. Isoamylases are α-1 $\to$ 6 starch debranching enzymes. They were first recognized in plants and were isolated from broad beans [23].

α-Amylases have also been found in plants, although they don't seem to be as widely distributed in plants as they are in the animal and microbial kingdoms. Barley malt α-amylase has been characterized and found to have a product specificity similar to that of *B. amyloliquefaciens* α-amylase and to produce maltotriose, maltohexaose, and maltoheptaose as the primary products [24].

Fungi, such as *Aspergillus oryzae*, elaborate an α-amylase [25]. Many fungi, such as *Aspergillus niger* [26], *Aspergillus awamori* [27,28], *Rhizopus delemar* [29], and *Rhizopus niveus* [30], elaborate another kind of starch-hydrolyzing enzyme called *glucoamylase*. Glucoamylase is an exo-acting enzyme that hydrolyzes glucose residues from the nonreducing ends of starch chains, giving β-D-glucopyranose. In contrast to other types of amylases, glucoamylases can catalyze the hydrolysis of both α-1 $\to$ 4 and α-1 $\to$ 6 glycosidic linkages, although at different rates. The α-1 $\to$ 4 linkage is hydrolyzed approximately 600 times faster than the α-1 $\to$ 6 linkage. Glucoamylases can, thus, completely convert starch into D-glucose, and they have attained a central role in the industrial conversion of starch into high-glucose syrups. The other α-amylases are also used industrially. *B. licheniformis* α-amylase is used at high temperatures to produce maltodextrins. *P. stutzeri* maltotetraose α-amylase is used to a limited extent to produce specialized maltotetraose syrups.

11.3 Biodegradation of Starch and Dextran to Cyclodextrins

While the formation of cyclodextrins might be considered a biosynthetic process, their formation is actually the degradation of starch and dextran. *Cyclodextrin glucanosyltransferase* (CGTase) catalyzes the breakdown of starch into cyclic, nonreducing α-1 $\to$ 4-linked maltodextrins (see section 8.1 in Chapter 8). There are several different enzymes that are elaborated by different species of bacteria. The first enzyme of this type to be studied was elaborated by *B. macerans* [31]. The enzyme is exo-acting and catalyzes an intramolecular transfer of α-1 $\to$ 4 glycosidic linkages of starch to form cyclic α-1 $\to$ 4–linked maltodextrins containing six, seven, and eight D-glucose residues [32]. The primary product of this enzyme is cyclomaltohexaose. CGTases have also been reported from *Bacillus megaterium* [33] and *B. circulans* [34]. Their primary product is cyclomaltoheptaose. A CGTase elaborated by *Brevibacterium* sp. produces cyclomaltooctaose as the primary product [35].

The CGTases also catalyze acceptor or transfer reactions between the cyclic maltodextrins and a carbohydrate acceptor [31,32]. In these acceptor reactions, the cyclodextrin ring is opened and the acceptor is specifically added to the reducing end of the maltodextrin chain that is formed from the cyclodextrin. The reaction is as follows:

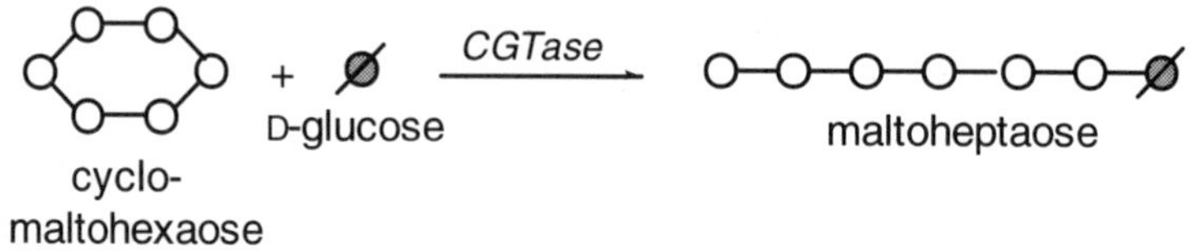

$$\text{cyclomaltohexaose} + \text{D-glucose} \xrightarrow{\text{CGTase}} \text{maltoheptaose} \qquad (11.5)$$

The maltodextrin acceptor product undergoes further disproportionation reactions in which two linear chains interact to transfer part of one chain to the other chain, as formulated in reaction (11.6).

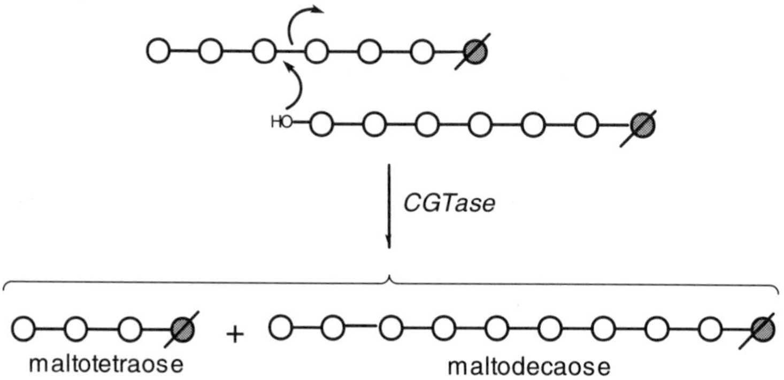

$$(11.6)$$

maltotetraose + maltodecaose

Several different disproportionation reactions can occur to give a wide range of maltodextrin products with different chain lengths, as shown in the following reactions:

$$G_7 + G_7 \xrightarrow{\text{CGTase}} \begin{cases} G_1 + G_{13} \\ G_2 + G_{12} \\ G_3 + G_{11} \\ G_4 + G_{10} \\ G_5 + G_9 \\ G_6 + G_8 \end{cases} \begin{array}{l} \text{various disproporionation} \\ \text{products from reaction of two} \\ \text{maltoheptaose molecules} \end{array} \qquad (11.7)$$

Each of the products of the reactions of 11.7 can itself undergo disproportionation to produce a complex mixture of homologous maltodextrins. A consequence of this disproportionation reaction is that, if the starting reactants are cyclomaltodextrins and ^{14}C-D-glucose, all of the resulting maltodextrins are specifically labeled in the reducing-end glucose residue [36]. Acceptors other than D-glucose can also be used [37]. For example, isomaltose, sucrose, or panose will give products with these acceptors attached at the reducing end of the maltodextrin chains.

The cycloisomaltodextrins (see section 8.4 in Chapter 8) are produced from B-512F dextran by the action of *cycloisomaltodextrin glucanosyltransferase* (CIG-Tase), elaborated by a *Bacillus* sp. Cycloisomaltodextrins with seven, eight, and nine glucose residues are formed, with cycloisomaltoheptaose as the primary product [38,39]. It is reported that this enzyme also catalyzes acceptor and disproportionation reactions, but they have not been described in any detail. The reaction mechanism seems to be similar to that of CGTase in which the CIGTase is exo-acting, catalyzing an intramolecular transfer of α-1 $\rightarrow$ 6 glucosidic linkages of dextran to form cyclic isomaltodextrins.

11.4 Biodegradation of Liver and Muscle Glycogen

Liver glycogen is degraded by liver phosphorylase that catalyzes the reaction between the glucose units of the nonreducing ends of glycogen and inorganic phosphate (Pi) to give α-G-1-P (reaction 11.8) [40]. The α-G-1-P is converted into D-glucose-6-phosphate (G-6-P) by the enzyme *phosphoglucomutase* (reaction 11.9). This is then hydrolyzed by liver *phosphatase* to give D-glucose (reaction 11.10), which goes into the blood stream to maintain normal blood glucose levels. The process is under strict hormone control (insulin and glucagon) to keep the concentration of glucose in the blood relatively constant.

Skeletal muscle glycogen is degraded in a similar manner, but the phosphatase enzyme is missing in skeletal muscle, and the end product is G-6-P, which is further broken down to give the muscles chemical energy in the form of ATP (see section 11.6).

$$G_n \ + \ m \ Pi \ \xrightarrow{\ phosphorylase\ } \ G_{n-m} \ + \ m \ \alpha\text{-G-1-P} \tag{11.8}$$

glycogen   glycogen chain
                                                             shortened by m
                                                             Glc units

$$\alpha\text{-G-1-P} \ \xrightarrow[\ mutase\ ]{\ phosphogluco\ } \ \text{G-6-P} \tag{11.9}$$

$$\text{G-6-P} \ + \ H_2O \ \xrightarrow[\ phosphatase\ ]{\ liver\ } \ \text{blood Glc} \tag{11.10}$$

Phosphorylase does not bypass α-1 $\rightarrow$ 6 branch linkages. It does not even remove all of the glucose residues close to the branch linkage. Phosphorylase action stops four glucose residues from the branch point on the two chains (the A-chain and the B-chain) [41,42]. See Fig. 11.2 for the structure of the phosphorylase limit dextrin. Another enzyme, *glycogen debranching enzyme,* is required for further action of phosphorylase. This is an unusual enzyme that has two types of activities. It first catalyzes the transfer of a maltotriosyl unit from the A-chain to the B-chain, leaving a single glucosyl residue linked α-1 $\rightarrow$ 6 to the B-chain. The enzyme then hydrolyzes the α-1 $\rightarrow$ 6 linkage of this glucose residue, removing the impediment for the action of phosphorylase [43,44] (see Fig. 11.2).

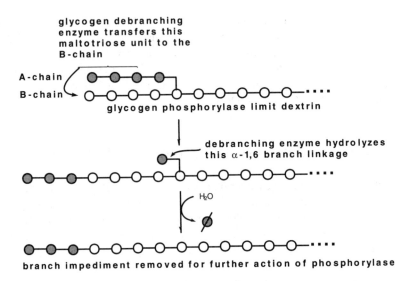

Figure 11.2. Phosphorylase limit dextrin of glycogen and the action of glycogen debranching enzyme.

11.5 Biodegradation of Cellulose and Related Polysaccharides

The highly intermolecular hydrogen-bonded cellulose is very water insoluble and resistant to enzyme degradation. Nevertheless, there are a number of fungi (*Aspergillus niger, Myrothecium verrucaria, Trichoderma viride,* and *Trich. ressi*) and some bacteria (*Cellvibrio gilvus* and *Pseudomonas* sp.) that secrete cellulases that hydrolyze cellulose [45–47]. Some of the microorganisms that secrete cellulose-digesting enzymes are symbiotic in the digestive tracts of herbivores and termites. The digestion of fibrous plants, such as grasses, by herbivores is complex and takes place in multichambered stomachs. The cellulose of dead plants is digested by fungi and other microorganisms in a slow process that can take several years.

Digestion of cellulose involves an endo-acting cellulase that attacks the less-ordered, amorphous regions of the cellulose fiber bundle. The ends of the hydrolyzed cellulose chains are then hydrolyzed by an exo-acting *cellobiohydrolase* that removes cellobiose from the ends of the chains [48,49]. The cellobiose is then hydrolyzed to D-glucose by a *β-1,4-glucosidase*. The weakening of the fibrous bundle by the action of these enzymes apparently produces an unraveling of the chains, providing further action of the enzymes.

Xylans and substituted xylans are hydrolyzed by endo-β-1 → 4-D-xylanase. Reaction of xylanase with xylan and arabino-, glucurono-, and arabinoglucuronoxylan gives D-xylose, xylodextrins (d.p. = 2–4), and a range of substituted xylodextrins containing various monosaccharide branch units [50–52].

The bacterial cell wall peptidomurein is hydrolyzed by hen egg-white lysozyme [53] to give di-, tetra-, and octasaccharides by the specific hydrolysis of the glycosidic bond of *N*-acetyl-D-muramic acid [53]. Lysozyme will also hydrolyze the β-1 → 4 GlcNAc bond of chitin, but at a much lower rate. There are specific chitinases that will also hydrolyze chitin to give chitobiose and chitodextrins.

11.6 Chemical Energy from Carbohydrates

Glucose is the primary source of chemical energy for all nonphotosynthesizing living organisms. Much of this energy comes from the breakdown of starch by amylases and α-glucosidases, and the breakdown of cellulose by cellulases, although an immediate source of energy in skeletal muscles is obtained from the breakdown of muscle glycogen by phosphorylase to give α-G-1-P.

Before the production of molecular oxygen into the atmosphere by photosynthesis, organisms, mostly bacteria, had developed a mechanism for obtaining energy from glucose that was anaerobic. Some of the descendents of these bacteria are still around and are anaerobic or facultatively anaerobic. The process is known as *anaerobic glycolysis* and produces L-lactic acid as an end product that is excreted from the cell into the environment. Other organisms, mainly simple eukaryotes such as yeasts, modified the end product by adding a couple of enzymes that produce CO_2 and alcohol (ethanol) in what is known as a fermentation process. To start anaerobic glycolysis from D-glucose, energy must first be expended by two phosphorylations. D-Glucose is phosphorylated by ATP, catalyzed by *hexokinase,* to give D-glucose-6-phosphate (G-6-P). G-6-P is isomerized to D-fructose-6-phosphate that is then phosphorylated by ATP, catalyzed by *phosphofructokinase,* to give D-fructose-1,6-bisphosphate (see Fig. 11.3). At this point the D-fructose-1,6-bisphosphate is broken into two trioses, 3-phospho-D-glyceraldehyde and dihydroxyacetone phosphate, catalyzed by *aldolase* [54,55]. The substrate for aldolase is the open chain form of D-fructose-1,6-bisphosphate. Dihydroxyacetone phosphate is converted into 3-phospho-D-glyceraldehyde by *triose phosphate isomerase,* and the 3-phospho-D-glyceraldehyde is oxidized by *3-phospho-D-glyceraldehyde dehydrogenase* to give D-glycerate-1,3-bisphosphate. The oxidation is coupled with inorganic phosphate to produce a mixed acid–anhydride of a carboxyl group and a phospho group. This phosphocarboxylate is a high-energy phosphate that is used to make ATP. The reactions for anaerobic glycolysis, starting with D-glucose are given in Fig. 11.3. ATP is formed in two reactions (reactions 7 and 10 of Fig. 11.3). The formation of ATP by these types of reactions is called *substrate phosphorylation.* Because the glycolytic reactions involve a triose, 3-phospho-D-glyceraldehyde, and two of them are effectively formed from D-glucose, each of the reactions of glycolysis occurs twice for each D-glucose residue. Thus, the glycolytic reactions generate 4 ATP per D-glucose molecule. But before the glucose can enter glycolysis, two ATP must be used, thus giving a net of 2 ATP formed per D-glucose molecule. To keep glycolysis going, NAD^+

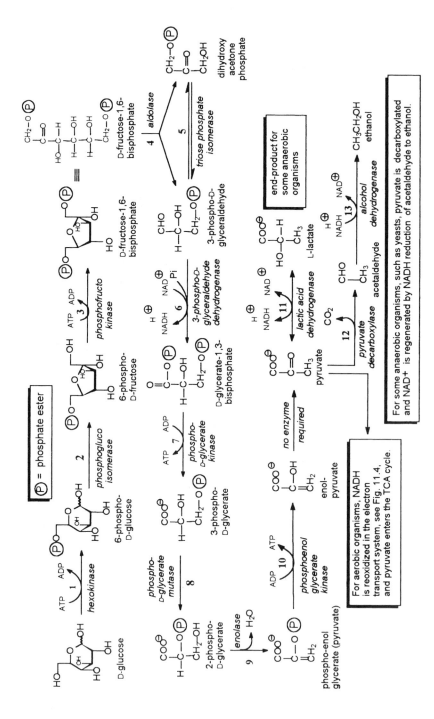

Figure 11.3. Reactions of anaerobic and aerobic glycolysis.

must be regenerated. This is accomplished by the reduction of pyruvate by NADH to give a waste product, L-lactic acid (reaction 11 of Fig. 11.3). In some anaerobic organisms, such as yeast, pyruvate is decarboxylated to produce acetaldehyde (reaction 12 of Fig. 11.3) and the NAD^+ regenerated by reduction of the acetaldehyde to ethanol (reaction 13 of Fig. 11.3).

As oxygen increased in the atmosphere due to an increase in photosynthesis, organisms developed an aerobic glycolytic process that was capable of producing 6 additional ATP by oxidizing NADH with molecular oxygen, for a net total of 8 ATP. This was four times the amount of chemical energy that could be obtained by anaerobic glycolysis. The increased amount of chemical energy that was available under aerobic conditions provided organisms a much broader scope for evolutionary expansion and gave more flexibility to what organisms could do.

To use molecular oxygen for the oxidation of NADH, the organisms developed an electron transport system, similar in many respects to the electron transport system that had been developed by photosynthetic organisms. This system took the reduced product, NADH, and reoxidized it with a flavoprotein enzyme that used a flavomononucleotide coenzyme (FMN). The oxidizing group has the same three-ring, fused heterocyclic system that FAD has (see Fig. 10.2B). The FMN becomes reduced to $FMNH_2$, a process similar to the reduction of FAD to $FADH_2$. The $FMNH_2$ is reoxidized by an enzyme using coenzyme Q (CoQ). This process passes electrons down the chain via oxidations and reductions. Two electrons are passed to two ferric ions of cytochrome b giving two ferrous ion, which are oxidized by two ferric ions of cytochrome $c1$, which in turn has its ferrous ions oxidized by two ferric ions of cytochrome c, and these ferrous ions are oxidized by two ferric ions of cytochrome oxidase. The last reaction is the oxidation of the two ferrous ions of cytochrome oxidase by molecular oxygen to give water. In the process of the oxidations and reductions of the electron transport chain, the two electrons that were removed from the triose phosphate substrate, slowly lose their energy. Some of this energy is coupled with ADP and Pi to form 3 ATP molecules that each have about 10 kcal/mol in the terminal pyrophosphate group. The reactions of the electron transport system are called *oxidative phosphorylation* when the energy of the electrons are coupled with the formation of ATP. The reactions of oxidative phosphorylation are summarized in Fig. 11.4. The ATP that is formed is used to run all of the various functions of the cell that are involved in the process of living. A fascinating aspect of this is that ATP is generated by these same reactions by all organisms, and ATP is the universal energy carrier and donor for all organisms. We have seen some of these energy-requiring functions in Chapter 10 in which we considered the biosynthesis of oligosaccharides and polysaccharides that required the formation of nucleotide diphospho sugars as the activated donors of monosaccharides for their biosynthesis. The energy for the formation of the nucleotide diphospho sugars is provided by ATP (see reactions 10.7 and 10.8 in Chapter 10).

Aerobic glycolysis, however, still had not oxidized the carbohydrate to CO_2 and H_2O. As the complexity of life-forms continued to increase, there was a need

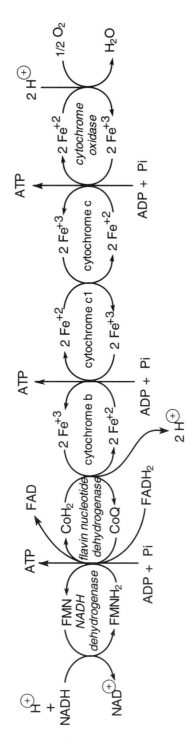

Figure 11.4. The reactions of electron transport and oxidative phosphorylation.

for additional energy. This energy could be provided by the complete oxidation of the carbohydrate. The end product of aerobic glycolysis is pyruvate. An enzyme system evolved that decarboxylated and oxidized pyruvate to give CO_2 and acetyl coenzyme A (acetyl CoA) [56–58]. In this reaction, the acetyl group forms a thioester with coenzyme A. This ester enters the *citric acid cycle,* sometimes called the *Kreb's cycle,* by condensing with oxaloacetate to give citrate [59,60]. In the process, the thioester is hydrolyzed, driving the reaction. A series of reactions then take place in which there are three more oxidations by NAD^+ requiring enzymes [60]. The reduced NADH coenzymes are reoxidized to NAD^+ in the electron transport system, each generating 3 ATP. Considering the oxidative decarboxylation reaction of pyruvate (reaction 1 of Fig. 11.5), there are four NADH that are reoxidized in the electron transport system. In the oxidative decarboxylation of α-ketoglutarate, a thioester, succinyl CoA, is formed [57]. The hydrolysis of this thioester provides energy for the formation of GTP, which in turn can produce ATP. There then is an FAD-requiring enzyme that oxidizes succinate to fumarate, giving $FADH_2$. $FADH_2$ enters the electron transport system and is reoxidized by a CoQ dehydrogenase to give 2 ATP. Oxaloacetate is ultimately regenerated by NAD^+ oxidation of L-malate (reaction 11 of Fig. 11.5) for entry into another turn of the cycle. From the entry of pyruvate into the cycle and the formation of succinyl-CoA, 3 CO_2 are produced from the various oxidations and decarboxylations [60]. The reactions of the citric acid cycle are summarized in Fig. 11.5.

Early in glycolysis (from the action of *aldolase*), there are two three-carbon fragments produced from D-glucose. For the complete oxidation of D-glucose to $CO_2 + H_2O$, the two three-carbon fragments must each go through the two processes of glycolysis and the citric acid cycle. Thus, by two turns of the citric acid cycle that involve two three-carbon fragments, 6 CO_2 are produced. Two three-carbon fragments generate 30 ATP from the oxidative decarboxylation of pyruvate and the four oxidations occurring in the cycle. There is a net of 8 ATP formed from D-glucose in aerobic glycolysis, thus giving a net total of 38 ATP formed for the complete oxidation of D-glucose to carbon dioxide and water.

As with the photosynthetic process, we see that the respiratory process is also more complicated than the simple reaction (reaction 1.3) that we wrote in Chapter 1. The energy contained in the carbon-carbon bonds of carbohydrates that are formed in photosynthesis are built up in a series of small steps (reactions) that conserve the energy from the sun. This conserved chemical energy is also obtained in a series of small chemical degradation reactions that make available utilizable chemical energy in the form of ATP.

The two processes of photosynthesis and respiration have common features such as the oxidation and reduction reactions that use NAD^+/NADH and FAD/$FADH_2$ and the electron transport systems, using CoQ and cytochromes to generate energy in the form of ATP, and, in the case of photosynthesis, the generation of NADPH for reducing CO_2. Although photosynthesis and respiration have different purposes, the one being to conserve the energy of the sun, and the other being to obtain chemical energy, they balance each other and are tied together.

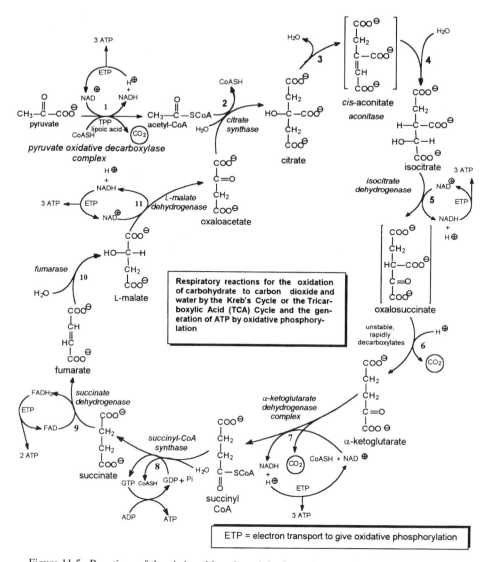

Figure 11.5. Reactions of the citric acid cycle and the formation of ATP and carbon dioxide.

The photosynthetic process uses carbon dioxide and produces molecular oxygen, while the respiration process uses molecular oxygen and produces carbon dioxide. Both processes involve carbohydrates, the one synthesizing and the other degrading.

11.7 Literature Cited

1. J. F. Robyt and D. French, *J. Biol. Chem.,* **245** (1970) 3917–3827.
2. K. Kainuma and D. French, *FEBS Lett.,* **5** (1969) 257–260.
3. K. Kainuma and D. French, *FEBS Lett.,* **6** (1970) 182–186.
4. G. M. Gray, B. C. Lally, and K. A. Conklin, *J. Biol. Chem.,* **254** (1978) 6038–6042.
5. J. F. Robyt and D. French, *Arch. Biochem. Biophys.,* **122** (1967) 8–16.
6. J. F. Robyt and D. French, *Arch. Biochem. Biophys.,* **138** (1970) 662–670.
7. B. Y. Tao, P. J. Reilly, and J. F. Robyt, *Biochim. Biophys. Acta,* **995** (1989) 214–220.
8. J. F. Robyt and D. French, *Arch. Biochem. Biophys.,* **100** (1963) 451–467.
9. J. F. Robyt and D. French, *Arch. Biochem. Biophys.,* **104** (1964) 338–345.
10. J. F. Robyt and R. J. Ackerman, *Arch. Biochem. Biophys.,* **145** (1971) 105–114.
11. K. Kainuma, S. Kobayashi, T. Ito, and S. Suzuki, *FEBS Lett.,* **26** (1972) 281–285.
12. K. Kainuma, K. Wako, S. Kobayashi, A. Nogami, and S. Suzuki, *Biochim. Biophys. Acta,* **410** (1975) 333–340.
13. Y. Takasaki, *Agric. Biol. Chem.,* **49** (1985) 1091–1097.
14. Y. Takasaki, *Agric. Biol. Chem.,* **47** (1983) 2193–2199.
15. F. J. Morgan and F. R. Priest, *J. Appl. Bacteriol.,* **50** (1981) 107–112.
16. T. Harada, K. Yokobayashi, and A. Misaki, *Appl. Microbiol.,* **16** (1968) 1493–1496.
17. K. Yokobayashi, A. Misaki, and T. Harada, *Agric. Biol. Chem.,* **33** (1969) 625–631.
18. K. Yokobayashi, A. Misaki, and T. Harada, *Biochim. Biophys. Acta,* **212** (1970) 458–464.
19. T. Harada, A. Misaki, H. Akai, K. Yokobayashi, and K. Sugimoto, *Biochim. Biophys. Acta,* **268** (1972) 497–505.
20. K. Kainuma, S. Kobayashi, and T. Harada, *Carbohydr. Res.,* **61** (1978) 345–357.
21. H. Bender and K. Wallenfels, *Biochem. Z.,* **334** (1961) 79–86.
22. B. J. Catley, J. F. Robyt, and W. J. Whelan, *Biochem. J.,* **100** (1966) 5p.
23. P. N. Hobson, W. J. Whelan, and S. Peat, *J. Chem. Soc.,* (1951) 1451–1455.
24. E. A. MacGregor and A. W. MacGregor, *Carbohydr. Res.,* **142** (1958) 223–236.
25. T. Suganuma, R. Matsuno, M. Ohinishi, and K. Hiromi, *J. Biochem.,* **84** (1978) 293–316.
26. J. H. Pazur and T. Ando, *J. Biol. Chem.,* **234** (1959) 1966–1971.
27. S. Ueda, *Bull. Agric. Chem. Soc. Jpn.,* **20** (1956) 148–155.
28. A. N. Savel'ev, V. R. Sergev, and L. M. Firsov, *Biochem. (USSR),* **47** (1982) 330–337.
29. K. Hiromi, *Biochem. Biophys. Res. Commun.,* **40** (1970) 1–11.
30. Y. Tsujisaka, J. Fukumoto, and T. Yamoto, *Nature,* **181** (1958) 94–96.
31. D. French, *Adv. Carbohydr. Chem.,* **12** (1975) 189–205.
32. J. F. Robyt, "Enzymes and Their Action on Starch", in *Starch: Chemistry and Technology,* 3rd edn. (R. L. Whistler and J. N. BeMiller, eds.) Academic, San Diego (1997).
33. S. Kitahata, N. Tsuyama, and S. Okada, *Agric. Biol. Chem.,* **38** (1974) 387–391.
34. S. Kitahata and S. Okada, *Denpun Kagaku,* **29** (1982) 13–18.

35. S. Mori, S. Hirose, O. Takaichi, and S. Kitahata, *Biosci. Biotech. Biochem.*, **58** (1994) 1968–1972.
36. J. F. Robyt, in *Biotechnology of Amylodextrin Oligosaccharides*, pp. 98–110 (R. B. Friedman, ed.) ACS Symposium Series 458, American Chemical Society, Washington, D.C. (1991).
37. E. Norberg and D. French, *J. Am. Chem. Soc.*, **72** (1950) 1202–1205.
38. T. Oguma, T. Horuichi, and M. Kobayashi, *Biosci. BioTech. Biochem.*, **57** (1993) 1225–1227.
39. T. Oguma, K. Tobe, and M. Kobayashi, *FEBS Lett.*, **345** (1994) 135–138.
40. E. G. Krebs and E. H. Fischer, *J. Biol. Chem.*, **216** (1955) 113–120.
41. W. J. Whelan and J. M. Bailey, *Biochem. J.*, **58** (1954) 560–566.
42. G. J. Walker and W. J. Whelan, *Biochem. J.*, **76** (1960) 264–268.
43. D. H. Brown and B. B. Illingworth, in *Control of Glycogen Metabolism*, pp. 139–150, (W. J. Whelan and M. P. Cameron, eds.) J & A Churchill, London (1964).
44. D. H. Brown and B. B. Illingworth, *Biochem. J.*, **100** (1966) 8p–9p.
45. S. P. Shoemaker and R. D. Brown, *Biochim. Biophys. Acta*, **523** (1978) 133–146; 147–161.
46. T. Kanda, K. Wakabayashi, and K. Nisizawa, *J. Biochem.*, **79** (1976) 977–988.
47. G. Okada, *J. Biochem.*, **80** (1976) 913–922.
48. K. E. Ericksson, B. Pettersson, and J. Westermark, *FEBS Lett.*, **49** (1974) 282–285.
49. M. Streamer, K. E. Ericksson, and B. Pettersson, *Eur. J. Biochem.*, **59** (1975) 607–613.
50. R. F. H. Dekker and G. N. Richards, *Adv. Carbohydr. Chem. Biochem.*, **32** (1976) 277–352.
51. T. E. Timell, *Adv. Carbohydr. Chem. Biochem.*, **20** (1965) 409–483.
52. K. C. B. Wilkie, *Adv. Carbohydr. Chem. Biochem.*, **36** (1979) 215–264.
53. D. M. Chipman and N. Sharon, *Science*, **165** (1969) 454–465.
54. A. Sols, in *Carbohydrate Metabolism and Its Disorders*, Vol. 1, Chap. 3 (F. Dickens, P. J. Randle, and W. J. Whelan, eds.) Academic, New York (1968).
55. E. B. Cunningham, *Biochemistry, Mechanisms of Metabolism*, Chap. 9, McGraw-Hill (1978).
56. L. J. Reed, *Acc. Chem. Res.*, **7** (1974) 40–46.
57. J. H. Collins and L. J. Reed, *Proc. Natl. Acad. Sci. USA*, **74** (1977) 4223–4227.
58. G. D. Grenville, in *Carbohydrate Metabolism and Its Disorders*, Vol. 1, Chap. 9 (F. Dickens, P. J. Randle, and W. J. Whelan, eds.) pp. 279–335, Academic New York (1968).
59. H. A. Krebs and W. A. Johnson, *Enzymologia*, **4** (1937) 148–156.
60. J. M. Lowenstein, in *Metabolic Pathways*, Vol. 1, 3rd edn., pp. 146–270 (D. M. Greenberg, ed.) Academic, New York (1967).

Chapter 12

Determinations

12.1 Determination of the Presence of Carbohydrate

The determination of the presence of carbohydrate in a sample is obtained by performing the Molisch test [1] (see section 3.4 in Chapter 3 for the chemistry involved in the Molisch test). The test is sensitive down to 10 μg/mL and is relatively broad but specific for all types of carbohydrates, with the exception of sugar alcohols, 2-deoxy sugars, and 2-amino-2-deoxy sugars or 2-acetamido-2-deoxy sugars.

12.2 Purification of Carbohydrates

The purification of carbohydrate in a sample depends on what else might be in the sample and how the sample was obtained. If the sample has been prepared by chemically modifying a known carbohydrate, the possible contaminants are the starting material and any side products or intermediates that might have been produced in the synthetic reactions. This type of sample can usually be purified by silica-gel column chromatography, using "flash chromatography" [2] or filtering column chromatography [3] and/or charcoal-celite column chromatography [4]. The latter is usually used for water-soluble compounds and the former for organic-solvent-soluble compounds, although silica-gel chromatography also can be used for water-soluble compounds. Purification is also possible by crystallization of the synthesized product.

If the sample has been obtained from a natural source, such as a cell or tissue extract, protein and/or nucleic acid might be present. The presence of protein and nucleic acid can be determined by measuring the absorbance at 280 nm and 254 nm, respectively. The nature of the carbohydrate is also important, that is, is it a low molecular weight material such as a mono-, di-, or trisaccharide, or is it a high molecular weight material such as an oligo- or polysaccharide? The presence of low molecular weight saccharide can be detected by thin-layer chromatography (TLC) [5] (see Fig. 12.1). If it is a low molecular weight saccharide, it can be separated from proteins and nucleic acids (polymeric substances) by precipitation of the polymers with two volumes of alcohol. The precipitate should be washed with 50% (v/v) alcohol to remove any occluded saccharide. The alcohol supernatant and washings are combined and evaporated to dryness, often resulting in a syrup. The product can then be dissolved in water for further study, or it can be treated with dry acetone several times to remove water, followed by a final treatment with anhydrous ethanol and drying in a vacuum oven to give a dry powder that can be easily weighed and handled.

If the sample is a polysaccharide or oligosaccharide of appreciable size (d.p. 20 or greater), it can be precipitated with two volumes of alcohol along with protein and nucleic acid. This precipitation removes the polysaccharide or oligosaccharide from salts and low molecular weight materials such as amino acids, peptides, and nucleotides. The precipitate can then be dissolved in water. A certain amount of protein can be denatured in the precipitation process, especially if the temperature is above 4°C, and can be removed. The soluble protein and nucleic acid can be removed from a neutral polysaccharide by ion-exchange chromatography on DEAE-cellulose or CM-cellulose. The neutral polysaccharide would pass through the column and be obtained free of the protein and nucleic acid contaminants. Protein and nucleic acid can also be removed by treatment with proteinases, for example, *pronase,* and nucleases, for example, *ribonuclease* for RNA and *deoxyribonuclease* for DNA. The polysaccharide or oligosaccharide can then be separated from the low molecular weight proteinase and nuclease products by precipitation with two volumes of alcohol.

If the polysaccharide is composed of uronic acid residues, it can be specifically precipitated by lowering the pH. Different pH values between 5 and 1 can be used to precipitate polysaccharides with different amounts and kinds of uronic acids [6]. Likewise, mixtures of neutral polysaccharides can sometimes be precipitated by differential alcohol precipitation in which the polysaccharides have different water solubilities due to differences in their degrees of branching [7]. Saccharides with certain specific structures can be selectively separated by using lectins (see section 12.10).

12.3 Monosaccharide Composition of Carbohydrates

The monosaccharide composition of the oligosaccharide or polysaccharide is determined by first hydrolyzing the sample with 10% (v/v) concentrated hydrochloric acid or trifluoroacetic acid at 100–120°C in a sealed ampule for 30–240 min. The monosaccharide composition is then analyzed by TLC [5] (see Fig. 12.1). Some laboratories might use high-performance liquid chromatography (HPLC) [8], although analysis by TLC uses much less sample and much simpler equip-

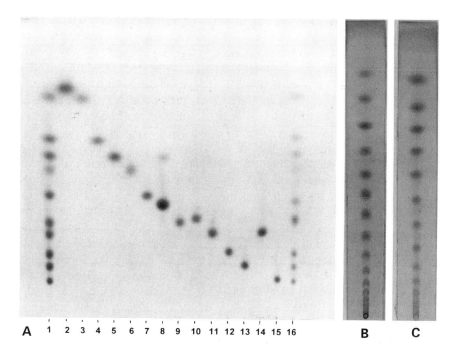

Figure 12.1. Thin layer chromatography of the separation of carbohydrates: **A,** chromatography of mono-, di-, and trisaccharides on Whatman K5 silica gel, using four ascents of acetonitrile/H$_2$O (85:15, v/v); (1 & 16), standards, (2), D-ribose, (3) D-xylose, (4) D-fructose, (5) D-xylose, (6) D-galactose, (7) sucrose, (8) turanose, (9) maltose, (10) cellobiose, (11) lactose, (12) α,α-trehalose, (13) melibiose, (14) melezitose, (15) raffinose; **B,** chromatography of maltodextrins (d.p. 1–13) on Whatman K5, using three ascents of acetonitrile/ethyl acetate/propanol-1/H$_2$O (85:20:50:50); **C,** chromatography of isomaltodextrins (d.p. 1–11) on Whatman K5, using acetonitrile/ethyl acetate/propanol-1/H$_2$O (85:20: 50:90).

ment and is much less expensive. Carbohydrates can be detected in the 100- to 200-ng range on silica gel TLC using 0.3% (w/v) *N*-naphthyl ethylenediamine in methanol containing 5% (v/v) sulfuric acid [9].

12.4 Determination of the Positions of the Glycosidic Linkage

The best method that has been developed for determining the positions of attachment of the glycosidic linkages of monosaccharide residues in oligo- and polysaccharides is methylation analysis [10]. Methylation analysis can also provide information about the type and degree of branching. It has been used for about 75 years. Initially the methylation reaction employed dimethyl sulfate as the methylating reagent. This resulted in a number of false determinations due to incomplete methylation. The samples had to be methylated three and four times, and even then methylation was sometimes incomplete. Methylation analysis was greatly improved by the development of the Hakomori procedure [11], which provides complete methylation in one reaction. The Hakomori reagent is prepared by treating dry dimethylsulfoxide with sodium hydride. The saccharide is dissolved in the Hakomori reagent, which contains methyl sulphinyl carbanions and forms polyalkoxide ions with the free hydroxyl groups on the saccharide. This mixture is then methylated with methyl iodide. The Hakomori reagent solubilizes most oligo- and polysaccharides. The methylated saccharide is then acid hydrolyzed, and the methylated monosaccharides are analyzed. In the 1950s, the primary method available for analysis of carbohydrate derivatives was gas chromatography (GS) [12]. To use GS, the carbohydrate sample has to be volatile. To convert the methylated monosaccharides into volatile derivatives, they were reduced and acetylated [13–15] and analyzed by GC. Later, when mass spectrometry (MS) was developed, the samples were analyzed by gas chromatography–mass spectrometry (GCMS), which still required volatile derivatives [13]. Because of the relatively large number of reactions involved, the initial sample had to be on a macro scale, since losses occurred with every step. Further, the instrumentation had to be dedicated and was expensive and not readily available to many laboratories. An alternate method of analysis has recently been developed in which the methylated monosaccharides are examined directly by TLC [16], eliminating the reduction and acetylation reactions and the need for expensive, dedicated equipment. Micro samples can be analyzed, and the procedure is relatively simple, fast, and inexpensive.

The kinds of glycosidic linkages in the saccharide are deduced by the types of methylated monosaccharides that are obtained. For example, if the saccharide contains glucose residues linked 1 → 4 with 1 → 6 branch linkages, three types of methylated glucose residues are obtained: 2,3,6-tri-O-methyl-D-glucose from the inner parts of the main chains; 2,3-di-O-methyl-D-glucose from the branched glucose residues; and 2,3,4,6-tetra-O-methyl-D-glucose from the ends of the chains. Usually the amount of di-O-methyl and tetra-O-methyl glucoses will equal each other. Some examples of the types of methylated glucoses that are obtained from saccharides linked in different ways are given in Fig. 12.2.

12.5 Determination of the Position of Substitution of Monosaccharide Residue(s) by Periodate Oxidation

Periodate oxidation has been used to determine structure, especially the position of the glycosidic linkage or the position of substitution on a monosaccharide residue. In Chapter 2, section 2.5 we saw how periodate oxidation was used to determine the size of the carbohydrate rings that were produced by the formation of intramolecular hemiacetals. As indicated in that section, periodate stoichiometrically cleaves carbon-carbon bonds of carbons that each have a hydroxyl group.

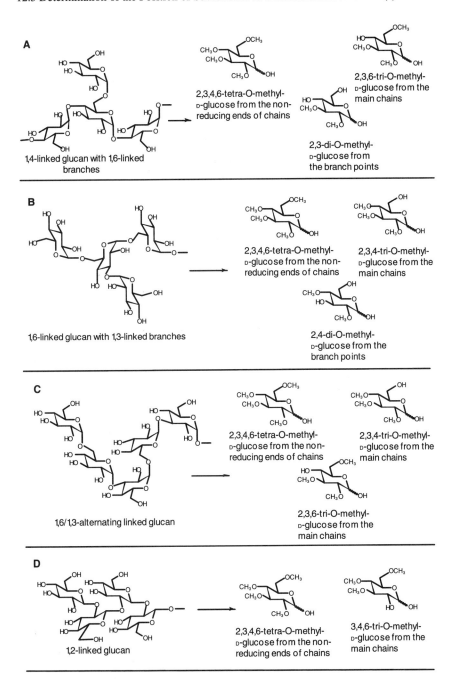

Figure 12.2. Products from the methylation analysis of glucans linked in different ways: **A**, 1 → 4-linked glucan with 1 → 6-linked branches; **B**, 1 → 6-linked glucan with 1 → 3-linked branches; **C**, 1 → 6/1 → 3-alternating-linked glucan; **D**, 1 → 2-linked glucan.

Periodate oxidation is relatively easy to perform, and the amount of periodate consumed is easy to determine spectrophotometrically by measuring the absorbance at 290 nm, using an absorbance coefficient of 0.22 mM/cm [17]. Sometimes formic acid is a product of the reaction, and it too is relatively easy to measure by acid/base titration. Different kinds of linkages in a polysaccharide or in an oligosaccharide consume different amounts of periodate and form different amounts of formic acid. From this, deductions can be made about the kinds of glycosidic linkage(s) present.

The amounts of periodate consumed and the amount of formic acid formed by glucans with different kinds of linkages are given as examples in Fig. 12.3. Glucose residues with $1 \rightarrow 3$ linkages do not undergo periodate oxidation because there are no vicinal hydroxyl groups; glucose residues with $1 \rightarrow 4$ linkages consume one mole of periodate per mole of glucose; glucose residues with $1 \rightarrow 2$ linkages also consume one mole of periodate per mole of glucose; and glucose residues with $1 \rightarrow 6$ linkages consume two moles of periodate and produce one mole of formic acid per mole of glucose. A glucan with an alternating structure of $1 \rightarrow 6$ and $1 \rightarrow 3$ glycosidic linkages would consume one mole of periodate per glucose residue (two from the glucose residue with the $1 \rightarrow 6$-linked residue and none from the $1 \rightarrow 3$-linked residue) and would produce one mole of formic acid for every two glucose residues. From this, one can deduce a structure containing the alternating $1 \rightarrow 6/1 \rightarrow 3$-linked glucan. A glucan with a sequence of two $1 \rightarrow 4$ linkages and one $1 \rightarrow 6$ linkage would consume four moles of periodate and produce one formic acid for every three glucose residues.

As can be seen from these examples, the periodate method is not an absolute method and gives us only an approximate structure. There are instances where glycans with different kinds of linkages or combination of linkages might give the same result. But, because the method is relatively fast and easy to perform, it is used in two principal ways: (1) to obtain a presumptive structure before other methods are employed and (2) to confirm a structure that has been obtained by using other methods.

12.6 Configuration of Glycosidic Linkages

The configuration of a glycosidic linkage is not obtained from a methylation analysis. The configuration can be estimated by the determination of the specific optical rotation of the saccharide. If the saccharide has a relatively high positive rotation, $[\alpha] \geq +100°$, the saccharide probably has a high number of α-linkages, and if the specific rotation is relatively low, $[\alpha] \leq 10°$, it probably has a high number of β-linkages. Maltose, for example, with one α-glycosidic linkage, has an equilibrium specific rotation of $+130.4°$, while cellobiose, with one β-glycosidic linkage, has an equilibrium specific rotation of $+35°$. The addition of one α-linkage (maltotriose) increases the specific rotation to $+160°$, and the addition of one β-linkage (cellotriose) decreases the specific rotation to $+22°$. Sucrose, a structure that has both an α-linkage and a β-linkage, has an intermediate specific rota-

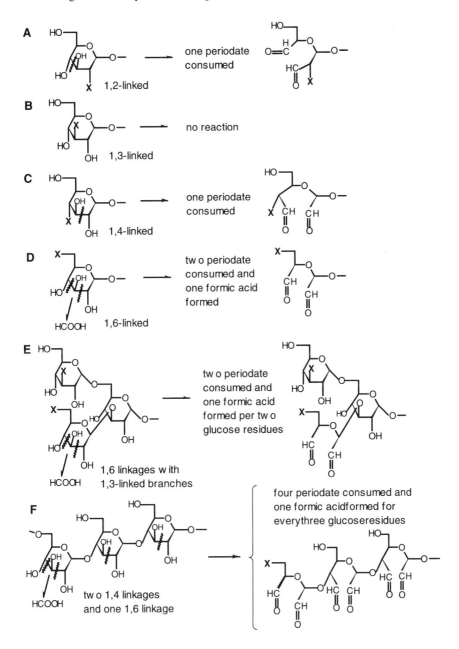

Figure 12.3. Products from the periodate oxidation of different linked or substituted monosaccharide residues. X, position of linkage or substitution; *wavy line* indicates carbon bonds cleaved by periodate.

tion of $+66.5°$. Polymers with a high number of α-linkages, such as starch, have a high specific rotation of $+220°$, and polymers with a high number of β-linkages, such as C-6–oxidized chitin or cellulose, have negative specific rotations of $-199°$ and $-45°$, respectively.

The absolute determination of the presence of α- or β-linkages can be made by ^{13}C-NMR because the carbons of each anomeric form give specific chemical signals [18]. The determination can sometimes also be made by using enzymes that are specific for the α- or β -linkages, if they are available and their specificity is known.

12.7 Sequence of Monosaccharides and Their Linkages

The sequence of monosaccharides and linkages in a saccharide is determined by obtaining and analyzing partially hydrolyzed fragments. The fragments can be obtained by partial acid hydrolysis and enzyme hydrolysis. Different kinds of linkages have different susceptibilities to acid hydrolysis. For example, $1 \rightarrow 6$ linkages are more stable than the secondary linkages of $1 \rightarrow 2$, $1 \rightarrow 3$, or $1 \rightarrow 4$, and β-linkages are more stable than α-linkages. The reverse is true for acetolysis in which $1 \rightarrow 6$ linkages are cleaved faster than the secondary linkages [19]. The formation of fragments by enzyme-catalyzed hydrolysis is important because the reactions take place under very mild conditions, with a certain specificity for hydrolyzing specific linkages, resulting in fragments whose structures are characteristic of the structure of the original saccharide. See the following section in which we give specific examples of the use of enzymes in determining saccharide structure.

12.8 Use of Enzymes in Determining Structure

The use of enzymes to characterize a polysaccharide or oligosaccharide involves the formation of specific products whose structures are further determined using other methods such as composition analysis, methylation analysis, NMR spectroscopy, and other enzymes. The products can be both of low molecular weight (mono-, di-, and trisaccharides) and high molecular weight (enzyme-resistant fragments and limit dextrins). Sometimes two or more enzymes can be used in a sequential manner and the products analyzed after each reaction. The same set of enzymes can also be used, but in a different sequence to give different kinds of products. Other times, two enzymes can be used together to determine some structural aspect that might not be evident by using one enzyme or even by using two enzymes in sequence. Frequently, enzymes are used to determine the "fine" structure, after the general structural features have been determined by other methods. Besides the specificity that enzymes can provide, another advantage is that enzyme reactions can be performed with relatively small samples (milligram amounts).

Enzymes may be used to detect particular types of linkages in polysaccharides because of their high degree of specificity in hydrolyzing a specific kind of linkage, and in so doing can also elucidate the sequence of the linkages in the polysaccharide. They can be used to purify specific polysaccharides from a mixture of polysaccharides by using an enzyme that selectively hydrolyzes a contaminating polysaccharide but does not hydrolyze the polysaccharide to be purified, similar to the strategy of removal of proteins and nucleic acids with proteinases and nucleases. As an example, starch can be purified from cellulose by hydrolyzing the cellulose with cellulase. Enzymes specific for the polysaccharide under investigation can produce oligosaccharides in high yields. They can cleave polysaccharides that have acid-resistant linkages, and they can produce oligosaccharides that have acid-labile linkages. Biosynthetic enzymes can also contribute to the understanding of the structures of polysaccharides by synthesizing intermediates whose structures can be determined.

Enzymes can be used to give both qualitative and quantitative determinations, depending on the types of analyses made. Quantitative data can be obtained by the measurement of the increase in reducing values with time of hydrolysis and/or the determination of a specific product, such as glucose, using the glucose oxidase method of analysis (see section 3.5 in Chapter 3). Other specific products can be determined using qualitative and quantitative TLC or HPLC.

Enzymes have played an important role in the determination of the structure of starch. Acid hydrolysis of starch provided some controversy about the possible presence of α-1 $\rightarrow$ 3, β-1 $\rightarrow$ 4, and β-1 $\rightarrow$ 6 linkages in starch. This was apparently due to the formation of acid reversion products. Definitive proof that starch consisted exclusively of α-1 $\rightarrow$ 4-linked D-glucose residues with α-1 $\rightarrow$ 6-linked branch chains was obtained by hydrolysis with highly purified human salivary and porcine pancreatic α-amylase to produce α-limit dextrins [20]. These α-amylase limit dextrins were isolated from their action on amylopectin as described in Chapter 11. The structures of these limit dextrins were shown to have only α-1 $\rightarrow$ 6 branch linkages [21,22] and helped to define the structure of the amylopectin component of starch. The doubly branched dextrins, 6^4, 6^6-α-di-D-glucopyranosyl maltohexaose and 6^3, 6^5-α-di-D-glucopyranosyl maltopentaose isolated from the action of porcine pancreatic α-amylase with waxymaize starch, established that the α-1 $\rightarrow$ 6 branch linkages in waxymaize starch was as close as one nonbranched glucose residue between the two branch linkages. No saccharides were obtained that had α-1 $\rightarrow$ 6 branch linkages on consecutive glucose residues.

Crystalline human salivary α-amylase produced 6^3-phosphomaltotetraose from potato amylopectin, demonstrating that the covalently linked phosphate in potato starch was attached to the C-6 hydroxyl group by a phosphoester linkage in amylopectin [23].

Initially, the action of plant β-amylase with amylose gave complete conversion to β-maltose [24], indicating that the amylose component of starch was a linear α-1 $\rightarrow$ 4-linked glucan. However, when highly purified, crystalline sweet potato β-amylase was used, the amylose was only converted to 70% β-maltose [25–27].

The resistant fraction gave a deep blue color with triiodide, similar to that given by amylose. Treatment of the original amylose or the resistant fragment with pullulanase increased its conversion to 100% β-maltose, indicating that some of the amylose molecules had α-1 → 6 branch linkages that prevented complete conversion to β-maltose by β-amylase [28,29].

The structures of the slightly branched amyloses have been determined using pullulanase and *Pseudomonas amyloderamosa* isoamylase [30–32]. The size distribution of the branch chains were determined by hydrolysis of the α-1 → 6 branch linkage with isoamylase, followed by HPLC analysis of the released chains [33].

Endodextranase hydrolysis of *Leuconostoc mesenteroides* B-512F dextran has provided similar information about the branch linkages of B-512F dextran by producing a series of branch dextrins (B4–B8) that were analogous to those produced by α-amylase acting on amylopectin [34]. The branched dextran oligosaccharides had an α-1 → 3-glucopyranosyl unit attached to the third α-1 → 6-linked glucosyl residue from the reducing end, giving 3^3-α-glucopyranosyl-substituted isomaltodextrins [34]. A difference was that doubly branched dextrins were not formed, indicating that the distance between the branch linkages of B-512F dextran was much greater than the distance between the branch linkages found in waxymaize starch. The more highly branched dextrans produced by *L. mesenteroides* B-742 and *Streptococcus mutans* 6715 were resistant to endodextranase due to the high percentage of single α-1 → 3-linked glucose units that were adjacent to each other or had only one glucose unit between them. An α-1 → 3-dextran debranching enzyme has yet to be reported.

The unbranched structure of β-1 → 4 D-mannans from ivory nuts has been confirmed by hydrolysis with endo-β-1 → 4 mannanase in which no branched dextrins were formed [35,36]. The structure of pullulan (see Chapter 6, Fig. 6.17) was established by hydrolysis with pullulanase [37], an endohydrolase that gave maltotriose, a hexasaccharide (6^3-α-maltoriosyl maltotriose), a saccharide of nine glucose residues with two α-1 → 6 linkages, and a saccharide of 12 glucose residues with three α-1 → 6 linkages. These saccharides could be hydrolyzed by pullulanase to give maltotriose and the next lower saccharide, showing that each of the saccharides had a repeating structure of maltotriose units linked α-1 → 6 end to end [37]. In addition, there were some minor saccharides with 7, 10, and 13 glucose residues. These minor saccharides were hydrolyzed by pullulanase to give maltotriose and maltotetraose, indicating that there was a small amount of maltotetraose in the pullulan structure [37,38]. Salivary α-amylase hydrolyzed the maltotetraose units at the α-1 → 4 glycosidic bond that was adjacent and on the nonreducing side of the glucose residue linked α-1 → 6, leaving a maltotriose unit at the reducing end of the fragment and a B4 structure (see Fig. 11.1 in Chapter 11 for the structure of B4) at the nonreducing end of the other fragment [38]. The sizes of the fragments released by salivary α-amylase were determined by gel chromatography, and the distribution of the fragments indicated that the maltotetraose units were randomly distributed in pullulan [39].

12.9 Determination of Carbohydrate Structure by Using NMR

The proton-decoupled ^{13}C-NMR spectrum of carbohydrates gives a signal for each of the specific types of carbons present. For highly asymmetric carbohydrate units, this usually means a signal for each of the carbons in the asymmetric unit. This has been demonstrated for a number of saccharides, such as amylose [40], cyclomaltodextrins [41], dextrans [42], and many others [18]. It is most clearly demonstrated for sucrose in which 12 distinct signals are produced, one for each of the 12 carbons (see Fig. 12.4A).

Substitution for the hydroxyl groups attached to the carbohydrate carbons usually produces a shift in the signal. When chlorine was substituted for hydroxyl groups of sucrose, the signals were significantly shifted upfield 15–16 ppm [43]. These shifts for substitutions on various carbons are shown in Fig. 12.4B–C. Substitutions onto the hydroxyl groups, as in forming esters, however, does not produce significant shifts in the signals. A sucrose compound that had presumptive ester groups on C-6 and C-6′ had the structure proven by chlorination of the presumptive ester, followed by a proton-decoupled ^{13}C-NMR spectrum (see Fig. 12.4D) [44]. The signals for the C-6 and C-6′ of the chlorinated sucrose ester were not shifted upfield, indicating that they were already substituted (esterified). The signals for C-4 and C-1′ were shifted upfield as expected for chlorine substitution (compare the spectrum for the 4,6,1′,6′-tetrachlorosucrose with the spectrum for the chlorinated sucrose ester, Fig. 12.4B,D). The chlorination and ^{13}C-NMR spectrum established that the esterifications were at C-6 and C-6′ [44].

Proton-decoupled ^{13}C-NMR spectra have been obtained for a number of polysaccharides. They also show distinctive signals for the various carbons that are characteristic for the structures of the polysaccharides. In general, the C-1 signals for the α-configured glycosidic linkages range between 99 and 102 ppm, while the C-1 signals for the β-configured glycosidic linkages are a little more downfield at 102–105 ppm [18].

The proton-decoupled ^{13}C-NMR spectrum for *L. mesenteroides* B-1355L dextran (the same structure as B-512F dextran) gave a single signal at 98.72 ppm for the C-1 of the α-1 → 6 linkage [42]. The signals for the six carbons of B-1355L and B-512F dextran are identified on Fig. 12.5A. The ^{13}C-NMR spectrum for *L. mesenteroides* B-742 regular comb dextran (see Chapter 6, Fig. 6.16 for the structure) is given in Fig. 12.5B. This spectrum shows that there are two signals for C-1, the first one at 100.3 ppm for the C-1 making the α-1 → 3 branch linkage, and the second one at 98.84 ppm for the C-1 making the α-1 → 6 linkage of the main chain. In this spectrum there are two signals for C-6, the one at 66.5 ppm for the *O*-substituted, glycosidically linked C-6, and the other at 61.5 ppm for the unsubstituted C-6 produced by the single-branched glucose residues. The presence of a significant signal of approximately the same magnitude as the *O*-substituted C-6 signal indicates that the branching is by a single glucosyl

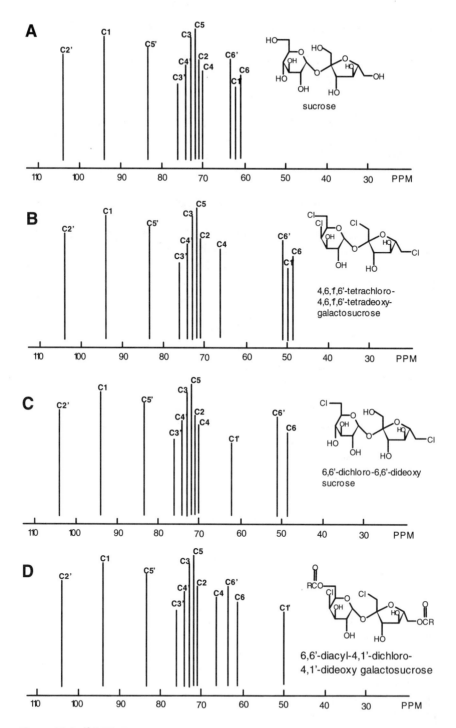

Figure 12.4. ^{13}C-NMR proton-decoupled spectra of sucrose and substituted sucroses in D$_2$O. PPM are relative to tetramethylsilane. From ref. [44].

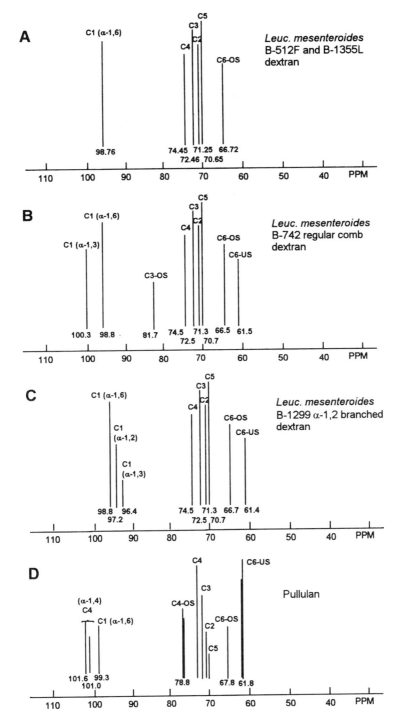

Figure 12.5. ^{13}C-NMR proton-decoupled spectra of glucans in D_2O. PPM are relative to tetramethylsilane. From refs. [42,45]. US, unsubstituted carbons; OS O-substitute carbons.

residue. In the B-1355L dextran spectrum, the unsubstituted C-6 signal was not apparent due to its relatively low amount in comparison with the glycosidically O-substituted C-6. In the B-742 dextran spectrum, there is also a signal at 81.7 ppm for the C-3 O-substituted branch linkage, which is in high amount in B-742 dextran. A very similar spectrum was observed for *S. mutans* dextran [45], which is an alternating comb dextran whose structure is similar to that of *L. mesenteroides* B-742 regular comb dextran but with approximately 50% fewer α-1 → 3 branch linkages. The ^{13}C-NMR spectrum for *L. mesenteroides* B-1299 dextran (Fig. 12.5C) gave three C-1 signals for the α-1 → 6, α-1 → 2, and α-1 → 3 linkages at 98.7, 97.2, and 96.4 ppm, respectively. There also were two signals for C-6 at 66.5 ppm for the O-substituted C-6, and at 61.5 ppm for the unsubstituted C-6 single-branch glucose residue. One of the several signals in the range of 76–70 ppm probably represents the C-2 O-substituted carbon that is in this dextran [42].

The ^{13}C-NMR of pullulan (Fig. 12.5D) gave three C-1 signals at 101.6, 101.0, and 99.3 ppm [40,46]. The first two are for the two C-1 signals from the two α-1 → 4 glycosidic linkages, and the last one is for the C-1 from the α-1 → 6 glycosidic linkage. There are three signals for C-6, the one at 67.8 ppm for the O-substituted C-6, and the other two at 62.0 and 61.8 ppm for the two unsubstituted but structurally distinct C-6 carbons. There also are two signals for the two O-substituted C-4 carbons at 79.2 and 78.8 ppm. This spectrum shows that the ^{13}C-NMR signals are very sensitive to small differences in structure.

The proton-decoupled ^{13}C-NMR spectra of polysaccharides whose methylene hydroxyl group had presumptively been oxidized by TEMPO showed signals at 175 ppm and no signals in the C-6 area of 66 ppm, indicating that the primary alcohol groups of the polysaccharides had been oxidized to carboxyl groups. Chemical shifts in the range of 198–205 ppm were absent, indicating the absence of keto groups that could have been formed by the oxidation of the secondary alcohols. These spectra confirmed that the oxidation had exclusively resulted in the formation of carboxyl groups from primary alcohol groups [47].

NMR spectra of atoms other than ^{13}C that are substituted onto carbohydrates, such as ^{19}F or ^{31}P, can also be used to establish structures. One such example is the substitution of fluorine for hydroxyl groups at C-6 and C-6′ of sucrose. In this case, the ^{19}F spectrum (Fig. 12.6) showed two distinct sextets centered at 235.21 and 227.74 ppm that were due to signals arising from a fluorine atom on C-6 coupled to the two hydrogen atoms on C-6, giving a triplet. Each of the triplets was further split by the hydrogen on C-5, resulting in the sextet. Similarly, the signals arising from fluorine on C-6′ indicates a triplet that is split by C-5′-H, giving a sextet. Substitution of fluorine on no other carbon atoms would have given these two sextets. If substitution had occurred on C-1′ (a possible substitution site), a unique triplet would have resulted and would not have been split into a sextet because the requisite hydrogen on C-2′ is not present [48].

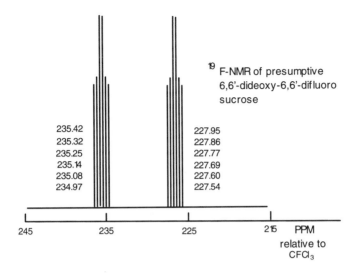

¹⁹F-NMR of presumptive
6,6'-dideoxy-6,6'-difluoro
sucrose

235.42	227.95
235.32	227.86
235.25	227.77
235.14	227.69
235.08	227.60
234.97	227.54

245   235   225   215   PPM
relative to
CFCl$_3$

Figure 12.6. ^{19}F-NMR spectrum of a fluoro-substituted sucrose. PPM are relative to CFCl$_3$. From ref. [48].

12.10 Separation and Determination of Carbohydrate Structure by Using Lectins

Lectins are proteins that bind specific carbohydrates and agglutinate cells. The first lectin to be recognized was *concanavalin A,* a protein from jack bean that precipitated glycogen and starch from solution [49]. In the late 1940s, it was reported that certain seeds contained proteins that agglutinated some human blood group antigens [50]. Using lectins and glucosidases, Watkins and Morgan showed that carbohydrates were the determinants in the four human blood types [51]. As we saw in Chapter 9, animal cell surfaces are coated with carbohydrates, and these carbohydrates are specific for different kinds of cells. The agglutinating power of lectins has been used to investigate cell surface carbohydrate structures and the changes in structure that occur as these cells undergo growth, differentiation, and the formation of malignancy [52,53].

Lectins may be considered as pseudoenzymes that have specific carbohydrate binding sites but are devoid of catalytic activity. But, because of their multivalent character, more than one carbohydrate binding site per protein molecule, and their ability to produce precipitates and agglutinate cells, they are more properly analogous to antibodies. Concanavalin A has specificity for D-mannose and D-glucose residues. Other lectins from peanuts, pokeweed, soybean, and wheat germ have specificities for *N*-acetyl-D-glucosamine, *N*-acetyl-D-galactosamine, D-galactose, L-fucose, and *N*-acetyl-D-neuraminic acid. See Table 12.1 for a listing of lectins

Table 12.1 Lectins and Their Carbohydrate-Binding Specificities

Lectin	Carbohydrate binding
Arbin	D-Gal
Asparagus pea	α-L-Fuc
Broad bean	D-Man, D-Glc
Castor bean	D-GalNAc, β-D-Gal
Chick pea	Fetuin (see Chap. 9)
Concanavalin A	D-Man, D-Glc
Eel	α-L-Fuc
Elder	α-D-NeuNAc-(2 → 6)-D-Gal/D-GalNAc
Gorse II	(D-GlcNAc)$_2$
Hairy vetch	D-GalNAc
Jimson weed	(D-GlcNAc)$_2$
Lima bean	D-GalNAc
Limulin	D-NeuNAc
Mistletoe, European	β-D-Gal
Mung bean	α-D-Gal
Mushroom	β-D-Gal-(1 → 3)-D-GalNAc
Osage orange	α-D-Gal, α-D-GalNAc
Pagoda tree	β-D-GalNAc
Garden pea	α-D-Man
Peanut	β-D-Gal(1 → 3)-D-GalNAc
Pokeweed	(D-GlcNAc)$_3$
Potato	(D-GalNAc)$_3$
Garden snail	D-GalNAc
Soybean	D-GalNAc
Sweet pea	α-D-Man
Wheat germ	D-GlcNAc, α-D-NeuNAc

and their carbohydrate binding specificities. The specificities that lectins show for binding different carbohydrate structures can be used for the isolation and separation of glycoproteins and glycolipids by affinity chromatography [54] and in studying their structures.

Radioactive and fluorescent lectins have been prepared and are used to detect specific glycoproteins on electrophoresis gels [54,55]. Lectins can be coupled on solid supports (see section 7.8 in Chapter 7) and used in preparative affinity chromatography of glycoproteins [56,57].

Lectins have also found specialized use in the identification and isolation of specific microorganisms by binding to specific surface carbohydrates, such as teichoic acids, lipopolysaccharides, and capsular polysaccharides [58,59]. The agglutination of a microorganism by a particular lectin can be used as a confirmatory identification of the microorganism, eliminating the necessity of performing a series of time-consuming biochemical and serological tests.

12.11 Removal of Glycosides from Glycoproteins for Structure Determination

The determination of the structures of glycosides of glycoproteins are obtained by first removing them from the protein. *O*-Glycosides attached to serine and threonine and *S*-glycosides attached to cysteine can be released by a β-elimination using a mild alkaline treatment (see Fig. 9.2B,C,F in Chapter 9) [60–62]. The reaction is usually conducted in the presence of sodium borohydride, which reduces the reducing end of the released oligosaccharide. This reduction prevents alkaline degradation of the oligosaccharide by the "peeling" reaction, and it determines the carbohydrate residue that is attached to the protein as an alditol.

To achieve β-elimination, various strengths of alkali (0.05–0.5 N NaOH), temperatures of 0–45°C, and lengths of 15–216 hr are used. The sodium borohydride is 0.15–1.0 M. A standard procedure uses 0.1 N NaOH and 0.3 M NaBH$_4$ at 37°C for 48 hr. The conditions, however, should be established for each glycoprotein or glycopeptide [61]. The β-elimination reaction does not proceed satisfactorily if the glycosylated amino acid is the C- or N-terminal residue. In such cases, the amino or carboxyl group has to be derivatized to eliminate the charge [62].

The asparagine *N*-linked glycosides can be cleaved by hydrazinolysis [63]. The glycoprotein is heated at 100°C with anhydrous hydrazine for 8–12 hr. A series of reactions take place (Fig. 12.7) The first and second reactions proceed relatively quickly. The third reaction is much slower. The procedure is carried out by suspending 0.2–1 mg of glycoprotein in 0.5–1.0 mL of freshly distilled anhydrous hydrazine. The solution is heated in a sealed tube at 100°C for 8–12 hr. The glycoprotein sample usually dissolved after 1 hr.

Various endoglycosidases such as endo-β-*N*-acetylglucosaminidase can be used to liberate asparagine *N*-linked oligosaccharide chains [64].

Chemical methods that liberate both *O*- and *N*-linked oligosaccharides from glycoproteins involve anhydrous trifluoromethane sulfonic acid (TFMS) or anhy-

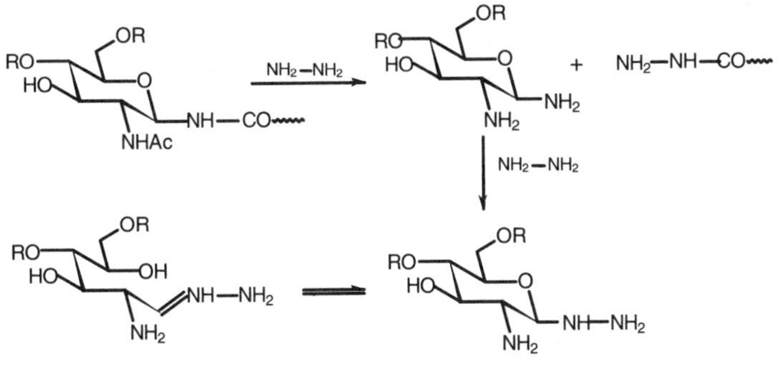

Figure 12.7. Hydrazinolysis of asparagine *N*-linked oligosaccharide chains of glycoproteins. R represents hydrogen or carbohydrate residues.

drous hydrogen fluoride [65]. TFMS hydrolysis is performed at 0°C for 0.5–2 hr under nitrogen. After reaction the mixture is cooled below −20°C in a dry ice–ethanol bath and slowly neutralized by 60% (v/v) aqueous pyridine that is previously cooled to −20°C. The TFMS reaction is easier and less hazardous to perform than the HF procedure and does not require a special apparatus [65].

The structure of the released oligosaccharide can be determined by using a number of specific glycosidases (N-acetylneuraminidase, α-L-fucosidase, α-mannosidase, and β-N-acetylglucosaminidase) sequentially or in a mixture to hydrolyze the oligosaccharide chain.

12.12 Literature Cited

1. H. Molisch, *Monatsch. Chem.,* **7** (1886) 108–111.
2. W. C. Still, M. Kahn, and A. Mitra, *J. Org. Chem.,* **43** (1978) 2923–2926.
3. E. K. Yau and J. K. Coward, *Aldrichim. Acta,* **21(4)** (1988) 106–107.
4. R. L. Whistler and J. N. BeMiller, *Methods Carbohydr. Chem.,* **1** (1962) 42–44.
5. R. Gauch, U. Leuenberger, and E. Baumgartner, *J. Chromatog.,* **174** (1979) 195–200.
6. A. Haug, B. Larsen, and O. Smidsrød, *Acta Chemica Scand.,* **20** (1966) 183–190.
7. C. A. Wilham, B. H. Alexander, and A. Jeanes, *Arch. Biochem. Biophys.,* **59** (1955) 61–68.
8. S. C. Churms, in *Handbook of Chromatography, Carbohydrates,* Vol. 1, pp. 69–129 and pp. 175–187 (G. Zweig and J. Sherma, eds.) CRC, Boca Raton (1982).
9. M. Bounias, *Anal. Biochem.,* **106** (1980) 291–295.
10. H. Bouveng and B. Lindberg, *Adv. Carbohydr. Chem.,* **15** (1960) 58–68.
11. S. Hakomori, *J. Biochem.,* **55** (1964) 205–208.
12. G. G. S. Dutton, *Adv. Carbohydr. Chem. Biochem.,* **30** (1974) 9–110.
13. H. Bjorndal, C. G. Hellerquist, B. Lindberg, and S. Svensson, *Angew. Chem. Internat. Ed.,* **9** (1970) 610–619.
14. K. Stellner, H. Saito, and S. Hakomori, *Arch. Biochem. Biophys.,* **155** (1973) 464–472.
15. M. E. Slodki, R. E. England, R. D. Plattner, and W. E. Dick, Jr., *Carbohydr. Res.,* **156** (1986) 199–206.
16. R. Mukerjea, D. Kim, and J. F. Robyt, *Carbohydr. Res.,* **292** (1996) 11–20.
17. T. Ikenaka, *J. Biochem.,* **54** (1963) 328–333.
18. P. A. J. Gorrin, *Adv. Carbohydr. Chem. Biochem.,* **38** (1981) 13–104.
19. H. Suzuki and E. J. Hehre, *Arch. Biochem. Biophys.,* **104** (1964) 305–315.
20. D. French, *Biochem. J.,* **100** (1966) 2p.
21. K. Kainuma and D. French, *FEBS Lett.,* **5** (1968) 257–261.
22. K. Kainuma and D. French, *FEBS Lett.,* **6** (1969) 182–186.
23. F. W. Parrish and W. J. Whelan, *Stärke,* **13** (1961) 231–236.
24. W. Z. Hassid and R. M. McCready, *J. Am. Chem. Soc.,* **65** (1943) 1157–1162.
25. S. Peat, W. J. Whelan, and S. J. Pirt, *Nature,* **164** (1949) 499–500.
26. G. J. Thomas, W. J. Whelan, and S. Peat, *Biochem. J.,* **47** (1950) 1x.
27. S. Peat, S. J. Pirt, and W. J. Whelan, *J. Chem. Soc.* (1952) 705–710.
28. O. Kjølberg and D. J. Manners, *Biochem. J.,* **86** (1963) 258–264.
29. W. Banks and C. T. Greenwood, *Arch. Biochem. Biophys.,* **117** (1966) 674–680.

30. S. Hizukuri, Y. Takeda, M. Yasuda, and A. Suzuki, *Carbohydr. Res.,* **94** (1981) 205–213.

31. Y. Takeda, K. Shiraska, and S. Hizukuri, *Carbohydr. Res.,* **132** (1984) 83–92.

32. Y. Takeda, S. Hizukuri, C. Takeda, and A. Suzuki, *Carbohydr. Res.,* **165** (1987) 139–145.

33. S. Hizukuri, *Carbohydr. Res.,* **141** (1985) 295–306.

34. G. J. Walker and A. Pulkownik, *Carbohydr. Res.,* **36** (1974) 53–59.

35. G. O. Aspinall, E. L. Hirst, E. G. V. Percival, and I. R. Williamson, *J. Chem. Soc.,* (1953) 3184–3188.

36. G. O. Aspinall, R. B. Rashbrook, and G. Kessler, *J. Chem. Soc.,* (1958) 215–221.

37. B. J. Catley, J. F. Robyt, and W. J. Whelan, *Biochem. J.,* **100** (1966) 5p.

38. B. J. Catley and W. J. Whelan, *Arch. Biochem. Biophys.,* **143** (1971) 138–144.

39. G. Carolan, B. J. Catley, and F. J. McDougal, *Carbohydr. Res.,* **114** (1983) 237–243.

40. H. J. Jennings and I. C. P. Smith, *J. Am. Chem. Soc.,* **95** (1973) 606–608.

41. K. Takeo, K. Hirose, and T. Kuge, *Chem. Lett.,* (1973) 1233–1236.

42. F. R. Seymour, R. D. Knapp, S. H. Bishop, and A. Jeanes, *Carbohydr. Res.,* **68** (1979) 123–140.

43. L. Hough, in *Sucochemistry,* pp.9–21, (J. L. Hickson, ed.) ACS Symposium Series 41, American Chemical Society, Washington, D.C. (1977).

44. R. Mukerjea and J. F. Robyt, unpublished data.

45. P. Colson, H. C. Jarrell, B. L. Lamberts, and I. C. P. Smith, *Carbohydr. Res.,* **71** (1979) 265–272.

46. P. Colson, H. J. Jennings, and I. C. P. Smith, *J. Am. Chem. Soc.,* **96** (1974) 8081–8087.

47. P. S. Chang and J. F. Robyt, *J. Carbohydr. Chem.,* **15** (1996) 667–677.

48. J. N. Zikopoulos, S. H. Eklund, and J. F. Robyt, *Carbohydr. Res.,* **104** (1982) 245–251.

49. J. B. Sumner and S. F. Howell, *J. Bacteriol.,* **32** (1936) 227–237.

50. W. C. Boyd and R. M. Reguera, *J. Immunol.,* **62** (1949) 333–339.

51. W. M. Watkins and W. T. J. Morgan, *Nature,* **169** (1952) 825–826.

52. G. L. Nicolson, *Int. Rev. Cytol.,* **39** (1974) 89–190.

53. G. L. Nicolson, *Biochim. Biophys. Acta,* **457** (1976) 57–108; **458** (1976) 1–72.

54. K. Burridge, *Methods Enzymol.,* **50** (1978) 54–64.

55. M. Furlan, B. A. Perret, and E. A. Beck, *Anal. Biochem.,* **96** (1979) 208–214.

56. S. Narasimhan, J. R. Wilson, E. Martin, and H. Schachter, *Can. J. Biochem.,* **57** (1979) 83–96.

57. K. Kornfeld, M. L. Reitman, and R. Kornfeld, *J. Biol. Chem.,* **256** (1981) 6633–6640.

58. T. G. Pistole, *Ann. Rev. Microbiol.,* **35** (1981) 85–112.

59. P. Z. Allen, M. C. Connelly, and M. A. Apicella, *Can. J. Microbiol.,* **26** (1980) 468–474.

60. B. Anderson, N. Seno, P. Sampson, J. G. Riley, P. Hoffman, and K. Meyer, *J. Biol. Chem.,* **239** (1964) 2716–2719.

61. R. G. Spiro and V. D. Bhoyroo, *Fed. Proc. Fed. Amer. Soc. Exp. Biol.,* **30** (1971) 1223–1225.

62. R. G. Spiro, *Methods Enzymol.,* **28** (1972) 35–40.

63. S. Takasaki, T. Mizuochi, and A. Kobata, *Methods Enzymol.,* **83** (1982) 263–268.

64. T. Muramatsu, *Methods Enzymol.,* **50** (1978) 555–559; A. Kobata, *Methods Enzymol.,* **50** (1978) 560–574; A. L. Tarentino, R. B. Trimble, and F. Maley, *Methods Enzymol.,* **50** (1978) 574–584.

65. H. T. Sojar and O. P. Bahl, *Methods Enzymol.,* **138** (1987) 341–344.

12.13 References for Further Study

"Methods in structural polysaccharides," H. Bouvend and B. Lindberg, *Adv. Carbohydr. Chem. Biochem.* **15** (1960) 58–68.

"Carbon-13 nuclear magnetic resonance spectroscopy of polysaccharides," P. A. J. Gorin, *Adv. Carbohydr. Chem. Biochem.,* **38** (1981) 13–104.

"Enzymic analysis of polysaccharide structure", B. V. McCleary and N. K. Matheson, *Adv. Carbohydr. Chem. Biochem.,* **44** (1986) 147–276.

Receptor-Specific Proteins: Plant and Animal Lectins, E. R. Gold and P. Balding, Excerpta Medica, Amsterdam (1975).

The Lectins: Properties, Functions, and Applications in Biology and Medicine, I. E. Liener, N. Sharon, and I. J. Goldstein, eds., Academic, New York (1986).

"Spectroscopic Methods in the Determination of Carbohydrate Structures," A. S. Perlin and B. Casu, in *The Polysaccharides,* Vol. 1, Chap. 4, pp. 135–195, Academic, New York (1982).

Appendix A

Primer on Carbohydrate Nomenclature

A.1 Names of Simple Sugars

Carbohydrates are defined as polyhydroxy aldehydes or ketones. They usually can be identified as carbohydrates by the ending *-ose* at the end of their names. The smallest carbohydrate has three carbons and is called a *triose;* a carbohydrate with four carbons is called a *tetrose;* one with five carbons, a *pentose;* one with six carbons, a *hexose,* and so on.

Carbohydrates with an aldehyde group are called *aldoses,* and those with a ketone group are called *ketoses.* So, an *aldopentose* would be a carbohydrate with an aldehyde group and five carbons, and a *ketohexose* would be a carbohydrate with a ketone group and six carbons.

The names of specific trioses, tetroses, pentoses, and so forth were given to the sugars by nineteenth- and early twentieth-century chemists, often by using some aspect of the source from which they were isolated. The following names refer to specific sugars: *glyceraldehyde, erythrose, arabinose, xylose, glucose, fructose, mannose* (see Chapter 1).

The carbohydrates are also divided into two classes, called "D" and "L" based on whether they are derived from the trioses *D-glyceraldehyde* or *L-glyceraldehyde,* which have their asymmetric hydroxyl groups to the right and left, respectively. Most of the naturally occurring sugars belong to the D-class.

Ketoses can be formed from each of the aldoses, and aldoses can be formed from ketoses by isomerization of the carbons at positions 1 and 2 (see reactions 1.6 and 1.7 in Chapter 1). A naturally occurring ketose that was first isolated from honey was *D-fructose* (see Chapter 2). It can be formed by isomerization of either D-glucose or D-mannose (see reaction 1.7 in Chapter 1). Other ketoses can be formed from other aldoses. They are usually named by replacidng *-ose* with *-ulose* at the end of the name of the sugar from which they are formed. Thus,

D-xylose gives *D-xylulose,* D-ribose gives *D-ribulose,* D-allose gives *D-allulose,* and so on.

Hexoses and pentoses can form intramolecular hemiacetals or hemiketals. The newly created asymmetric carbon at C-1 is called the *anomeric carbon,* and the hemiacetal (or hemiketal) hydroxyl is called the *anomeric hydroxyl* group. When C-1 is to the right, the anomeric hydroxyl can be below or above the plane of the ring and is called α or β, respectively. The ring that is formed can have six atoms or five atoms and is called *pyranose* or *furanose,* respectively (see Chapters 2 and 3). Thus, an aldohexose such as D-glucose can have four ring forms, depending on whether the anomeric hydroxyl group is α or β and whether the ring is pyranose or furanose. The four ring forms of D-glucose are called α-*D-glucopyranose,* β-*D-glucopyranose,* α-*D-glucofuranose,* and β-*D-glucofuranose.* D-Xylose likewise can have four ring forms that are called α-*D-xylopyranose,* β-*D-xylopyranose,* α-*D-xylofuranose,* and β-*D-xylofuranose.* Each of the four possible D-pentoses, eight D-hexoses, and four D-ketohexoses can theoretically exist in the same types of four-ring forms.

A.2 Sugars Derived from Polyhydroxy Aldehydes or Ketones

a. Sugar Alcohols

Reduction of the aldehyde group results in a single sugar alcohol, and reduction of a keto group results in two sugar alcohols. The sugar alcohols have the general name, *alditols.* The names of the sugar alcohols are obtained by replacing -*ose* on the name of the parent aldose with -*itol.* Thus D-erythrose gives *D-erythritol,* D-glucose gives *D-glucitol,* D-mannose gives *D-mannitol,* and D-xylose gives *D-xylitol.* Reduction of D-fructose gives D-glucitol and D-mannitol.

b. Aldonic Acids

Oxidation of the aldehyde group to a carboxyl group produces an aldonic acid. These are named by replacing -*ose* on the name of the parent aldose with -*onic acid.* Thus, the oxidation of C-1 of D-glucose gives *D-gluconic acid;* oxidation of C-1 of D-mannose gives *D-mannonic acid,* and so on.

c. Uronic Acids

The oxidation of the primary alcohol group (CH_2OH) to a carboxyl group (COOH) forms a *uronic acid.* Uronic acids are named by replacing -*ose* on the name of the parent aldose with -*uronic acid.* Thus, the oxidation of C-6 of D-glucose gives *D-glucuronic acid;* oxidation of D-mannose gives *D-mannuronic acid,* and so on.

d. Aldaric Acids

The oxidation of both the aldehyde and primary alcohol groups (CHO and CH_2OH) to carboxyl groups gives an aldaric acid. Aldaric acids are named by replacing *-ose* with *-aric acid.* Thus, the oxidation of C-1 and C-6 of D-glucose gives *D-glucaric acid,* oxidation of D-mannose gives *D-mannaric acid,* and so forth.

e. Lactones

The oxidation of C-1 of a hemiacetal forms a lactone (intramolecular carboxylic acid ester). Lactones are named by replacing *-ose* on the name of the parent aldose with *-onolactone.* Thus, oxidation of C-1 of β-D-glucopyranose gives *D-glucono-lactone,* specifically, *1,5-D-gluconolactone;* oxidation of C-1 of α-D-mannose gives *1,5-D-mannolactone,* and so forth.

f. Glycosides

The reaction of a hemiacetal with an alcohol to split out water produces an *acetal.* When the hemiacetal is a carbohydrate and the alcohol is not a carbohydrate, the resulting product is called a *glycoside.* A glycoside is named by considering the configuration of the part resulting from the hemiacetal and replacing *-ose* on the name of the parent aldose with *-oside.* Thus, the reaction of α-D-glucopyranose with methanol gives *methyl-α-D-glucopyranoside.* The methyl group is called the *aglycone,* and its name comes first, followed by the name of the carbohydrate (the *glycone*), replacing *-ose* on the name of the parent aldose by *-oside.* The reaction of *p*-nitro-phenol with β-D-mannose would give *p-nitrophenyl-β-D-manno-pyranoside.*

g. Oligosaccharides

A special kind of glycoside results when an alcohol group of one monosaccharide reacts with the hemiacetal hydroxyl of another carbohydrate unit to split out water and form an acetal linkage between the two. The bond joining the two units is given the special name *glycosidic linkage.* Oligosaccharides are named by taking the name of the monosaccharide that is joined to the other monosaccharide unit, adding α- or β-, depending on the configuration of the acetal linkage, and replacing *-ose* with *-osyl,* followed by the position of the linkage and the name of the monosaccharide unit to which the attachment is made. Thus, maltose, which is composed of a D-glucopyranose unit linked α-1 → 4 to another D-glucopyranose unit, would be named *α-D-glucopyranosyl-(1 → 4)-D-glucopyranose.* Lactose, which is composed of β-D-galactopyranose linked β-1 → 4 to D-glucopyranose, would be named *β-D-galactopyranosyl-(1 → 4)-D-glucopyranose.*

More than two monosaccharide units can be joined together by forming glycosidic linkages to give oligosaccharides. Oligosaccharides are classified and named according to the number of residues that are joined together. An oligosaccharide with two monosaccharide units is a *disaccharide*, one with three is a *trisaccharide*, one with four is a *tetrasaccharide*, one with five is a *pentasaccharide*, and so on. An oligosaccharide is defined as a saccharide that is homogenous and has a specified number of monosaccharide units. Oligosaccharides can consist of a single kind of monosaccharide residue that is joined together by a single type of glycosidic linkage to give a homologous series of oligosaccharides with different numbers of monosaccharides residues (see Fig. 3.10 in Chapter 3 for the formation of eight kinds of D-glucose disaccharides, each joined by different kinds of glycosidic linkages). The following are the names of the oligosaccharides with two to six D-glucose residues in the eight families of oligosaccharides joined together by α-1 $\rightarrow$ 2, α-1 $\rightarrow$ 3, α-1 $\rightarrow$ 4, and so forth glycosidic linkages:

α-Linked Families

No. Glc	α-1 $\rightarrow$ 2	α-1 $\rightarrow$ 3	α-1 $\rightarrow$ 4	α-1 $\rightarrow$ 6
2	Kojiobiose	Nigerose	Maltose	Isomaltose
3	Kojiotriose	Nigerotriose	Maltotriose	Isomaltotriose
4	Kojiotetraose	Nigerotetraose	Maltotetraose	Isomaltotetraose
5	Kojiopentaose	Nigeropentaose	Maltopentaose	Isomaltopentaose
6	Kojiohexaose	Nigerohexaose	Maltohexaose	Isomaltohexaose

β-linked families

	β-1 $\rightarrow$ 2	β-1 $\rightarrow$ 3	β-1 $\rightarrow$ 4	β-1 $\rightarrow$ 6
2	Sophorose	Laminaribiose	Cellobiose	Gentiobiose
3	Sophorotriose	Laminaritriose	Cellotriose	Gentiotriose
4	Sophorotetraose	Laminaritetraose	Cellotetraose	Gentiotetraose
5	Sophoropentaose	Laminaripentaose	Cellopentaose	Gentiopentaose
6	Sophorohexaose	Laminarihexaose	Cellohexaose	Gentiohexaose

Oligosaccharides can also be formed by the reaction of two hemiacetal hydroxyl groups with each other (or two hemiketal hydroxyl groups, or a hemiacetal hydroxyl group and a hemiketal hydroxyl group) to split out water and form an acetal-acetal (or a ketal-ketal, or an acetal-ketal) linkage. These kinds of saccharides are called *nonreducing saccharides* because the potential reducing aldehyde has been removed from both of the monosaccharide residues. This is in contrast to the previously discussed oligosaccharides that each have one potential reducing aldehyde group and are called *reducing saccharides*.

A well-known nonreducing disaccharide is *sucrose;* it results from the reaction of the hemiacetal hydroxyl group of α-D-glucopyranose and the hemiketal hydroxyl group of β-D-fructofuranose. The formal name of sucrose is β-*D-fructofuranosyl*-α-*D-glucopyranoside* or, very frequently, α-*D-glucopyranosyl*-β-*D-fructo-*

furanoside. Another well-known, naturally occurring nonreducing disaccharide is α,α-*trehalose,* which is formed when the hemiacetal hydroxyl of α-D-glucopyranose reacts with the hemiacetal hydroxyl of another α-D-glucopyranose (see Chapters 2 and 3 and Fig. 3.11). Nonreducing oligosaccharides having more than two monosaccharide residues are also possible, for example, the trisaccharide *raffinose* (see Fig. 3.12 in Chapter 3). Polysaccharides usually have a reducing end, but nonreducing polysaccharides do occur in which the reducing end is capped with a sucrose unit (see section 6.2.f. in Chapter 6 and Fig. 6.11).

h. Polysaccharides

Polysaccharides are polymers composed of many monosaccharides joined together by glycosidic linkages. The general name used for polysaccharides is *glycan.* The ending *-an* is added to the name to designate a polysaccharide such as dextr*an,* pullul*an,* xanth*an,* and so forth. There are many exceptions such as *starch, cellulose, agar, algin, amylose, amylopectin,* and *chitin.* Many of these polysaccharides were named before their chemical nature was fully understood or before a systematic naming system had been developed. Starch was known very early, and cellulose was recognized very early as a carbohydrate and hence the *-ose* ending was used. In the above examples, the ending *-in* is common. This apparently was given to high molecular weight polysaccharides that gave viscous, gellike solutions or suspensions.

A.3 Numbering of Carbons in a Monosaccharide

The carbon atoms in a monosaccharide are numbered consecutively starting with the aldehyde group or potential aldehyde group (hemiacetal carbon) as number 1.

A.4 Naming Monosaccharides That Have Added Substituents

Substituents may be added in two ways: (1) they may be added to a hydroxyl group or (2) they may be added in place of a hydroxyl group. When a substituent is added to a hydroxyl group, the letter *O* is usually used along with the number of the carbon atom that has the substitution. The *O* indicates that the substituent is attached to the oxygen atom attached to the carbon. Hence, if there was an acetyl group attached to position 4 of D-glucopyranose, the derivative would be named *4-O-acetyl-D-glucopyranose.* If there were acetyl groups at positions 4 and 6, the name would be *4,6-di-O-acetyl-D-glucopyranose,* and if there were acetyl groups at positions 2, 3, 4, and 6, the name would be *2,3,4,6-tetra-O-acetyl-D-glucopyranose.* The order of the numbering starts with the lowest number.

When a hydroxyl group has been replaced by a substituent, the number of the carbon holding the substituent is used with the name of the substituent, and the term *deoxy* is added. Thus, if we substitute a chlorine for a hydroxyl group at C-4 of D-glucopyranose, the analogue would be named *4-chloro-4-deoxy-D-gluco-pyranose.* But if we substitute an iodine atom, the name would be *4-deoxy-4-iodo-D-glucopyranose.* The order of the names of substituents and *deoxy* is alphabetical. If we substitute an iodine atom at C-4 and a chlorine atom at C-6 of D-glucopyranose, the name would be *4-deoxy-4-iodo-6-chloro-6-deoxy-D-gluco-pyranose.* In this case the numbering takes precedence over putting *chloro* before *iodo.*

There are some specific substitutions where a name is given without using this type of numbering and naming. This occurs especially for monosaccharides that are substituted at C-2 with an amino group. Names such as *D-glucosamine, D-galactosamine,* and so forth are common in place of *2-amino-2-deoxy-D-gluco-pyranose* and *2-amino-2-deoxy-D-galactopyranose,* probably because substitution of an amino group at C-2 is a frequently encountered and naturally occurring type of amino sugar. Even the disaccharide, lactose, that has an amino group substituted onto the D-glucopyranose unit is called *lactosamine.* Substitution of a hydrogen for a hydroxyl group at C-6 to give certain 6-deoxyhexoses is relatively common for some naturally occurring sugars. They too have been given trivial names: 6-deoxy-D-glucose is called *D-quinovose,* 6-deoxy-L-mannose is called *L-rhamnose,* and 6-deoxy-L-galactose is called *L-fucose.*

Carbohydrates that are substituted on the hydroxyl groups are usually thought of as *derivatives,* whereas those that replace the hydroxyl group are usually considered *analogues.*

A.5 Naming and Numbering
of Substituted Oligosaccharides

Naming of substituted oligosaccharides presents special problems and possibilities for confusion. One reason is that a full systematic name is often long and cumbersome. Some examples of the full systematic names are *α-D-glucopy-ranosyl-(1 → 4)-β-D-glucopyranose* for β-maltose, and *α-D-glucopyranosyl-(1 → 4)-α-D-glucopyranosyl-(1 → 4)-β-D-glucopyranose* for β-maltotriose. Hence, the shorter, less formal names, β-maltose and β-maltotriose, are more frequently used. We have already seen that the name sucrose is much easier to use than the formal name. The naming of substituted oligosaccharides uses the shorter, less formal name whenever possible, to make the name simpler and easier to use.

The numbering of the positions of substitution of oligosaccharides uses the numbering of the carbons of individual monosaccharide units in the oligosaccharide, with the first unit in the systematic name being given numbers with primes. This is especially so for the nonreducing disaccharides. The order is unprimed numbers with the lowest numbers first, followed by the primed numbers with the lowest numbers first. Thus, the substitution of chlorine for hydroxyl groups of sucrose at C-6, of the glucose moiety, and C-1 and C-6 of the fructose would be *6,1',6'-trichloro-6,1',6'-trideoxysucrose.* If another chlorine is substituted at C-4 of the glucose moiety with inversion, the name is *4,6,1',6'-tetrachloro-4,6,1',6'-tetradeoxygalactosucrose.* The inversion of the configuration at C-4 of the glucose gives D-galactose, and hence the trivial name galactosucrose is used (see Fig. 5.7 for halo-substituted sucroses). Thus, the substitution of chlorine at the C-6 position of the D-galactose moiety of raffinose, the C-4 position of the glucose moiety, and the C-1 and C-6 positions of fructose can be named as *6-[6-chloro-6-deoxy-α-D-galactopyranosyl]-4,1',6'-trichloro-4,1',6'-trideoxysucrose.* Sometimes the substitution in a nonreducing trisaccharide, such as raffinose, is most clearly and simply named by using a three-letter superscript for the monosaccharide unit that is substituted. For example if position 2 of the glucose moiety of raffinose is substituted by an amino group, it could be named *2^{Glc}-amino-2^{Glc}-deoxyraffinose.* This type of nomenclature was introduced by Whelan in 1960 [1] and has further use in more complex saccharides with mixed linkages and branch linkages, as shown in the following discussion.

Substitutions in a reducing oligosaccharide that contains only one kind of monosaccharide unit can be done in a similar manner by numbering each of the carbons of the monosaccharide units with a numeral indicating the particular residue from the reducing end. Thus, substitution of a hydroxyl group by a hydrogen at the C-4 of the nonreducing residue of maltose (residue number two from the reducing end) would be named *4^2-deoxymaltose.* If the substitution occurred at C-6 of the reducing residue, the name would be *6^1-deoxy maltose,* and if the substitution occurred at both C-4 of the nonreducing residue and C-6 of the reducing residue, the name would be *$4^2,6^1$-dideoxymaltose.*

This type of naming and numbering can be easily extended to oligosaccharides that contain one kind of monosaccharide but with two or more kinds of glycosidic linkages. Thus, the trisaccharide with an α-glucopyranosyl unit linked 1 → 6 to the nonreducing glucose unit of maltose (trivial name of *panose*) could be named *6^2-α-D-glucopyranosylmaltose,* and the isomer *isopanose* with the α-D-glucose unit linked 1 → 6 to the reducing end of maltose would be *6^1-α-D-glucopyranosylmaltose.* The α-limit dextrin, tetrasaccharide, that is obtained by human salivary α-amylase hydrolysis of amylopectin (see Chapter 11) and given the trivial designation of B4, would be named *6^3-α-D-glucopyranosylmaltotriose.* Similarly, B5, an α-limit dextrin pentasaccharide, would be *6^3-α-glucopyranosylmaltotetraose.*

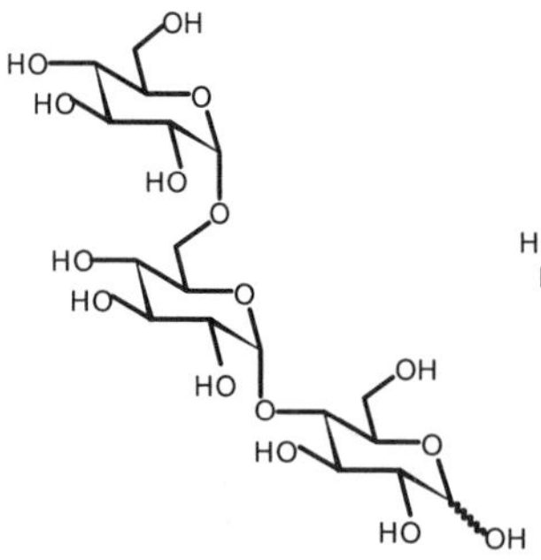

6^2-α-D-glucopyranosyl maltose 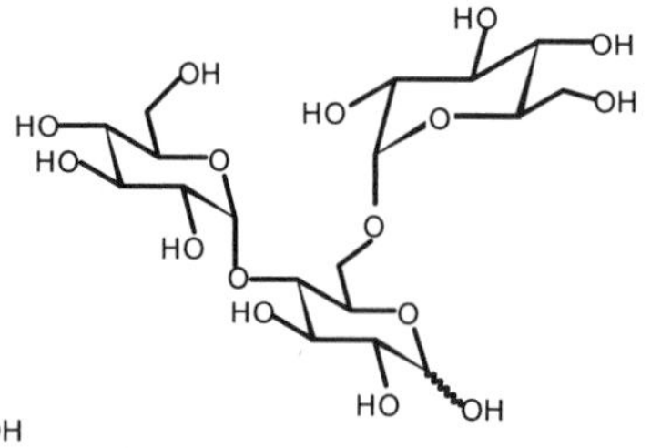 6^1-α-D-glucopyranosyl maltose

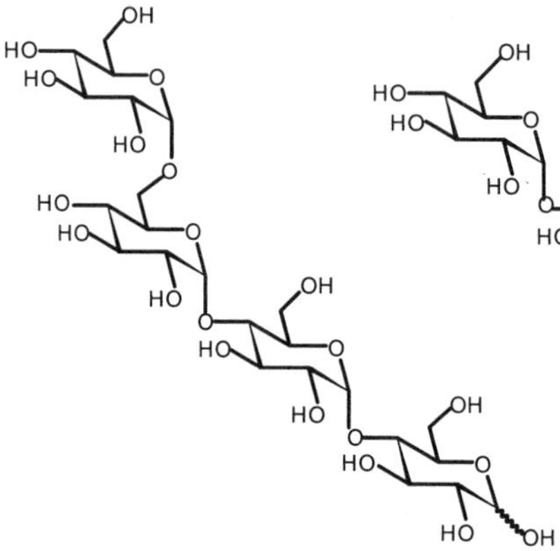

6^3-α-D-glucopyranosyl maltotriose 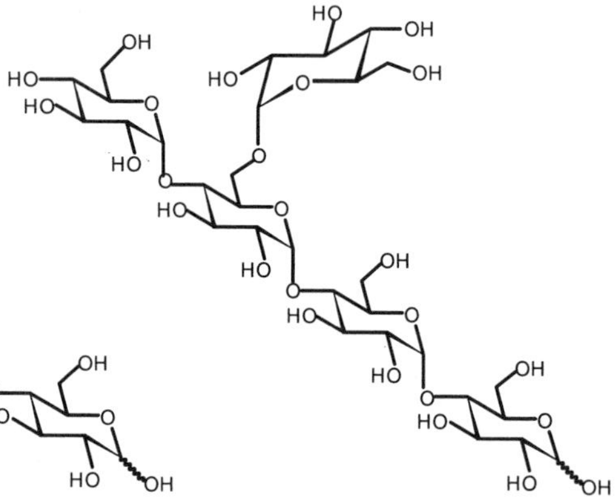 6^3-α-D-glucopyranosyl maltotetraose

A tetrasaccharide with a sequence of three different kinds of linkages, for example, α-1 → 3, α-1 → 6, and α-1 → 4, would be named *6^2-α-nigerosyl maltose.* One of the isomers with the sequence of α-1 → 3, α-1 → 4, and α-1 → 6 would be named *4^2-nigerosyl isomaltose,* and another isomer with the sequence α-1 → 6, α-1 → 3, and α-1 → 4 would be named *3^2-isomaltosyl maltose,* and another isomer with the sequence α-1 → 4, α-1 → 3, α-1 → 6 would be named *3^2-α-maltosyl isomaltose.* A hexasaccharide with an α-isomaltotriosyl unit linked 1 → 6 to the second residue of a maltotriose unit would be named *6^2-α-isomaltotriosyl maltotriose.* In summary, the carbon of substitution is an ordinary numeral, and the residue on which the carbon is substituted is designated by a superscript.

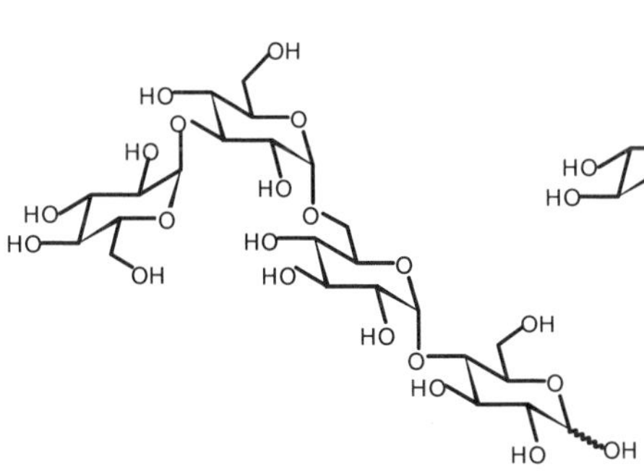

6^2-α-nigerosyl maltose

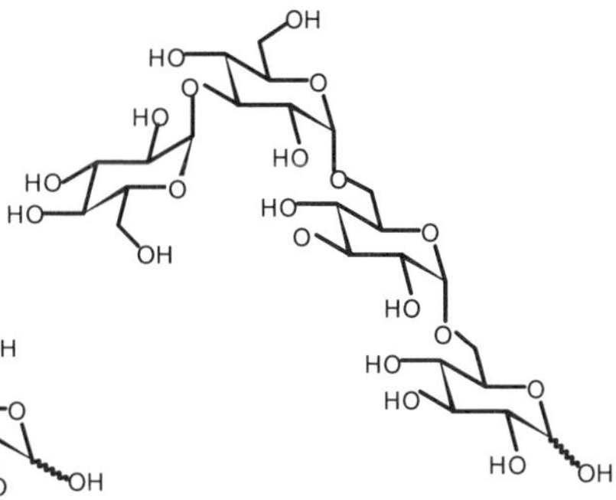

6^2-α-nigerosyl isomaltose

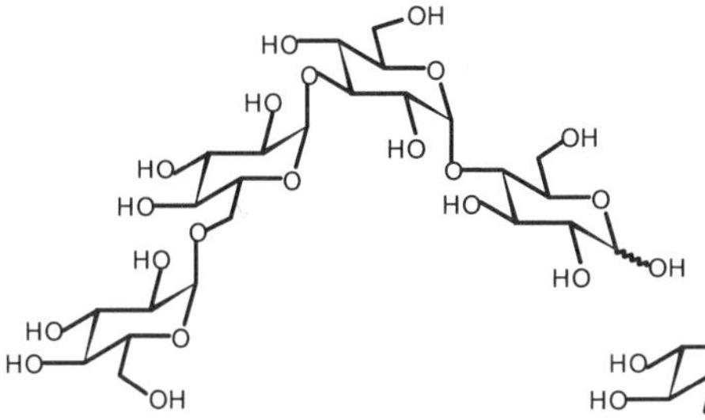

3^2-α-isomaltosyl maltose

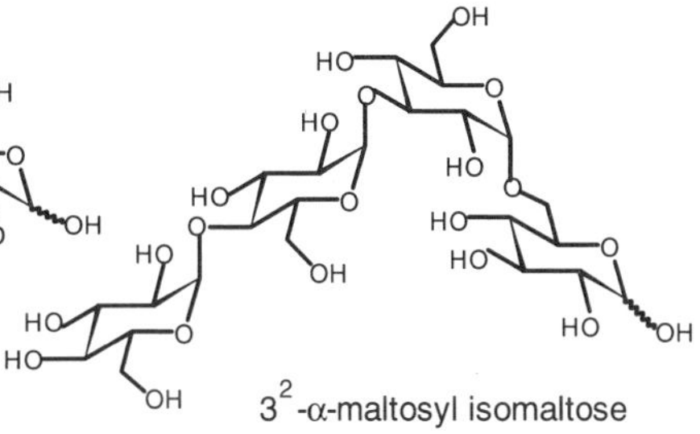

3^2-α-maltosyl isomaltose

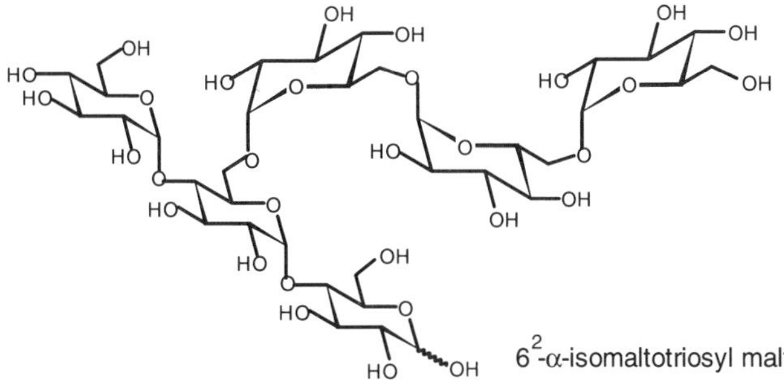

6^2-α-isomaltotriosyl maltotriose

Chemical substitution onto these types of oligosaccharides can be named in a similar manner. Substitution of a tosyl group onto the C-6 of the third glucose residue of maltotetraose would be 6^3-O-*tosylmaltotetraose*. It should be noted that the Joint Commission on Biochemical Nomenclature of the International Union of Pure and Applied Chemistry and the International Union of Biochemistry and Molecular Biology has recommended the use of roman numeral superscripts to designate the particular residue from the reducing end [2]. Thus, the above substituted maltotetraose would be 6^{III}-O-*tosylmaltotetraose*. The author prefers the more established and less cumbersome use of arabic numerals to indicate both the position of substitution and the position of different kinds of monosaccharide substituted units. Thus, the substitution of a tosyl group on the branched glucose unit of the B5 α-limit dextrin would be named 6^3-[6-O-*tosyl-α-D-gluco-pyranosyl*]-*maltotetraose*.

Substitution of a sulfur atom for a glycosidic oxygen in the α-1 → 6 linkage of panose would be named 6^2-α-D-*glucopyranosyl-6²-thiomaltose,* and substitution of a sulfur atom for the α-1 → 6 glycosidic oxygen of isopanose would be 6^1-α-D-*glucopyranosyl-6¹-thiomaltose.* Substitution of a sulfur atom for the glycosidic linkage of the α-1- → 4 linkage of panose would be named 6^2-α-D-*glucopyra-nosyl-4¹-thiomaltose.*

6^3-[6-O-tosyl-α-D-glucopyranosyl]-
maltotetraose

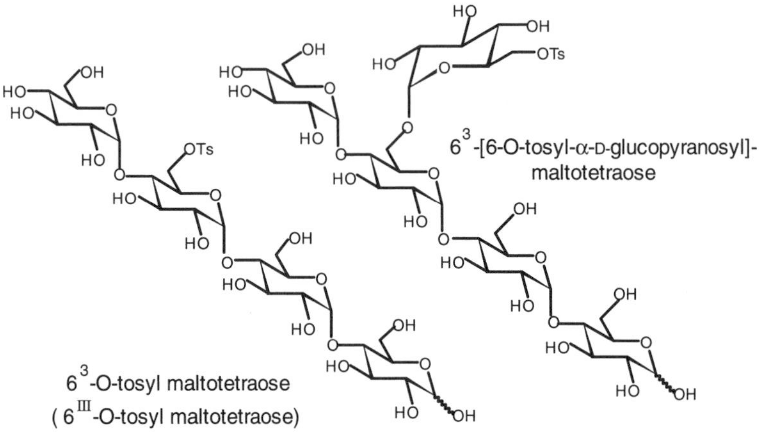

6^3-O-tosyl maltotetraose
(6^{III}-O-tosyl maltotetraose)

Substitution of a sulfur atom for a glycosidic oxygen in the α-1 → 6 linkage of panose would be named 6^2-α-D-*glucopyranosyl-6²-thiomaltose,* and substitution of a sulfur atom for the α-1 → 6 glycosidic oxygen of isopanose would be 6^1-α-D-*glucopyranosyl-6¹-thiomaltose.* Substitution of a sulfur atom for the glycosidic linkage of the α-1 → 4 linkage of panose would be named 6^2-α-D-*glucopyra-nosyl-4¹-thiomaltose.*

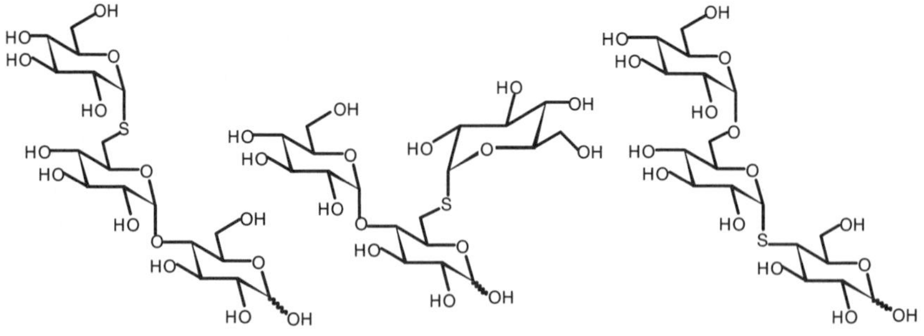

6^2-α-D-glucopyranosyl-6^2thio-
maltose

6^1-α-D-glucopyranosyl-6^1-thio-
maltose

6^2-α-D-glucopyranosyl-4^1-thio-
maltose

A.6 Abbreviations of Monosaccharide Residues

The various monosaccharide residues have been assigned three- to six-letter abbreviations that are similar to those assigned to amino acids. The following are the accepted abbreviations for several monosaccharide residues.

Abequose	Abe
Allose	All
Altrose	Alt
Arabinose	Ara
2-Deoxyribose	dRib
Dihydroxyacetone phosphate	DHAP
Fructose	Fru
Fucose	Fuc
Galactose	Gal
Galactosamine	GalN
N-Acetyl galactosamine	GalNAc
Glucose	Glc
Glucosamine	GlcN
Glucitol	Glc-ol
N-Acetyl glucosamine	GlcNAc
Glucuronic acid	GlcUA
Glyceraldehyde-3-phosphate	GAP
Gulose	Gul
Idose	Ido
Mannose	Man
Muramic acid	Mur
N-Acetyl muramic acid	MurNAc
Neuraminic acid	Neu
N-Acetyl neuraminic acid	NeuNAc
2-keto-3-deoxymannoctulosonic acid	Kdo
Rhamnose	Rha
Quinovose	Qui
Ribose	Rib
Xylose	Xyl
Xylulose	Xul

It should be kept in mind that in developing a name for a carbohydrate, the basic rules should be followed, but one should not get bogged down by the rules. The name should be correct; it should reflect something about the structure; and it should be as simple as possible. This latter aspect is probably the most important factor in that the names of compounds are meant to convey information so that the compounds they refer to can be easily identified, used, and discussed.

For more details on the rules for naming carbohydrates, consult ref. [2].

A.7 Literature Cited

1. W. J. Whelan, *Annu. Rev. Biochem.*, **29** (1960) 105–106.
2. *Carbohydr. Res.*, **297** (1997) 1–92.

Appendix B

Primer on Enzyme Names and Their Catalyzed Reactions

Enzymes usually have names that end in -ase. There are some exceptions, primarily for some of the first enzymes that were recognized, for example, trypsin, chymotrypsin, papain, and elastin. These are enzymes that catalyze the hydrolysis of peptide bonds of proteins. The more systematic naming of enzymes can be divided into three parts: (1) the name of the substrate on which the enzyme acts is indicated; (2) the type of reaction that is catalyzed is indicated; and (3) the biological source of the enzyme (such as a particular animal or species of bacteria or plant and/or a particular organ or secretion fluid, for example, pancreas, liver, heart, saliva, etc.) is given. This third category is used when enzymes from different biological sources act on the same substrate and catalyze the same reaction but have different rates, optimum temperature or pH, and/or produce different kinds of products.

The number of kinds of reactions catalyzed by enzymes is finite. There are approximately 1,500 known enzymes. We will consider some of the major kinds of enzymes that catalyze reactions involved in the metabolism of carbohydrates.

B.1 Hydrolytic Reactions

Hydrolases are enzymes that catalyze the hydrolysis of a particular bond, such as an ester, peptide, or acetal bond. These kinds of enzymes are widely distributed and were some of the first enzymes to be recognized. Many of these enzymes have common names that do not conform to the pattern of naming both the substrate and the reaction catalyzed.

There are several hydrolases that are important in catalyzing the hydrolysis of glycosidic bonds. In some cases, the names of these enzymes generally suggest or indicate the substrates. *Amylases* catalyze the hydrolysis of α-1 $\to$ 4 glycosidic

linkages of starch, glycogen, and maltodextrins. There are several classes of amylases (see Chapter 11). *Glycosidases* are enzymes that catalyze the hydrolysis of
the glycosidic linkage of glycosides. There are α- and β-glycosidases that hydrolyze specific glycosidic configurations. There are also glycosidases that are
specific for different kinds of linkages, such as α-1 → 4 or α-1 → 6. An enzyme
that specifically hydrolyzes the glycosidic bond of sucrose and which was recognized early on is *invertase.* It was so named because its action produced an inversion of the specific optical rotation from +62° to −62°. The enzyme is actually a
sucrase or *sucrose hydrolase* that catalyzes the hydrolysis of the linkage between
D-glucopyranose and D-fructofuranose.

Phosphatases hydrolyze phosphate esters. Sucrose-6′-phosphate is hydrolyzed
by *phosphosucrose phosphatase. Pyrophosphatases* catalyze the hydrolysis of the
acid anhydride linkage of pyrophosphates (see Chapter 10).

B.2 Transfer Reactions

The general name for the enzymes that catalyze transfer reactions of one group to
another is *transferases. Kinases* are a specific kind of transferase that are important in carbohydrate metabolism. All kinases use ATP (or ADP in the reverse direction) as a substrate and phosphorylate other substrates to give phosphate esters.
For example, *hexokinase* or *glucokinase* phosphorylates D-glucose to give D-
glucose-6-phosphate which begins the reactions of glycolysis (see Chapter 11).
Another important carbohydrate kinase is *phosphofructokinase* which catalyzes
the phosphorylation of D-fructose-6-phosphate to D-fructose-1,6-bisphosphate. A
kinase reaction in the reverse direction, in which phosphate is transferred from
phosphoenol pyruvate to ADP to give ATP, is catalyzed by *phosphoenol pyruvate
kinase* (see Chapter 11).

Another common transfer reaction in carbohydrate enzymology is the transfer
of an organic unit to inorganic phosphate (Pi). This type of reaction is catalyzed
by *phosphorylases* such as *starch phosphorylase* or *glycogen phosphorylase,* and
D-glucopyranosyl groups are transferred to Pi.

There are enzymes called *glycosyltransferases* that catalyze the transfer of a
carbohydrate unit from one compound to another. These kinds of enzymes are
particularly found in the transfer of monosaccharide from a nucleotide disphos-
phomonosaccharide to some carbohydrate acceptor, such as in the transfer of
GlcNAc from UDP-GlcNAc to dolichol phosphate by *UDP-GlcNAc/dolichol
phosphate glycosyl transferase,* or the transfer of D-mannose from GDP-Man to
another carbohydrate to give an α-1 → 4 or α-1 → 6 linkage, for example, catalyzed by *GDP-Man-α-1 → 6-glycosyl transferase* (see Chapter 10, Fig.
10.17).

Starch branching enzyme is a *glucanosyltransferase* that catalyzes the cleavage
of an α-1 → 4 linkage to give an α-1 → 6 branch linkage (see Chapter 10, Fig.
10.16A, B).

B.3 Isomerizations

Isomerases catalyze the isomerization of one compound into another. There are many important isomerization reactions in the metabolism of carbohydrates. D-Glucose-6-phosphate is converted into D-fructose-6-phosphate by *phosphoglucoisomerase*. Dihydroxyacetone phosphate is converted into 3-phospho-D-glyceraldehyde by the enzyme *triose phosphate isomerase*. In the Calvin cycle of photosynthesis, this same enzyme converts 3-phospho-D-glyceraldehyde into dihydroxyacetone phosphate.

A special kind of isomerization moves a group from one position to another position in the same compound. These kinds of isomerases are called *mutases*. The conversion of α-Glc-1-P into Glc-6-P is catalyzed by *phosphoglucomutase*. Another example is the transfer of the phosphate group from position 3 of 3-phospho-D-glycerate to position 2 to give 2-phospho-D-glycerate, catalyzed by *phospho-D-glycerate mutase*.

Isomerization of the anomeric carbon is catalyzed by *mutarotase* (see Chapter 3). The inversion of the configuration at C-4 or C-5 is catalyzed by *C-4 epimerase* and *C-5 epimerase*, respectively.

B.4 Oxidation-Reduction Reactions

There are two classes of enzymes that catalyze oxidation reactions. The first class uses molecular oxygen as the oxidizing agent; there are only a few of these enzymes. They are called *oxidases*. One oxidase is *glucose oxidase* which catalyzes the oxidation of β-D-glucose to D-gluconic acid (see Chapter 3). Another enzyme that uses molecular oxygen as an oxidizing agent is *cytochrome oxidase*. This enzyme terminates the oxidative phosphorylation/electron transport chain in the process of respiration and returns ferrous ion back to ferric ion (see Chapter 11).

The second class of enzyme catalyzing oxidation is much more frequent than the first class. The enzymes in this class are known as *dehydrogenases*. They require one of two coenzymes, NAD^+ (or $NADP^+$) and FAD (see Chapters 10 and 11). Both of these enzyme-coenzyme complexes remove a hydrogen ion and a hydride ion from their substrates. The enzymes using NAD^+ as the oxidizing coenzyme remove a hydrogen ion from a hydroxyl group, and a hydride ion from a carbon. A classic reaction is the oxidation of L-lactic acid to pyruvic acid giving NADH, catalyzed by *lactic acid dehydrogenase*. *Alcohol dehydrogenase* catalyzes a similar oxidation of ethanol to acetaldehyde. Both enzymes can act reversibly.

The other types of dehydrogenases use FAD as the oxidizing agent. They oxidize a carbon-carbon arrangement by removing a hydrogen ion from one carbon and a hydride ion from the other. A typical reaction is the oxidation of succinate to fumarate by *succinate dehydrogenase* (see Fig. 11.5).

The reverse reaction of reducing a substrate is also catalyzed by dehydrogenases. The reducing coenzyme is frequently NADPH. There are exceptions in which FADH$_2$ is the reducing coenzyme. The FADH$_2$ reduces a carbon-carbon double bond. A classical example of NADPH reduction is the reduction of D-glycerate-1,3-biphosphate to D-glyceraldehyde-3-phosphate, catalyzed by *D-glyceraldehyde-3-phosphate dehydrogenase* in the Calvin cycle of photosynthesis (see reaction 5 in Fig. 10.3).

B.5 Synthetic Reactions

Enzymatically catalyzed syntheses primarily involve three kinds: (1) formation of carbon-carbon bonds; (2) formation of glycosidic bonds; and (3) formation of pyrophosphates or phosphate esters. A carbon-carbon bond is formed when carbon dioxide is "fixed" in the photosynthetic Calvin cycle. This reaction is catalyzed by *ribulose-1,5-bisphosphate carboxylase* (see Chapter 10). Another important reaction forming a carbon-carbon bond is the aldol condensation of D-glyceraldehyde-3-phosphate with dihydroxyacetone phosphate to give D-fructose-1,6-bisphosphate, catalyzed by *aldolase* in the Calvin cycle. The reverse reaction of aldolase is an important reaction in glycolysis in which D-fructose-1,6-bisphosphate is cleaved into two parts, D-glyceraldehyde-3-phosphate and dihydroxyacetone phosphate (see Chapter 11). A carbon-carbon bond is formed in the synthesis of citric acid, catalyzed by *citric acid synthase.* In this reaction the enzyme forms a carbanion on the methyl group of acetyl CoA that then adds by an aldol-type condensation to oxaloacetate to give citrate.

Glycosidic linkages are formed in a wide variety of reactions such as the addition of monosaccharide units to form sucrose, starch, glycogen, cellulose, peptidoglycan, and so forth, catalyzed by *synthases* using nucleotide diphosphomonosaccharides. Glycosidic bonds are also formed by the transfer of monosaccharide residues to an oligosaccharide chain from nucleotide diphosphomonosaccharides, catalyzed by *glycosyl transferases.*

Nucleotide diphosphomonosaccharides are synthesized by the reaction of α-monosaccharide-1-phosphate with nucleotide triphosphate, catalyzed by *pyrophosphorylases.*

Phosphodiesters are synthesized in the polymerization of DNA and RNA from nucleotide triphosphates catalyzed by *DNA and RNA polymerases.* Teichoic acids are synthesized from nucleotide diphosphoalditols (and in some instances by nucleotide diphosphomonosaccharides), catalyzed by *teichoic acid synthases.*

B.6 Dehydrases and Hydrases

There are enzymes that eliminate water and add water. They have been given special names, primarily indicating the substrate but not the reaction. *Enolase* is an important enzyme in glycolysis that splits out water in D-glyceraldehyde-2-phosphate to give phosphoenol pyruvate. An enzyme that catalyzes the addition of water to a carbon-carbon double bond of fumarate to give L-malate is *fu-marase*. An enzyme that removes water from citrate and then adds it back, with the hydrogen and hydroxyl groups in the opposite positions to give isocitrate, is *aconitase*.

B.7 Glossary of Enzyme Names and Their Reactions

1. *Transferases*
   a. *Hydrolases* transfer organic units to water
      i. *Amylases* hydrolyze α-1 → 4 glycosidic linkages of starch, glycogen, and maltodextrins
      ii. *Isoamylases* hydrolyze α-1 → 6 branch linkages of amylopectin and glycogen
      iii. *Cellulase* hydrolyzes β-1 → 4 linkages of cellulose
      iv. *α-1 → 4 Glucosidase* hydrolyzes α-1 → 4 linkages of low molecular weight maltodextrins
      v. *α-1 → 6 Glycosidase* hydrolyzes α-1 → 6 linkages of low molecular weight maltodextrins
      vi. *α-Glycosidase* hydrolyzes α-linkages of glycosides
      vii. *β-Glycosidase* hydrolyzes β-linkages of glycosides
      viii. *Lactase* or *β-galactosidase* hydrolyze β-1 → 4 linkage of lactose
      ix. *Invertase* hydrolyzes the acetal-ketal linkage of sucrose
      x. *Phosphatases* hydrolyze organic phosphate esters
      xi. *Pyrophosphatase* hydrolyzes the phosphoric acid anhydride linkage of pyrophosphates
      xii. *Esterases* hydrolyze ester linkages
   b. *Kinases* phosphorylate organic alcohols with ATP
      i. *Hexokinase* phosphorylates C-6 hydroxyl of D-glucose
      ii. *Phosphofructokinase* phosphorylates C-1 of 6-phospho-D-fructofuranose
      iii. *Phosphoenol pyruvate kinase* phosphorylates ADP by phosphoenol pyruvate
   c. *Phosphorylase* transfers an organic unit with a glycosidic linkage to inorganic phosphate
      i. *Starch phosphorylase* forms α-Glc-1-P from the ends of starch chains
      ii. *Glycogen phosphorylase* forms α-Glc-1-P from the ends of glycogen chains
      iii. *Sucrose phosphorylase* forms α-Glc-1-P from sucrose
      iv. *Maltose phosphorylase* forms α-Glc-1-P from maltose
      v. *α,α-Trehalose phosphorylase* forms α-Glc-1-P from α,α-trehalose
   d. *Glycosyltransferases* transfer a carbohydrate unit from one component to another
      i. *UDP-GlcNAc/dolichol phosphate transferase* transfers GlcNAc-P to dolichol-P from UDP-GlcNAc

    ii. *UDP-MurNAc-pentapeptide/bactoprenol phosphate transferase* transfers MurNAc-pentapeptide-P to bactoprenol-P from UDP-MurNAc-pentapeptide

    iii. *GDP-Man-(1,4 or 1,6)-glycosyl transferases)* transfers Man from GDP-Man to a carbohydrate acceptor to form α-1 $\rightarrow$ 4 or α-1 $\rightarrow$ 6 linkages

    iv. *Starch (glycogen) branching enzymes* cleave an α-1 $\rightarrow$ 4 linkage and transfer the chain to form an α-1 $\rightarrow$ 6 branch linkage

2. *Isomerases* catalyze the isomerization of one compound into another

  a. *Triose phosphate isomerase* catalyzes the reversible conversion of D-glyceraldehyde-3-phosphate into dihydroxyacetone phosphate

  b. *Phosphoglucoisomerase* catalyzes the conversion of D-glucose-6-phosphate into D-fructose-6-phosphate

  c. *Mutases* catalyze the reversible movement of a group from one position in a compound to another

    i. *Phosphoglucomutase* catalyzes the movement of phosphate from C-1 to C-6 of D-glucose or from C-6 to C-1

    ii. *Phosphoglycerate mutase* catalyzes the movement of phosphate from C-3 to C-2

  d. *Mutarotase* catalyzes the reversible interconversion of α- and β-anomers

  e. *Epimerases* catalyze the inversion of configuration of a chiral carbon

    i. *C-4 epimerase* inverts C-4 hydroxyl of D-glucose to give D-galactose

    ii. *C-5 epimerase* inverts carboxyl group of D-mannuronate to give L-guluronate in alginate biosynthesis, and D-glucuronate in heparin to give L-iduronate

3. *Oxidoreductases* catalyze the oxidation or reduction of a substrate

  a. *Oxidases* use molecular oxygen as an oxidizing agent

    i. *Glucose oxidase* oxidizes β-D-glucopyranose

    ii. *Cytochrome oxidase* oxidizes ferrous ion in the terminal cytochrome in the oxidative phosphorylation/electron transport chain

  b. *Dehydrogenases* catalyze the oxidation of a substrate by removing a proton and a hydride ion, using a coenzyme oxidant

    i. *Alcohol dehydrogenase* removes a proton from the alcohol and a hydride ion from the carbon, using NAD^+, to give acetaldehyde

    ii. *L-lactic acid dehydrogenase* removes a proton from the alcohol and a hydride ion from the carbon, using NAD^+, to give pyruvic acid

    iii. *Succinate dehydrogenase* removes a proton from a carbon and a hydride from an adjacent carbon, using FAD, to give fumarate

    iv. *D-Glyceraldehyde-3-phosphate dehydrogenase* catalyzes the reduction of the carboxyl-phospho acid anhydride of D-glycerate-1,3-bisphosphate by NADPH to give D-glyceraldehyde-3-phosphate in the Calvin cycle of photosynthesis

4. *Synthases* catalyze the formation of carbon-carbon bonds or the formation of glycosidic linkages

    a. *Ribulose-1,5-bisphosphate carboxylase* catalyzes the fixing of carbon dioxide in the Calvin cycle of photosynthesis

    b. *Aldolase* catalyzes the condensation of D-glyceraldehyde with dihydroxyacetone phosphate to give D-fructose-1,6-bisphosphate

    c. *Citric acid synthase* catalyzes the aldol-type condensation of acetyl CoA with oxaloacetate to give citric acid

    d. *Pyrophosphorylases* catalyze the formation of nucleotide diphosphomonosaccharides from α-monosaccharide-1-phosphate and nucleotide triphosphate

    e. *Glycosyltransferases* catalyze the transfer of glycosyl units, primarily from nucleotide diphosphomonosaccharides to an acceptor to form a glycosidic linkage, but also from isoprenolpyrophosphate monosaccharide or an oligosaccharide

       i. *Starch synthase* elongates starch chains, using ADP-Glc

      ii. *Glycogen synthase* elongates glycogen chains, using UDP-Glc

     iii. *Cellulose synthase* elongates cellulose chains, using GDP-Glc

     iv. *Peptidoglycan synthase* elongates murein chains by transferring NAG-NAM-decapeptide from bactoprenol pyrophosphate-NAG-NAM-decapeptide

      v. *Salmonella O-antigen synthase* elongates the O-antigen polysaccharide by transferring a tetrasaccharide from bactoprenol pyrophosphate tetrasaccharide

     vi. *Teichoic acid synthase* elongates the teichoic acid chain by transfering the repeating unit from isoprenol pyrophosphate–repeating unit

    f. *DNA (or RNA) polymerase* catalyzes the polymerization of DNA (or RNA) from nucleotide triphosphates

5. *Dehydrases and Hydrases*

    a. *Enolase* catalyzes the removal of water from D-glyceraldehyde-2-phosphate to give phosphoenol pyruvate

    b. *Fumarase* catalyzes the addition of water to the carbon-carbon double bond of fumarate to give L-malate

    c. *Aconitase* catalyzes the removal of water from citrate and the addition of water back to *cis*-aconitate to give isocitrate.

Index

The following abbreviations are used with the page numbers in the index: *f* indicates the item appears in a figure, *s* indicates the item is a structure, *ss* indicates the item is a structure whose chemical synthesis is given, and *t* indicates the item appears in a table. The parent compound name has been used in alphabetizing, and chemical descriptors such as N-acetyl-, 2-deoxy-, α-, D-, and so forth are ignored in alphabetizing.